Topics in Applied Physics

Volume 143

Topics in Applied Physics is a well-established series of review books, each of which presents a comprehensive survey of a selected topic within the domain of applied physics. Since 1973 it has served a broad readership across academia and industry, providing both newcomers and seasoned scholars easy but comprehensive access to the state of the art of a number of diverse research topics.

Edited and written by leading international scientists, each volume contains high-quality review contributions, extending from an introduction to the subject right up to the frontiers of contemporary research.

Topics in Applied Physics strives to provide its readership with a diverse and interdisciplinary collection of some of the most current topics across the full spectrum of applied physics research, including but not limited to:

- Quantum computation and information
- Photonics, optoelectronics and device physics
- Nanoscale science and technology
- Ultrafast physics
- Microscopy and advanced imaging
- Biomaterials and biophysics
- Liquids and soft matter
- Materials for energy
- Geophysics
- Computational physics and numerical methods
- Interdisciplinary physics and engineering

We welcome any suggestions for topics coming from the community of applied physicists, no matter what the field, and encourage prospective book editors to approach us with ideas. Potential authors who wish to submit a book proposal should contact Zach Evenson, Publishing Editor:

zachary.evenson@springer.com

Topics in Applied Physics is included in Web of Science (2020 Impact Factor: 0.643), and is indexed by Scopus.

More information about this series at https://link.springer.com/bookseries/560

Noé Jiménez · Olga Umnova · Jean-Philippe Groby
Editors

Acoustic Waves in Periodic Structures, Metamaterials, and Porous Media

From Fundamentals to Industrial Applications

Springer

Editors
Noé Jiménez
I3M
Universitat Politècnica de València
Valencia, Spain

Olga Umnova
Acoustics Research Centre
University of Salford
Salford, UK

Jean-Philippe Groby
Laboratoire d'Acoustique de l'Université du Mans (LAUM)
Institut d'Acoustique - Graduate School (IA - GS)
Le Mans, France

ISSN 0303-4216 ISSN 1437-0859 (electronic)
Topics in Applied Physics
ISBN 978-3-030-84299-4 ISBN 978-3-030-84300-7 (eBook)
https://doi.org/10.1007/978-3-030-84300-7

This Springer imprint is published by the registered company Springer Nature Switzerland AG
The registered company address is: Gewerbestrasse 11, 6330 Cham, Switzerland

To our families

Preface

Noise, especially at low frequencies, is a major environmental problem across Europe. Increasingly more information is becoming available about the health impacts of noise. The latest publication of the World Health Organization (WHO) and the Joint Research Center of the European Commission shows that traffic-related noise may account for over 1 million healthy years of life lost annually in the European Union (EU) Member States and other Western European countries. In addition, the Guidelines for EU Noise acknowledge effects of environmental noise, including annoyance, as a serious health problem. According to the European Environment Agency, more than 30% of the EU population may be exposed to excessive noise levels causing annoyance, fatigue, and sleep disturbance.

Urbanization, growing demand for motorized transport, and inefficient urban planning are the main driving forces for environmental noise exposure. There is a pressing need for lighter, thinner, and more efficient structures for the absorption of low frequency sound. Until now, porous materials have been the common choice for noise and vibration control due to their ability to dissipate vibro-acoustic energy through thermal and viscous losses. However, bulky and heavy porous material treatments are required to absorb low frequency sound and mitigate low frequency elastic energy. In addition, in many engineered systems (such as aircraft) the multifunctionality of the noise reducing components, which need to carry mechanical loads and provide thermal or electromagnetic insulation, is essential. This cannot be achieved using conventional porous materials.

For many years the development of noise reducing treatments has been the subject purely of acoustics research. However, recent scientific advances provide a unique and timely opportunity to bring about significant improvements in the design of noise treatments. Phononic and sonic crystals, acoustic metamaterials, and metasurfaces can revolutionize noise and vibration control and in many cases replace traditional porous material. The major breakthroughs are expected in the areas where the traditional acoustics overlaps with new branches of physics and mechanics. Moreover, it is expected that not just attenuation, but also manipulation of sound and vibration by compact devices and structures will be an important next step in addressing this issue. It is therefore necessary to unite the efforts of all the scientific communities involved. The first action is to provide a common theoretical background to the different communities, researchers and engineers, working on acoustic metamaterials, metasurfaces, and sonic crystals as well as conventional acoustic materials.

The Training School *Sound waves in metamaterials and porous media* has been organized by the DENORMS COST Action (CA15125) in order to facilitate this. The Trainers decided to write the present book with the aim of providing the theoretical background on acoustic materials for the researchers from different communities. It is thus organized into three parts. Each part comprises a theoretical part illustrated by examples. The Part I (Chaps. 1–4) focuses on the wave propagation in periodic media and describes the commonly used modeling techniques such as Plane Wave Expansion, Multiple Scattering theory, and Transfer Matrix Method. The illustrating example considers their application to the analysis of the sonic crystal performance. The subject of the Part II (Chaps. 5–8) is the acoustic wave propagation in metamaterial and porous absorbers with viscothermal losses. The recent advances in the design of acoustic metamaterials are first reviewed. The acoustic wave propagation in viscothermal effective fluids, i.e., porous media, and the extension of this theory to non-local models for fluid saturated metamaterials are then considered. Numerical methods, relevant to this problem, are described in detail. Finally, the Part III (Chaps. 9–12) offers a review of industrial applications targeted at building, automotive, and aeronautic industry. This part is thought as a white book for these three industries.

In this book, we have tried to cover theoretical background of the subject, related solution methods, and applications, in order to equip the reader with the skills essential for a successful researcher. The Editors would like to thank the participants, both the Trainers and the Trainees, of the Training School *Sound waves in metamaterials and porous media* and the authors of each chapter. This book is the result of a huge collective effort over years and we hope it will be useful for the current and future generations of researchers in the field of acoustic materials.

Le Mans, France
Salford, UK
Valencia, Spain
November 2019

Jean-Philippe Groby
Olga Umnova
Noé Jiménez

Contents

Part I Wave Propagation in Periodic and Structured Media

1 Periodic Structures, Irreducible Brillouin Zone, Dispersion Relations and the Plane Wave Expansion Method 3
Jérôme O. Vasseur
1.1 Preamble 3
1.2 One-Dimensional Atomic Chains 4
1.2.1 One Dimensional Atomic Chain With One Atom by Unit Cell 4
1.2.2 One Dimensional Atomic Chain With Two Atoms Per Unit Cell 6
1.3 Elements of Crystallography 8
1.3.1 Bravais Lattice, Primitive Vectors, Wigner-Seitz Cell 8
1.3.2 Reciprocal Lattice, Irreducible Brillouin Zone 9
1.3.3 Examples 11
1.4 The Plane Wave Expansion Method 15
1.4.1 Plane Wave Expansion Method for Bulk Phononic Crystals 15
1.4.2 Limitations of the PWE Method 28
1.4.3 Modified PWE Method for Complex Band Structures 34
1.4.4 PWE Method for 2D Phononic Crystal Plates 37
1.5 Conclusions 40
References 41

2 Introduction to Multiple Scattering Theory for Scalar Waves 43
Dani Torrent
2.1 Introduction 43
2.2 Multiple Scattering of Acoustic Waves 45
2.2.1 Incident Fields 46
2.2.2 T-Matrix of a Fluid Cylinder 48

2.2.3 Multiple Scattering 50
2.3 Multiple Scattering of Flexural Waves 53
2.3.1 Effective T-Matrix 56
2.3.2 Scattering by a Cluster of Mass-Spring Resonators 57
2.4 Numerical Examples 60
References 63

3 Sound Wave Propagation in Sonic Crystals 65
Vicent Romero-García
3.1 Introduction: Origins of Sonic Crystals 65
3.2 The Physical Origin of Bandgaps 67
3.2.1 Transfer Matrix Method 68
3.2.2 Discussion 68
3.3 Dispersion Relation of Sonic Crystals 71
3.3.1 Plane Wave Expansion 71
3.3.2 Extended Plane Wave Expansion 73
3.3.3 Structure Factor 74
3.3.4 Supercell Approximation 76
3.4 Multiple Scattering Theory 77
3.4.1 Cylindrical Waves 77
3.4.2 Boundary Conditions 82
3.5 Research on Sonic Crystals 85
3.5.1 Long Wavelength Regime 86
3.5.2 Diffraction Regime 88
3.6 Technological Application: Sonic Crystal Acoustic Barrier 93
3.6.1 Design Process 94
3.6.2 Acoustic Standardization 96
3.6.3 Wind Tunnel Analysis: Structural Load 96
References 97

4 The Transfer Matrix Method in Acoustics 103
Noé Jiménez, Jean-Philippe Groby, and Vicent Romero-García
4.1 Introduction 103
4.2 The Transfer Matrix Method 104
4.2.1 Total Transfer Matrix 106
4.2.2 Effective Parameters 108
4.2.3 The Scattering Matrix 109
4.2.4 Reflection, Transmission and Absorption Coefficients 110
4.3 Review of the Usual Transfer Matrices 115
4.3.1 Particle Velocity Verses Flow Formulation 115
4.3.2 Fluid Layers: Particle Velocity Formulation 116
4.3.3 Ducts: Flow Formulation 117
4.3.4 Porous Media Layers 121
4.3.5 Locally Resonant Elements 126
4.3.6 Side Resonating Ducts 128

4.3.7 Helmholtz Resonators ... 130
4.3.8 Rigid Micro-perforated Plates ... 132
4.3.9 Elastic Plates ... 133
4.3.10 Membranes ... 136
4.3.11 Infinite Elastic Vibrating Wall ... 138
4.4 Examples of Application ... 140
4.4.1 Absorption of Multilayered Porous Structure ... 140
4.4.2 Noise Transmission Through a Double Wall ... 144
4.4.3 Phononic Crystals ... 151
4.4.4 Metamaterial Modelling Using TMM ... 156
4.5 Conclusions ... 160
4.6 Appendix—End Corrections ... 160
4.6.1 Change of Section in a Waveguide ... 161
4.6.2 Side Branch ... 161
4.6.3 Periodic Array of Slits ... 161
References ... 162

Part II Wave Propagation in Absorbing Metamaterials and Porous Media

5 Acoustic Metamaterial Absorbers ... 167
Jean-Philippe Groby, Noé Jiménez, and Vicent Romero-García
5.1 Introduction ... 168
5.1.1 The Scattering Matrix ... 169
5.1.2 Complex Frequency Plane Analysis ... 171
5.1.3 Perfect Absorption and Critical Coupling ... 171
5.2 Reflection Problems ... 172
5.2.1 Rigidly-Backed Porous Materials ... 173
5.2.2 Metaporous Absorbers ... 175
5.2.3 Metamaterials Based on Resonant Air Cavities ... 178
5.2.4 Metamaterials Based in Resonant Elastic Plates and Membranes ... 181
5.3 Transmission Problems ... 183
5.3.1 Absorption and Fabry–Pérot Modes: Limits of Absorption by Single Resonators in the Transmission Problem ... 184
5.3.2 Monopolar and Dipolar Resonances ... 186
5.3.3 Overlapping of Resonances in Porous Layers ... 188
5.3.4 Accumulation of Resonances Due to Dispersion ... 190
5.3.5 Symmetry Breaking Materials ... 193
5.3.6 Combinations of Monopolar and Dipolar Resonances ... 196
5.4 Summary and Conclusions ... 199
References ... 201

6 Acoustic Wave Propagation in Viscothermal Fluids 205
Denis Lafarge
6.1 Introduction 205
6.2 Molecular Constitution of Fluids 207
6.3 Wave Propagation in a Viscothermal Fluid: *Use of Near-Equilibrium Thermodynamics Because of Scale Separation* 210
6.4 Governing Navier-Stokes-Fourier Equations for Small Amplitude Wave Propagation 212
6.5 Equilibrium Thermodynamics: *Brief Recap on Trivariate Fluids* 214
6.6 Derivation of the Equations of Motion 217
6.6.1 Conservation Laws and Equation of State 218
6.6.2 Constitutive Laws: Stokes and Fourier 221
6.6.3 Inherent Thermodynamic Simplifications in Navier-Stokes-Fourier's Framework 225
6.7 The Different Normal Modes at Fixed Real Wavevector 226
6.8 Some Explicit Calculations for the Longitudinal Modes 228
6.9 General Representation of the Fields in Harmonic Regime Without Source 231
6.10 Revisiting the Longitudinal Acoustic Equations Using an Electromagnetic Analogy 233
6.11 Macroscopic Electromagnetics 234
6.11.1 Nonlocal Maxwellian Pattern of the Macroscopic Equations 234
6.11.2 Passage from Microscopics to Macroscopics 237
6.11.3 Ambiguities and a Suggested Way to Resolve Them 240
6.11.4 Ambiguities and the Customary Way 242
6.11.5 Discussion of Our Proposal 243
6.12 Nonlocal Maxwellian Pattern of Longitudinal Acoustics 244
6.12.1 Electromagnetic Analogy 245
6.12.2 Acoustic Maxwell Fields 248
6.12.3 Acoustic Nonlocal Operators 251
6.12.4 Summary: Action-Response Problem to Determine the Acoustic Kernels $\rho(\omega, \boldsymbol{k})$ and $\chi^{-1}(\omega, \boldsymbol{k})$ 256
6.12.5 Acoustics Translation of the Customary Viewpoint 257
Appendix: Electromagnetic-Acoustic Analogy 259
References 271

7 Nonlocal Dynamic Homogenization of Fluid-Saturated Metamaterials 273
Denis Lafarge
7.1 Sound Propagation in Fluid-Saturated Rigid-Framed Porous Media 273

7.2 Statement of the Problem ... 275
7.3 The Operations of Macroscopic Averaging ... 278
7.3.1 The Well-Defined Case of Stationary-Random Media ... 279
7.3.2 The Ambiguous Case of Periodic Media ... 288
7.4 Macroscopic Equations and Definition of the Acoustic H-Field from Electromagnetic Analogy ... 295
7.4.1 Unbounded Fluid (Longitudinal Motions) ... 296
7.4.2 Macroscopic Electromagnetics ... 297
7.4.3 Macroscopic Acoustics (Fluid-Saturated Rigid-Framed Porous Medium) ... 298
7.5 Macroscopic Equations: Other Points of View ... 300
7.5.1 Macroscopic Electromagnetics—Customary Point of View ... 300
7.5.2 Acoustics Translation of the Customary Point of View in Electromagnetics ... 301
7.5.3 Acoustics Formulation in Terms of Volume-Averaged Pressure ... 301
7.6 Derivation of the Macroscopic Equations by a Micro-Macro Passage ... 303
7.7 Action-Response Problem to Determine the Nonlocal Dynamic Operators from Microstructure ... 307
7.8 Other Points of View ... 311
7.8.1 Acoustics Translation of the Customary Point of View in Electromagnetics ... 311
7.8.2 Acoustics Formulation in Terms of Volume-Averaged Pressure ... 313
7.9 Characteristic Wavenumbers and Impedances ... 314
7.10 Conclusions ... 315
Appendix: Local Dynamic Homogenization of Rigid-Framed Fluid-Saturated Porous Materials ... 316
References ... 330

8 Numerical Methods for Modelling and Simulation of Porous Materials ... 333
Peter Goransson and Olivier Dazel
8.1 Introduction ... 333
8.2 Biot's Equations ... 335
8.2.1 Constitutive Laws ... 336
8.2.2 Momentum Equations ... 337
8.3 Weak Forms ... 339
8.3.1 PEM1 Weak Formulation ... 339
8.3.2 PEM2 Weak Formulation ... 339
8.3.3 Elastic Solid ... 340
8.3.4 Air ... 341

8.4 Discrete Linear System 341
8.4.1 Elementary Matrices 341
8.4.2 Discretisation of the Weak Forms in One Element 342
8.4.3 Assembly 344
8.5 Coupling Between Domains 345
8.5.1 Coupling with an Air Domain 345
8.5.2 Coupling with an Elastic Solid 347
8.5.3 Coupling Through a Thin Airgap Interface 348
8.5.4 Coupling with Bloch Waves 350
8.6 Application Examples 352
8.6.1 Convergence Aspects 353
8.6.2 FE Cases in 1D 354
8.6.3 Simulation of a Metaporous Material 359
8.6.4 3D Anisotropic Modelling 362
Appendix - Periodic Boundary Conditions 364
References 365

Part III Industrial Applications of Porous Media and Acoustic Metamaterials

9 Industrial Applications I 369
Arnaud Duval and Fabien Chevillotte
9.1 Introduction 369
9.2 Absorption Optimization 370
9.2.1 The Coupling Parameters: Johnson-Champoux-Allard (JCA) Model 371
9.2.2 Optimizing Felts or Fibrous Materials for Absorption 373
9.2.3 Optimizing Foam Materials for Absorption 375
9.2.4 Airflow Resistive Screens: Tunable Absorption 379
9.3 Insulation Optimization 383
9.3.1 Insulation Optimization: Fibrous Porous Material Case 384
9.3.2 Insulation Optimization: Foam Porous Material Case 384
9.3.3 The Standard Mass—Poroelastic Spring Insulator 384
9.3.4 The Hybrid Stiff Insulator Concept 385
9.4 Damping Optimization 386
9.4.1 Damping Optimization: Poroelastic Material Itself 386
9.4.2 Damping Optimization: Poroelastic Material Boundary Conditions 387
9.5 Conclusions 388
References 388

10 Industrial Applications II 391
Fabien Chevillotte
10.1 Introduction 391
10.2 Acoustical Applications in the Building Industry 392
10.2.1 Acoustical Correction 392
10.2.2 Air-Borne Insulation 393
10.2.3 Solid-Borne Insulation (Impact Noise) 400
10.2.4 Ceilings 401
10.3 New Trends in Building Acoustics 402
10.3.1 Thin and Aesthetic Absorbers 402
10.3.2 Low Frequency Performances and Non-conventional Phenomena 402
10.3.3 Rolling Noise 404
10.4 Conclusions 405
References 405

11 Industrial Applications III 407
Arnaud Duval
11.1 Introduction 407
11.2 Acoustic Package Optimization Methods in the Automotive Industry 408
11.2.1 Automotive Vehicle Main Noise Sources and Airborne Transfer Paths 408
11.2.2 Multilayer Noise Treatments Properties 412
11.3 Fast Broadband Highly Curved Insertion Loss Simulation of Automotive Trims 418
11.3.1 Trim Modelling Curvature Issue 418
11.3.2 Cylindrical Transfer Matrix Method: A Spectral Approach 419
11.3.3 Curvature Radius Influence on Trim Insertion Loss Slopes 421
11.3.4 Trim Thickness Influence on Curved Insertion Loss Breathing Frequencies 421
11.3.5 Multi-thickness Simplified 3D Dash Insulator Broadband Insertion Loss Simulation 423
11.3.6 Summary of Curved Insertion Loss Simulations 426
11.4 Conclusions 426
References 427

12 Industrial Applications IV 429
Israel Pereira, Sideto Futatsugi, and Maria L. V. Rodrigues
12.1 Introduction 429
12.2 Description of Noise Sources 430
12.2.1 Engine Sources 430
12.2.2 Turbulent Boundary Layer (TBL) 431
12.2.3 Systems 432

12.3 The Use of Porous Materials 433
12.4 Aeronautic Requirements 435
12.5 New Developments 436
12.6 Case Study 437
References 440

Correction to: Acoustic Wave Propagation in Viscothermal Fluids C1
Denis Lafarge

Index 441

Contributors

Fabien Chevillotte MATELYS - Research Lab., Bât. B, 7 Rue des Maraîchers, Vaulx-En-Velin, France

Olivier Dazel University of Le Mans, UMR CNRS 6613, Le Mans, France

Arnaud Duval Trèves Products, Services & Innovation, 2-4 rue Emile Arquès, Reims, France;
TREVES Product, Services & Innovation, 2–4 rue Emile Arquès, Reims, France

Sideto Futatsugi Embraer S.A, Av. Brigadeiro Faria Lima, São José dos Campos, Brazil

Peter Goransson KTH Royal Institute of Technology, Stockholm, Sweden

Jean-Philippe Groby Laboratoire d'Acoustique de l'Université du Mans, UMR CNRS 6613, Le Mans Université, Le Mans, France

Noé Jiménez Instituto de Instrumentación para Imagen Molecular, Universitat Politècnica de València, CSIC, Camino de Vera S/N, València, Spain

Denis Lafarge Laboratoire d'Acoustique de l'Université du Mans, UMR CNRS 6613, Le Mans Université, Le Mans, France

Israel Pereira Embraer S.A, Av. Brigadeiro Faria Lima, São José dos Campos, Brazil

Maria L. V. Rodrigues Embraer R&D Portugal, Parque da Indústria Aeronáutica de Évora, Lote A-I, Herdade do Pinheiro e Casa Branca, Évora, Portugal

Vicent Romero-García Laboratoire d'Acoustique de l'Université du Mans, UMR CNRS 6613, Le Mans Université, Le Mans, France

Dani Torrent Grup de Recerca d'òptica (GROC), Institut de Noves Tecnologies de la Imatge (INIT), Universitat Jaume I, Castelló, Spain

Jérôme O. Vasseur Institute of Electronic, Microelectronic and Nanotechnology, UMR CNRS 8520, Université de Lille, Villeneuve-d'Ascq, France

Part I
Wave Propagation in Periodic and Structured Media

Chapter 1
Periodic Structures, Irreducible Brillouin Zone, Dispersion Relations and the Plane Wave Expansion Method

Jérôme O. Vasseur

Abstract The Plane Wave Expansion (PWE) method allows the calculation of dispersion curves, i.e., the relation linking the frequency to the wave number for any propagating mode of periodic structures made of elastic materials such as phononic crystals. The method is relatively easy to implement numerically but presents some limitations. After recalling some fundamental aspects of crystallography that are necessary to the study of periodic structures, the PWE method described in detail for the case of bulk phononic crystals, i.e., structures of infinite extent, and its advantages and drawbacks are discussed. It is also shown that the method can be used for calculating the band structure of phononic crystals of finite thickness and for analysing the evanescent waves within the phononic band gaps.

1.1 Preamble

Propagation of elastic waves in composite materials exhibiting a periodic structure constitutes a very old topic in physics. One can mention the work, among others, of Lord Rayleigh in 1887 where has been demonstrated the existence of band gaps in periodically stratified media [1]. However since the beginning of the 1990s and the pioneering works of Sigalas et al. [2] and Kushwaha et al. [3] on phononic crystals, this topic received a renewed interest. These artificial material composites whose physical characteristics (density, elastic moduli, ...) are periodic functions of the position have been proven to exhibit very peculiar propagation properties such as frequency band gaps, negative refraction or self-collimation phenomena [4]. Studies of the propagation of elastic waves in periodic structures necessitate using theoretical tools that were initially developed in the frame of solid state physics such as the unit cell, the direct lattice, the reciprocal lattice, the irreducible Brillouin zone or dispersion relations. Moreover, these studies also require solving, with a high level

J. O. Vasseur (✉)
Institute of Electronic, Microelectronic and Nanotechnology, UMR CNRS 8520, Université de Lille, Villeneuve-d'Ascq, France
e-mail: jerome.vasseur@univ-lille.fr

N. Jiménez et al. (eds.), *Acoustic Waves in Periodic Structures, Metamaterials, and Porous Media*, Topics in Applied Physics 143,
https://doi.org/10.1007/978-3-030-84300-7_1

of accuracy, the equations of elastic waves propagation. Different theoretical tools were proposed for this. One can mention the Plane Wave Expansion (PWE) method, the Finite Difference Time Domain (FDTD) method, the Multiple Scattering (MS) method, the Finite Element (FE) method, and many others [4].

With the aim of introducing in a pedagogical way most of the solid state physics concepts listed above, very simple periodic structures such as one dimensional infinite atomic chains are considered first. In the second part of this Chapter, we recall in a more formal way, the elements of crystallography that are necessary for the study of periodic structures. The third part of the Chapter focuses on the PWE method. Basic principles of the method are first presented and its application to two-dimensional periodic structures is reported with many details. Limitations of the PWE method are discussed. Finally, it is also shown that the method can be used for calculating the band structure of phononic crystals of finite thickness and for analysing the evanescence of waves inside the phononic band gaps.

1.2 One-Dimensional Atomic Chains

1.2.1 One Dimensional Atomic Chain With One Atom by Unit Cell

We consider first a very simple periodic structure, namely an infinite one-dimensional linear chain of identical atoms with mass m, connected by springs with constant stiffness β and oriented along the x direction. The equilibrium position of atom n is $x_{n,eq} = na$, where a is the distance between two adjacent atoms in equilibrium. Atoms are assumed free to move slightly around their respective equilibrium position and their position, at any time t, is given as $x_n(t) = na + u_n(t)$ with $|u_n(t)| \ll |x_n(t)|$ and $u_n = x_n - x_{n,eq}$ is the displacement of the nth atom from the equilibrium position. In that case, the unit cell, shown in Fig. 1.1, that can be repeated along direction x with periodicity a, contains only one atom and the lattice spacing a defines the periodicity of the chain along the x axis. Newton's second law applied to atom n considering interaction between nearest neighbours leads to

$$m\frac{\partial^2 u_n}{\partial t^2} = -\beta\left(u_n - u_{n-1}\right) + \beta\left(u_{n+1} - u_n\right) = \beta\left(u_{n+1} + u_{n-1} - 2u_n\right). \quad (1.1)$$

Seeking solutions of Eq. (1.1) in the form of sinusoidal propagating waves of amplitude U_0 such as $u_n(t) = U_0 e^{i(kna-\omega t)}$ where k is the wave number and ω the circular frequency, Eq. (1.1) becomes

$$-m\omega^2 = \beta\left(e^{ika} + e^{-ika} - 2\right) = 2\beta\left(\cos\left(ka\right) - 1\right) = -4\beta\sin^2\left(\frac{ka}{2}\right). \quad (1.2)$$

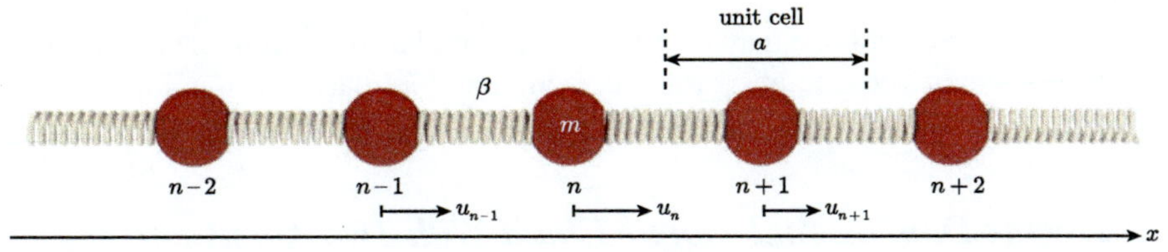

Fig. 1.1 Schematic illustration of the infinite atomic chain made of identical atoms of mass m with a lattice parameter a. β is the stiffness of the spring linking atoms

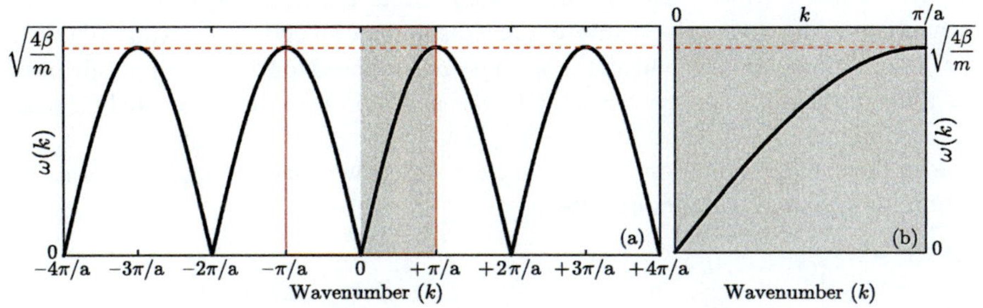

Fig. 1.2 **a** Dispersion relation of the infinite atomic chain made of identical atoms. The red and shaded boxes represent the first Brillouin zone and the irreducible Brillouin zone, respectively; **b** Dispersion relation plotted in the irreducible Brillouin zone

One deduces from Eq. (1.2), the dispersion relation of the atomic chain, i.e., the relation linking the circular frequency ω to the wave number k in the form

$$\omega(k) = \sqrt{\frac{4\beta}{m}} \left| \sin\left(\frac{ka}{2}\right) \right| . \tag{1.3}$$

Figure 1.2a shows the dispersion relation $\omega(k)$. Note that $|\sin(ka/2)|$ is a π-periodic function,

$$\left| \sin\left(\frac{ka}{2}\right) \right| = \left| \sin\left(\frac{ka}{2} + \pi\right) \right| = \left| \sin\left(\frac{a}{2}\left[k + \frac{2\pi}{a}\right]\right) \right| . \tag{1.4}$$

Then, $\omega(k)$ is a periodic function of k with periodicity $G = 2\pi/a$ and $\omega(k + nG) = \omega(k)$ where n is an integer. One deduces that a propagation mode of wave number k and a mode with wave number $(k + G)$ are exactly the same modes. The periodicity $G = 2\pi/a$ in the wave number space is associated with the *reciprocal lattice* of the chain while the lattice parameter a characterizes its *direct lattice*.

Due to the periodicity of the dispersion relation in the reciprocal space, the useful information concerning the vibration modes that can propagate in the chain, is contained in the waves with wave numbers lying between the limits $-\pi/a$ and $+\pi/a$. This range of wave numbers centred at $k = 0$ is named the *first Brillouin zone* of the reciprocal lattice. Therefore, the dispersion relation is also symmetric with respect of the plane $k = 0$, and one may restrict the study to the *irreducible Brillouin zone*, i.e., the domain of wave numbers ranging from 0 to $+\pi/a$, as shown in Fig. 1.2b.

1.2.2 One Dimensional Atomic Chain With Two Atoms Per Unit Cell

We can now turn to a little bit more complicated structure: an infinite one-dimensional linear chain with two atoms of different masses in the unit cell, as shown in Fig. 1.3. The lattice parameter is $2a$ and all the springs are supposed to have the same stiffness β. Atoms of mass m_1 and m_2 are named *even* and *odd* atoms and are labelled with integers $2n$ and $2n+1$ respectively.

With the same assumptions as that of Sect. 1.2.1, we can write the equations of motion for *even* and *odd* atoms in the form

$$\begin{cases} m_1 \dfrac{\partial^2 u_{2n}}{\partial t^2} &= -\beta\left(u_{2n} - u_{2n-1}\right) + \beta\left(u_{2n+1} - u_{2n}\right) \\ &= \beta\left(u_{2n+1} + u_{2n-1} - 2u_{2n}\right), \\ m_2 \dfrac{\partial^2 u_{2n+1}}{\partial t^2} &= -\beta\left(u_{2n+1} - u_{2n}\right) + \beta\left(u_{2n+2} - u_{2n+1}\right) \\ &= \beta\left(u_{2n+2} + u_{2n} - 2u_{2n+1}\right). \end{cases} \tag{1.5}$$

Seeking solutions of Eq. (1.5) in the form

$$\begin{cases} u_{2n}(t) = Ae^{i(k(2n)a-\omega t)}, \\ u_{2n+1}(t) = Be^{i(k(2n+1)a-\omega t)}, \end{cases} \tag{1.6}$$

where A and B are amplitude terms, one obtains a set of two equations that can be recast in the following matrix form

$$\begin{bmatrix} (2\beta - m_1\omega^2) & -2\beta\cos(ka) \\ 2\beta\cos(ka) & -(2\beta - m_2\omega^2) \end{bmatrix} \begin{bmatrix} A \\ B \end{bmatrix} = \begin{bmatrix} 0 \\ 0 \end{bmatrix}. \tag{1.7}$$

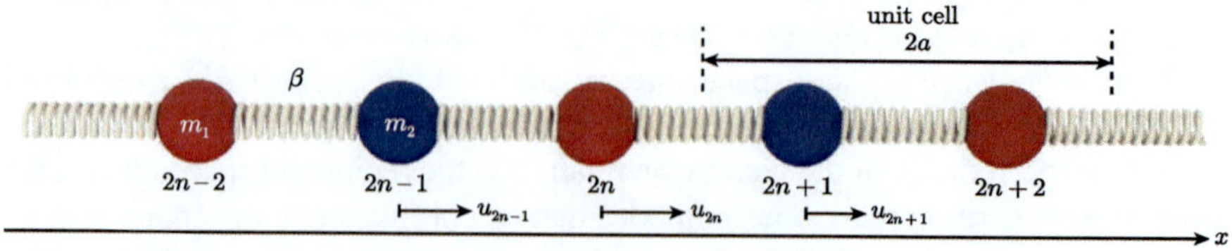

Fig. 1.3 Schematic illustration of the infinite atomic chain made of two atoms of masses m_1 and m_2 in the unit cell with a lattice parameter $2a$. β is the stiffness of the spring linking atoms

Equations (1.7) admit non-trivial solutions if the determinant of the matrix vanishes. This leads to

$$\omega^4 - 2\beta\left(\frac{m_1 + m_2}{m_1 m_2}\right) + \frac{4\beta^2 \sin^2(ka)}{m_1 m_2} = 0, \tag{1.8}$$

and one deduces

$$\omega(k) = \sqrt{\beta\frac{m_1 + m_2}{m_1 m_2}\left(1 \pm \sqrt{1 - 4\frac{m_1 m_2 \sin^2(ka)}{(m_1 + m_2)^2}}\right)}. \tag{1.9}$$

Consequently, Eq. (1.8) admits two real solutions $\omega_-(k)$ and $\omega_+(k)$ that are periodic in wave number, k, with a period of $+\pi/a$ and the first Brillouin zone corresponds to the wave numbers varying between $-\pi/2a$ and $+\pi/2a$. One notes that because the unit cell in the direct lattice of the chain is two times larger than that of the monoatomic chain, the first Brillouin zone is two times smaller. Figure 1.4 shows the dispersion relations plotted in the irreducible Brillouin zone (k between 0 and $+\pi/2a$) as a function of the ratio m_2/m_1 greater than or equal to 1. One observes that for $m_2 = m_1$, the dispersion relation of the infinite monoatomic chain is recovered but the band is folded in a smaller irreducible Brillouin zone. Moreover, for increasing mass ratio, a band gap appears at the edge of the irreducible Brillouin zone and higher is the mass ratio, the larger is the band gap.

In this Section, considering very simple one dimensional periodic structures, we have introduced the notions that are of fundamental importance in the study of periodic structures namely the unit cell, the direct and reciprocal lattices and the irreducible Brillouin zone. We will see in Sect. 1.3.1 of this Chapter, how these concepts can be generalized of much more complicated periodic structures such as the phononic crystals.

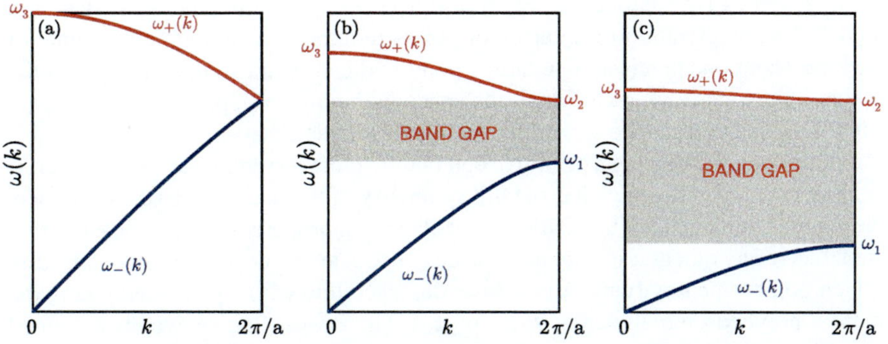

Fig. 1.4 Dispersion relations of the infinite atomic chain made of two atoms of masses m_1 and m_2 in the unit cell with a lattice parameter $2a$ plotted in the irreducible Brillouin zone for **a** $m_2 = m_1$, **b** $m_2 = 2m_1$, **c** $m_2 = 10m_1$. Circular frequencies ω_1, ω_2 and ω_3 are equal to $\sqrt{2\beta/m_2}$, $\sqrt{2\beta/m_1}$, $\sqrt{2\beta(m_1 + m_2)/m_1 m_2}$, respectively

1.3 Elements of Crystallography

In this Section, we recall the elements of crystallography that are necessary for the study of periodic structures. We limit ourselves to some simple structures. Complete reports on solid state physics are available in text books such as Refs. [5, 6]. These references are strongly recommended to the reader of the present Chapter.

1.3.1 Bravais Lattice, Primitive Vectors, Wigner-Seitz Cell

In solid state physics, solids possessing a crystalline structure are periodic arrays of atoms that are modelled by a combination of a basis and a Bravais lattice. In Ref. [6], a Bravais lattice is defined as: *an infinite array of discrete points with an arrangement and orientation that appear exactly the same, from whichever of the points the array is viewed (all the points have the same environment).*

In three dimensions (3D), there exist a total of fourteen different Bravais lattices [5, 6], five in 2D and one in 1D. The symmetry of any physical crystal is described by one of the Bravais lattices plus a basis. The basis consists of identical units, usually made by group of atoms, which are attached to every point of the underlying Bravais lattice. A crystal, whose basis consists of a single atom or ion, is said to have a monatomic Bravais lattice.

In 3D, from the mathematical point of view, a Bravais lattice is defined as a collection of points with position vectors $\vec{R}$ of the form $\vec{R} = \ell\vec{a}_1 + m\vec{a}_2 + n\vec{a}_3$ where $\vec{a}_1$, $\vec{a}_2$ and $\vec{a}_3$ are three vectors (named *the primitive vectors of the Bravais lattice*) not in the same plane and ℓ, m and n are three integers. For the sake of simplicity, the notion of Bravais lattice is illustrated in 2D on Fig. 1.5. In this figure, one observes that from a point chosen at the origin of the array, any other point can be obtained by a translation equals to a linear combination of the primitive vectors. Moreover for any given Bravais lattice, the set of primitive vectors is not unique, as shown in Fig. 1.5. Characterization of an array of points requires also to define a volume of space that contains precisely one lattice point and can be translated through all the vectors of a Bravais lattice to fill all the space without overlapping itself or leaving voids. This space is named a primitive cell and is depicted also in Fig. 1.5. There is no unique way of choosing a primitive cell but the most common choice, however, is the Wigner-Seitz cell, which has the full symmetry of the underlying Bravais lattice. The Wigner-Seitz cell about a lattice point also has a property of being closer to that point than to any other lattice point. It can be constructed by drawing lines connecting a given point to nearby lying points, bisecting each line with a plane and taking the smallest polyhedron bounded by these planes. The Bravais lattice, which is defined in real space, is sometimes referred to as a direct lattice.

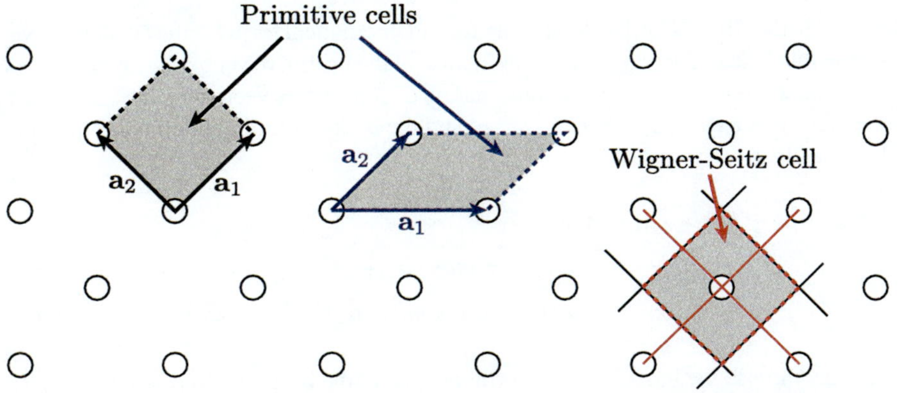

Fig. 1.5 A 2D Bravais lattice. Two possible choices of the primitive vectors $\vec{a}_1$ and $\vec{a}_2$ are indicated. Black and blue parallelograms represent the primitive unit cells associated with each set of primitive vectors. Red parallelogram is the Wigner-Seitz cell of the Bravais lattice. It has been constructed by drawing red lines connecting a given point to nearby lying points, drawing black lines bisecting each red line and considering the area bounded by red dashed lines

1.3.2 Reciprocal Lattice, Irreducible Brillouin Zone

With the direct lattice being defined in the real space, there exists a dual space named the reciprocal lattice. This concept is very important when studying wave propagation, diffraction and other wave phenomena in crystals. We know that in the direct lattice, a periodic function in space $f(\vec{r})$ satisfies $f(\vec{r}) = f(\vec{r} + \vec{R})$ where $\vec{R} = \ell\vec{a}_1 + m\vec{a}_2 + n\vec{a}_3$, see Sect. 1.3.1. For example, the function $f(\vec{r})$ in phononic crystals can be the mass density or the elastic moduli. This function being $\vec{R}$-periodic can be developed in Fourier series such as

$$f(\vec{r}) = \sum_{\vec{G}} f\left(\vec{G}\right) e^{i\vec{G}\cdot\vec{r}}, \tag{1.10}$$

where $\vec{G}$ are named the *reciprocal lattice vectors* and $f(\vec{G})$ are the Fourier coefficients of $f(\vec{r})$. Then one can write

$$f\left(\vec{r} + \vec{R}\right) = \sum_{\vec{G}} f\left(\vec{G}\right) e^{i\vec{G}\cdot(\vec{r}+\vec{R})} = f\left(\vec{r}\right) = \sum_{\vec{G}} f\left(\vec{G}\right) e^{i\vec{G}\cdot\vec{r}}, \tag{1.11}$$

and one deduces that $e^{i\vec{G}\cdot\vec{R}} = 1$ and $\vec{G}\cdot\vec{R} = 2\pi \cdot N$ where N is an integer. Consequently, one can define the reciprocal lattice as a set of points whose positions are given by a set of vectors $\vec{G}$ satisfying the condition

$$\vec{G}\cdot\vec{R} = 2\pi \cdot N; \quad N \in \mathbb{Z}, \tag{1.12}$$

for all $\vec{R}$ in the Bravais lattice. Searching for a mathematical form for the $\vec{G}$ vectors, we assume that because $\vec{R}$ is a linear combination of the primitive vectors $\vec{a}_i$, $i = 1, 2, 3$, $\vec{G}$ may be written also as a linear combination of some basis vectors $\vec{b}_i$, $i = 1, 2, 3$ as $\vec{G} = \ell'\vec{b}_1 + m'\vec{b}_2 + n'\vec{b}_3$ where ℓ', m', n' and the $\vec{b}_i$ are initially undefined. Equation (1.12) leads to

$$\begin{aligned}\ell'\ell\vec{b}_1 \cdot \vec{a}_1 + m'\ell\vec{b}_2 \cdot \vec{a}_1 + n'\ell\vec{b}_3 \cdot \vec{a}_1 + \\ + \ell' m\vec{b}_1 \cdot \vec{a}_2 + m'm\vec{b}_2 \cdot \vec{a}_2 + m'n\vec{b}_2 \cdot \vec{a}_3 + \\ + n'\ell\vec{b}_3 \cdot \vec{a}_1 + n'm\vec{b}_3 \cdot \vec{a}_2 + n'n\vec{b}_3 \cdot \vec{a}_3 = 2\pi N.\end{aligned} \tag{1.13}$$

One may impose the basis $\{\vec{b}_i\}$ to be orthonormal to the basis $\{\vec{a}_i\}$ and write $\vec{b}_i \cdot \vec{a}_j = 2\pi\delta_{ij}$ where δ_{ij} is the Kronecker's symbol and the multiplicative factor 2π has been introduced for simplification. Then Eq. (1.13) reduces to

$$\ell'\ell + m'm + n'n = N. \tag{1.14}$$

The left-hand side of Eq. (1.14) must be an integer as the right-hand side and consequently ℓ', m' and n' must be also integers. This implies that a reciprocal lattice of a direct lattice is also a Bravais lattice. We can now define vectors $\vec{b}_i$. Consider first $\vec{b}_1$. Because $\vec{b}_1 \perp \vec{a}_2$ and $\vec{b}_1 \perp \vec{a}_3$, one may write $\vec{b}_1 = \lambda \cdot \vec{a}_2 \times \vec{a}_3$ where λ is a constant to be determined. Moreover, $\vec{a}_1 \cdot \vec{b}_1 = 2\pi = \lambda\vec{a}_1 \cdot (\vec{a}_2 \times \vec{a}_3)$ and one deduces $\lambda = 2\pi/[\vec{a}_1 \cdot (\vec{a}_2 \times \vec{a}_3)]$. The same can be done for vectors $\vec{b}_2$ and $\vec{b}_3$ and one obtains $\vec{b}_2 = 2\pi\vec{a}_3 \times \vec{a}_1/[\vec{a}_2 \cdot (\vec{a}_3 \times \vec{a}_1)]$ and $\vec{b}_3 = 2\pi\vec{a}_1 \times \vec{a}_2/[\vec{a}_3 \cdot (\vec{a}_1 \times \vec{a}_2)]$. One notes that $\vec{a}_1 \cdot (\vec{a}_2 \times \vec{a}_3) = \vec{a}_2 \cdot (\vec{a}_3 \times \vec{a}_1) = \vec{a}_3 \cdot (\vec{a}_1 \times \vec{a}_2)$ and the basis vectors of the reciprocal lattice depend on the basis vectors of the direct lattice as

$$\begin{cases} \vec{b}_1 = 2\pi \dfrac{\vec{a}_2 \times \vec{a}_3}{\vec{a}_1 \cdot (\vec{a}_2 \times \vec{a}_3)}, & (1.15a) \\ \vec{b}_2 = 2\pi \dfrac{\vec{a}_3 \times \vec{a}_1}{\vec{a}_1 \cdot (\vec{a}_2 \times \vec{a}_3)}, & (1.15b) \\ \vec{b}_3 = 2\pi \dfrac{\vec{a}_1 \times \vec{a}_2}{\vec{a}_1 \cdot (\vec{a}_2 \times \vec{a}_3)}. & (1.15c) \end{cases}$$

In Eqs. (1.15), the scalar quantity $\vec{a}_1 \cdot (\vec{a}_2 \times \vec{a}_3)$ corresponds to the volume of the parallelepiped constructed from the three primitive vectors $\{\vec{a}_i\}$ of the original direct (Bravais) lattice, i.e., the volume of the unit cell [7]. Moreover one observes that the length of the reciprocal lattice vectors is proportional to the reciprocal of the length of the direct lattice vectors and this is the origin of the term *reciprocal lattice*.

As an example, Fig. 1.6 shows a simple-cubic Bravais lattice with a lattice constant a as well as its reciprocal lattice, which is also a simple-cubic one with a lattice constant $2\pi/a$, as follows from relations (1.15). Since the reciprocal lattice is a Bravais lattice, one can also find its Wigner-Seitz cell. The Wigner-Seitz cell of a reciprocal lattice is conventionally called a first Brillouin zone. Planes in reciprocal

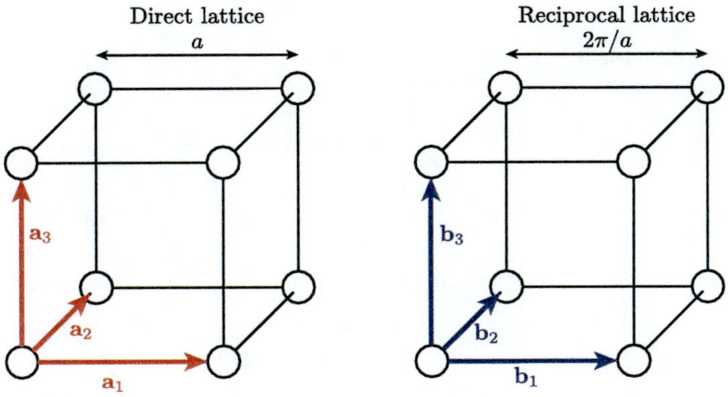

Fig. 1.6 Simple-cubic direct lattice and its reciprocal lattice. The primitive vectors of both lattices are also indicated

space, which bisect the lines joining a particular point of a reciprocal lattice with all other points, are known as Bragg planes. Therefore, the first Brillouin zone can also be defined as the set of all points in the reciprocal space that can be reached from the origin without crossing any Bragg plane. Symmetry properties may allow to reduce the dimensions of the first Brillouin zone and to define the smallest Brillouin zone also named the irreducible Brillouin zone.

In what follows we will consider, for the sake of simplicity, two examples of Bravais lattices in two dimensions, the square and the hexagonal one, and we will illustrate the concepts previously introduced.

1.3.3 Examples

1.3.3.1 The Square Bravais Lattice

We first consider the case of the square Bravais lattice with lattice parameter a. The points (the "atoms") are located at the vertices of a square as depicted in Fig. 1.7. The space is refereed to an orthonormal basis $(0, \vec{e}_1, \vec{e}_2, \vec{e}_3)$ with Cartesian coordinates (x_1, x_2, x_3) and O is a point chosen as origin. The array being two-dimensional in the plane (x_1, O, x_2), the primitive vectors of the direct lattice have components only in this plane and are $\vec{a}_1 = a\vec{e}_1$ and $\vec{a}_2 = a\vec{e}_2$ and vectors $\vec{R}$ write $\vec{R} = a\,(\ell\vec{e}_1 + m\vec{e}_2)$. In Fig. 1.7, the red lines correspond to lines connecting the origin to its nearest neighbours and blue lines bisect the red dashed ones. One deduces easily from this drawing that the Wigner-Seitz cell is the grey square.

One may define now the basis vectors of the reciprocal lattice applying relations (1.15). One obtains

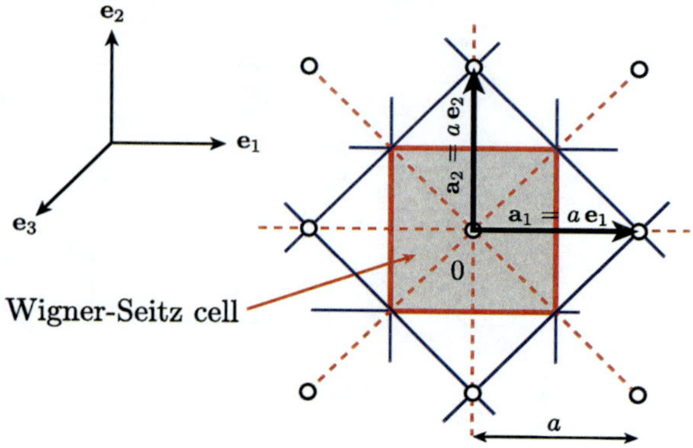

Fig. 1.7 The square direct lattice of lattice parameter a and its Wigner-Seitz cell

$$\begin{cases} \vec{b}_1 = 2\pi \dfrac{a\vec{e}_2 \times \vec{e}_3}{a\vec{e}_1 \cdot (a\vec{e}_2 \times \vec{e}_3)} = \dfrac{2\pi}{a}\vec{e}_1, & (1.16a) \\ \vec{b}_2 = 2\pi \dfrac{\vec{e}_3 \times a\vec{e}_1}{a\vec{e}_1 \cdot (a\vec{e}_2 \times \vec{e}_3)} = \dfrac{2\pi}{a}\vec{e}_2, & (1.16b) \end{cases}$$

where $\vec{e}_3$ is introduced only for constructing the different cross products.

In Fig. 1.8a, the dots represent points whose positions are given by two-dimensional vectors $\vec{G} = \ell'\vec{b}_1 + m'\vec{b}_2$. One chooses point Γ as the origin of the reciprocal lattice and draws red dashed lines connecting Γ to its nearest neighbours. With the help of the blue lines bisecting the red ones, the grey square is defined as

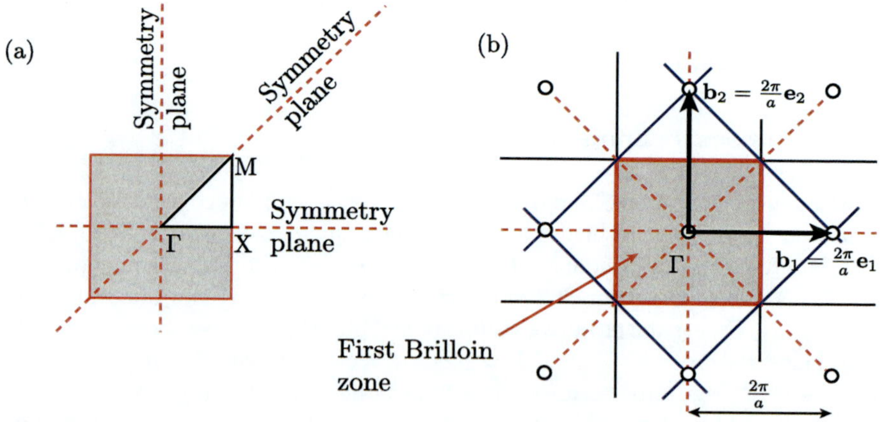

Fig. 1.8 **a** The reciprocal lattice of the square Bravais lattice and the first Brillouin zone (grey square), **b** the first Brillouin zone and the irreducible Brillouin zone (ΓXM) with $\Gamma : \frac{2\pi}{a}(0, 0)$, $X : \frac{2\pi}{a}(\frac{1}{2}, 0)$ and $M : \frac{2\pi}{a}(\frac{1}{2}, \frac{1}{2})$. Γ, X, and M are the points of highest symmetry in the irreducible Brillouin zone. In terms of wave propagation, vectors $\vec{\Gamma X}$, $\vec{XM}$ and $\vec{\Gamma M}$ represent the principal directions of propagation

the first Brillouin zone. Due to the symmetries of the grey square, shown in Fig. 1.8b, one reduces the study to the triangle (ΓXM) where the components of point Γ in the basis $(\vec{e}_1, \vec{e}_2)$ are $\frac{2\pi}{a}(0, 0)$. Because $\vec{\Gamma X} = \vec{b}_1/2$ and $\vec{\Gamma M} = (\vec{b}_1 + \vec{b}_2)/2$, the components of points X and M are $(\frac{2\pi}{a})(\frac{1}{2}, 0)$ and $\frac{2\pi}{a}(\frac{1}{2}, \frac{1}{2})$, respectively. The triangle (ΓXM) is the irreducible Brillouin zone of the square Bravais lattice.

When considering propagation of waves in two-dimensional phononic crystals, any wave vector $\vec{k}$ (which belongs to the reciprocal space), can be written as $\vec{k} = \vec{G} + \vec{K}_{\mathrm{IBZ}}$ where $\vec{K}_{\mathrm{IBZ}}$ belongs to the irreducible Brillouin zone. Consequently study of the propagation of waves can be limited to waves with wave vectors belonging to the irreducible Brillouin zone (see Sect. 1.2).

1.3.3.2 The Hexagonal Bravais Lattice

As depicted in Fig. 1.9, in the case of the hexagonal Bravais lattice, the "atoms" are located on the vertices and at the centre of a regular hexagon of side length a. One can construct the Wigner-Seitz cell following the same processes than that used for the square array. Around point O' the Wigner-Seitz cell has exactly the same symmetry as the direct lattice and is represented by the red hexagon. Nevertheless, due to the rather complicated geometry of the array, one may choose a simpler primitive unit cell such as the blue parallelogram constructed from the set of primitive lattice vectors

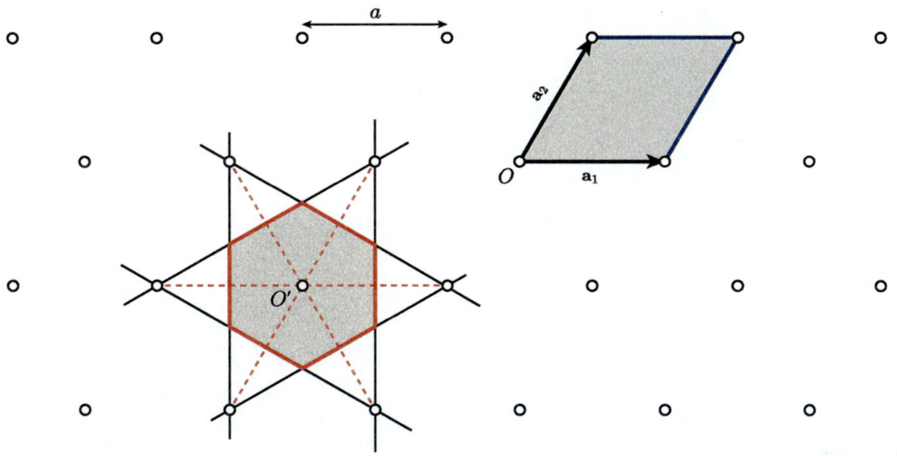

Fig. 1.9 The hexagonal direct lattice with lattice parameter a and its Wigner- Seitz cell around point O' (red hexagon). It is convenient to choose as primitive unit cell the blue parallelogram constructed from the primitive vectors $\vec{a}_1 = a\vec{e}_1$ and $\vec{a}_2 = \frac{a}{2}\vec{e}_1 + \frac{a\sqrt{3}}{2}\vec{e}_2$

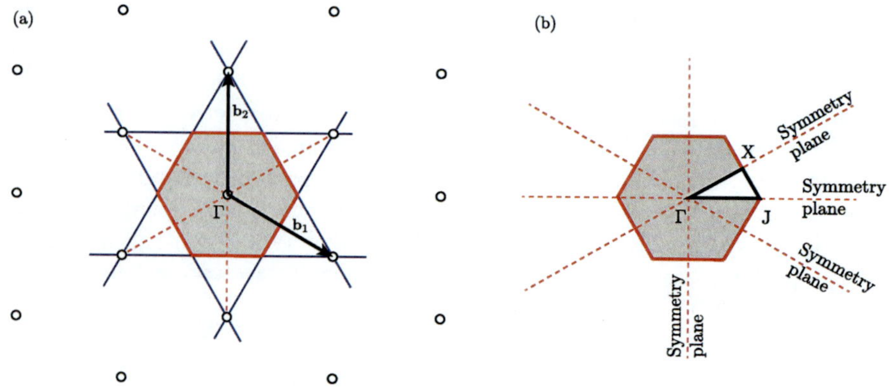

Fig. 1.10 **a** The reciprocal lattice of the hexagonal Bravais lattice and the first Brillouin zone (grey square), **b** the first Brillouin zone and the irreducible Brillouin zone $(\Gamma J X)$ with $\Gamma : \frac{2\pi}{a}(0, 0)$, $J : \frac{2\pi}{a}(\frac{2}{3}, 0)$ and $X : \frac{2\pi}{a}(\frac{1}{2}, \frac{1}{2\sqrt{3}})$. Points Γ, J, and X are the points of highest symmetry in the irreducible Brillouin zone

$\{\vec{a}_1 = a\vec{e}_1, \vec{a}_2 = \frac{a}{2}\vec{e}_1 + \frac{a\sqrt{3}}{2}\vec{e}_2\}$. The basis vectors of the reciprocal lattice are then obtained using relations (1.15) as

$$\begin{cases} \vec{b}_1 = 2\pi\frac{\vec{a}_2 \times \vec{e}_3}{\vec{a}_1 \cdot (\vec{a}_2 \times \vec{e}_3)} = \frac{2\pi}{a}\left(\vec{e}_1 - \frac{1}{\sqrt{3}}\vec{e}_2\right), & (1.17a) \\ \vec{b}_2 = 2\pi\frac{\vec{e}_3 \times \vec{a}_1}{\vec{a}_1 \cdot (\vec{a}_2 \times \vec{e}_3)} = \frac{2\pi}{a}\left(\frac{2}{\sqrt{3}}\right)\vec{e}_2, & (1.17b) \end{cases}$$

and the reciprocal lattice vectors are given in the basis $(O, \vec{e}_1, \vec{e}_2)$ as $\vec{G} = \ell'\vec{b}_1 + m'\vec{b}_2 = \frac{2\pi}{a}[\ell'\vec{e}_1 + \frac{1}{\sqrt{3}}(-\ell' + 2m')\vec{e}_2]$.

The reciprocal lattice of the hexagonal array together with its first Brillouin zone are depicted in Fig. 1.10a. Due to the symmetries of the grey hexagon, shown in Fig. 1.10b, one reduces the study to the triangle $(\Gamma J X)$ where the components of point Γ in the basis $(\vec{e}_1, \vec{e}_2)$ are $\frac{2\pi}{a}(0, 0)$. Because $\vec{\Gamma X} = (\vec{b}_1 + \vec{b}_2)/2$, the components of point X are $\frac{2\pi}{a}(\frac{1}{2}, \frac{1}{2\sqrt{3}})$. Finally, triangle $(\Gamma J X)$ being right-angled on X, one may write $\Gamma X^2 + X J^2 = \Gamma J^2$ and one deduces that point J has components $\frac{2\pi}{a}(\frac{2}{3}, 0)$. The area of the triangle $(\Gamma J X)$ is the irreducible Brillouin zone of the hexagonal Bravais lattice.

1.4 The Plane Wave Expansion Method

1.4.1 Plane Wave Expansion Method for Bulk Phononic Crystals

1.4.1.1 Equations of Propagation of Elastic Waves in Heterogeneous Media

One considers an heterogeneous elastic medium of infinite extent along the 3 spatial directions (x_1, x_2, x_3), made of constituent materials of specific crystallographic symmetry (isotropic, cubic, …). The Cartesian coordinates system is referred to an orthonormal basis $(O, \vec{e}_1, \vec{e}_2, \vec{e}_3)$. At every point, $\vec{r}$, the medium is characterized by the material parameters namely the mass density $\rho(\vec{r})$ and the elastic moduli $C_{ijk\ell}(\vec{r})$. The elements of the stress tensor T_{ij} and those of the strain tensor $S_{k\ell}$ are related through the Hooke's law [8]

$$T_{ij}(\vec{r}) = \sum_{kl} C_{ijk\ell}(\vec{r}) S_{k\ell}(\vec{r}), \tag{1.18}$$

where each index, i, j, k, ℓ varies between 1 and 3. One notes that in this relation the stress and the strain tensors are of rank 2 with $3^2 = 9$ elements while the tensor of elastic moduli is of rank 4 with $3^4 = 81$ elements. Constituent materials are assumed to be linear materials (limit of small strains) and the elements of the strain tensor are expressed as $S_{k\ell}(\vec{r}) = \frac{1}{2}(\partial u_k(\vec{r})/\partial x_\ell + \partial u_\ell(\vec{r})/\partial x_k)$ where $u_i(\vec{r})$, $i = 1, 2, 3$ refer to the components of the displacement vector $\vec{u}$ in the (x_1, x_2, x_3) Cartesian coordinates system. The elements of the elastic moduli tensor $C_{ijk\ell}$ must satisfy $C_{ijk\ell} = C_{jik\ell}$ because $T_{ij} = T_{ji}$ and $C_{ijk\ell} = C_{ij\ell k}$ because $S_{k\ell} = S_{\ell k}$ and thermodynamic laws impose $C_{ijk\ell} = C_{k\ell ij}$ [8]. Then, in terms of displacements, Hooke's law writes

$$\begin{aligned} T_{ij}(\vec{r}) &= \frac{1}{2} \sum_{kl} C_{ijk\ell}(\vec{r}) \frac{\partial u_k(\vec{r})}{\partial x_\ell} + \frac{1}{2} \sum_{kl} C_{ijk\ell} \frac{\partial u_\ell(\vec{r})}{\partial x_k}, \\ &= \frac{1}{2} \sum_{kl} C_{ijk\ell}(\vec{r}) \frac{\partial u_k(\vec{r})}{\partial x_\ell} + \frac{1}{2} \sum_{kl} C_{ij\ell k} \frac{\partial u_\ell(\vec{r})}{\partial x_k}, \end{aligned} \tag{1.19}$$

and since $C_{ijk\ell} = C_{ij\ell k}$, the two summations on the right are equal so that

$$T_{ij}(\vec{r}) = \sum_{k\ell} C_{ijk\ell}(\vec{r}) \frac{\partial u_k(\vec{r})}{\partial x_\ell}. \tag{1.20}$$

In absence of external forces, Newton's second law leads to the equations of motion

$$\rho(\vec{r})\frac{\partial^2 u_i(\vec{r})}{\partial t^2} = \sum_j \frac{\partial T_{ij}(\vec{r})}{\partial x_j} = \sum_j \frac{\partial}{\partial x_j}\left[\sum_{k\ell} C_{ijk\ell}(\vec{r})\frac{\partial u_k(\vec{r})}{\partial x_\ell}\right]. \tag{1.21}$$

The tensor of elastic moduli contains, *a priori*, 81 elements but due to the symmetries of this tensor ($C_{ijk\ell} = C_{jik\ell}$, $C_{ijk\ell} = C_{ij\ell k}$, $C_{ijk\ell} = C_{k\ell ij}$), this number reduces to 21 and the Hooke's law can be rewritten in the following matrix form

$$\begin{bmatrix} T_{11} \\ T_{22} \\ T_{33} \\ T_{23} \\ T_{31} \\ T_{12} \end{bmatrix} = \begin{bmatrix} C_{1111} & C_{1122} & C_{1133} & C_{1123} & C_{1131} & C_{1112} \\ C_{1122} & C_{2222} & C_{2233} & C_{2223} & C_{2231} & C_{2212} \\ C_{1133} & C_{2233} & C_{3333} & C_{3323} & C_{3331} & C_{3312} \\ C_{1123} & C_{2223} & C_{3323} & C_{2323} & C_{2331} & C_{2312} \\ C_{1131} & C_{2231} & C_{3331} & C_{2331} & C_{3131} & C_{3112} \\ C_{1112} & C_{2212} & C_{3312} & C_{2312} & C_{3112} & C_{1212} \end{bmatrix} \begin{bmatrix} S_{11} \\ S_{22} \\ S_{33} \\ 2S_{23} \\ 2S_{31} \\ 2S_{12} \end{bmatrix}. \tag{1.22}$$

This equation can be rewritten using the Voigt notation: *a pair of indices ij is replaced by a single index m* as follows

$$\begin{aligned} (11) &\leftrightarrow 1, \\ (22) &\leftrightarrow 2, \\ (33) &\leftrightarrow 3, \\ (23) \text{ or } (32) &\leftrightarrow 4, \\ (31) \text{ or } (13) &\leftrightarrow 5, \\ (12) \text{ or } (21) &\leftrightarrow 6, \end{aligned} \tag{1.23}$$

and Eq. (1.22) becomes

$$\begin{bmatrix} T_1 \\ T_2 \\ T_3 \\ T_4 \\ T_5 \\ T_6 \end{bmatrix} = \begin{bmatrix} C_{11} & C_{12} & C_{13} & C_{14} & C_{15} & C_{16} \\ C_{12} & C_{22} & C_{23} & C_{24} & C_{25} & C_{26} \\ C_{13} & C_{23} & C_{33} & C_{34} & C_{35} & C_{36} \\ C_{14} & C_{24} & C_{34} & C_{44} & C_{45} & C_{46} \\ C_{15} & C_{25} & C_{35} & C_{45} & C_{55} & C_{56} \\ C_{16} & C_{26} & C_{36} & C_{46} & C_{56} & C_{66} \end{bmatrix} \begin{bmatrix} S_1 \\ S_2 \\ S_3 \\ 2S_4 \\ 2S_5 \\ 2S_6 \end{bmatrix}. \tag{1.24}$$

In the particular case of constituent materials of cubic crystallographic symmetry, only three independent elastic moduli C_{11}, C_{12} and C_{44} are involved and Hooke's law writes in the matrix form

$$\begin{bmatrix} T_1 \\ T_2 \\ T_3 \\ T_4 \\ T_5 \\ T_6 \end{bmatrix} = \begin{bmatrix} C_{11} & C_{12} & C_{12} & 0 & 0 & 0 \\ C_{12} & C_{11} & C_{12} & 0 & 0 & 0 \\ C_{12} & C_{12} & C_{11} & 0 & 0 & 0 \\ 0 & 0 & 0 & C_{44} & 0 & 0 \\ 0 & 0 & 0 & 0 & C_{44} & 0 \\ 0 & 0 & 0 & 0 & 0 & C_{44} \end{bmatrix} \begin{bmatrix} S_1 \\ S_2 \\ S_3 \\ 2S_4 \\ 2S_5 \\ 2S_6 \end{bmatrix}. \tag{1.25}$$

The isotropic material can be considered as a particular case of cubic material imposing $C_{12} = C_{11} - 2C_{44}$ and only two independent elastic moduli, C_{11} and C_{44} are necessary to describe the elastic behaviour of the material. In what follows, we will limit ourself to constituent materials of cubic crystallographic symmetry and combining Eqs. (1.21) and (1.25), we can write the equation of motion in this case as follows

$$
\begin{aligned}
\rho \frac{\partial^2 u_1}{\partial t^2} &= \frac{\partial T_{11}}{\partial x_1} + \frac{\partial T_{12}}{\partial x_2} + \frac{\partial T_{13}}{\partial x_3}, \\
&= \frac{\partial}{\partial x_1}[C_{11}S_1 + C_{12}(S_2 + S_3)] + \frac{\partial}{\partial x_2}(C_{44}2S_6) + \frac{\partial}{\partial x_3}(C_{44}2S_5), \\
&= \frac{\partial}{\partial x_1}\left[C_{11}\frac{\partial u_1}{\partial x_1} + C_{12}\left(\frac{\partial u_2}{\partial x_2} + \frac{\partial u_3}{\partial x_3}\right)\right] + \cdots \\
&\quad + \frac{\partial}{\partial x_2}\left[C_{44}\left(\frac{\partial u_1}{\partial x_2} + \frac{\partial u_2}{\partial x_1}\right)\right] + \frac{\partial}{\partial x_3}\left[C_{44}\left(\frac{\partial u_1}{\partial x_3} + \frac{\partial u_3}{\partial x_1}\right)\right],
\end{aligned} \tag{1.26}
$$

$$
\begin{aligned}
\rho \frac{\partial^2 u_2}{\partial t^2} &= \frac{\partial T_{21}}{\partial x_1} + \frac{\partial T_{22}}{\partial x_2} + \frac{\partial T_{23}}{\partial x_3}, \\
&= \frac{\partial}{\partial x_1}\left[C_{44}\left(\frac{\partial u_1}{\partial x_2} + \frac{\partial u_2}{\partial x_1}\right)\right] + \frac{\partial}{\partial x_2}\left[C_{11}\frac{\partial u_2}{\partial x_2} + C_{12}\left(\frac{\partial u_1}{\partial x_1} + \frac{\partial u_3}{\partial x_3}\right)\right] + \\
&\quad + \frac{\partial}{\partial x_3}\left[C_{44}\left(\frac{\partial u_2}{\partial x_3} + \frac{\partial u_3}{\partial x_2}\right)\right],
\end{aligned} \tag{1.27}
$$

and

$$
\begin{aligned}
\rho \frac{\partial^2 u_3}{\partial t^2} &= \frac{\partial T_{31}}{\partial x_1} + \frac{\partial T_{32}}{\partial x_2} + \frac{\partial T_{33}}{\partial x_3}, \\
&= \frac{\partial}{\partial x_1}\left[C_{44}\left(\frac{\partial u_1}{\partial x_3} + \frac{\partial u_3}{\partial x_1}\right)\right] + \frac{\partial}{\partial x_2}\left[C_{44}\left(\frac{\partial u_2}{\partial x_3} + \frac{\partial u_3}{\partial x_2}\right)\right] + \cdots \\
&\quad + \frac{\partial}{\partial x_3}\left[C_{11}\frac{\partial u_3}{\partial x_3} + C_{12}\left(\frac{\partial u_1}{\partial x_1} + \frac{\partial u_2}{\partial x_2}\right)\right].
\end{aligned} \tag{1.28}
$$

In Eqs. (1.26), (1.27) and (1.28), for the sake of simplicity, dependence of ρ, u_i and C_{mn} on $\vec{r}$ has been omitted. Elastic wave propagation in a heterogeneous material of infinite extent is governed by three coupled second order partial differential equations (1.26–1.28). When the inhomogeneities are distributed periodically in space, e.g. phononic crystals, these three coupled equations can be solved using the plane wave expansion (PWE) method.

1.4.1.2 Basic Principles of the PWE Method for Bulk Phononic Crystals

One considers a periodic structure with the direct lattice (DL), of specific geometry, characterized by its unit cell (UC). The reciprocal lattice (RL) vectors are $\vec{G}(G_1, G_2, G_3)$ with respect to the orthonormal basis $(O, \vec{e}_1, \vec{e}_2, \vec{e}_3)$. The solutions of the equations of motion are assumed to be pure tones in the form $\vec{u}(\vec{r}, t) = \vec{u}(\vec{r})e^{-i\omega t}$ where ω is the circular frequency. Due to the periodicity of the structure, the Bloch-Floquet theorem states that $\vec{u}(\vec{r})$ can be written in the form

$$\vec{u}(\vec{r}) = e^{i\vec{K}\cdot\vec{r}}\vec{U}_{\vec{K}}(\vec{r}), \tag{1.29}$$

where $\vec{K}(K_1, K_2, K_3)$ is the Bloch wave vector and $\vec{U}_{\vec{K}}(\vec{r})$ has the periodicity of the direct lattice. Then $\vec{U}_{\vec{K}}(\vec{r})$ can be developed in Fourier series as

$$\vec{U}_{\vec{K}}(\vec{r}) = \sum_{\vec{G}'} \vec{U}_{\vec{K}}\left(\vec{G}'\right) e^{i\vec{G}'\cdot\vec{r}}, \quad \text{where} \quad \vec{G}' \in (RL) \tag{1.30}$$

and

$$\vec{u}(\vec{r}, t) = e^{-i\omega t} \sum_{\vec{G}'} \vec{U}_{\vec{K}}(\vec{G}')e^{i(\vec{G}'+\vec{K})\cdot\vec{r}}. \tag{1.31}$$

The material parameters, mass density $\rho(\vec{r})$ and elastic moduli C_{mn} are periodic functions of the position, i.e., $\rho\left(\vec{r} + \vec{R}\right) = \rho(\vec{r})$ and $C_{mn}(\vec{r} + \vec{R}) = C_{mn}(\vec{r})$ where $\vec{R} \in (DL)$ and can be expanded in Fourier series such as

$$\begin{cases} \rho(\vec{r}) = \sum_{\vec{G}''} \rho(\vec{G}'')e^{i\vec{G}''\cdot\vec{r}} & \text{(1.32a)} \\ C_{mn}(\vec{r}) = \sum_{\vec{G}f''} C_{mn}(\vec{G}'')e^{i\vec{G}''\cdot\vec{r}} & \text{(1.32b)} \end{cases}$$

where $\vec{G}'' \in (RL)$. Note that the Fourier coefficients $\rho(\vec{G}'')$ and $C_{mn}(\vec{G}'')$ in Eqs. (1.32) are defined as

$$\begin{cases} \rho(\vec{G}'') = \dfrac{1}{V_{(U.C.)}} \displaystyle\iiint_{(U.C.)} \rho(\vec{r})e^{-i\vec{G}''\cdot\vec{r}}d^3\vec{r}, & \text{(1.33a)} \\ C_{mn}(\vec{G}'') = \dfrac{1}{V_{(U.C.)}} \displaystyle\iiint_{(U.C.)} C_{mn}(\vec{r})e^{-i\vec{G}''\cdot\vec{r}}d^3\vec{r}, & \text{(1.33b)} \end{cases}$$

where the integration is performed over the volume $V_{(U.C.)}$ of the UC of the direct lattice.

Inserting Eqs. (1.31, 1.32) in Eqs. (1.26–1.28) results in the Fourier transform of the equations of motion. For example, performing these substitutions on the left hand side of Eq. (1.26), one obtains

$$\rho(\vec{r})\frac{\partial^2 u_1(\vec{r})}{\partial t^2} = -\omega^2 e^{i(\vec{K}\cdot\vec{r}-\omega t)} \sum_{\vec{G}',\vec{G}''} \rho(\vec{G}'')U_{1,\vec{K}}(\vec{G}')e^{i(\vec{G}'+\vec{G}'')\cdot\vec{r}} \tag{1.34}$$

where $U_{1,\vec{K}}$ is the component of vector $\vec{U}_{\vec{K}}$ along $\vec{e}_1$. Subsequently, the first term of the right hand side of Eq. (1.26) is transformed as

$$\begin{aligned}
&\frac{\partial}{\partial x_1}\left[C_{11}(\vec{r})\frac{\partial u_1(\vec{r})}{\partial x_1}\right] = \\
&= e^{-i\omega t}\frac{\partial}{\partial x_1}\left[\sum_{\vec{G}',\vec{G}''} C_{11}(\vec{G}'')e^{i\vec{G}''\cdot\vec{r}}[i(K_1+G_1')]e^{i(\vec{K}+\vec{G}')\cdot\vec{r}}U_{1,\vec{K}}(\vec{G}')\right] \\
&= e^{-i\omega t}\frac{\partial}{\partial x_1}\left[\sum_{\vec{G}',\vec{G}''} C_{11}(\vec{G}'')[i(K_1+G_1')]e^{i(\vec{K}+\vec{G}'+\vec{G}'')\cdot\vec{r}}U_{1,\vec{K}}(\vec{G}')\right] \\
&= -e^{i(\vec{K}\cdot\vec{r}-\omega t)}\left[\sum_{\vec{G}',\vec{G}''} C_{11}(\vec{G}'')[(K_1+G_1')(K_1+G_1'+G_1'')]e^{i(\vec{G}'+\vec{G}'')\vec{r}} \times \cdots \right. \\
&\qquad\qquad \left. U_{1,\vec{K}}(\vec{G}')\right],
\end{aligned} \tag{1.35}$$

similarly to the other terms of the right-hand side of Eq. (1.26). Finally, the Fourier transform of Eq. (1.26) writes as

$$\begin{aligned}
&-\omega^2 e^{i(\vec{K}\cdot\vec{r}-\omega t)}\sum_{\vec{G}',\vec{G}''}\rho(\vec{G}'')U_{1,\vec{K}}(\vec{G}')e^{i(\vec{G}'+\vec{G}'')\cdot\vec{r}} = \cdots \\
&= -e^{i(\vec{K}\cdot\vec{r}-\omega t)}\cdot\sum_{\vec{G}',\vec{G}''} e^{i(\vec{G}'+\vec{G}'')\cdot\vec{r}}\Big\{\Big[\Big[C_{11}(\vec{G}'')(K_1+G_1')(K_1+G_1'+G_1'')+\cdots \\
&+ C_{44}(\vec{G}'')\Big[(K_2+G_2')(K_2+G_2'+G_2'')+(K_3+G_3')(K_3+G_3'+G_3'')\Big]\Big]U_{1,\vec{K}}(\vec{G}')+\cdots \\
&\quad + \Big[C_{12}(\vec{G}'')(K_2+G_2')(K_1+G_1'+G_1'')+\cdots \\
&\qquad\qquad + C_{44}(\vec{G}'')(K_1+G_1')(K_2+G_2'+G_2'')\Big]U_{2,\vec{K}}(\vec{G}')+\cdots \\
&\quad + \Big[C_{12}(\vec{G}'')(K_3+G_3')(K_1+G_1'+G_1'')+\cdots \\
&\qquad\qquad + C_{44}(\vec{G}'')(K_1+G_1')(K_3+G_3'+G_3'')\Big]U_{3,\vec{K}}(\vec{G}')\Big\}.
\end{aligned} \tag{1.36}$$

Eliminating $-e^{i(\vec{K}\cdot\vec{r}-\omega t)}$ in both parts of Eq. (1.36) and then multiplying both parts by $e^{-i\vec{G}\cdot\vec{r}}$ where $\vec{G}\in(RL)$ allows representation of the terms in the form $e^{i(\vec{G}'+\vec{G}''-\vec{G})\cdot\vec{r}}$. Because

$$\frac{1}{V_{(U.C.)}}\iiint_{(U.C.)} e^{i(\vec{G}'+\vec{G}''-\vec{G})\cdot\vec{r}}d^3\vec{r} = \cdots$$

$$= \delta_{(\vec{G}'+\vec{G}''-\vec{G}),\vec{0}} = \begin{cases} 1, & \text{if } \vec{G}'+\vec{G}''-\vec{G}=\vec{O} \\ 0, & \text{if } \vec{G}'+\vec{G}''-\vec{G}\neq\vec{O} \end{cases} \tag{1.37}$$

integration over $V_{(U.C.)}$ of Eq. (1.36) leads to

$$\begin{aligned}
\omega^2 \textstyle\sum_{\vec{G}'} \rho(\vec{G}-\vec{G}')U_{1,\vec{K}}(\vec{G}') = \textstyle\sum_{\vec{G}'} \Big\{ \Big[& C_{11}(\vec{G}-\vec{G}')(K_1+G'_1).(K_1+G_1)+\cdots \\
+C_{44}(\vec{G}-\vec{G}')\Big[(K_2+G'_2)(K_2+G_2)&+(K_3+G'_3)(K_3+G_3)\Big]\Big]\cdot U_{1,\vec{K}}(\vec{G}')+\cdots \\
&+\Big[C_{12}(\vec{G}-\vec{G}')(K_2+G'_2)(K_1+G_1)+\cdots \\
&+C_{44}(\vec{G}-\vec{G}')(K_1+G'_1)(K_2+G_2)\Big]\cdot U_{2,\vec{K}}(\vec{G}')+\cdots \\
&+\Big[C_{12}(\vec{G}-\vec{G}')(K_3+G'_3)(K_1+G_1)+\cdots \\
&+C_{44}(\vec{G}-\vec{G}')(K_1+G'_1)(K_3+G_3)\Big]\cdot U_{3,\vec{K}}(\vec{G}')\Big\}.
\end{aligned} \tag{1.38}$$

Applying the same algebra for Eqs. (1.27) and (1.28), the Fourier transforms of the equations of motion are derived in the form of 3 coupled equations given as

$$\begin{cases}
\omega^2 \sum_{\vec{G}'} B^{(11)}_{\vec{G},\vec{G}'} U_{1,\vec{K}}(\vec{G}') = \\
\quad = \sum_{\vec{G}'} \left\{ A^{(11)}_{\vec{G},\vec{G}'} U_{1,\vec{K}}(\vec{G}') + A^{(12)}_{\vec{G},\vec{G}'} U_{2,\vec{K}}(\vec{G}') + A^{(13)}_{\vec{G},\vec{G}'} U_{3,\vec{K}}(\vec{G}') \right\}, & (1.39a) \\
\omega^2 \sum_{\vec{G}'} B^{(22)}_{\vec{G},\vec{G}'} U_{2,\vec{K}}(\vec{G}') = \\
\quad = \sum_{\vec{G}'} \left\{ A^{(21)}_{\vec{G},\vec{G}'} U_{1,\vec{K}}(\vec{G}') + A^{(22)}_{\vec{G},\vec{G}'} U_{2,\vec{K}}(\vec{G}') + A^{(23)}_{\vec{G},\vec{G}'} U_{3,\vec{K}}(\vec{G}') \right\}, & (1.39b) \\
\omega^2 \sum_{\vec{G}'} B^{(33)}_{\vec{G},\vec{G}'} U_{3,\vec{K}}(\vec{G}') = \\
\quad = \sum_{\vec{G}'} \left\{ A^{(31)}_{\vec{G},\vec{G}'} U_{1,\vec{K}}(\vec{G}') + A^{(32)}_{\vec{G},\vec{G}'} U_{2,\vec{K}}(\vec{G}') + A^{(33)}_{\vec{G},\vec{G}'} U_{3,\vec{K}}(\vec{G}') \right\}, & (1.39c)
\end{cases}$$

where

$$
\begin{cases}
B^{(11)}_{\vec{G},\vec{G}'} = B^{(22)}_{\vec{G},\vec{G}'} = B^{(33)}_{\vec{G},\vec{G}'} = \rho(\vec{G}-\vec{G}'), & (1.40a)\\
A^{(11)}_{\vec{G},\vec{G}'} = C_{11}(\vec{G}-\vec{G}')(G_1+K_1)(G'_1+K_1)+\cdots \\
\quad +C_{44}(\vec{G}-\vec{G}')\Big[(G_2+K_2)(G'_2+K_2)+(G_3+K_3)(G'_3+K_3)\Big], & (1.40b)\\
A^{(12)}_{\vec{G},\vec{G}'} = C_{12}(\vec{G}-\vec{G}')(G_1+K_1)(G'_2+K_2)+\cdots \\
\quad +C_{44}(\vec{G}-\vec{G}')(G'_1+K_1)(G_2+K_2), & (1.40c)\\
A^{(13)}_{\vec{G},\vec{G}'} = C_{12}(\vec{G}-\vec{G}')(G_1+K_1)(G'_3+K_3)+\cdots \\
\quad +C_{44}(\vec{G}-\vec{G}')(G'_1+K_1)(G_3+K_3), & (1.40d)\\
A^{(21)}_{\vec{G},\vec{G}'} = C_{12}(\vec{G}-\vec{G}')(G'_1+K_1)(G_2+K_2)+\cdots \\
\quad +C_{44}(\vec{G}-\vec{G}')(G'_2+K_2)(G_1+K_1), & (1.40e)\\
A^{(22)}_{\vec{G},\vec{G}'} = C_{11}(\vec{G}-\vec{G}')(G_2+K_2)(G'_2+K_2)+\cdots \\
\quad +C_{44}(\vec{G}-\vec{G}')\Big[(G_1+K_1)(G'_1+K_1)+(G_3+K_3)(G'_3+K_3)\Big], & (1.40f)\\
A^{(23)}_{\vec{G},\vec{G}'} = C_{12}(\vec{G}-\vec{G}')(G_2+K_2)(G'_3+K_3)+\cdots \\
\quad +C_{44}(\vec{G}-\vec{G}')(G'_2+K_2)(G_3+K_3), & (1.40g)\\
A^{(31)}_{\vec{G},\vec{G}'} = C_{12}(\vec{G}-\vec{G}')(G'_1+K_1)(G_3+K_3)+\cdots \\
\quad +C_{44}(\vec{G}-\vec{G}')(G_1+K_1)(G'_3+K_3), & (1.40h)\\
A^{(32)}_{\vec{G},\vec{G}'} = C_{12}(\vec{G}-\vec{G}')(G'_2+K_2)(G_3+K_3)+\cdots \\
\quad +C_{44}(\vec{G}-\vec{G}')(G_2+K_2)(G'_3+K_3), & (1.40i)\\
A^{(33)}_{\vec{G},\vec{G}'} = C_{11}(\vec{G}-\vec{G}')(G_3+K_3)(G'_3+K_3)+\cdots \\
\quad +C_{44}(\vec{G}-\vec{G}')\Big[(G_1+K_1)(G'_1+K_1)+(G_2+K_2)(G'_2+K_2)\Big]. & (1.40j)
\end{cases}
$$

Equation (1.39) can be recast in the following matrix form

$$
\omega^2 \begin{bmatrix} B^{(11)}_{\vec{G},\vec{G}'} & 0 & 0 \\ 0 & B^{(22)}_{\vec{G},\vec{G}'} & 0 \\ 0 & 0 & B^{(33)}_{\vec{G},\vec{G}'} \end{bmatrix} \begin{bmatrix} U_{1,\vec{K}}(\vec{G}') \\ U_{2,\vec{K}}(\vec{G}') \\ U_{3,\vec{K}}(\vec{G}') \end{bmatrix} =
$$

$$
= \begin{bmatrix} A^{(11)}_{\vec{G},\vec{G}'} & A^{(12)}_{\vec{G},\vec{G}'} & A^{(13)}_{\vec{G},\vec{G}'} \\ A^{(21)}_{\vec{G},\vec{G}'} & A^{(22)}_{\vec{G},\vec{G}'} & A^{(23)}_{\vec{G},\vec{G}'} \\ A^{(31)}_{\vec{G},\vec{G}'} & A^{(32)}_{\vec{G},\vec{G}'} & A^{(33)}_{\vec{G},\vec{G}'} \end{bmatrix} \begin{bmatrix} U_{1,\vec{K}}(\vec{G}') \\ U_{2,\vec{K}}(\vec{G}') \\ U_{3,\vec{K}}(\vec{G}') \end{bmatrix}, \qquad (1.41)
$$

or

$$\omega^2 \overleftrightarrow{B} \vec{\mathcal{U}} = \overleftrightarrow{A} \vec{\mathcal{U}}, \tag{1.42}$$

where $\overleftrightarrow{A}$ and $\overleftrightarrow{B}$ are square matrices and $\vec{\mathcal{U}}$ is a vector whose sizes depend on the number of reciprocal lattice vectors taken into account in the Fourier series. The numerical resolution of this generalized eigenvalue problem is performed for a fixed value of the wave vector $\vec{K} = (K_1, K_2, K_3)$ describing the contour of the irreducible Brillouin zone of the array of inclusions. Then one obtains a set of eigenfrequencies $\omega(\vec{K})$.

Equation (1.41) is general and constitutes the basis equation for applying the PWE method to the calculation of the dispersion curves of a three-dimensional periodic structure with a specific geometry. For lower dimensions, this *master* equation can be simplified by cancelling some components of the wave vectors and of the reciprocal lattice vectors. This is the object of the next part in which we will consider the particular case of bulk two-dimensional phononic crystals.

1.4.1.3 PWE Method for Bulk Two-Dimensional Phononic Crystals

A bulk two-dimensional phononic crystal consists of an array of infinitely long parallel cylinders, of specific cross-section, e.g., circular, square, elliptical, hexagonal, made of an elastic material A embedded in an elastic matrix B, as shown in Fig. 1.11. Materials A and B are supposed to be of cubic crystallographic symmetry.

Due to the infinite length of the cylinders along the x_3 axis, there exists an invariance by translation along x_3 and the material parameters (density and elastic moduli) and hence the displacement field does not depend on x_3. This means that G_3 and G'_3 vanish in Eq. (1.39). Moreover, one can limit the propagation of waves to the transverse plane $(x_1 O x_2)$ and hence assume $K_3 = O$. Due to this, in Eq. (1.40), terms

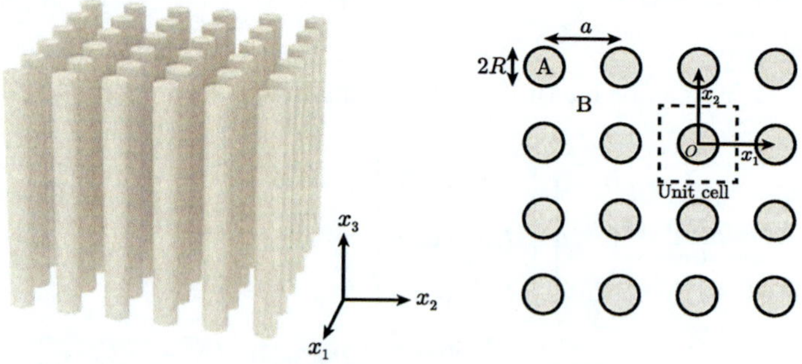

Fig. 1.11 Left panel: Two-dimensional phononic crystal made of a square array (lattice parameter a) of circular cylinders A of radius R embedded in a matrix B. The cylinders are oriented along the x_3 axis and are assumed infinite along this direction. The structure is periodic along x_1 and x_2; Right panel: Cross section in the transverse $(x_1, 0, x_2)$ plane of the 2D phononic crystal

$A^{(13)}_{\vec{G},\vec{G}'}$, $A^{(23)}_{\vec{G},\vec{G}'}$, $A^{(31)}_{\vec{G},\vec{G}'}$ and $A^{(32)}_{\vec{G},\vec{G}'}$ vanish and Eq. (1.41) becomes

$$\omega^2 \begin{bmatrix} B^{(11)}_{\vec{G},\vec{G}'} & 0 & 0 \\ 0 & B^{(22)}_{\vec{G},\vec{G}'} & 0 \\ 0 & 0 & B^{(33)}_{\vec{G},\vec{G}'} \end{bmatrix} \begin{bmatrix} U_{1,\vec{K}}(\vec{G}') \\ U_{2,\vec{K}}(\vec{G}') \\ U_{3,\vec{K}}(\vec{G}') \end{bmatrix} = \\ = \begin{bmatrix} A^{(11)}_{\vec{G},\vec{G}'} & A^{(12)}_{\vec{G},\vec{G}'} & 0 \\ A^{(21)}_{\vec{G},\vec{G}'} & A^{(22)}_{\vec{G},\vec{G}'} & 0 \\ 0 & 0 & A^{(33)}_{\vec{G},\vec{G}'} \end{bmatrix} \begin{bmatrix} U_{1,\vec{K}}(\vec{G}') \\ U_{2,\vec{K}}(\vec{G}') \\ U_{3,\vec{K}}(\vec{G}') \end{bmatrix}, \tag{1.43}$$

where

$$\begin{cases} B^{(11)}_{\vec{G},\vec{G}'} = B^{(22)}_{\vec{G},\vec{G}'} = B^{(33)}_{\vec{G},\vec{G}'} = \rho(\vec{G} - \vec{G}'), & (1.44a) \\ A^{(11)}_{\vec{G},\vec{G}'} = C_{11}(\vec{G} - \vec{G}')(G_1 + K_1)(G'_1 + K_1) + \cdots \\ \qquad + C_{44}(\vec{G} - \vec{G}')(G_2 + K_2)(G'_2 + K_2), & (1.44b) \\ A^{(12)}_{\vec{G},\vec{G}'} = C_{12}(\vec{G} - \vec{G}')(G_1 + K_1)(G'_2 + K_2) + \cdots \\ \qquad + C_{44}(\vec{G} - \vec{G}')(G'_1 + K_1)(G_2 + K_2), & (1.44c) \\ A^{(21)}_{\vec{G},\vec{G}'} = C_{12}(\vec{G} - \vec{G}')(G'_1 + K_1)(G_2 + K_2) + \cdots \\ \qquad + C_{44}(\vec{G} - \vec{G}')(G'_2 + K_2)(G_1 + K_1), & (1.44d) \\ A^{(22)}_{\vec{G},\vec{G}'} = C_{11}(\vec{G} - \vec{G}')(G_2 + K_2)(G'_2 + K_2) + \cdots \\ \qquad + C_{44}(\vec{G} - \vec{G}')(G_1 + K_1)(G'_1 + K_1), & (1.44e) \\ A^{(33)}_{\vec{G},\vec{G}'} = C_{44}(\vec{G} - \vec{G}')\Big[(G_1 + K_1)(G'_1 + K_1) + \cdots \\ \qquad + (G_2 + K_2)(G'_2 + K_2)\Big]. & (1.44f) \end{cases}$$

Equation (1.43) involves *super-diagonal matrices* and can be split into two independent matrix equations

$$\omega^2 \begin{bmatrix} B^{(11)}_{\vec{G},\vec{G}'} & 0 \\ 0 & B^{(22)}_{\vec{G},\vec{G}'} \end{bmatrix} \begin{bmatrix} U_{1,\vec{K}}(\vec{G}') \\ U_{2,\vec{K}}(\vec{G}') \end{bmatrix} = \begin{bmatrix} A^{(11)}_{\vec{G},\vec{G}'} & A^{(12)}_{\vec{G},\vec{G}'} \\ A^{(21)}_{\vec{G},\vec{G}'} & A^{(22)}_{\vec{G},\vec{G}'} \end{bmatrix} \begin{bmatrix} U_{1,\vec{K}}(\vec{G}') \\ U_{2,\vec{K}}(\vec{G}') \end{bmatrix}, \tag{1.45}$$

and

$$\omega^2 \sum_{\vec{G}'} B^{(33)}_{\vec{G},\vec{G}'} U_{3,\vec{K}}(\vec{G}') = \sum_{\vec{G}'} A^{(33)}_{\vec{G},\vec{G}'} U_{3,\vec{K}}(\vec{G}'). \tag{1.46}$$

Equations (1.45, 1.46) show that propagation modes in the 2D bulk phononic crystal decouple. Equation (1.45) corresponds to modes polarized in the transverse plane $(x_1 O x_2)$ (often named XY modes) while Eq. (1.46) characterizes modes (often named Z modes) with a displacement vector oriented along the x_3 direction.

Equations (1.44) contain the Fourier coefficients $\rho(\vec{G}-\vec{G}')$ and $C_{mn}(\vec{G}-\vec{G}')$, with $(mn) \equiv (11)$, (44) and (12) defined by Eqs. (1.33). For two-dimensional phononic crystals, Eqs. (1.33) must be rewritten as

$$\eta(\vec{G}-\vec{G}') = \frac{1}{\Sigma_{(U.C.)}} \iint_{(U.C.)} \eta(\vec{r}) e^{-i(\vec{G}-\vec{G}')\cdot\vec{r}} d^2\vec{r}, \; \eta \equiv \rho \text{ or } C_{mn}, \tag{1.47}$$

where $\Sigma_{(U.C.)}$ is the area of the two-dimensional unit cell in the $(x_1 O x_2)$ plane and

$$\eta(\vec{G}-\vec{G}') = \frac{1}{\Sigma_{(U.C.)}} \iint_{(A_{(UC)})} \eta_A \cdot e^{-i(\vec{G}-\vec{G}')\cdot\vec{r}} d^2\vec{r} + \cdots + \frac{1}{\Sigma_{(U.C.)}} \iint_{(B_{(UC)})} \eta_B \cdot e^{-i(\vec{G}-\vec{G}')\cdot\vec{r}} d^2\vec{r}, \tag{1.48}$$

where the integrals are performed over the areas filled by materials A and B inside the UC and η_A (resp. η_B) denotes the value of parameter η for material A (resp. B).

Equation (1.48) can be rewritten as

$$\begin{aligned}\eta(\vec{G}-\vec{G}') = & \frac{1}{\Sigma_{(U.C.)}} \iint_{(A_{(UC)})} \eta_A \cdot e^{-i(\vec{G}-\vec{G}')\cdot\vec{r}} d^2\vec{r} + \cdots \\ & - \frac{1}{\Sigma_{(U.C.)}} \iint_{(A_{(UC)})} \eta_B \cdot e^{-i(\vec{G}-\vec{G}')\cdot\vec{r}} d^2\vec{r} + \cdots \\ & + \frac{1}{\Sigma_{(U.C.)}} \iint_{(A_{(UC)})} \eta_B \cdot e^{-i(\vec{G}-\vec{G}')\cdot\vec{r}} d^2\vec{r} + \cdots \\ & + \frac{1}{\Sigma_{(U.C.)}} \iint_{(B_{(UC)})} \eta_B \cdot e^{-i(\vec{G}-\vec{G}')\cdot\vec{r}} d^2\vec{r} \\ = & (\eta_A - \eta_B)\left[\frac{1}{\Sigma_{(U.C.)}} \iint_{(A_{(UC)})} e^{-i(\vec{G}-\vec{G}')\cdot\vec{r}} d^2\vec{r}\right] + \cdots \\ & + \eta_B \left[\frac{1}{\Sigma_{(U.C.)}} \iint_{(UC)} e^{-i(\vec{G}-\vec{G}')\cdot\vec{r}} d^2\vec{r}\right].\end{aligned} \tag{1.49}$$

Because

$$\frac{1}{\Sigma_{(U.C.)}} \iint_{(UC)} e^{-i(\vec{G}-\vec{G}')\cdot\vec{r}} d^2\vec{r} = \delta_{(\vec{G}-\vec{G}'),\vec{0}} = \begin{cases} 1 \text{ if } (\vec{G}-\vec{G}') = \vec{0}, \\ 0 \text{ if } (\vec{G}-\vec{G}') \neq \vec{0}, \end{cases} \tag{1.50}$$

and defining the quantity $F(\vec{G}-\vec{G}')$ as

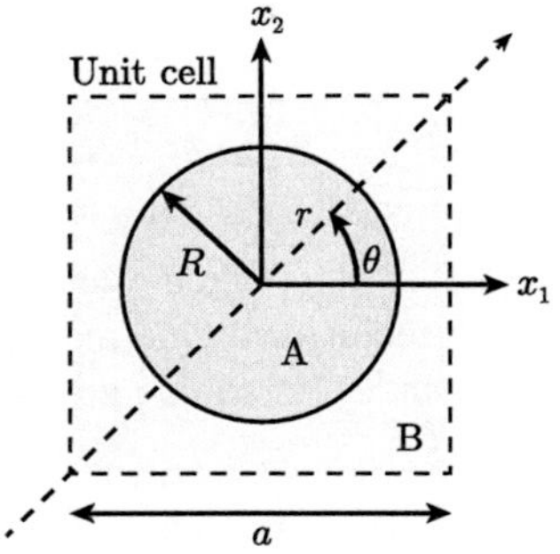

Fig. 1.12 Unit cell in the transverse plane ($x_1 0 x_2$) of the two-dimensional phononic crystal made of a square array of cylinders with circular cross-section. The area of the unit cell is a^2

$$F(\vec{G}-\vec{G}') = \frac{1}{\Sigma_{(U.C.)}} \iint_{(A_{(UC)})} e^{-i(\vec{G}-\vec{G}')\cdot\vec{r}} d^2\vec{r}, \tag{1.51}$$

equation (1.48) becomes

$$\eta(\vec{G}-\vec{G}') = (\eta_A - \eta_B) \cdot F(\vec{G}-\vec{G}') + \eta_B \cdot \delta_{(\vec{G}-\vec{G}'),\vec{0}}. \tag{1.52}$$

where $F(\vec{G}-\vec{G}')$ is the structure factor which depends on the cross section shape of the cylindrical inclusion.

For example if one considers inclusions of circular cross-section, see Fig. 1.12, the structure factor can be calculated using polar coordinates (r, θ) as

$$\begin{aligned} F(\vec{G}-\vec{G}') &= \frac{1}{\Sigma_{(U.C.)}} \iint_{(A_{(UC)})} e^{-i(\vec{G}-\vec{G}')\cdot\vec{r}} d^2\vec{r} \\ &= \frac{1}{a^2} \int_0^R \int_0^{2\pi} e^{-i|\vec{G}-\vec{G}'|r\cos\theta} r dr d\theta = \frac{1}{a^2} \int_0^R 2\pi r dr J_0(|\vec{G}-\vec{G}'|r) \\ &= \frac{2\pi}{(a^2|\vec{G}-\vec{G}'|^2)} \int_0^{|\vec{G}-\vec{G}'|R} (|\vec{G}-\vec{G}'|r) J_0(|\vec{G}-\vec{G}'|r) d(|\vec{G}-\vec{G}'|r) \\ &= \frac{2\pi}{(a^2|\vec{G}-\vec{G}'|^2)} (|\vec{G}-\vec{G}'|R) J_1(|\vec{G}-\vec{G}'|R) = f \frac{2J_1(|\vec{G}-\vec{G}'|R)}{|\vec{G}-\vec{G}'|R}, \end{aligned} \tag{1.53}$$

where $f = \frac{\pi R^2}{a^2}$ $(0 \le f \le \frac{\pi}{4})$ is the filling fraction of inclusions, i.e., the ratio between the area of the cross-section of the cylinder and that of the unit cell, and J_0 and J_1 are the Bessel's functions of the first kind of orders 0 and 1, respectively. When $(\vec{G}-\vec{G}') = \vec{0}$,

$$F(\vec{0}) = \frac{1}{\Sigma_{(U.C.)}} \iint_{(A_{(UC)})} d^2\vec{r} = \frac{\pi R^2}{a^2} = f, \tag{1.54}$$

and one can rewrite Eq. (1.52) as

$$\eta(\vec{G}-\vec{G}') = \begin{cases} f\eta_A + (1-f)\eta_B = \overline{\eta} & \text{if } (\vec{G}-\vec{G}') = \vec{O}, \\ (\eta_A - \eta_B)F(\vec{G}-\vec{G}') & \text{if } (\vec{G}-\vec{G}') \neq \vec{O}, \end{cases} \tag{1.55}$$

where $\overline{\eta}$ is an average value on the unit cell of parameter η.

The structure factor depends on the reciprocal lattice vectors and on the geometry of the inclusions. One may consider cross-section section of other geometries and for cylinders of square cross-section with side ℓ, for example,

$$F(\vec{G}-\vec{G}') = \frac{1}{\Sigma_{(U.C.)}} \iint_{(A_{(UC)})} e^{-i(\vec{G}-\vec{G}')\cdot\vec{r}} d^2\vec{r}, \tag{1.56}$$

$$= \frac{1}{a^2} \int_{-\frac{\ell}{2}}^{+\frac{\ell}{2}} e^{-i(G_1-G_1')\cdot x_1} dx_1 \int_{-\frac{\ell}{2}}^{+\frac{\ell}{2}} e^{-i(G_2-G_2')\cdot x_2} dx_2, \tag{1.57}$$

$$= f \left[\frac{\sin\left[(G_1-G_1')\frac{\ell}{2}\right]}{(G_1-G_1')\frac{\ell}{2}} \right] \left[\frac{\sin\left[(G_2-G_2')\frac{\ell}{2}\right]}{(G_2-G_2')\frac{\ell}{2}} \right]. \tag{1.58}$$

Other authors have considered elliptical or hexagonal [9] cross-sections and analytical expressions of the structure factor can be derived in these cases. For much more complicated geometries, the calculation of the structure factor can be performed numerically by a numerical resolution of the surface integral involved in Eq. (1.51). However, computational time could be an issue. On the other hand, some complicated geometries could still be treated analytically. This applies to concentrically layered circular cylinders. Their treatment requires rewriting Eq. (1.49).

Equation (1.46) governs the Z modes propagating in the bulk 2D phononic crystal and can be rewritten as

$$\omega^2 \sum_{\vec{G}'} \rho(\vec{G}-\vec{G}') U_{3,\vec{K}}(\vec{G}') = \tag{1.59}$$

$$= \sum_{\vec{G}'} C_{44}(\vec{G}-\vec{G}') \Big[(G_1+K_1)(G_1'+K_1) + (G_2+K_2)(G_2'+K_2) \Big] U_{3,\vec{K}}(\vec{G}').$$

If in Eq. (1.59), we single out the term $\vec{G} = \vec{G}'$ in the summation, then this equation can be recast as

$$\omega^2 \Big\{ \rho(\vec{0}) U_{3,\vec{K}}(\vec{G}) + \sum_{\vec{G}' \neq \vec{G}} \rho(\vec{G}-\vec{G}') U_{3,\vec{K}}(\vec{G}') \Big\} =$$

$$= C_{44}(\vec{0}) \Big[(G_1+K_1)(G_1+K_1) + (G_2+K_2)(G_2+K_2) \Big] U_{3,\vec{K}}(\vec{G}) + \cdots$$

$$\sum_{\vec{G}' \neq \vec{G}} C_{44}(\vec{G}-\vec{G}') \Big[(G_1+K_1)(G_1'+K_1) + (G_2+K_2)(G_2'+K_2) \Big] U_{3,\vec{K}}(\vec{G}'), \tag{1.60}$$

and

$$\omega^2\Big[\overline{\rho}U_{3,\vec{K}}(\vec{G}) + (\rho_A - \rho_B)\sum_{\vec{G}'\neq\vec{G}} F(\vec{G}-\vec{G}')U_{3,\vec{K}}(\vec{G}')\Big] =$$
$$= \overline{C_{44}}(\vec{G}+\vec{K})^2 U_{3,\vec{K}}(\vec{G}) + \cdots$$
$$+ (C_{44A} - C_{44B})\sum_{\vec{G}'\neq\vec{G}} F(\vec{G}-\vec{G}')\Big[(G_1+K_1)(G_1'+K_1) + \cdots$$
$$+ (G_2+K_2)(G_2'+K_2)\Big]U_{3,\vec{K}}(\vec{G}'). \quad (1.61)$$

Introducing the dimensionless vectors $\vec{g} = \frac{a}{2\pi}\vec{G}$, $\vec{g}' = \frac{a}{2\pi}\vec{G}'$ and $\vec{k} = \frac{a}{2\pi}\vec{K}$, Eq. (1.61) is transformed as

$$\omega^2\overline{\rho}\left[U_{3,\vec{k}}(\vec{g}) + \frac{(\rho_A-\rho_B)}{\overline{\rho}}\sum_{\vec{g}'\neq\vec{g}} F(\vec{g}-\vec{g}')U_{3,\vec{k}}(\vec{g}')\right] =$$
$$= \overline{C_{44}}\left(\frac{2\pi}{a}\right)^2\Bigg[(\vec{g}+\vec{k})^2 U_{3,\vec{k}}(\vec{g}) + \cdots$$
$$+ \frac{(C_{44A}-C_{44B})}{\overline{C_{44}}}\sum_{\vec{g}'\neq\vec{g}} F(\vec{g}-\vec{g}')(\vec{g}+\vec{k})(\vec{g}'+\vec{k})U_{3,\vec{k}}(\vec{g}')\Bigg], \quad (1.62)$$

and, eventually, can be rewritten as

$$\Omega^2\left[U_{3,\vec{k}}(\vec{g}) + \Delta\rho\sum_{\vec{g}'\neq\vec{g}} F(\vec{g}-\vec{g}')U_{3,\vec{k}}(\vec{g}')\right] =$$
$$= (\vec{g}+\vec{k})^2 U_{3,\vec{k}}(\vec{g}) + \Delta C_{44}\sum_{\vec{g}'\neq\vec{g}} F(\vec{g}-\vec{g}')(\vec{g}+\vec{k})(\vec{g}'+\vec{k})U_{3,\vec{k}}(\vec{g}'), \quad (1.63)$$

where

$$\Omega = \frac{\omega}{\frac{2\pi}{a}\sqrt{\frac{\overline{C_{44}}}{\overline{\rho}}}}, \quad \Delta C_{44} = \frac{C_{44A}-C_{44B}}{\overline{C_{44}}}, \quad \text{and} \quad \Delta\rho = \frac{\rho_A-\rho_B}{\overline{\rho}}$$

are dimensionless quantities. The same transformations can be performed for Eq. (1.45) governing the XY modes of propagation. Equation (1.63) shows that it is convenient to compute the dimensionless frequency Ω as a function of the dimensionless Bloch wave vector $\vec{k}$. For example, in the case of a square array of cylindrical inclusions, the $2D$ dimensionless vectors $\vec{g}$ (resp. $\vec{g}'$) are $\vec{g} = \ell\vec{e}_1 + m\vec{e}_2$ (resp. $\vec{g}' = \ell'\vec{e}_1 + m'\vec{e}_2$) where ℓ and m (resp. ℓ' and m') are integers, see Sect. 1.3.3.1. In the course of the numerical resolution of Eqs. (1.45, 1.46), we consider

$-MT \leq (\ell, \ell') \leq +MT$ and $-MT \leq (m, m') \leq +MT$ where MT is a positive integer, i.e., $(2MT+1)^2$ $\vec{g}$ or $\vec{g}'$ vectors are taken into account in the truncated Fourier series. This gives $(2MT+1)^2$ (resp. $2(2MT+1)^2$) real eigenfrequencies for the Z modes (resp. the XY modes) for a given reduced wave vector $\vec{k}$ describing the principal directions of propagation in the irreducible Brillouin zone (at points Γ, X and M of the irreducible Brillouin zone of the square array, $\vec{k}$ is $\vec{k_\Gamma} = (0, 0)$, $\vec{k_X} = (\frac{1}{2}, 0)$ and $\vec{k_M} = (\frac{1}{2}, \frac{1}{2})$, respectively). Solving for Eq. (1.63) requires to resolve a generalized eigenvalues problem for each value of $\vec{k}$. The size of the matrices involved in this problem and then the choice of the values of the integer MT is of crucial importance for insuring the convergence of the Fourier series and consequently optimizing the precision of the numerical solutions for the eigenfrequencies Ω. This will be discussed in the next subsection.

1.4.2 Limitations of the PWE Method

The PWE method is a useful tool for calculating band structures of phononic crystals. It is relatively easy to implement numerically while the main difficulty is the correct representation of the matrices involved in the generalized eigenvalues problem to solve. There exists many numerical codes (e.g., in Fortran and C languages, or in Matlab software) of band structure calculations based on the PWE method that are freely available online. Many of them were developed for photonic crystals but could be easily adapted to phononic crystals. The method is general in the sense that it can be applied for 1D, 2D and 3D structures, with inclusions of different shapes distributed in arrays of various geometries. Inclusions and matrix can be made of materials of much more complicated crystallographic symmetry than the isotropic or the cubic ones [10]. Moreover, not only passive elastic constituent materials but also active ones such as piezoelectric [11] or magneto-elastic [12] materials may be considered. However, the method presents some limitations linked to convergence problems of the truncated Fourier series and the choice of constitutive materials.

1.4.2.1 Convergence of the Truncated Fourier Series

As already mentioned at the end of Sect. 1.4.1.3, while the Fourier series are assumed infinite theoretically, a finite number of reciprocal lattice vectors must be taken into account in the course of the numerical calculations. We analyse here the effect of this truncation of the series on the calculated eigenvalues. For that we consider a particular case of a phononic crystal made of highly contrasted constituent materials namely a square array of steel cylinders embedded in an epoxy matrix with a filling factor of inclusions $f = 0.55$. The study is limited to the Z modes of propagation defined by Eq. (1.63). Figures 1.13 present the first 10 bands of these band structures calculated along the principal directions of propagation of the irreducible Brillouin

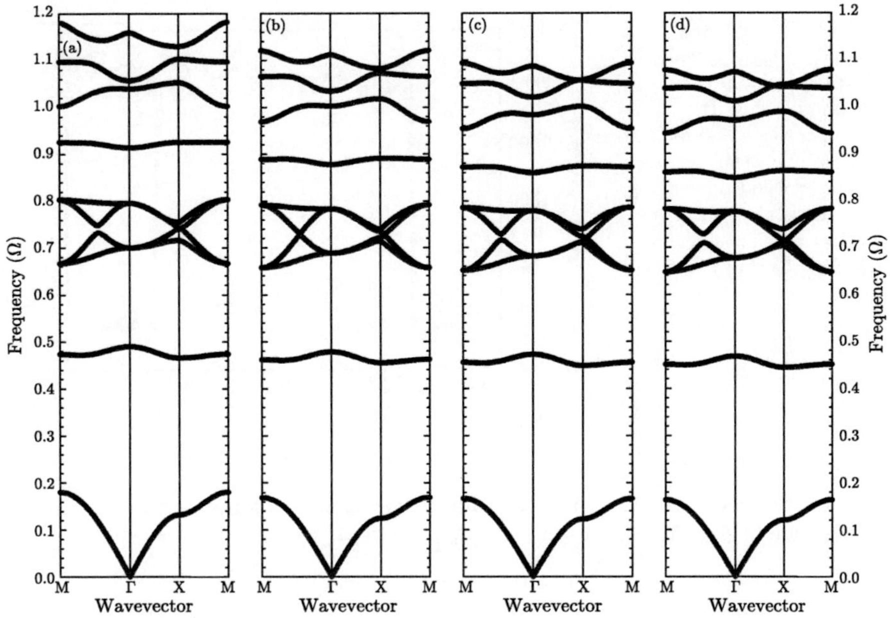

Fig. 1.13 Z modes band structure for a square array of steel cylinders embedded in an epoxy matrix for $MT = 6$ (**a**), $MT = 8$ (**b**), $MT = 10$ (**c**), $MT = 12$ (**d**). $\rho_A = 7780\ kg.m^{-3}$, $C_{44}^A = 8.1.10^{10}\ N.m^{-2}$, $\rho_B = 1142\ kg.m^{-3}$, $C_{44}^B = 0.148.10^{10}\ N.m^{-2}$ and $f = 0.55$

zone for different values of the integer MT namely for $MT = 6, 8, 10$ and 12. While the overall shape of the band structure remains almost the same, one observes that the location in frequency of some bands (for Ω around 0.2, 0.45 and 1.0 for examples) is strongly influenced by the value of MT, this effect being stronger for larger Ω. This slow convergence results mainly from the difficulty to reproduce accurately a discontinuous function such as the density or an elastic modulus by the sum of a finite number of sinusoidal continuous functions [13]. This is often referred to the *Gibbs phenomenon* and can be clearly viewed in Fig. 1.14 where the function

$$\rho_{\text{truncated}}(\vec{r}) = \sum_{|\vec{G}| \leq G_{max}} \rho(\vec{G}) e^{i\vec{G}\cdot\vec{r}} \tag{1.64}$$

(with $G_{max} = \frac{2\pi}{a}\sqrt{MT^2 + MT^2} = \frac{2\pi}{a} MT\sqrt{2}$) has been plotted for different numbers of reciprocal lattice vectors, i.e., for $MT = 6, 8, 10$ and 12). One observes that whatever is the value of MT, the function $\rho_{truncated}$ is rather different from $\rho(\vec{r})$. Then when using the PWE method, it is necessary to fix the value of MT allowing a good compromise between convergence and calculation time. Convergence of the truncated Fourier series is the main concern when considering very different constituent materials such as steel and epoxy. In this case a value of MT equal or larger than 10 is necessary for obtaining the values of the eigenfrequencies with reasonable

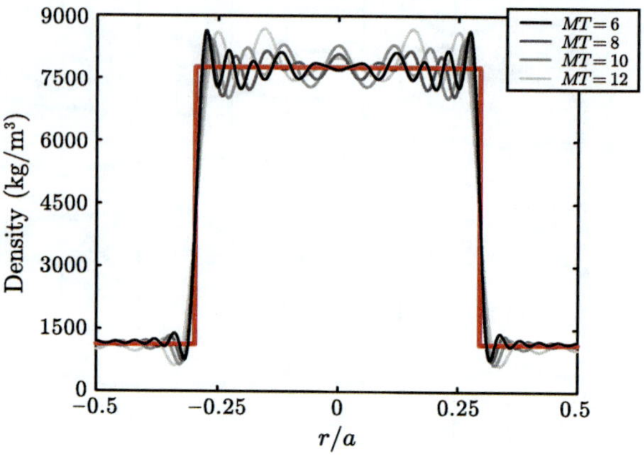

Fig. 1.14 Density, $\rho(\vec{r})$ (red) and $\rho(\vec{r})_{\text{truncated}}$ with $MT = 6$ (black), $MT = 8$ (dark grey), $MT = 10$, $MT = 12$ (clear grey) for a square array of steel cylinders embedded in an epoxy matrix along the path $r = x_1 = x_2$ in the unit cell, see Fig. 1.12, using $f = 0.55$

precision. Some authors have proposed alternative scheme aiming at speeding up the convergence of the PWE algorithm [4, Chap. 11] [14].

1.4.2.2 Choice of the Materials

In the preceding sections, it was assumed that all the constituent materials are purely elastic solids. However, one may consider phononic crystals made of materials of different nature for example a fluid (liquid or gas) and a solid. The reliability of the PWE method in this case could be questioned. Let us consider a 2D phononic crystal made of hollow cylindrical voids drilled in a solid matrix and filled with a liquid. Intuitively, the fluid could be thought as an isotropic *solid* material with $C_{44} = 0$ because of the absence of transverse displacements. However, the PWE method still assumes a finite non-vanishing displacement amplitude of the transverse mode in the cylinders and considering $C_{44} = 0$ in the fluid will introduce numerical instabilities in the PWE code. As an example, Fig. 1.15 shows the XY modes band structure of a square array of mercury filled cylindrical voids in an aluminium matrix. Liquid mercury was modelled as a solid isotropic material with $C_{44} = 0$, values of ρ and C_{11} being those of *real* mercury. PWE calculations reveal the existence of flat bands, whose number increases when the number of reciprocal lattice vectors taken into account in the Fourier series increases, alongside with the bands without clear physical meaning. The absence of these modes in the band structure using other methods (for example, the finite element method via the COMSOL™ software which models solid material and fluid constituents with their real elastic characteristics, i.e., two elastic moduli for the solid and the compressibility modulus for the fluid) shows unambiguously that the PWE additional modes are fictitious [15]. Hou et al. [16] argued that these fictitious modes result from an incorrect use of the Bloch theorem in the application of the PWE method in such mixed solid/fluid phononic crystals. Notes that the boundary conditions at the interface between the solid and the liquid

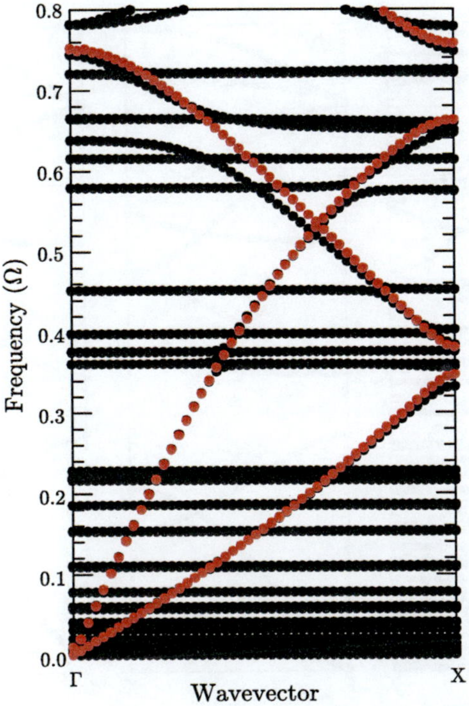

Fig. 1.15 XY band structure along the ΓX direction of the irreducible Brillouin zone of a square array of hollow cylinders drilled in an Aluminium matrix and filled with liquid mercury for $f = 0.4$; black dots: PWE calculations where mercury has been considered as an isotropic solid with $C_{44} = 0$ and $MT = 8$, red dots: finite element results where mercury has been considered has a *real* fluid. The PWE method leads to unphysical modes in this case and fails to predict accurately the propagating modes in the mixed fluid/solid 2D structure. $\rho_A = 13600\ \mathrm{kg\cdot m^{-3}}$, $C_{11}^A = 2.86\ 10^{10}\ \mathrm{N\cdot m^{-2}}$, $\rho_B = 2700\ \mathrm{kg\cdot m^{-3}}$, $C_{44}^B = 2.61\ 10^{10}\ \mathrm{N\cdot m^{-2}}$, $C_{11}^B = 11.09\ 10^{10}\ \mathrm{N\cdot m^{-2}}$

are strictly satisfied in the finite element calculations while the PWE method does not take into account these conditions.

Moreover, phononic crystals made of voids in a solid matrix present many advantages due to the ease in manufacturing. For the reasons reported previously, if one models air inside the voids as a solid material with vanishing C_{44}, the fictitious flat bands in the band structure are obtained. This is demonstrated in Fig. 1.16a where the XY band structure for a square array of holes drilled in an Aluminium matrix is plotted. However, it was shown [17] that for an accurate use of the PWE method in this case, it is preferable to replace air inside the cylinders by vacuum and to model vacuum as a pseudo-solid material with very low elastic moduli and density. Indeed, modelling vacuum by a material with vanishing density and elastic moduli leads to unphysical solutions of the eigenvalue problem. For the sake of simplicity, this low impedance medium (LIM) was supposed elastically isotropic and was characterized by a longitudinal speed of sound C_ℓ, and a transversal speed of sound

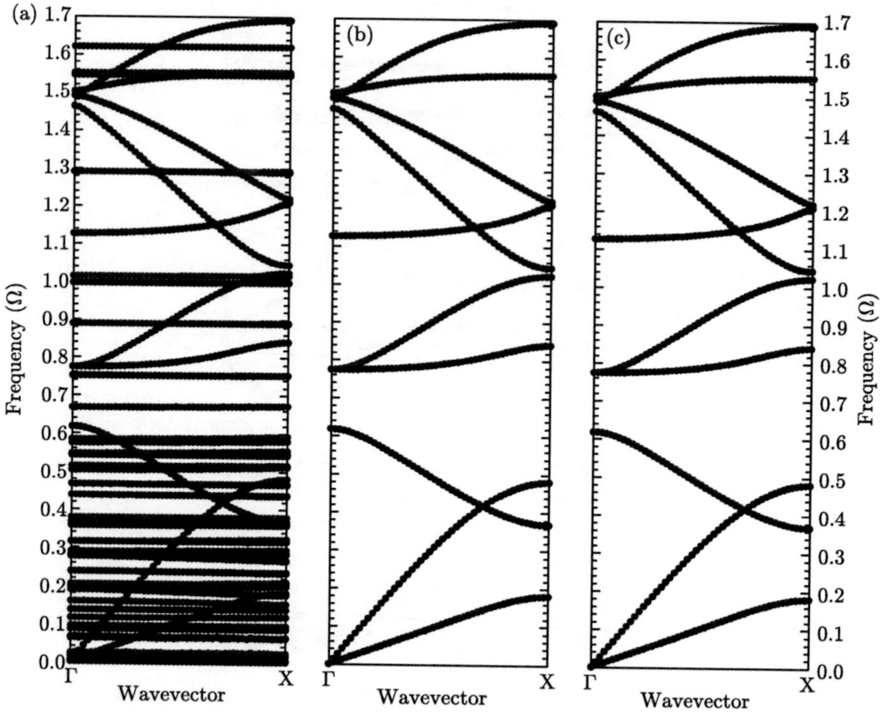

Fig. 1.16 XY band structure along the ΓX direction of the irreducible Brillouin zone of a square array of hollow cylinders drilled in an Aluminium matrix for $f = 0.4$; **a** PWE calculations where the medium inside the cylinders is air with $C_{44} = 0$, $C_{11} = 1.49\ 10^5 \mathrm{N} \cdot \mathrm{m}^{-2}$, $\rho = 1.3\ \mathrm{kg} \cdot \mathrm{m}^{-3}$ and $MT = 6$; **b** PWE calculations where the medium inside the cylinders is the LIM with $C_{11} = C_{44} = 10^6\ \mathrm{N} \cdot \mathrm{m}^{-2}$, $\rho = 10^{-4}\ \mathrm{kg} \cdot \mathrm{m}^{-3}$ and $MT = 6$; **c** Finite elements method (see text for details)

C_t or equivalently by two elastic moduli, $C_{11} = \rho.C_\ell{}^2$ and $C_{44} = \rho C_t{}^2$. The choice of the values for these parameters is governed by the boundary condition between any solid material and vacuum. Indeed, this interface must be free of stress and this requires that $C_{11} = 0$ and $C_{44} = 0$ in vacuum. Then, using the LIM to model vacuum in the PWE computations, the non vanishing values of these parameters must be as small as possible and the ratio between the elastic moduli of the LIM and those of any other solid material constituting the phononic crystal must approach zero. We imposed C_ℓ and C_t to be much larger than the speeds of sound in usual solid materials in order to limit propagation of acoustic waves to the solid. Large speeds of sound and small elastic moduli impose a choice of a very low mass density for the LIM. More specifically, we choose $\rho = 10^{-4}\ \mathrm{kg} \cdot \mathrm{m}^{-3}$ and $C_\ell = C_t = 10^5 \mathrm{m} \cdot \mathrm{s}^{-1}$, i.e., the acoustic impedances of the LIM are equal to $10\ \mathrm{kg} \cdot \mathrm{m}^{-2} \cdot \mathrm{s}^{-1}$. With these values $C_{11} = C_{44} = 10^6\ \mathrm{N} \cdot \mathrm{m}^{-2}$ and the elastic constants of the LIM are approximately 10^4 times lower than those of any usual solid material that are typically on the order of $10^{10}\ \mathrm{N} \cdot \mathrm{m}^{-2}$. The values we have chosen for C_{11} and C_{44} are a compromise to achieve satisfactory convergence of the PWE method and still satisfy boundary conditions.

C_{11} and C_{44} were assumed the same for convenience. Figure 1.16b presents the same band structure as that of Fig. 1.16a but the medium inside the hollow cylinders is modelled with the LIM. One observes that the flat bands that appeared in Fig. 1.16a are removed. Results obtained with the finite element method where only the space occupied by the Aluminium matrix has been discretized, reported in Fig. 1.16c, show a very good agreement with those of Fig. 1.16b. This shows that the PWE method is suitable for calculating the band structure of phononic crystals made of holes drilled inside a solid matrix provided the medium inside the holes is replaced by the LIM.

Many experimental studies are dealing with phononic crystals, often named as sonic crystals, where solid inclusions are surrounded by air. Again considering air as a solid with $C_{44} = 0$ leads to unphysical results. Nevertheless, due to the huge contrast between the physical characteristics of the solid and those of air, the solid inclusions can be assumed infinitely hard with a high density and high elastic moduli. This implies that the sound does not penetrate such inclusions, and hence the propagation of acoustic waves is predominant in air where only longitudinal waves exist. The periodic structure made of rigid inclusions in air can then be considered as an inhomogeneous fluid and the equation of propagation of longitudinal acoustic waves in such medium [18] is

$$-\frac{1}{C_{11}(\vec{r})}\frac{\partial^2 p(\vec{r},t)}{\partial t^2} = -\frac{\omega^2}{C_{11}(\vec{r})}p(\vec{r}) = \vec{\nabla}\cdot\left(\frac{1}{\rho(\vec{r})}\vec{\nabla}p(\vec{r},t)\right), \tag{1.65}$$

where $p(\vec{r},t) = e^{i\omega t}p(\vec{r})$ is the acoustic pressure field inside the heterogeneous fluid. In a periodic fluid medium, Eq. (1.65) can be Fourier transformed to give

$$\omega^2\sum_{\vec{G}'} C_{11}^{-1}(\vec{G}-\vec{G}')p_{\vec{K}}(\vec{G}') = \sum_{\vec{G}'}\rho^{-1}(\vec{G}-\vec{G}')\Big[(G_1+K_1)(G_1'+K_1)+\cdots + (G_2+K_2)(G_2'+K_2)+(G_3+K_3)(G_3'+K_3)\Big]p_{\vec{K}}(\vec{G}'). \tag{1.66}$$

Note that in the case of a 2D array of rigid cylindrical inclusions surrounded with air, i.e., for $G_3 = G_3' = K_3 = 0$, Eq. (1.66) becomes

$$\omega^2\sum_{\vec{G}'} C_{11}^{-1}(\vec{G}-\vec{G}')p_{\vec{K}}(\vec{G}') = \sum_{\vec{G}'}\rho^{-1}(\vec{G}-\vec{G}')\Big[(G_1+K_1)(G_1'+K_1)+\cdots + (G_2+K_2)(G_2'+K_2)\Big]p_{\vec{K}}(\vec{G}'), \tag{1.67}$$

and it coincides with the equation of propagation of Z modes in an elastic solid 2D phononic crystal (provided ρ, C_{44} and $U_{3,\vec{K}}$ in Eq. (1.59) play the roles of C_{11}^{-1}, ρ^{-1} and $p_{\vec{k}}$ in Eq. (1.67)). Considering these analogies and using a numerical code written for the Z modes, it is very easy to calculate the band structure shown in Fig. 1.17 for a square array of steel cylinders in air.

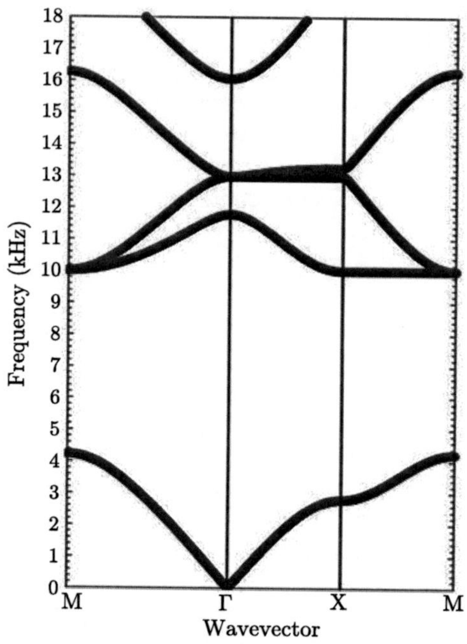

Fig. 1.17 PWE band structure of a square array of circular steel cylinders placed in air with $a = 2.7$ cm and $R = 1.29$ cm. PWE calculations were performed considering the assumption of infinitely rigid solid inclusions and with $MT = 10$. Notes the existence of a large absolute stop band in the audible frequency range

Due to the existence of a large acoustic stop band in the audible frequency range, application of these sonic crystals as sound barriers was largely investigated by many authors [19, 20].

The reliability of the assumption of infinitely rigid inclusions when the matrix of the phononic crystal is made of air was validated (see Fig. (2) of Ref. [21]). One may note that this assumption leads to unreliable results in most of the cases (array of inclusions, filling factor of inclusions, ...) when considering a fluid matrix with the density and elastic moduli values higher than those of air, such as water. One can also remark that Eq. (1.66) can be used for calculating band structures of phononic crystals composed of fluid constituents only such as periodic arrays of air inclusions in water [18, 22].

1.4.3 *Modified PWE Method for Complex Band Structures*

In classical PWE expansion methods (see Sect. 1.4.1.2), one calculates a set of real eigenfrequencies $\omega(\vec{K})$ for a specific wave vector $\vec{K}$. That means that only propagating modes with a real wave vector can be deduced from the $\omega(\vec{K})$ PWE method. Then a modified PWE method has been proposed that allows the calculation of not only the propagating modes but also the evanescent modes. The wave vector for evanescent waves has a non-vanishing imaginary part. We have seen previously that the Fourier transform of the equation of propagation of elastic waves in a phononic crystal leads to the resolution of a generalized eigenvalue equation in the form $\omega^2 \overleftrightarrow{B}\vec{\mathcal{U}} = \overleftrightarrow{A}\vec{\mathcal{U}}$. The matrix elements of $\overleftrightarrow{A}$ and $\overleftrightarrow{B}$ involve terms dependent on the components of the wave vector $\vec{K}$. It is always possible to rewrite matrix $\overleftrightarrow{A}$ as $\overleftrightarrow{A} = K_\alpha{}^2 \overleftrightarrow{A_1} + K_\alpha \overleftrightarrow{A_2} + \overleftrightarrow{A_3}$, where K_α is one of the components of the wave vector, and $\overleftrightarrow{A_1}$, $\overleftrightarrow{A_2}$ and $\overleftrightarrow{A_3}$ are matrices of the same size as $\overleftrightarrow{A}$. The generalized eigenvalue equation $\omega^2 \overleftrightarrow{B}\vec{\mathcal{U}} = \overleftrightarrow{A}\vec{\mathcal{U}}$ may be recast as $K_\alpha{}^2 \overleftrightarrow{A_1}\vec{\mathcal{U}} = (\omega^2 \overleftrightarrow{B} - K_\alpha \overleftrightarrow{A_2} - \overleftrightarrow{A_3}).\vec{\mathcal{U}}$ and one can write

$$K_\alpha \begin{bmatrix} \overleftrightarrow{I} & \overleftrightarrow{0} \\ \overleftrightarrow{0} & \overleftrightarrow{A_1} \end{bmatrix} \begin{bmatrix} \vec{\mathcal{U}} \\ K_\alpha \vec{\mathcal{U}} \end{bmatrix} = \begin{bmatrix} 0 & \overleftrightarrow{I} \\ \omega^2 \overleftrightarrow{B} - \overleftrightarrow{A_3} & -\overleftrightarrow{A_2} \end{bmatrix} \begin{bmatrix} \vec{\mathcal{U}} \\ K_\alpha \vec{\mathcal{U}} \end{bmatrix}, \tag{1.68}$$

where $\overleftrightarrow{I}$ is the identity matrix. Equation (1.68) is nothing else but a generalized eigenvalue equation where the eigenvalues are the component K_α of the wave vector. For a specific value of the circular frequency ω, one calculates a set of complex eigenvalues K_α. The size of the matrices present on the left and right-hand sides of Eq. (1.68) is twice that of matrices $\overleftrightarrow{A}$ and $\overleftrightarrow{B}$. One may illustrate these general ideas by considering the peculiar case of the Z elastic modes propagating in a bulk $2D$ phononic crystal made of a square array of lattice parameter a, of cylindrical inclusions embedded in a solid matrix. If one assumes $K_3 = 0$, then these modes are given by Eq. (1.63), where ω depends on the two variables K_1 and K_2. Consider the propagation of elastic waves along the ΓX direction of the irreducible Brillouin zone for which $K_2 = 0$ and $0 \leq \mathfrak{Re}(K_1) \leq \frac{\pi}{a}$. Equation (1.63) leads to

$$\begin{aligned} &\omega^2 \sum_{\vec{G}'} \rho(\vec{G} - \vec{G}') U_{3,\vec{K}}(\vec{G}') = \\ &= \sum_{\vec{G}'} C_{44}(\vec{G} - \vec{G}') \Big[(G_1 + K_1)(G_1' + K_1) + G_2 G_2' \Big] U_{3,\vec{K}}(\vec{G}'), \end{aligned} \tag{1.69}$$

and this can be rewritten as

$$\begin{aligned} &K_1{}^2 \sum_{\vec{G}'} C_{44}(\vec{G} - \vec{G}') U_{3,\vec{K}}(\vec{G}') = \\ &\sum_{\vec{G}'} \Big[\omega^2 \rho(\vec{G} - \vec{G}') - (G_1 G_1' + G_2 G_2') C_{44}(\vec{G} - \vec{G}') \Big] U_{3,\vec{K}}(\vec{G}') + \cdots \\ &\qquad - K_1 \sum_{\vec{G}'} (G_1 + G_1') C_{44}(\vec{G} - \vec{G}') U_{3,\vec{K}}(\vec{G}'), \end{aligned} \tag{1.70}$$

or in matrix form

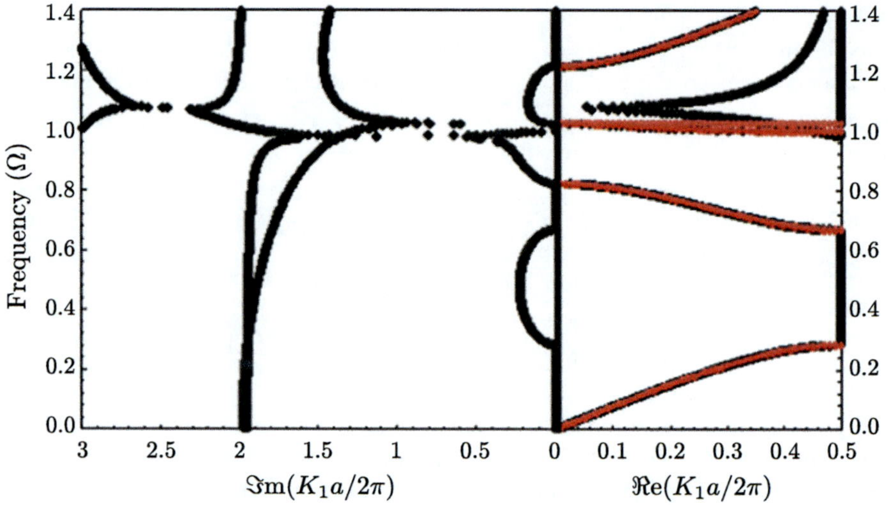

Fig. 1.18 *Z* modes band structures along the ΓX direction of the irreducible Brillouin zone for a square array of holes drilled in a Silicon matrix: Red dots: $\omega(\vec{K})$ method; Black dots: $\vec{K}(\omega)$ method

$$K_1 \begin{bmatrix} \overleftrightarrow{I} & \overleftrightarrow{0} \\ \overleftrightarrow{0} & \overleftrightarrow{A_1} \end{bmatrix} \begin{bmatrix} \vec{\mathcal{U}} \\ K_1\vec{\mathcal{U}} \end{bmatrix} = \begin{bmatrix} 0 & \overleftrightarrow{I} \\ \omega^2 \overleftrightarrow{B} - \overleftrightarrow{A_3} & -\overleftrightarrow{A_2} \end{bmatrix} \begin{bmatrix} \vec{\mathcal{U}} \\ K_1\vec{\mathcal{U}} \end{bmatrix}, \tag{1.71}$$

where

$$\begin{cases} B_{\vec{G},\vec{G}'} = \rho(\vec{G} - \vec{G}'), & (1.72a) \\ A_{1\vec{G},\vec{G}'} = C_{44}(\vec{G} - \vec{G}'), & (1.72b) \\ A_{2\vec{G},\vec{G}'} = C_{44}(\vec{G} - \vec{G}')(G_1 + G_1'), & (1.72c) \\ A_{3\vec{G},\vec{G}'} = C_{44}(\vec{G} - \vec{G}')(G_1G_1' + G_2G_2'). & (1.72d) \end{cases}$$

Numerical resolution of Eq. (1.71) leads to $2N$ (if $N \times N$ is the size of matrices $\overleftrightarrow{A}$ and $\overleftrightarrow{B}$) complex values of $K_1 = \Re e\{K_1\} - i\Im m\{K_1\}$ for any value of ω. Eigenvalues belonging to the irreducible Brillouin zone and corresponding to waves with a vanishing amplitude when $x_1 \to \infty$ may be taken into account, i.e., $0 \le \Re e\{K_1\} \le \frac{\pi}{a}$ and $\Im m\{K_1\} \ge 0$. Figure 1.18 demonstrates the band structures calculated by both $\omega(\vec{K})$ and $\vec{K}(\omega)$ methods. This figure shows the ability of the $\vec{K}(\omega)$ method to calculate the evanescent modes. Of particular interest is the existence of additional bands (see right panel of Fig. 1.18 for $\Omega \approx 1.1$) not predicted by the classical $\omega(\vec{K})$ method (red dots). These vibrational modes are characterized by a non-vanishing $\Im m\{K_1\}$. $\vec{K}(\omega)$ method requires considering only one component of the wave vector $\vec{K}$ as eigenvalue. The other components need to be fixed or a linear relation between them assumed. For example, along the ΓM direction in the irreducible Brillouin zone of the square array, one can write $K_1 = K_2$ and consider K_1 as the eigenvalue. In the same way, one can deal with any direction of propagation and not only with the high-symmetry directions. Plotting all the values of K_1 and K_2 corresponding to a specific frequency leads to the equi-frequency contour (EFC) of the phononic crystal. Knowing precisely the shape of these EFC's is of fundamental interest for study of focusing or self-collimating of elastic waves by phononic crystals [23]. Moreover, the $\vec{K}(\omega)$ PWE method allows consideration of the frequency dependent elastic moduli and should be applied for calculating the band structures of phononic crystals made of viscoelastic materials [24].

1.4.4 PWE Method for 2D Phononic Crystal Plates

The PWE method, presented in Sect. 1.4.1.2, can be applied for calculating the elastic band structure of phononic crystal plates, i.e., phononic crystals of finite extent along one specific direction of the Cartesian coordinates system (O, x_1, x_2, x_3). For the sake of clarity, we consider two-dimensional phononic crystals with a finite thickness h_2 along the x_3 direction and of infinite extent along x_1 and x_2. We limit ourselves to the case of a phononic crystal made of a square array (lattice parameter a) of cylindrical inclusions A embedded in a matrix B as in Sect. 1.4.1.3 but the method can be generalized to other geometries of the array of inclusions. It is assumed that the plate is sandwiched between two slabs of thickness h_1 and h_2, made of elastic homogeneous materials C and D, as shown in Fig. 1.19a. Materials A, B, C and D are supposed to be of cubic symmetry. Then, one may define a super-cell in the form of a parallelepiped whose basis in the (x_1, O, x_2) plane includes that of the two-dimensional primitive unit cell of the square array of inclusions and its height along the x_3 direction is $\ell = h_1 + h_2 + h_3$, see Fig. 1.19b.

This super-cell is repeated periodically along the 3 spatial directions and the overall structure is ℓ-periodic along x_3 and has the periodicity of the square array of

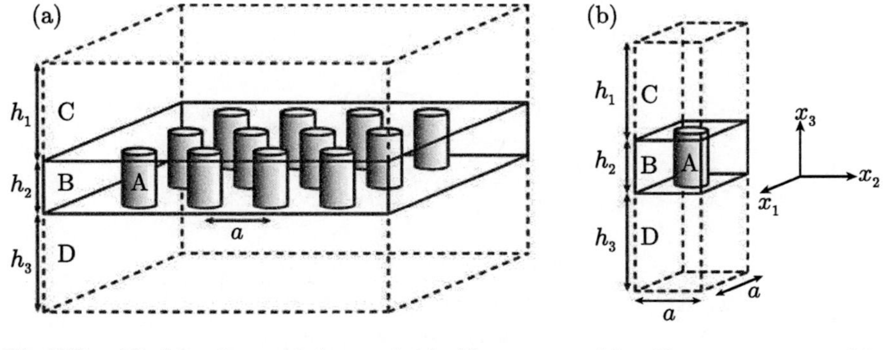

Fig. 1.19 **a** 2D phononic crystal plate sandwiched between two slabs of homogeneous materials, and **b** three-dimensional supercell considered in the course of the supercell-PWE computation

inclusions in the (x_1, O, x_2) plane. In this way, solving for the eigenfrequencies of the plate using the PWE method is a full three-dimensional problem and Eqs. (1.41) and (1.42) apply. In these equations, the reciprocal lattice vectors $\vec{G} = (G_1, G_2, G_3)$ may be written as $\vec{G} = \vec{G}_{//} + \vec{G}_3$ where $\vec{G}_{//} = G_1\vec{e}_1 + G_2\vec{e}_2$ and $\vec{G}_3 = G_3\vec{e}_3$. Similarly, the wave vector $\vec{K} = (K_1, K_2, K_3)$ associated with a propagating mode is $\vec{K} = \vec{K}_{//} + \vec{K}_3$ where $\vec{K}_{//}$ describes the periphery of the two-dimensional irreducible Brillouin zone of the square array of inclusions and $\vec{K}_3 = K_3\vec{e}_3$ where K_3 takes a fixed value between 0 and π/ℓ. Fourier coefficients in Eqs. (1.39) and (1.40) are now given as:

$$\eta(\vec{G} - \vec{G}') = \frac{1}{V_u} \iiint_{\text{(Super Cell)}} \eta(\vec{r}) e^{-i(\vec{G}-\vec{G}')\cdot\vec{r}} d^3\vec{r}, \tag{1.73}$$

where $V_u = \Sigma_{(U.C.)} \cdot \ell$ is the volume of the super-cell, and $\Sigma_{(U.C.)}$ is the area of the two-dimensional unit cell in the $(x_1 O x_2)$ plane. For a square array of inclusions, the Fourier coefficients become

$$\eta(\vec{G}-\vec{G}') = \begin{cases} f\eta_A\left(\frac{h_2}{\ell}\right) + (1-f)\eta_B\left(\frac{h_2}{\ell}\right) + \cdots \\ \qquad + \eta_C\left(\frac{h_1}{\ell}\right) + \eta_D\left(\frac{h_3}{\ell}\right), \quad \text{if } \vec{G}-\vec{G}' = \vec{0}, \\ (\eta_A - \eta_B)F_I^s(\vec{G}-\vec{G}') + (\eta_C - \eta_B)F_{II}^s(\vec{G}-\vec{G}') + \cdots \\ \qquad + (\eta_D - \eta_B)F_{III}^s(\vec{G}-\vec{G}'), \quad \text{if } \vec{G}-\vec{G}' \neq \vec{0}, \end{cases} \tag{1.74}$$

with

$$\begin{cases} F_I^s(\vec{G}-\vec{G}') = \frac{1}{V_u} \iiint_{(A)} e^{-i(\vec{G}-\vec{G}')\cdot\vec{r}} d^3\vec{r} = \\ \qquad = F(\vec{G}_{//} - \vec{G}'_{//}) \left(\frac{\sin[(G_3 - G'_3)\frac{h_2}{2}]}{(G_3 - G'_3)\frac{h_2}{2}} \right) \frac{h_2}{\ell}, & (1.75\text{a}) \\ F_{II}^s(\vec{G}-\vec{G}') = \frac{1}{V_u} \iiint_{(C)} e^{-i(\vec{G}-\vec{G}')\cdot\vec{r}} d^3\vec{r} = \\ \qquad = \left(\frac{\sin[(G_1 - G'_1)\frac{a}{2}]}{(G_1 - G'_1)\frac{a}{2}} \right) \left(\frac{\sin[(G_2 - G'_2)\frac{a}{2}]}{(G_2 - G'_2)\frac{a}{2}} \right) \times \cdots \\ \qquad \left(\frac{\sin[(G_3 - G'_3)\frac{h_1}{2}]}{(G_3 - G'_3)\frac{h_1}{2}} \right) \frac{h_1}{\ell} e^{-i(G_3 - G'_3)\frac{h_1+h_2}{2}}, & (1.75\text{b}) \\ F_{III}^s(\vec{G}-\vec{G}') = \frac{1}{V_u} \iiint_{(D)} e^{-i(\vec{G}-\vec{G}')\cdot\vec{r}} d^3\vec{r} = \\ \qquad = \left(\frac{\sin[(G_1 - G'_1)\frac{a}{2}]}{(G_1 - G'_1)\frac{a}{2})} \right) \left(\frac{\sin[(G_2 - G'_2)\frac{a}{2}]}{(G_2 - G'_2)\frac{a}{2}} \right) \times \cdots \\ \qquad \left(\frac{\sin[(G_3 - G'_3)\frac{h_3}{2}]}{(G_3 - G'_3)\frac{h_3}{2}} \right) \frac{h_3}{\ell} e^{+i(G_3 - G'_3)\frac{h_2+h_3}{2}}. & (1.75\text{c}) \end{cases}$$

In Eqs. (1.75), (1.75b), (1.75c), the integration is performed over the volume occupied by each material A, C, or D inside the super cell. In Eq. (1.75a), $F(\vec{G}_{//} - \vec{G}'_{//})$ is the structure factor defined by Eq. (1.53) for cylindrical inclusions.

The two-dimensional $\vec{G}_{//}$ (resp. $\vec{G}'_{//}$) vectors are $\vec{G}_{//} = \frac{2\pi}{a}(m\vec{e}_1 + n\vec{e}_2)$ (resp. $\vec{G}'_{//} = \frac{2\pi}{a}(m'\vec{e}_1 + n'\vec{e}_2)$) where m and n (resp. m' and n') are integers. The third component of the reciprocal lattice vectors G_3 and G'_3 are defined as $G_3 = p\frac{2\pi}{\ell}$ and $G_3 = p'\frac{2\pi}{\ell}$ where p and p' are integers. In the course of the numerical resolution of Eqs. (1.41), we consider $-M_1 \leq (m, m') \leq +M_1$, $-M_2 \leq (n, n') \leq +M_2$ and $-M_3 \leq (p, p') \leq +M_3$ (M_1, M_2, M_3 are three positive integers). This leads to $3(2M_1 + 1)(2M_2 + 1)(2M_3 + 1)$ eigenfrequencies $\omega(\vec{K})$ for a given wave vector $\vec{K}$.

The super-cell method requires the weak interaction between the vibrational modes of neighbouring periodically repeated, along the x_3 direction, phononic crystal plates. Then, in order to allow the top surface of the plate to be free of stress, medium C should behave, similar to vacuum [25]. For alleviating numerical instabilities, vacuum is modelled with the low impedance medium defined in Sect. 1.4.2.2. On the other hand, medium D can be either vacuum or a homogeneous material depending on whether one wants to model a phononic crystal plate or a structure made of a phononic crystal plate deposited on a substrate of finite thickness. Computations of dispersion curves of phononic crystal plates with $K_3 = 0$ and with any other non-vanishing value of K_3, lower than $\frac{\pi}{\ell}$, lead to nearly the same result. Indeed, the eigenvalues computed with $K_3 = 0$ and $K_3 \neq 0$ differ only in their third decimal. This indicates that the homogeneous slabs C and D made of the LIM modelling vacuum are adequate to provide appropriate decoupling of the plate modes of vibration in the x_3 direction. Then, the value of K_3 may be fixed to zero. Due to this three-dimensional nature, the numerical convergence of the super-cell PWE method is relatively slow and it has been shown that this method is suitable for voids/solid matrix plates but is not reliable for constituent materials with very different physical properties [17]. The super-cell PWE method does not to require formulating and explicitly satisfying the boundary conditions at the free surfaces. Nevertheless, other authors have proposed PWE schemes for phononic crystals plates where these boundary conditions are satisfied but these methods also suffer from convergence difficulties [26].

Figure 1.20 shows the elastic band structure of square arrays of cylindrical holes in steel for three different thickness of the phononic crystal plate. In PWE numerical calculations, the material inside the holes, i.e., air was modelled by the LIM depicted in Sect. 1.4.2.2. We consider a high filling factor of inclusion $f = 0.7$. Calculations have been performed with $M_1 = M_2 = 4$ and $M_3 = 2$ and a good convergence of the super-cell PWE method was obtained. One observes that the bulk band structure exhibits a complete band gap centred on $\Omega \approx 0.5$ and a complete band gap only appears for a thickness of the plate approaching the value of the lattice parameter, namely $h_2/a = 1.0$, see Fig. 1.20b.

The super-cell PWE method is an interesting tool for calculating the band structure of free-standing or deposited on a substrate, 2D phononic crystal plate especially

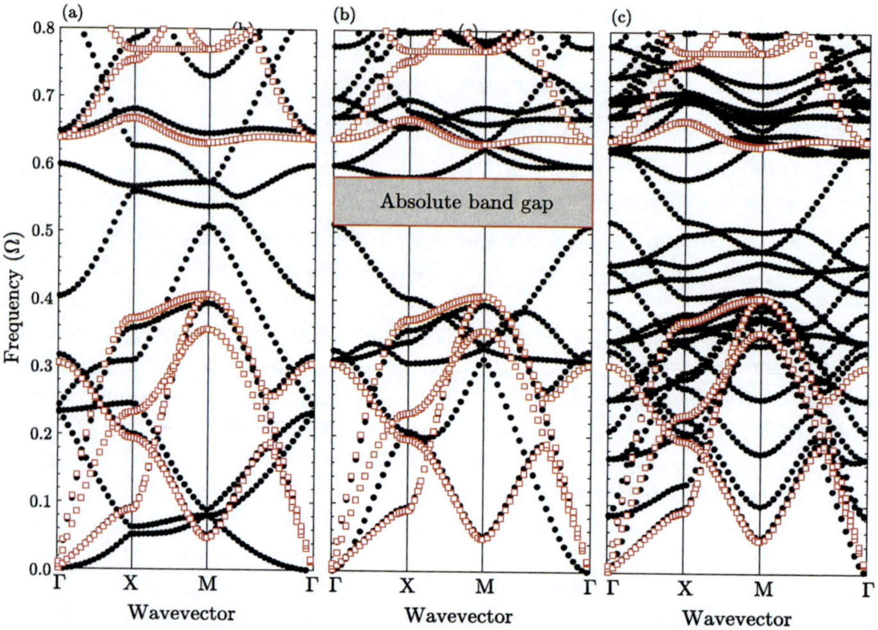

Fig. 1.20 PWE elastic band structures for the bulk 2D phononic crystal (red squares) and the phononic crystal plate of thickness h_2 (black filled circles) made of a square array of circular holes drilled in a steel matrix with $f = 0.7$. **a** $h_2 = 0.1a$; **b** $h_2 = 0.7a$; **c** $h_2 = 4a$

when the periodic structure is made of holes drilled in a matrix. For other structures, it is preferable to choose an alternate method of calculation such as for example the finite element method.

1.5 Conclusions

We have presented a detailed explanation of PWE method which is considered to be a useful tool for computing dispersion curves of the periodic structures. The method is quite easy to implement but presents some limitations regarding the convergence of the Fourier series and the choice of the constituent materials. The method is reliable in the case of phononic crystals made of solid or fluid components but is less so for most of the mixed structures where a fluid component is associated with a solid one. In these cases other methods of calculation such as the finite difference time domain method or the finite element method may be preferred. Nevertheless, the tips presented here might help to increase the accuracy of PWE method when applied to structures made of holes drilled in a solid matrix or constituted of solid inclusions surrounded with air. The classical PWE method can also be extended for considering phononic crystals of finite extent, for analysing the evanescence of waves inside the band gaps and for calculating the equi-frequency surfaces of any periodic structure.

Acknowledgements I would like to thank C. Croënne (IEMN, Villeneuve d'Ascq, France) for his help with some numerical calculations.

References

1. J.W. Strutt (Lord Rayleigh), XVII. On the maintenance of vibrations by forces of double frequency, and on the propagation of waves through a medium endowed with a periodic structure. Phil. Mag. S. 5. **24**(147), 145–159 (1887)
2. M.M. Sigalas, E.N. Economou, Elastic and acoustic wave band structure. J. Sound Vib. **158**(2), 377–382 (1992)
3. M.S. Kushwaha, P. Halevi, L. Dobrzynski, B. Djafari-Rouhani, Acoustic band structure of periodic elastic composites. Phys. Rev. Lett. **71**(13), 2022–2025 (1993)
4. P.A. Deymier, *Acoustic Metamaterials and Phononic Crystals* Springer Series in Solid-State Sciences. (Springer, Berlin, 2013) Springer Series in Solid-State Sciences 173 Springer Series in Solid-State Sciences. (Springer, Berlin, 2013)
5. C. Kittel, *Introduction to Solid State Physics*, 8th edn. 1 Wiley Drive Somerset (NJ 08875-1272, USA, Wiley, 2004)
6. N.W. Ashcroft, N.D. Mermin, *Solid State Physics*, 8th edn. (Saunders College Publishing, Fort Worth, 1976)
7. T. Wegener, "Physics in a nutshell." "http://www.physics-in-a-nutshell.com/article/1", 2017
8. D. Royer, E. Dieulesaint, *Elastic Waves in Solids I: Free and Guided Propagation* (Springer, Berlin, 1999)
9. Y.-Z. Wang, F.-M. Li, W.-H. Huang, Y.-S. Wang, Effects of inclusion shapes on the band gaps in two-dimensional piezoelectric phononic crystals. J. Phys.: Condens. Matter **19**(49), 496204 (2007)
10. S.-C.S. Lin, T.J. Huang, Tunable phononic crystals with anisotropic inclusions. Phys. Rev. B **83**(17), 174303 (2011)
11. M. Wilm, S. Ballandras, V. Laude, T. Pastureaud, A full 3D plane-wave-expansion model for 1–3 piezoelectric composite structures. J. Acoust. Soc. Am. **112**(3), 943–952 (2002)
12. O. Boumatar, J.F. Robillard, J.O. Vasseur, A.-C. Hladky-Hennion, P.A. Deymier, P. Pernod, V. Preobrazhensky, Band gap tunability of magneto-elastic phononic crystal. J. Appl. Phys. **111**(5), 054901 (2012)
13. H.S. Sözüer, J.W. Haus, R. Inguva, Photonic bands: convergence problems with the plane-wave method. Phys. Rev. B **45**(24), 13962–13972 (1992)
14. Y. Cao, Z. Hou, Y. Liu, Convergence problem of plane-wave expansion method for phononic crystals. Phys. Lett. A **327**(2), 247–253 (2004)
15. Y. Tanaka, Y. Tomoyasu, S. Tamura, Band structure of acoustic waves in phononic lattices: two-dimensional composites with large acoustic mismatch. Phys. Rev. B **62**(11), 7387–7392 (2000)
16. Z. Hou, X. Fu, Y. Liu, Singularity of the Bloch theorem in the fluid/solid phononic crystal. Phys. Rev. B **73**(2), 024304 (2006)
17. J.O. Vasseur, P.A. Deymier, B. Djafari-Rouhani, Y. Pennec, A.-C. Hladky-Hennion, Absolute forbidden bands and waveguiding in two-dimensional phononic crystal plates. Phys. Rev. B **77**(8), 085415 (2008)
18. M.S. Kushwaha, B. Djafari-Rouhani, Giant sonic stop bands in two-dimensional periodic system of fluids. J. Appl. Phys. **84**(9), 4677–4683 (1998)

19. C. Goffaux, F. Maseri, J.O. Vasseur, B. Djafari-Rouhani, P. Lambin, Measurements and calculations of the sound attenuation by a phononic band gap structure suitable for an insulating partition application. Applied Physics Letters **83**(2), 281–283 (2003)
20. M.P. Peiró-Torres, J. Redondo, J.M. Bravo, J.V.S. Pérez, Open noise barriers based on sonic crystals: advances in noise control in transport infrastructures. Transp. Res. Procedia **18**(Supplement C), 392–398 (2016)
21. J. Vasseur, P.A. Deymier, M. Beaugeois, Y. Pennec, B. Djafari-Rouhani, D. Prevost, Experimental observation of resonant filtering in a two-dimensional phononic crystal waveguide. Zeitschrift für Kristallographie **220**(9–10), 829–835 (2009)
22. M. Kushwaha, B. Djafari-Rouhani, L. Dobrzynski, Sound isolation from cubic arrays of air bubbles in water. Phys. Lett. A **248**(2), 252–256 (1998)
23. C. Croenne, E.D. Manga, B. Morvan, A. Tinel, B. Dubus, J. Vasseur, A.-C. Hladky-Hennion, Negative refraction of longitudinal waves in a two-dimensional solid-solid phononic crystal. Phys. Rev. B **83**(5), 054301 (2011)
24. R.P. Moiseyenko, V. Laude, Material loss influence on the complex band structure and group velocity in phononic crystals. Phys. Rev. B **83**(6), 064301 (2011)
25. B. Manzanares-Martínez, F. Ramos-Mendieta, Surface elastic waves in solid composites of two-dimensional periodicity. Phys. Rev. B **68**(13), 134303 (2003)
26. C. Charles, B. Bonello, F. Ganot, Propagation of guided elastic waves in two-dimensional phononic crystals. Ultrasonics **44**(Supplement), e1209–e1213 (2006)

Chapter 2
Introduction to Multiple Scattering Theory for Scalar Waves

Dani Torrent

Abstract In this chapter some fundamentals of the theory of multiple scattering between objects is presented. The chapter is focused on the scalar case and in two dimensions, so that essentially two fields are considered: the acoustic field and flexural waves in thin plates. The scattering by one single circular object is first presented and then the theory is generalized to multiple objects. For flexural waves an expression is derived for the effective scattering matrix of a given cluster, as well as the expression for the scattering of point-like objects. Some numerical examples are given at the end of the chapter.

2.1 Introduction

Multiple scattering theory [1–4] is a very efficient method, from the numerical point of view, to compute the interaction of a set of N objects with a given incident field. The theory has been applied to electromagnetic and mechanical waves, and it offers a complementary tool for more numerical methods. Although commercially available software packages can treat more or less accurately the interaction of waves with arbitrarily shaped scatterers, analytical methods continue being a reliable tool for research, since they allow us to obtain a deeper knowledge of the different physical processes involved in the interaction. However, it has to be pointed out that multiple scattering can be numerically more efficient than purely numerical methods, although our objective here is not to discuss when multiple scattering theory is better than other methods. The aim of this chapter is to present multiple scattering as a theoretical tool to other methods, and we do not pretend to present a more efficient theory but only a theory whose analytical manipulation is easy and which can be useful to understand the nature of the interaction of waves with a cluster of objects. Given the pedagogical nature of this chapter, we will focus on scalar problems in two dimensions, since

D. Torrent (✉)
Grup de Recerca d'òptica (GROC), Institut de Noves Tecnologies de la Imatge (INIT), Universitat Jaume I, Castelló, Spain
e-mail: dtorrent@uji.es

N. Jiménez et al. (eds.), *Acoustic Waves in Periodic Structures, Metamaterials, and Porous Media*, Topics in Applied Physics 143,
https://doi.org/10.1007/978-3-030-84300-7_2

once this situation is properly understood it is just a mater of technical work to learn the three-dimensional and vectorial situation.

Before going to a detailed explanation, let us see a pedagogical introduction to the problem of multiple scattering. Let us assume that some incident field ψ_0, e.g. a plane wave or something similar, arrives to a region of space where there is an object which scatters waves. The total field ψ_T in space now will be

$$\psi_T = \psi_0 + \psi_{sc}, \tag{2.1}$$

where ψ_{sc} is the scattered field by the object. This scattered field depends on the incident field, and it is proportional to it, with the constant of proportionality called the T matrix, thus we can write

$$\psi_{sc} = T\psi_0, \tag{2.2}$$

and the total field is therefore given by (we will see later that this is just a symbolic solution)

$$\psi_T = (1 + T)\psi_0. \tag{2.3}$$

Now, let us assume that we have instead a cluster of N objects located in positions $\boldsymbol{R}_\alpha$, for $\alpha = 1, 2, \ldots, N$, each one scatters a field ψ_{sc}^α, thus

$$\psi_T = \psi_0 + \sum_{\alpha=1}^{N} \psi_{sc}^\alpha. \tag{2.4}$$

We know the T matrix of each scatterer, labeled T_α, however the T matrix relates the response of the scatterer to the incident field, but now this incident field is not only ψ_0, since the field that arrives to the scatterer α, labeled $\psi_{0\alpha}$, is the incident field ψ_0 plus the scattered field by all the other scatterers $\beta \neq \alpha$. Then we can write

$$\psi_{sc}^\alpha = T_\alpha \psi_{0\alpha} = T_\alpha \left(\psi_0 + \sum_{\beta \neq \alpha} \psi_{sc}^\beta \right), \tag{2.5}$$

which is equivalent to

$$\psi_{sc}^\alpha - T_\alpha \sum_{\beta \neq \alpha} \psi_{sc}^\beta = T_\alpha \psi_0. \tag{2.6}$$

The above equation is actually a set of N equations from which we can obtain the N unknowns ψ_{sc}^α. Let us assume for instance that we have only two scatterers, then $\alpha = 1, 2$ and we have the two equations

$$\psi_{sc}^1 - T_1 \psi_{sc}^2 = T_1 \psi_0 \tag{2.7}$$

and

$$\psi_{sc}^2 - T_2\psi_{sc}^1 = T_2\psi_0, \tag{2.8}$$

from which we solve the scattered fields as

$$\psi_{sc}^1 = T_1 \frac{1 + T_2}{1 - T_1 T_2}\psi_0 \tag{2.9}$$

and a similar solution is found for ψ_{sc}^2. We see then the formal mechanism of multiple scattering, and how the interaction between scatterers appears by means of the crossing products $T_1 T_2$. The strength of the interaction between the different scatterers will depend in general on this product, among other variables like the distance between them or the cluster configuration in the case of large clusters. Within the next sections we will derive a more formal theory to study this complex process.

2.2 Multiple Scattering of Acoustic Waves

When an external acoustic field reaches a closed region defined by some boundary $\partial\Omega$ having acoustic parameters different to that of the surrounding medium a scattered field is excited. In two dimensions the external field can always be expressed in terms of regular Bessel functions $J_q(\cdot)$ in polar coordinates $\boldsymbol{r} = (r, \theta)$ and the field wave number k,

$$P^0(\boldsymbol{r}) = \sum_{q=-\infty}^{\infty} A_q^0 J_q(kr)e^{iq\theta}. \tag{2.10}$$

The scattered field cannot be constructed with regular Bessel functions only, but an irregular part must be added. This irregular part is formed with the Bessel functions of the second kind $Y_q(\cdot)$, and the resulting functions are called Hankel functions $H_q(\cdot) = J_q(\cdot) + iY_q(\cdot)$. In terms of Hankel functions the scattered field is expressed as

$$P^{\text{sc}}(\boldsymbol{r}) = \sum_{q=-\infty}^{\infty} A_q H_q(kr)e^{iq\theta}. \tag{2.11}$$

In a general problem, the coefficients A_q^0 of the incident field are the inputs and the scattered field coefficients A_q are the output. The coefficients A_q depend on both the physical nature of the scatterer and the incident field. The dependence of the physical nature of the scatterer is obtained mathematically by means of the boundary conditions, which are different for each type of cylinders, as will be shown later, but finally a linear relationship is found by means of the T matrix [5],

$$A_q = \sum_s T_{qs}(k, \eta_i)A_s^0, \tag{2.12}$$

where the argument η_i indicates that the matrix elements T_{qs} are dependent on some set of physical constants η_i. Nevertheless, in the present work this notation is simplified and only the wavenumber dependence will be explicitly indicated in some special cases. The expression of this T matrix depends not only on the physical nature of the cylinders but also on the external shape of them.

In this section, after showing the expressions for the most commonly used incident fields, the T matrix of a fluid cylinder (circular and with arbitrary cross-section) and of an elastic cylinder (only circular) will be computed.

2.2.1 *Incident Fields*

Here, only two types of 2D incident fields are used: the plane wave and the point source. Both are basic fields and can be used to understand the response of a given system to more complicated sources. Independently of the functional form of the incident fields, when working in a multiple scattering environment these fields have to be expanded in terms of regular Bessel functions in translated reference frames. The goal of this section is not only introduce the fields, but also show how they can be expanded in Bessel functions.

2.2.1.1 The Plane Wave

The plane wave is a field distribution defined by some wavenumber $\boldsymbol{k}_0 = k_{0x}\hat{\boldsymbol{x}} + k_{0y}\hat{\boldsymbol{y}} = k(\cos\theta_0, \sin\theta_0)$ and some complex amplitude C_0, and has a functional form

$$P^0(\boldsymbol{r}) = C_0 e^{i\boldsymbol{k}_0\cdot\boldsymbol{r}}, \tag{2.13}$$

the integral definition of the Bessel functions [6] allow us to express this field as

$$P^0(\boldsymbol{r}) = C_0 e^{i\boldsymbol{k}_0\cdot\boldsymbol{r}} = C_0 \sum_q i^q e^{-iq\theta_0} J_q(kr) e^{iq\theta}. \tag{2.14}$$

The coefficients A_q^0 in (2.10) are

$$A_q^0 = C_0 i^q e^{-iq\theta_0}. \tag{2.15}$$

Now let us assume that the same field has to be expressed in another reference frame $\boldsymbol{r}'$ such that, as shown in Fig. 2.1

$$\boldsymbol{r}' = \boldsymbol{r} - \boldsymbol{R}', \tag{2.16}$$

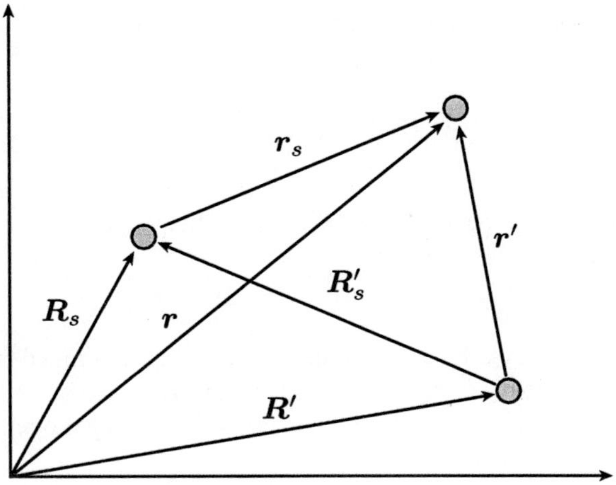

Fig. 2.1 Coordinate definitions for the incident fields

where $\boldsymbol{R}'$ is a vector which goes from the origin in the frame $\boldsymbol{r}$ to that of the frame $\boldsymbol{r}'$. The expression needed could be obtained by Graf's addition theorem [6]. For the case of plane waves a simpler approach can be used,

$$\begin{aligned} P^0(\boldsymbol{r}) &= C_0 e^{i\boldsymbol{k}_0\cdot\boldsymbol{r}} = C_0 e^{i\boldsymbol{k}_0\cdot\boldsymbol{R}'} e^{i\boldsymbol{k}_0\cdot\boldsymbol{r}'} \\ &= C_0 e^{i\boldsymbol{k}_0\cdot\boldsymbol{R}'} \sum_q i^q e^{-iq\theta_0} J_q(kr') e^{iq\theta'}, \end{aligned} \tag{2.17}$$

thus,

$$A'^0_q = C_0 e^{i\boldsymbol{k}_0\cdot\boldsymbol{R}'} i^q e^{-iq\theta_0} = C_0 e^{i\boldsymbol{k}_0\cdot\boldsymbol{R}'} A^0_q. \tag{2.18}$$

In multiple scattering theory this operation moves the origin of coordinates to the center of new cylindrical system where the vector $\boldsymbol{R}'$ is the origin.

Any incident field can be expanded in a sum (discrete or continuous) of plane waves, the resulting field being the sum of the response to all these plane waves. In general, the incident field will be

$$P^0(\boldsymbol{r}) = \sum_{\theta_0} C_{\theta_0} e^{i\boldsymbol{k}_0\cdot\boldsymbol{r}}. \tag{2.19}$$

2.2.1.2 The Point Source

The point source of order s is defined by a Hankel function of the same order. For a source located in $\boldsymbol{R}_s$

$$P^0(\boldsymbol{r}) = C_s H_s(kr_s) e^{is\theta_s}, \tag{2.20}$$

where $\boldsymbol{r}_s = \boldsymbol{r} - \boldsymbol{R}_s$ and C_s is a complex constant (see Fig. 2.1).

This field can be expressed in the frame $\boldsymbol{r}'$ of the previous section by means of the Graf's addition theorem,

$$P^0(\boldsymbol{r}) = C_s \sum_q H_{s-q}(kR'_s)e^{i(s-q)\Phi'_s} J_q(kr')e^{iq\theta'}, \tag{2.21}$$

where the vector $\boldsymbol{R}'_s = (R'_s, \Phi'_s)$ is the position vector of the source in the primed frame, and is given by

$$\boldsymbol{R}'_s = \boldsymbol{R}_s - \boldsymbol{R}'. \tag{2.22}$$

When working with multiple scattering the primed refers to some cylinder, so that $\boldsymbol{R}'$ is the position vector of this cylinder.

If a more general source is needed, it can be modelled with a linear combination of all the s-sources

$$P^0(\boldsymbol{r}) = \sum_s C_s H_s(kr_s)e^{is\theta_s}, \tag{2.23}$$

and the response of the system will be the addition of the responses to all the individual sources.

2.2.2 *T-Matrix of a Fluid Cylinder*

Let us consider a fluid cylinder of radius R_a, and a density and sound speed given by v_a and ρ_a, respectively. Let us also assume that some external field, defined by the coefficients A_q^0, impinges the cylinder. In the region outside the cylinder, as shown previously, the total acoustic field is

$$P^+(\boldsymbol{r}) = \sum_q A_q^0 J_q(kr)e^{iq\theta} + \sum_q A_q H_q(kr)e^{iq\theta}. \tag{2.24}$$

Inside the cylinder there are no sources, and the total field can be expressed in terms of regular Bessel functions with associated wavenumber k_a

$$P^-(\boldsymbol{r}) = \sum_q B_q^{\text{in}} J_q(k_a r)e^{iq\theta}. \tag{2.25}$$

Boundary conditions are applied at the circular surface, which are the continuity of the pressure field and the normal component of the velocity field [7], thus,

$$P^+(\boldsymbol{R}_a) = P^+(\boldsymbol{R}_b), \tag{2.26}$$

$$\frac{1}{\rho_b}\frac{\partial P^+(\boldsymbol{r})}{\partial r}\bigg|_{r=R_a} = \frac{1}{\rho_a}\frac{\partial P^+(\boldsymbol{r})}{\partial r}\bigg|_{r=R_a}. \tag{2.27}$$

Substituting (2.24) and (2.25) into these boundary conditions we obtain

$$\sum_q A_q^0 J_q(kR_a)e^{iq\theta} + \sum_q A_q H_q(kR_a)e^{iq\theta} = \sum_q B_q^{\text{in}} J_q(k_a R_a)e^{iq\theta}, \tag{2.28}$$

and

$$\frac{k}{\rho_b}\sum_q A_q^0 J_q'(kR_a)e^{iq\theta} + \frac{k}{\rho_b}\sum_q A_q H_q'(kR_a)e^{iq\theta} = \frac{k_a}{\rho_a}\sum_q B_q^{\text{in}} J_q'(k_a R_a)e^{iq\theta}, \tag{2.29}$$

where the $'$ implies derivatives respect to the argument.

The sums in q can be eliminated by multiplying the equations by $e^{is\theta}$ and integrating from 0 to 2π. In this case only those terms such that $s = q$ are different from zero and the final form of the equations

$$A_q^0 J_q(kR_a)e^{iq\theta} + A_q H_q(kR_a)e^{iq\theta} = B_q^{\text{in}} J_q(k_a R_a)e^{iq\theta}, \tag{2.30}$$

$$\frac{k}{\rho_b} A_q^0 J_q'(kR_a)e^{iq\theta} + \frac{k}{\rho_b} A_q H_q'(kR_a)e^{iq\theta} = \frac{k_a}{\rho_a} B_q^{\text{in}} J_q'(k_a R_a)e^{iq\theta}. \tag{2.31}$$

Now it is easy to obtain the relations between the coefficients A_q and A_q^0, which defines the T matrix. Note that in the obtained relation the coefficient A_q only depends on the coefficient A_q^0; i.e., the T matrix is diagonal, $T_{sq} = T_q\delta_{sq}$, with T_q given by

$$T_q = -\frac{\rho_q J_q'(kR_a) - J_q(kR_a)}{\rho_q H_q'(kR_a) - H_q(kR_a)}, \tag{2.32}$$

where

$$\rho_q = \frac{\rho_a v_a}{\rho_b v_b}\frac{J_q(k_a R_a)}{J_q'(k_a R_a)}. \tag{2.33}$$

The T matrix of two special cases can now be derived. The first one is the rigid cylinder, where the density of the cylinder ρ_a, in relation to the background ρ_b, approaches to infinity. In this situation also the quantity ρ_q approaches to infinity, so that the T matrix for the rigid cylinder is

$$T_q = -\frac{J_q'(kR_a)}{H_q'(kR_a)}. \tag{2.34}$$

The second case is the void cylinder, which corresponds to the opposite situation: now the density of the cylinder, in relation to the background, approaches to zero, then the T matrix for the void is

$$T_q = -\frac{J_q(kR_a)}{H_q(kR_a)}. \tag{2.35}$$

The T matrix allows one to obtain the field outside the cylinder. When the field inside the cylinder is needed, it can be computed with the help of (2.30), which solving for B_q^{in} gives

$$B_q^{\text{in}} = \frac{J_q(kR_a) + T_q H_q(kR_a)}{J_q(k_a R_a)} A_q^0, \tag{2.36}$$

for both the rigid and the void cylinders there is no field inside them.

2.2.3 Multiple Scattering

Consider a cluster of N parallel cylinders with arbitrary cross-section shape located at $\boldsymbol{R}_\alpha$, with $\alpha = 1, 2, \ldots, N$. If an incident field $P^0(r, \theta)$ impinges the cluster, the total scattered field will be given by the sum of the scattered fields by each individual cylinder, that is

$$P^{\text{sc}}(r, \theta) = \sum_\alpha \sum_{q=-\infty}^{\infty} (A_\alpha)_q H_q(kr_\alpha) e^{iq\theta_\alpha}, \tag{2.37}$$

where $H_q(\cdot)$ is the q-Th order Hankel function of first kind and $(r_\alpha, \theta_\alpha)$ are the polar coordinates with the origin translated to the center of the α-cylinder, i.e, $\boldsymbol{r}_\alpha = \boldsymbol{r} - \boldsymbol{R}_\alpha$, as shown in Fig. 2.2. Here, $k = \omega/c_b$ and $(A_\alpha)_q$ are the coefficients to be determined.

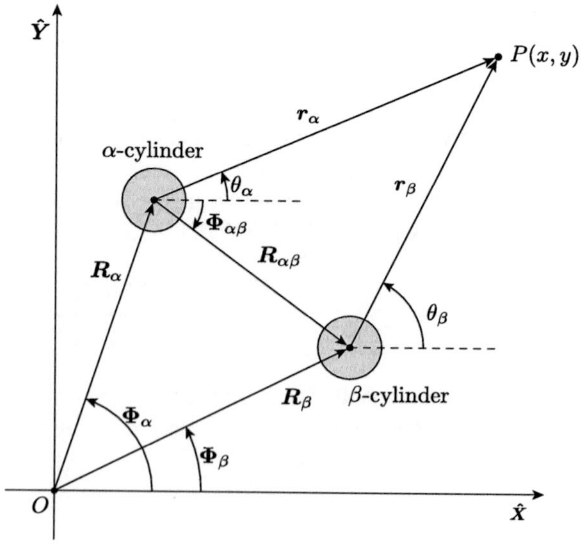

Fig. 2.2 Definition of variables employed in the multiple scattering theory

The total field incident on the α-cylinder can be expressed as a linear combination of Bessel functions

$$P^0_\alpha(r_\alpha, \theta_\alpha) = \sum_s (B_\alpha)_s J_s(kr_\alpha)e^{is\theta_\alpha}. \tag{2.38}$$

These coefficients are related to the $(A_\alpha)_q$ by means of the T matrix

$$(A_\alpha)_q = \sum_s (T_\alpha)_{qs}(B_\alpha)_s, \tag{2.39}$$

with T_α the T matrix of the α-cylinder.

The total field (2.38) incident upon the α-cylinder is the sum of the external field $P^0(r, \theta)$ and the field scattered by all the cylinders except α. The external field can be expressed in the α frame. The scattered field can also be expressed in the α reference frame by means of the Graft's addition theorem. Thus,

$$P^0_\alpha(r_\alpha, \theta_\alpha) = \sum_q (A^0_\alpha)_q J_q(kr_\alpha)e^{iq\theta_\alpha} + \sum_{q,s}\sum_{\beta\neq\alpha} (A_\beta)_s H_{q-s}(kr_{\alpha\beta})e^{i(s-q)\theta_{\alpha\beta}} J_q(kr_\alpha)e^{iq\theta_\alpha}, \tag{2.40}$$

where the coefficients $(A^0_\alpha)_q$ are those of the external field in the α frame of reference. Details of the variables employed are given in Fig. 2.2. From (2.40) and (2.38), the relation between $(B_\alpha)_q$ and $(A_\beta)_s$ coefficients is

$$(B_\alpha)_q = (A^0_\alpha)_q + \sum_{\beta\neq\alpha} (A_\beta)_s H_{q-s}(kr_{\alpha\beta})e^{i(s-q)\theta_{\alpha\beta}}. \tag{2.41}$$

Multiplying this equation by $(T_\alpha)_{qr}$ and performing summation over q we get

$$(A_\alpha)_r - \sum_s \sum_\beta (G_{\alpha\beta})_{rs}(A_\beta)_s = \sum_q (T_\alpha)_{qr}(A^0_\alpha)_q, \tag{2.42}$$

where

$$(G_{\alpha\beta})_{rs} = \sum_q (1 - \delta_{\alpha\beta})(T_\alpha)_{qr} H_{q-s}(kr_{\alpha\beta})e^{i(s-q)\theta_{\alpha\beta}}. \tag{2.43}$$

Equation (2.43) has been derived for a cylinder of arbitrary cross-section with a non-diagonal T matrix. However, in the present work, the multiple scattering theory has been mainly applied to circular cylinders, where the T matrix is diagonal, $(T_\alpha)_{qr} = (T_\alpha)_q\delta_{qr}$. Thus (2.42) and (2.43) become, respectively,

$$(A_\alpha)_q - \sum_s \sum_\beta (G_{\alpha\beta})_{qs}(A_\beta)_s = (T_\alpha)_q (A^0_\alpha)_q, \tag{2.44}$$

$$(G_{\alpha\beta})_{qs} = (1-\delta_{\alpha\beta})(T_\alpha)_q H_{q-s}(kr_{\alpha\beta})e^{i(s-q)\theta_{\alpha\beta}}. \tag{2.45}$$

Returning to the general expression (2.42), in principle, the multi-polar expansions are infinite, but the angular momentum index is truncated to a maximum value $s_{\max}$ such that $|s| <= s_{\max}$. In this case, the set of equations (2.42) defines a system of $N(2s_{\max}+1)$ linear equations. The inversion of the matrix M defined by

$$(M_{\alpha\beta})_{rs} = \delta_{rs}\delta_{\alpha\beta} - (G_{\alpha\beta})_{rs}, \tag{2.46}$$

gives the solution of the problem

$$(A_\alpha)_q = \sum_\beta \sum_r \sum_s (M^{-1}_{\alpha\beta})_{qr}(T_\alpha)_{rs}(A^0_\alpha)_s. \tag{2.47}$$

This solution is obtained in terms of the position and properties of each cylinder and of the external field.

Note that, for the case of a plane wave of amplitude C_0 and propagation angle θ_0, the coefficients $(A^0_\alpha)_q$ are

$$(A^0_\alpha)_q = C_0 e^{i\boldsymbol{k}_0\cdot\boldsymbol{R}_\alpha} i^q e^{-iq\theta_0}, \tag{2.48}$$

and for the point source of order s and amplitude C_s

$$(A^0_\alpha)_q = C_s H_{s-q}(kr'_s)e^{i(s-q)\Phi'_s}. \tag{2.49}$$

Finally, the total pressure at any point of the plane is the addition of the incident field and the total scattered field,

$$P(r,\theta) = P^0(r,\theta) + \sum_\alpha \sum_{q=-\infty}^{\infty} (A_\alpha)_q H_q(kr_\alpha)e^{iq\theta_\alpha}. \tag{2.50}$$

The scattered far field is a very useful quantity to characterize a cluster of cylinders. This scattered far field is computed for $kr \gg 1$, then

$$P^{sc}(r,\theta) = \sum_\alpha \sum_{q=-\infty}^{\infty} (A_\alpha)_q H_q(kr_\alpha)e^{iq\theta_\alpha} \approx$$

$$\sqrt{\frac{2}{\pi kr}} e^{-i\pi/4} e^{ikr} \sum_\alpha \sum_{q=-\infty}^{\infty} (-i)^q (A_\alpha)_q e^{-ikr_\alpha} e^{iq\theta_\alpha}. \tag{2.51}$$

Thus the scattered far field is

$$\sigma_{sc}(k,\theta) = \left| \sqrt{\frac{2}{\pi k}} \sum_{\alpha} \sum_{q=-\infty}^{\infty} (-i)^q (A_\alpha)_q e^{-ikr_\alpha} e^{iq\theta_\alpha} \right|, \tag{2.52}$$

and the scattering form factor is

$$F(k) \equiv \int_0^{2\pi} \sigma_{sc}^2(k,\theta) d\theta = \frac{4}{k} \sum_{\alpha} \sum_{q=-\infty}^{\infty} |(A_\alpha)_q|^2. \tag{2.53}$$

2.3 Multiple Scattering of Flexural Waves

Flexural waves are a special case of elastic waves propagating in thin elastic plates. In the long wavelength limit, it can be shown that these waves propagate as a scalar field which satisfies a two-dimensional wave theory. From the pedagogical point of view this fact is interesting since it is a scalar field in two dimensions which, additionally, describes a true physical situation, so that it is a very instructive example to learn about the multiple scattering of waves. We will follow the theoretical introduction for these waves given in [8–11].

The equations describing flexural waves can be found in many textbooks (see for instance [12, 13]). If the wavelength of the field is larger than the thickness of the plate, the wave equation is the fourth order differential equation

$$\begin{aligned} -\frac{\partial^2}{\partial x^2}\left(D_b\left[\frac{\partial^2 W}{\partial x^2} + \nu_b \frac{\partial^2 W}{\partial y^2}\right]\right) \\ -\frac{\partial^2}{\partial y^2}\left(D_b\left[\frac{\partial^2 W}{\partial y^2} + \nu_b \frac{\partial^2 W}{\partial x^2}\right]\right) \\ -2\frac{\partial^2}{\partial x \partial y}\left(D_b(1-\nu_b)\frac{\partial^2 W}{\partial x \partial y}\right) = \rho h \frac{\partial^2 W}{\partial t^2}, \end{aligned} \tag{2.54}$$

being ρ_b, h_b and $D_b = E_b h_b^3/12(1-\nu_b^2)$ the mass density, thickness and bending stiffness of the plate, respectively, with E_b the Young's modulus and ν_b and Poisson's ratio. When the background's parameters are constant, and we assume harmonic time dependence of the field W, the above equation reduces to

$$(D_b \nabla^4 - \rho_b h_b \omega^2) W(x,y) = 0, \tag{2.55}$$

whose solution in polar coordinates is given by a linear combination of Bessel and modified Bessel functions [14] of argument k_b, such that

$$k_b^4 = \frac{\rho_b h_b}{D_b}\omega^2. \tag{2.56}$$

We limit here the study to the scattering by circular inclusions in the low frequency limit, defined for wavelengths such that $\lambda > 4a$, being a the typical distance between scatterers. Thus, as long as the thickness of the plate be smaller than a, the above equation is a good approximation.

For a scattering problem, the incident field is expressed as

$$W_0 = \sum_q \left[A_q^J J_q(k_b r) + A_q^I I_q(k_b r)\right] e^{iq\theta}, \tag{2.57}$$

while the scattered field is given by

$$W_{sc} = \sum_q \left[B_q^H H_q(k_b r) + B_q^K K_q(k_b r)\right] e^{iq\theta}. \tag{2.58}$$

If the scatterer is a circular inhomogeneity of radius R_a, inside the scatterer ($r < R_a$), since there are no sources, the field is expressed as

$$W_i = \sum_q \left[C_q^J J_q(k_a r) + C_q^I I_q(k_a r)\right] e^{iq\theta}. \tag{2.59}$$

Boundary conditions are explained for instance in [14], and they provide a system of four equations which solves for the four unknowns: two scattering coefficients B_q^H, B_q^K and the two internal coefficients C_q^J, C_q^I. The system of equations can be expressed as

$$\boldsymbol{X}_q^0 A_q + \boldsymbol{X}_q^{sc} B_q = \boldsymbol{X}_q^a C_q, \tag{2.60}$$

$$\boldsymbol{Y}_q^0 A_q + \boldsymbol{Y}_q^{sc} B_q = \boldsymbol{Y}_q^a C_q, \tag{2.61}$$

where the matrices $\boldsymbol{X}_q^i$ and $\boldsymbol{Y}_q^i$, with $i = 0, sc, a$, are 2×2 matrices and the coefficient vectors are $A_q = (A_q^J, A_q^I)$, $B_q = (B_q^H, B_q^K)$ and $C_q = (C_q^J, C_q^I)$. Solving (2.60) for C_q^i, and inserting the result into (2.61) gives

$$\boldsymbol{Y}_q^0 A_q + \boldsymbol{Y}_q^{sc} B_q = \boldsymbol{Y}_q^a (\boldsymbol{X}_q^a)^{-1} (\boldsymbol{X}_q^0 A_q + \boldsymbol{X}_q^{sc} B_q), \tag{2.62}$$

from which B_q can be found as a function of A_q

$$B_q = -\left(\boldsymbol{Y}_q^{sc} - \boldsymbol{Y}_q^a (\boldsymbol{X}_q^a)^{-1} \boldsymbol{X}_q^{sc}\right)^{-1} \times \left(\boldsymbol{Y}_q^0 - \boldsymbol{Y}_q^a (\boldsymbol{X}_q^a)^{-1} \boldsymbol{X}_q^0\right) A_q. \tag{2.63}$$

The above equation defines the T matrix of the scatterer, and it gives the scattering coefficients B_q in terms of A_q. It is a 2×2 matrix and each element of the matrix is related to the excitation of a different mode, that is, in full matrix form we have

$$\begin{bmatrix} B_q^H \\ B_q^K \end{bmatrix} = \begin{bmatrix} T_q^{HJ} & T_q^{HI} \\ T_q^{KJ} & T_q^{KI} \end{bmatrix} \begin{bmatrix} A_q^J \\ A_q^I \end{bmatrix}. \tag{2.64}$$

If the scatterer is a hole, the clamped free boundary conditions gives simply

$$B_q = -(\boldsymbol{Y}_q^{sc})^{-1}\boldsymbol{Y}_q^0 A_q. \tag{2.65}$$

Let us assume now that we have a cluster of scatterers located at positions $\boldsymbol{R}_\alpha$, for $\alpha = 1, 2, \ldots, N$, the total scattered field by the cluster is now given by

$$\Psi_{sc} = \sum_\alpha \sum_q \left[B_{q\alpha}^H H_q(k_b r_\alpha) + B_{q\alpha}^K K_q(k_b r_\alpha)\right] e^{iq\theta_\alpha} \tag{2.66}$$

where the B_α coefficients are related with the incident field on the α scatterer by the T matrix

$$B_{q\alpha}^H = T_{q\alpha}^{HJ} A_{q\alpha}^J + T_{q\alpha}^{HI} A_{q\alpha}^I, \tag{2.67}$$

$$B_{q\alpha}^K = T_{q\alpha}^{KJ} A_{q\alpha}^J + T_{q\alpha}^{KI} A_{q\alpha}^I. \tag{2.68}$$

However, the coefficients A_q have now two contributions: one due to the external incident field and the other given by the scattered field by the other scatterers, thus, after applying the addition theorem we have

$$A_{q\alpha}^J = \sum_s (G_{\alpha O}^0)_{qs}^J A_s^J + \sum_{\beta\neq\alpha}\sum_s (G_{\alpha\beta}^{sc})_{qs}^H B_{s\beta}^H, \tag{2.69}$$

$$A_{q\alpha}^I = \sum_s (G_{\alpha O}^0)_{qs}^I A_s^I + \sum_{\beta\neq\alpha}\sum_s (G_{\alpha\beta}^{sc})_{qs}^K B_{s\beta}^K \tag{2.70}$$

where

$$(G_{\alpha\beta}^{sc})_{qs}^H = H_{q-s}(k_b R_{\alpha\beta}) e^{i(s-q)\theta_{\alpha\beta}}, \tag{2.71}$$

$$(G_{\alpha\beta}^{sc})_{qs}^K = (-1)^s K_{q-s}(k_b R_{\alpha\beta}) e^{i(q-s)\theta_{\alpha\beta}}. \tag{2.72}$$

Then, we can obtain the equation for the B_α coefficients

$$\begin{aligned} B_{q\alpha}^H =& T_{q\alpha}^{HJ} \sum_s (G_{\alpha O}^0)_{qs}^J A_s^J + T_{q\alpha}^{HI} \sum_s (G_{\alpha O}^0)_{qs}^I A_s^I + \\ & T_{q\alpha}^{HJ} \sum_{\beta\neq\alpha}\sum_s (G_{\alpha\beta}^{sc})_{qs}^H B_{s\beta}^H + T_{q\alpha}^{HI} \sum_{\beta\neq\alpha}\sum_s (G_{\alpha\beta}^{sc})_{qs}^K B_{s\beta}^K \end{aligned} \tag{2.73}$$

and

$$B_{q\alpha}^{K}=T_{q\alpha}^{KJ}\sum_{s}(G_{\alpha O}^{0})_{qs}^{J}A_{s}^{J}+T_{q\alpha}^{KI}\sum_{s}(G_{\alpha O}^{0})_{qs}^{I}A_{s}^{I}+$$
$$T_{q\alpha}^{KJ}\sum_{\beta\neq\alpha}\sum_{s}(G_{\alpha\beta}^{sc})_{qs}^{H}B_{s\beta}^{H}+T_{q\alpha}^{KI}\sum_{\beta\neq\alpha}\sum_{s}(G_{\alpha\beta}^{sc})_{qs}^{K}B_{s\beta}^{K}. \tag{2.74}$$

The above two equations give us the coefficients B_q as functions of the A_q, after solution of the system of equations

$$MB=TG^{0}A, \tag{2.75}$$

where the matrix M is given by

$$(M_{\alpha\beta})_{qs}=\boldsymbol{I}\delta_{\alpha\beta}\delta_{qs}-\boldsymbol{T}_{\alpha q}(\boldsymbol{G}_{\alpha\beta}^{sc})_{qs}, \tag{2.76}$$

being $\boldsymbol{I}$ the 2×2 identity matrix, $\boldsymbol{T}_{\alpha q}$ the T matrix of the α-scatterer and

$$(\boldsymbol{G}_{\alpha\beta})_{qs}=\begin{bmatrix}(G_{\alpha\beta}^{sc})_{qs}^{H} & 0\\ 0 & (G_{\alpha\beta}^{sc})_{qs}^{K}\end{bmatrix}. \tag{2.77}$$

2.3.1 Effective T-Matrix

There is a very useful tool to describe the scattering of a cluster of objects, which is called the *effective T-matrix*. This effective T matrix has been properly computed for electromagnetic and acoustic scatterers [15–17], however for flexural waves an explicit expression is not found in the literature. The idea is to describe the scattering of one cluster of objects by means of a single T-matrix, as if it were just one complex objects. This method is specially useful for homogenization problems.

Let us expand the scattered fields given by (2.66) around a coordinate system located at the origin. By means of the addition theorem we get

$$\Psi_{sc}=\sum_{\alpha}\sum_{q}\left[B_{q\alpha}^{H}\sum_{s}(G_{\alpha O}^{O})_{qs}^{J}H_{s}(k_{b}r)+B_{q\alpha}^{K}\sum_{s}(G_{\alpha O}^{O})_{qs}^{I}K_{s}(k_{b}r)\right]e^{is\theta}, \tag{2.78}$$

which can be expressed as (as long as $r>R_{\alpha}$ for all α)

$$\Psi_{sc}=\sum_{s}B_{s}H_{s}(k_{b}r)e^{is\theta} \tag{2.79}$$

with

$$B_{s}=\sum_{\alpha}\sum_{q}\left[(G_{O\alpha}^{O})_{sq}^{J}B_{q\alpha}^{H}+(G_{O\alpha}^{O})_{sq}^{I}B_{q\alpha}^{K}\right] \tag{2.80}$$

or in matrix notation

$$B_0 = G^0 B_\alpha \tag{2.81}$$

given that

$$B_\alpha = M^{-1} T G^0 A_0 \tag{2.82}$$

we get

$$B_0 = G^0 M^{-1} T G^0 A_0 \tag{2.83}$$

which defines the effective T-matrix of the cluster as

$$T = G^0 M^{-1} T G^0 \tag{2.84}$$

or, more specifically,

$$T_{qs} = \sum_{\alpha,\beta} \sum_{n,m} (G^O_{O\alpha})_{qn} (M^{-1}_{\alpha\beta})_{nm} T_{\beta m} (G^O_{O\beta})_{ms}. \tag{2.85}$$

Then, the cluster of scatterers behaves as a single object with a non-diagonal matrix given by the above expression. The ideas of homogenization developed in [17–19] can now be applied here, although approximated expressions have been found in [11].

2.3.2 Scattering by a Cluster of Mass-Spring Resonators

A particular case is the scattering of flexural waves by a point-like scatterer, which is typically modelled as a spring-mass attachment to the plate (see [9, 20] for further details). The solution for the scattering of a point resonator can be obtained directly from the wave equation,

$$(D\nabla^4 - \omega^2 \rho h) W(\boldsymbol{r}) = f \delta(\boldsymbol{r}). \tag{2.86}$$

The force term is related with the movement of the mass-spring resonator as

$$f = -k_0 (W_0 - W), \tag{2.87}$$

being W_0 the displacement of the mass. Also, assuming time harmonic dependence of W_0, Newton's law reads

$$-\omega^2 m_0 W_0 = f. \tag{2.88}$$

From these two equations we can obtain, by solving for W_0, the relationship between f and W,being

$$f = \frac{\omega_0^2 m_0}{1 - \omega_0^2/\omega^2} W, \tag{2.89}$$

where $\omega_0^2 = k_0/m_0$ is the resonant frequency of the resonator. Thus, the wave equation is

$$(D\nabla^4 - \omega^2 \rho h) W(\boldsymbol{r}) = \frac{\omega^2 \omega_0^2 m_0}{\omega^2 - \omega_0^2} W(\boldsymbol{r}) \delta(\boldsymbol{r}). \tag{2.90}$$

To solve the above equation we need Green's function, defined as

$$\left(\nabla^4 - k_b^4\right) G_0(\boldsymbol{r}) = \delta(\boldsymbol{r}), \tag{2.91}$$

being

$$G_0(\boldsymbol{r}) = \frac{i}{8k_b^2} \left[H_0(k_b r) + \frac{2i}{\pi} K_0(k_b r) \right] \tag{2.92}$$

and $k_b^4 = \omega^2 \rho h / D$. Then we propose a self-consistent solution for W

$$W(\boldsymbol{r}) = \psi_0(\boldsymbol{r}) + T_0 \psi_0(0) G_0(\boldsymbol{r}), \tag{2.93}$$

where the incident field ψ_0 satisfies

$$\left(D\nabla^4 - k_b^4\right) \psi_0(\boldsymbol{r}) = 0. \tag{2.94}$$

Inserting W into the wave equation we get

$$T_0 \psi_0(0) \delta(\boldsymbol{r}) = \frac{\omega^2 \omega_0^2}{\omega^2 - \omega_0^2} \frac{m_0}{D} \delta(\boldsymbol{r}) (\psi_0(0) + T_0 \psi_0(0) G_0(0)), \tag{2.95}$$

from which we can obtain the T matrix of the resonator.

Once we know the response of one single scatterer, we can solve the multiple scattering of waves by a cluster of N resonators located at positions $\boldsymbol{R}_\alpha$, with $\alpha = 1, 2, \ldots N$, as was shown in [9], thus we know that

$$\psi(\boldsymbol{r}) = \psi_0(\boldsymbol{r}) + \sum_\alpha T_\alpha \psi_e(\boldsymbol{R}_\alpha) G_0(\boldsymbol{r} - \boldsymbol{R}_\alpha). \tag{2.96}$$

The quantity T_α can be defined as the T matrix of a resonator, an it is given by

$$T_\alpha = \frac{t_\alpha}{1 - i t_\alpha / (8k_b^2)} \tag{2.97}$$

with

$$t_\alpha = \frac{m_{R\alpha}}{D_b} \frac{\omega_{R\alpha}^2 \omega^2}{\omega_{R\alpha}^2 - \omega^2}. \tag{2.98}$$

The coefficients $\psi_e(\boldsymbol{R}_\alpha)$ are obtained from the multiple scattering equation

$$\psi_e(\boldsymbol{R}_\alpha) = \sum_\beta M^{-1}_{\alpha\beta}\psi_0(\boldsymbol{R}_\beta), \tag{2.99}$$

being the matrix M given by

$$M_{\alpha\beta} = \delta_{\alpha\beta} - (1-\delta_{\alpha\beta})T_\beta G_0(\boldsymbol{R}_\alpha - \boldsymbol{R}_\beta), \tag{2.100}$$

which solves the multiple scattering problem.

It is interesting to connect the above solution with the multiple scattering of flexural waves by finite scatterers. For this purpose, we can compute the T matrix of a cluster of point-like scatterers. First, the Green's function must be expanded in a set of Hankel functions and modified Bessel functions of second kind,

$$G_0(\boldsymbol{r} - \boldsymbol{R}_\alpha) = \sum_q \left[G^H_{\alpha q} H_q(k_b r) + G^K_{\alpha q} K_q(k_b r)\right] e^{iq\theta}, \tag{2.101}$$

being

$$G^H_{\alpha q} = \frac{i}{8k_b^2} J_q(k_b R_\alpha) e^{-iq\theta_\alpha}, \tag{2.102}$$

$$G^K_{\alpha q} = -\frac{1}{4\pi k_b^2} I_q(k_b R_\alpha) e^{-iq\theta_\alpha}, \tag{2.103}$$

which allow us to express the total field as

$$\psi(\boldsymbol{r}) = \psi_0(\boldsymbol{r}) + \sum_q \left[B^H_q H_q(k_b r) + B^K_q K_q(k_b r)\right] e^{iq\theta} \tag{2.104}$$

being

$$B^H_q = \sum_\alpha G^H_{q\alpha} T_\alpha \psi_e(\boldsymbol{R}_\alpha), \tag{2.105}$$

$$B^K_q = \sum_\alpha G^K_{q\alpha} T_\alpha \psi_e(\boldsymbol{R}_\alpha). \tag{2.106}$$

Given the solution for ψ_e, we find that

$$B^H_q = \sum_{\alpha\beta} G^H_{q\alpha} T_\alpha M^{-1}_{\alpha\beta} \psi_0(\boldsymbol{R}_\beta), \tag{2.107}$$

$$B^K_q = \sum_{\alpha\beta} G^K_{q\alpha} T_\alpha M^{-1}_{\alpha\beta} \psi_0(\boldsymbol{R}_\beta). \tag{2.108}$$

Also, we know that the incident field can be expanded in terms of Bessel and modified Bessel functions of first kind,

$$\psi_0(\boldsymbol{R}_\alpha) = \sum_s \left[A_s^J J_s(k_b r) + A_s^I I_s(k_b r)\right] e^{is\theta}, \tag{2.109}$$

thus, we get

$$B_q^H = \sum_s A_s^J \sum_{\alpha\beta} G_{q\alpha}^H T_\alpha M_{\alpha\beta}^{-1} J_s(k_b R_\beta) e^{is\theta_\beta} + \sum_s A_s^I \sum_{\alpha\beta} G_{q\alpha}^H T_\alpha M_{\alpha\beta}^{-1} I_s(k_b R_\beta) e^{is\theta_\beta} \tag{2.110}$$

and

$$B_q^K = \sum_s A_s^J \sum_{\alpha\beta} G_{q\alpha}^K T_\alpha M_{\alpha\beta}^{-1} J_s(k_b R_\beta) e^{is\theta_\beta} + \sum_s A_s^I \sum_{\alpha\beta} G_{q\alpha}^K T_\alpha M_{\alpha\beta}^{-1} I_s(k_b R_\beta) e^{is\theta_\beta}. \tag{2.111}$$

Therefore the above equations define the (non-diagonal) T matrix elements as

$$T_{qs}^{HJ} = \frac{i}{8k_b^2} \sum_{\alpha\beta} J_q(k_b R_\alpha) e^{-iq\theta_\alpha} T_\alpha M_{\alpha\beta}^{-1} J_s(k_b R_\beta) e^{is\theta_\beta}, \tag{2.112}$$

$$T_{qs}^{HI} = \frac{i}{8k_b^2} \sum_{\alpha\beta} J_q(k_b R_\alpha) e^{-iq\theta_\alpha} T_\alpha M_{\alpha\beta}^{-1} I_s(k_b R_\beta) e^{is\theta_\beta}, \tag{2.113}$$

$$T_{qs}^{KJ} = \frac{-1}{4\pi k_b^2} \sum_{\alpha\beta} I_q(k_b R_\alpha) e^{-iq\theta_\alpha} T_\alpha M_{\alpha\beta}^{-1} J_s(k_b R_\beta) e^{is\theta_\beta}, \tag{2.114}$$

$$T_{qs}^{KI} = \frac{-1}{4\pi k_b^2} \sum_{\alpha\beta} I_q(k_b R_\alpha) e^{-iq\theta_\alpha} T_\alpha M_{\alpha\beta}^{-1} I_s(k_b R_\beta) e^{is\theta_\beta}, \tag{2.115}$$

so that the cluster of resonators can be described as a finite-sized non-circular object.

2.4 Numerical Examples

Although numerous multiple scattering simulations can be found in the bibliography, we provide here some examples of simulations to illustrate the behaviour of waves. We will focus the examples in the domain of homogenization, in which the wavelength of the incident field is larger than the typical distance between scatterers and the cluster behaves as an effective material.

Figure 2.3a shows the interaction of a plane wave arriving from the left with a large slab of scatterers (in blue in the picture). As we see, there is a strong reflection at the surface of the slab, but diffraction effects occurs only at the borders of it, therefore

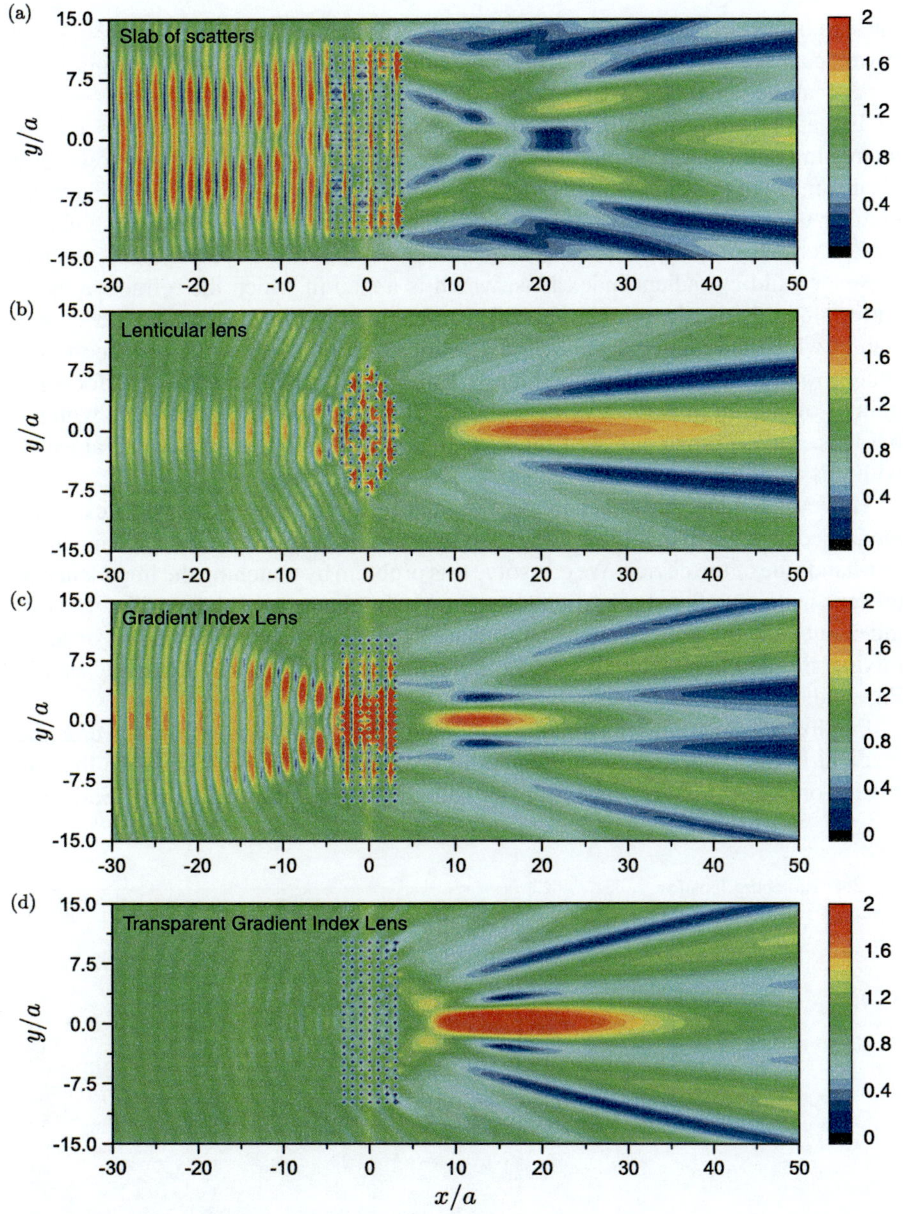

Fig. 2.3 Multiple scattering of waves by **a** a rectangular, **b** lenticular slabs. Due to the effective medium approximation, the material behaves as a refractive material, as shown in the "lens" formed in (**b**). **c** Gradient index lens. **d** Transparent gradient index lens designed with a mixture of soft and hard scatterers

the slab behaves as a homogeneous material. This can be seen from Fig. 2.3b, since if we "cut" the slab with a lenticular shape refraction occurs and a focusing point appears at the other side. We have built an acoustic lens [21].

The application of multiple scattering to the design of acoustic lenses is very interesting and a great amount of work has been devoted to it. There is however a disadvantage of this approach in terms of geometry. Since the slab is made of a small number of scattering units, the lenticular shape of the slab will have some abrupt surface effects: a continuous curved surface is not possible for these sizes. We can however build a gradient index lens, which is a slab in which the refractive index varies along the vertical direction. This can be done in Fig. 2.3c, where we show a slab in which the radius of the cylinders varies as a function of the distance to the center, following a well described lens to tailor the refractive index. The focusing point is of a better quality here due to the high contrast that can be achieved with the flat lens, and also the control of the focusing point is found to be in general easier with this approach.

However, a major drawback in the design of refractive lenses for acoustics is that the impedance mismatch is very high, so that a strong reflected field appears at the left-hand side of the device. We can solve this problem by matching the impedance of the lens with that of the background, so that we simultaneously change the refractive index but keep the impedance equal to that of the background. This can be done by mixing soft and hard scatterers, as explained in [22], and the result can be seen in Fig. 2.3d.

Finally, Fig. 2.4 shows a similar example but for flexural waves. In this case the gradient in the properties of the scatterers is along the radial direction, so that we have an omnidirectional refractive device called *Luneburg lens*. A plane wave arriving

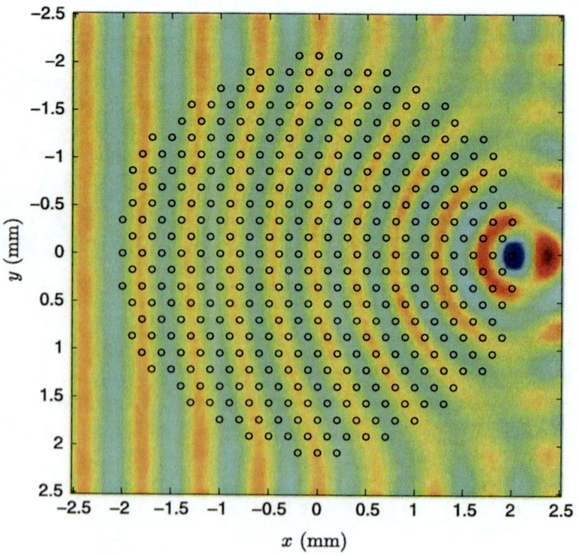

Fig. 2.4 Luneburg lens for flexural waves

from any direction will be focused at the surface of the device (see [23–26] for more devices and details). The refractive effect is more evident here, where we can see clearly the bending of the wavefront.

The above examples illustrate clearly the nature of the multiple scattering of waves: a large combination of scattering between objects gives a coherent field which travels at single wavelength which is different than that of the wavelength of the background, resulting in an effective medium which can be even position-dependent, to give different types of refractive devices.

Acknowledgements Daniel Torrent acknowledges José Sánchez-Dehesa, Bahram Djafari-Rouhani and Yan Pennec for relevant and fruitful discussions. Work supported through the "Ramón y Cajal" fellowship under grant number RYC-2016-21188.

References

1. L.L. Foldy, The multiple scattering of waves. I. General theory of isotropic scattering by randomly distributed scatterers. Phys. Rev. **67**(3–4), 107 (1945)
2. V. Twersky, Multiple scattering of radiation by an arbitrary configuration of parallel cylinders. J. Acoust. Soc. Amer. **24**(1), 42–46 (1952)
3. M. Kafesaki, E.N. Economou, Multiple-scattering theory for three-dimensional periodic acoustic composites. Phys. Rev. B **60**(17), 11993 (1999)
4. P.A. Martin, *Multiple Scattering: Interaction of Time-harmonic Waves with N Obstacles*, No. 107 (Cambridge University Press, Cambridge, 2006)
5. P. Waterman, New formulation of acoustic scattering. J. Acoust. Soc. Amer. **45**(6), 1417–1429 (1969)
6. M. Abramowitz, I.A. Stegun, *Handbook of Mathematical Functions: With Formulas, Graphs, and Mathematical Tables*, vol. 55 (Courier Corporation, 1965)
7. P.M. Morse, K.U. Ingard, *Theoretical Acoustics* (Princeton University Press, Princeton, 1986)
8. W.-M. Lee, J.-T. Chen, Scattering of flexural wave in a thin plate with multiple circular holes by using the multipole Trefftz method. Int. J. Solids Struct. **47**(9), 1118–1129 (2010)
9. D. Torrent, D. Mayou, J. Sánchez-Dehesa, Elastic analog of graphene: dirac cones and edge states for flexural waves in thin plates. Phys. Rev. B **87**(11), 115143 (2013)
10. P. Packo, A.N. Norris, D. Torrent, Inverse grating problem: efficient design of anomalous flexural wave reflectors and refractors. Phys. Rev. Appl. **11**(1), 014023 (2019)
11. D. Torrent, Y. Pennec, B. Djafari-Rouhani, Effective medium theory for elastic metamaterials in thin elastic plates. Phys. Rev. B **90**(10), 104110 (2014)
12. S.P. Timoshenko, S. Woinowsky-Krieger, *Theory of Plates and Shells* (McGraw-Hill, New York, 1959)
13. K.F. Graff, *Wave Motion in Elastic Solids* (Courier Corporation, 2012)
14. A. Norris, C. Vemula, Scattering of flexural waves on thin plates. J. Sound vibr. **181**(1), 115–125 (1995)
15. B. Peterson, S. Ström, T matrix for electromagnetic scattering from an arbitrary number of scatterers and representations of E (3). Phys. Rev. D **8**(10), 3661 (1973)
16. D. Torrent, A. Håkansson, F. Cervera, J. Sánchez-Dehesa, Homogenization of two-dimensional clusters of rigid rods in air. Phys. Rev. Lett. **96**(20), 204302 (2006)
17. D. Torrent, J. Sánchez-Dehesa, Effective parameters of clusters of cylinders embedded in a nonviscous fluid or gas. Phys. Rev. B **74**(22), 224305 (2006)
18. D. Torrent, J. Sánchez-Dehesa, Sound scattering by anisotropic metafluids based on two-dimensional sonic crystals. Phys. Rev. B **79**(17), 174104 (2009)

19. D. Torrent, J. Sánchez-Dehesa, Broadband acoustic cloaks based on the homogenization of layered materials. Wave Motion **48**(6), 497–504 (2011)
20. N. Lera, D. Torrent, P. San-Jose, J. Christensen, J.V. Alvarez, Valley hall phases in kagome lattices. Phys. Rev. B **99**(13), 134102 (2019)
21. F. Cervera, L. Sanchis, J. Sánchez-Pérez, R. Martinez-Sala, C. Rubio, F. Meseguer, C. López, D. Caballero, J. Sánchez-Dehesa, Refractive acoustic devices for airborne sound. Phys. Rev. Lett. **88**(2), 023902 (2001)
22. D. Torrent, J. Sánchez-Dehesa, Acoustic metamaterials for new two-dimensional sonic devices. New J. Phys. **9**(9), 323 (2007)
23. A. Climente, D. Torrent, J. Sánchez-Dehesa, Omnidirectional broadband insulating device for flexural waves in thin plates. J. Appl. Phys. **114**(21), 214903 (2013)
24. D. Torrent, Y. Pennec, B. Djafari-Rouhani, Omnidirectional refractive devices for flexural waves based on graded phononic crystals. J. Appl. Phys. **116**(22), 224902 (2014)
25. Y. Jin, D. Torrent, Y. Pennec, Y. Pan, B. Djafari-Rouhani, Gradient index devices for the full control of elastic waves in plates. Sci. Rep. **6**, 24437 (2016)
26. Y. Jin, B. Djafari-Rouhani, D. Torrent, Gradient index phononic crystals and metamaterials. Nanophotonics **8**(5), 685–701 (2019)

Chapter 3
Sound Wave Propagation in Sonic Crystals

From Long Wavelength Approximation to the Diffraction Regime

Vicent Romero-García

Abstract Motivated by the analogous photonic crystals acting on electromagnetic waves, periodic distribution of solid materials, i.e., phononic crystals, have been exploited during the last decades to control elastic and acoustic waves. This Chapter reviews phononic crystals in which the background medium is a fluid and the scatterers are rigid solids, i.e., sonic crystals. Following a bottom up approach, this Chapter presents the fundamentals of these periodic media for acoustic waves. First, by using the transfer matrix method we show the physical origin of most interesting properties of sonic crystals in one-dimensional structures. Then, we introduce the plane wave expansion method to obtain the dispersion relation of sonic crystals and the multiple scattering theory to deal with finite 2D distributions of N scatterers. Through an intense bibliographic review, we will show the different ways to control acoustic waves by using sonic crystals and locally resonant acoustic sonic materials, known as acoustic metamaterials. Finally, we show a technological application of sonic crystals for the broadband transmission loss of sound waves.

3.1 Introduction: Origins of Sonic Crystals

Wave propagation in periodic structured media has a long history [1]. The phenomenon of dispersion in a medium was described by Lord Kelvin by using a periodic spring-mass model [2], as it was illustrated in Chap. 1. Lord Rayleigh's early work studied the interaction of light with periodic layered media showing the presence of bandgaps in the frequency spectrum [3]. Later, the Brillouin's book [4] on periodic structures, first published in 1946, gave a detailed discussion of the early development in this field. In the late 80's Yablonovitch [5] and John [6] simultaneously triggered the primary emphasis in periodic systems due to their interesting propagation properties for the electromagnetic waves. Their proposal consisted of using a periodic

V. Romero-García (✉)
Laboratoire d'Acoustique de l'Université du Mans, UMR CNRS 6613, Le Mans Université, Avenue Olivier Messiaen, 72085 Le Mans, France
e-mail: vicente.romero@univ-lemans.fr

N. Jiménez et al. (eds.), *Acoustic Waves in Periodic Structures, Metamaterials, and Porous Media*, Topics in Applied Physics 143,
https://doi.org/10.1007/978-3-030-84300-7_3

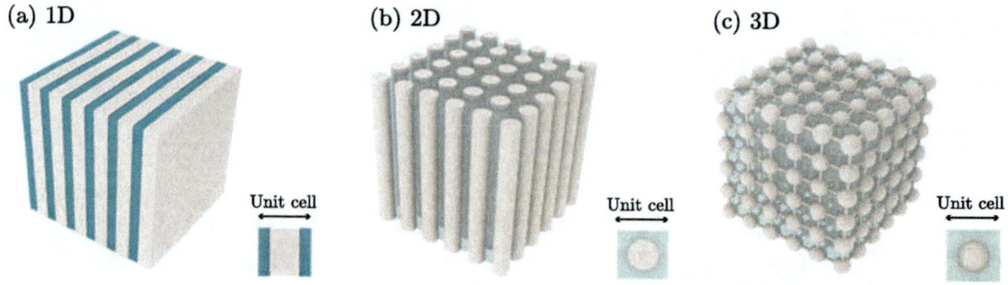

Fig. 3.1 Example of simple sonic/phononic crystals. **a** 1D configuration: layered media with alternating acoustic properties. **b** 2D configuration: periodic distribution of cylindrical scatters embedded in a fluid/solid matrix. **c** 3D configuration: period distribution of spherical scatters embedded in a fluid/solid matrix. Corresponding unit cells also are shown

distribution of dielectric scatterers embedded in a host medium with different dielectric properties. These periodic systems exhibit ranges of frequencies related to the periodicity of the structure where there is no wave propagation. By analogy with the electronic bandgap in semiconductor crystals, these ranges of frequencies were called bandgaps (BG) and these periodic structures were called photonic crystals. For an extended review of photonic band structures see reference [7].

At the beginning of the late 90's, an increasing interest in the comparable process of acoustic wave propagation in periodic arrays appeared. Motivated by the results of the photonic crystals, several theoretical works started the analysis of the periodic arrays made of isotropic solids embedded in an elastic background which was also isotropic [8–13]. By analogy with the photonic case, these periodic arrangements presented BG, defined here as frequency ranges where vibrations, sound and phonons were forbidden. Analogously, theses systems were called phononic crystals (PC).

Depending on the distribution of scatterers, one can obtain one-dimensional (1D), as shown in Fig. 3.1a, two-dimensional (2D), as shown in Fig. 3.1b, or three-dimensional (3D) PC, as shown in Fig. 3.1c. In each of these PC one can observe different combinations of transverse, longitudinal or mixed waves [14]. However, a drastic simplification arises in the case of fluids, which permits only longitudinal waves. It is said that if one of the elastic materials in the PC is a fluid, then PCs are called sonic crystals (SCs). Several studies discuss the similarities and differences between these periodic systems [13, 15].

The measurements of the sound attenuation by a sculpture, by Eusebio Sempere, exhibited at the Juan March Foundation in Madrid, constituted the first experimental evidence of the presence of BG in a SC [16]. The work of Martínez-Sala et al. [16] experimentally showed that the distribution of cylinder rods with a strong periodic modulation (2D), inhibited the sound transmission for certain frequency ranges related to this modulation, just as photonic crystals do with light. Immediate theoretical predictions [17–19] and experimental results [20] were motivated by these experimental results in order to explain the propagation properties of this sculpture that could filter noise.

Since these acoustical properties were measured in that minimalist sculpture, a large body of research, both experimental and theoretical, emphasized the existence of complete elastic/acoustic BG, opening possibilities to interesting applications such as elastic/acoustic filters [21–23] or noise control [24], as well as for the study of fundamental physics phenomena such as localization of waves [25–27]. The study of acoustic wave propagation in periodic binary composites shows that BG can exist under specific conditions concerned mainly the density and velocity contrast of the components of the composite, the volume fraction of one of the two components, the lattice structures and the topology [15]. The presence of BG in SC is due to Bragg's scattering, which in these systems is produced by an interplay between the sound speed and density of materials forming the composite as well as their spatial distribution. The emphasis in the acoustical properties of SC for frequencies high enough to distinguish the inner structure of the array marks the initial steps in the research on SC. A great research interest in the existence of spectral gaps in the PC made of several materials, shapes and distribution of scatterers were witnessed at the end of the 90s.

In this Chapter, using an extended review of the state of the art in the control of sound by SCs, we show both the fundamentals as well as the physical interpretation of the most relevant properties of SCs. In Sect. 3.2 the physical origin of the dispersion relation for sound waves in periodic structures is described by using the transfer matrix method applied to an academic case of a 1D periodic multilayer system. Sections 3.3 and 3.4 briefly demonstrate the modelling of the dispersion relation by using the plane wave expansion and the multiple scattering theory. We will focus on the particular features of each method to provide a deep understanding of each tool. Section 3.5 contains a bibliographic review of the main results obtained during the last years. Finally, Sect. 3.6 focuses on technological application of SCs as acoustic barriers.

3.2 The Physical Origin of Bandgaps

One of the most interesting properties of the periodic structures is the presence of bandgaps, i.e. ranges of frequencies in which waves cannot propagate. In this Section, the physical origin of the bandgaps for sound waves is described by using the transfer matrix method applied to an academic case. The analysed system consists of a 1D periodic array which unit cell is made of two fluid layers, A and B as shown in Fig. 3.2a. We note here that the general principles discussed in this Section will apply also to more complex periodic systems made of two- and three-dimensional structures.

3.2.1 Transfer Matrix Method

The transfer matrix between the two faces of the homogeneous and isotropic 1D material, extending from $x = 0$ to $x = l$, is used to relate the sound pressure, p, and normal acoustic particle velocity, v, at the two faces

$$\begin{bmatrix} p \\ v \end{bmatrix}_{x=0} = \begin{bmatrix} \cos(kl) & \imath Z \sin(kl) \\ \imath \sin(kl)/Z & \cos(kl) \end{bmatrix} \begin{bmatrix} p \\ v \end{bmatrix}_{x=l}, \tag{3.1}$$

where the wave vector of the material is defined as $k = \omega/c = \omega\sqrt{\rho/K}$, ω is the angular frequency, c is the sound speed, ρ is the density and K is the bulk modulus, the acoustic impedance is defined as $Z = \sqrt{\rho K}$, and $\imath = \sqrt{-1}$ is the imaginary unity. A Fourier time convention $e^{\imath\omega t}$ is assumed.

The transfer matrix, $\mathbf{T}$, of the unit cell made of two sublayers A and B, as shown in Fig. 3.2a, reads as

$$\mathbf{T} = \mathbf{T}_A \mathbf{T}_B, \tag{3.2}$$

where

$$\mathbf{T}_A = \begin{bmatrix} \cos(k_A l_A) & \imath Z_A \sin(k_A l_A) \\ \imath \sin(k_A l_A)/Z_A & \cos(k_A l_A) \end{bmatrix}, \tag{3.3}$$

$$\mathbf{T}_B = \begin{bmatrix} \cos(k_B l_B) & \imath Z_B \sin(k_B l_B) \\ \imath \sin(k_B l_B)/Z_B & \cos(k_B l_B) \end{bmatrix}. \tag{3.4}$$

Considering the periodic conditions at the boundaries of the unit cell, the Bloch parameters can then be obtained as follows,

$$q = \frac{1}{a} \cos^{-1}\left(\frac{T_{11} + T_{22}}{2}\right) = \mathrm{Tr}\,(\mathbf{T}), \tag{3.5}$$

$$Z_q = \sqrt{\frac{T_{12}}{T_{21}}}, \tag{3.6}$$

where q is the wavenumber and Z_q is the impedance, with $a = l_A + l_B$ being the periodicity of the system.

3.2.2 Discussion

In order to start the discussion we consider a plane wave $e^{\imath\omega t - \imath kx}$ propagating in a 1D medium, A, in which a fictitious periodicity with lattice constant a, defined as the distance between the fictitious scatterers in the periodic medium is imposed.

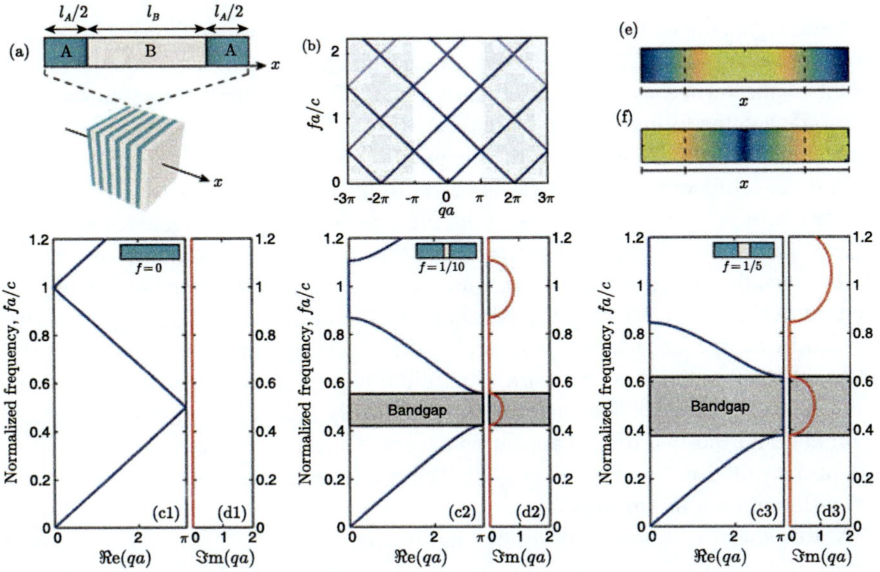

Fig. 3.2 Description of the dispersion relation in a 1D periodic system. **a** Scheme of the unit cell. **b** Dispersion relation in the extended Brillouin zone of a homogeneous material with artificial 1D periodicity with lattice constant a. **c1** and **d1** represent the complex dispersion relation of an homogeneous material, i.e., $ff = 0$, of a 1D periodic structure with $ff = 1/10$, **c2** and **d2**, and of a 1D periodic structure with $ff = 1/5$, **c3** and **d3**. In all cases, $\rho_B = 3\rho_A$ and $c_B = 3c_A$. **e** and **f** represent the pressure distribution in the unit cell for the lower and upper edges of the first bandgap for the case $ff = 1/10$ respectively

Therefore, the acoustic pressure should also follow this artificial periodicity, i.e., $p(x + na) = p(x)$. This condition implies that the wavenumber in the medium should also be periodic, following the relation $q_n = q + 2\pi n/a$ with n an integer. For free propagation in a fluid the dispersion relation is $q = \omega/c$. Figure 3.2b represents the dispersion relation in a homogeneous fluid medium with the artificial periodicity, therefore considering the n dispersion relations. In this case, q_n is periodic and the dispersion relation folds back into the well known Brillouin zone, white region in Fig. 3.2b, when it reaches an edge ($qa = \pi$).

We introduce now a second material in the unit cell, B. In this case, we consider that $\rho_b = 3\rho_A$ and $c_B = 2c_A$. We define here the filling fraction ff, as the volume occupied by the scatterer with respect to the total volume of the unit cell, in this case, $ff = l_B/a$. We start by a low filling fraction medium, $ff = 1/10$. Figures 3.2c1–c3 and Figs. 3.2d1–d3 represent the complex dispersion relation obtained form the transfer matrix method, (3.5). When small scatters are introduced, i.e., Figs. 3.2c2, d2 for $ff = 1/10$, the dispersion relations look like those for a homogeneous medium, however, presenting bands of frequencies where the wave number is complex, i.e., regions in which waves are evanescent. These bands correspond to the sonic bandgaps of the structure and are characterized by a strong dispersion around them. Moreover, in the bandgap, the wave amplitude decays exponentially through the crystal, i.e., the modes are evanescent. This results from the fact that the modes have a complex wave vector of the form $k = \pi/a + i\kappa$. Figure 3.2c3, d3, show the complex dispersion relation when the filling fraction increases to $ff = 1/5$. We notice that the width of the bandgaps increases as well as the imaginary part of the wavenumber.

In order to understand the physical origin of the sonic bandgap, we consider the pressure profiles in the band edges of the first bandgap, i.e., the eigenvectors of the eigenvalue problem for the case $qa = \pi$. For this case, the modes have a wavelength of $2a$, twice the lattice constant. There are two ways to center a mode of this type in the unit cell without violating the symmetry of the unit cell: positioning the nodes in each low impedance layer, or in each high impedance layer. In the case of the homogeneous material, at point $qa = \pi$, there is double degeneracy as shown in Fig. 3.2b. Once the second material is periodically placed, then this degeneracy is broken and the two modes appear as described before. Figure 3.2e, f show the pressure profile for the case $ff = 1/10$. The two different modes with nodes in the two different regions are shown.

At this stage, it is worth noting that the low frequency modes concentrate their energy in the high density regions, and the high frequency modes have a larger fraction of the energy in the low density regions [7]. Figures 3.2e, f show the node of the pressure profile in the high and, respectively, low density region. With this in mind, it is understandable why there is a frequency difference between the two cases and why the degeneracy is broken, i.e., why the bandgap is opened.

In the perturbative case of weak filling fraction and low contrast of impedances we can derive a simple formula from (3.5), for the size of the bandgap. We consider that the two materials have the same sound speed and densities differ as $\rho_B = \rho_A + \Delta\rho_A$. If either the density contrast is weak ($\Delta\rho_A/\rho_A \ll 1$) or the filling fraction l_B/a is small, then the bandwidth of the bandgap, $\Delta\omega$ reads as

$$\frac{\Delta\omega}{\omega_m} \approx \frac{\Delta\rho_A}{\rho_A}\frac{\sin(\pi l_B/a)}{\pi}, \tag{3.7}$$

where ω_m is the frequency of the middle of the gap.

As we have seen from the solutions given by the transfer matrix method, no purely real wave vector exists for any mode in the bandgap. The imaginary component of the wave vector causes the decay on a length scale of $1/\kappa$. But, what is the origin of this imaginary part? This can be understood by examining the bands in the immediate vicinity of the gap. At this stage we must have in mind that the time reversal symmetry should be preserved, i.e., the band structure should be even symmetric with respect to the limits of the Brillouin zone. Therefore, if we expand the upper band edge, ω_2 in powers of q around the edge $q = \pi/a$, the expansion cannot contain odd powers of q, and thus

$$\Delta\omega = \omega_2(q) - \omega_2\left(\frac{\pi}{a}\right) \propto \left(q - \frac{\pi}{a}\right)^2 = (\Delta q)^2. \tag{3.8}$$

For frequencies slightly higher than the top of the gap, $\Delta\omega > 0$, then Δq is purely real, and the wave propagates in the second propagating band. However, for $\Delta\omega < 0$, Δq is purely imaginary, and the wave is within the bandgap. The states decay exponentially since $\Delta q = \imath\kappa$. As we traverse the gap, the decay constant κ grows as the frequency reaches the gaps centre, then disappears again at the lower gap edge. This behaviour corresponds to the one shown in Fig. 3.2d. In this sense, larger gaps usually result in a larger κ at the middle of the bandgap, and thus less penetration of the wave into the crystal. It is worth noting here, that there is no physical way to excite the evanescent wave within an idealized crystal of infinite extent. However, a defect or an edge state in an otherwise perfect crystal can terminate this exponential decrease and thereby sustain an evanescent mode.

3.3 Dispersion Relation of Sonic Crystals

The periodicity of the system and the Bloch's theorem are exploited in this Section to solve the wave equation, obtaining a simple eigenvalue problem relating the wave vector and the frequency of the incident wave, $\omega(k)$ [12, 28]. This procedure is known as the Plane Wave Expansion (PWE) method. This model only allows the analysis of the propagative part of the dispersion relation, i.e. the real wave numbers of the dispersion relation. To obtain the complex wave numbers, that is to obtain the attenuation properties of the dispersion relation, the equations can be modified to solve the $k(\omega)$ problem instead of the $\omega(k)$ one. In this case obtaining both the real and the imaginary parts of the wave vector, k. This model is known as Extended Plane Wave Expansion (EPWE) [29–31]. In this Section we show the details of each expansion to obtain the band structures of a sonic crystal. Particular examples for 2D cases are shown.

3.3.1 *Plane Wave Expansion*

Propagation of sound is described by the equation

$$\frac{1}{\rho c_0^2}\frac{\partial^2 p}{\partial t^2} = \nabla\left(\frac{1}{\rho}\nabla p\right), \tag{3.9}$$

where c_0 is the sound speed, ρ is the density of the medium and p is the pressure. In this Section a system made of a periodic distribution of an isotropic solid A, embedded in an acoustic isotropic background B is considered. Due to periodicity, it is possible to expand the physical properties of the medium in Fourier series,

$$\sigma = \frac{1}{\rho(\mathbf{r})} = \sum_{\mathbf{G}} \sigma_{\mathbf{k}}(\mathbf{G}) e^{\imath \mathbf{G}\mathbf{r}}, \tag{3.10}$$

$$\eta = \frac{1}{K(\mathbf{r})} = \sum_{\mathbf{G}} \eta_{\mathbf{k}}(\mathbf{G}) e^{\imath \mathbf{G}\mathbf{r}}, \tag{3.11}$$

where $K(\mathbf{r}) = \rho(\mathbf{r})c(\mathbf{r})^2$ is the spatially varying bulk modulus. $\mathbf{G} = n_i \mathbf{b}_i$ (with n_i an integer) is a vector in the reciprocal space made of a linear combination of the primitive reciprocal lattice vectors defined as follows: if the primitive lattice vectors are $\mathbf{a}_i$ (with $i = 1, 2, 3$), the primitive lattice vectors are then

$$\mathbf{b}_i = 2\pi \frac{\varepsilon_{ijk} \mathbf{a}_j \times \mathbf{a}_k}{\mathbf{a}_1 \cdot (\mathbf{a}_2 \times \mathbf{a}_3)}, \tag{3.12}$$

where ε_{ijk} is the three-dimensional Levi-Civita completely anti-symmetric symbol. For the pressure p we use the Bloch theorem and harmonic temporal dependence,

$$p(\mathbf{r}, t) = e^{\imath(\mathbf{k}\mathbf{r} - \omega t)} \sum_{\mathbf{G}} p_k(\mathbf{G}) e^{\imath \mathbf{G}\mathbf{r}}. \tag{3.13}$$

Then, it is easy to show that [12]

$$\beta(\mathbf{G}) = \begin{cases} \beta_A ff + \beta_B(1 - ff) & \text{if } \mathbf{G} = \mathbf{0}, \\ (\beta_A - \beta_B)\, F(\mathbf{G}) & \text{if } \mathbf{G} \neq \mathbf{0}, \end{cases} \tag{3.14}$$

where $\beta = (\sigma, \eta)$, and $F(\mathbf{G})$ is the structure factor, described in the Section 3.3.3. Using Equations (3.10, 3.11), (3.13) and (3.9) we obtain [12]

$$\sum_{\mathbf{G}'} \left((\mathbf{k} + \mathbf{G})\sigma_k(\mathbf{G} - \mathbf{G}')(\mathbf{k} + \mathbf{G}') - \omega^2 \eta_{\mathbf{k}}(\mathbf{G} - \mathbf{G}') \right) p_{\mathbf{k}}(\mathbf{G}') = 0. \tag{3.15}$$

For $\mathbf{G}$ taking all the possible values, (3.15) constitutes a set of linear, homogeneous equations for the eigenvectors $p_{\mathbf{k}(\mathbf{G})}$ and the eigenfrequencies $\omega(\mathbf{k})$. We obtain the band structures letting $\mathbf{k}$ scan the area of the irreducible region of the first Brillouin zone.

Equation (3.15) can be expressed by the following matrix formulation

$$\sum_{i=1}^{3} \Gamma_i \Sigma \Gamma_i P = \omega^2 \Omega P, \tag{3.16}$$

where $i = 1, 2, 3$. Considering

$$(\Gamma_i)_{mn} = \delta_{mn}(k_i + G_i^m), \tag{3.17}$$

the matrices Γ_i, Σ and Ω are defined as

$$\Gamma_i = \begin{bmatrix} k_i + G_i & 0 & \dots & 0 \\ 0 & k_i + G_i & \dots & 0 \\ \vdots & \vdots & \ddots & \vdots \\ 0 & \dots & \dots & k_i + G_i \end{bmatrix}, \tag{3.18}$$

$$\Sigma = \begin{bmatrix} \sigma(\mathbf{G}_1 - \mathbf{G}_1) & \dots & \sigma(\mathbf{G}_1 - \mathbf{G}_{N\times N}) \\ \vdots & \ddots & \vdots \\ \sigma(\mathbf{G}_{N\times N} - \mathbf{G}_1) & \dots & \sigma(\mathbf{G}_{N\times N} - \mathbf{G}_{N\times N}) \end{bmatrix}, \tag{3.19}$$

$$\Omega = \begin{bmatrix} \eta(\mathbf{G}_1 - \mathbf{G}_1) & \dots & \eta(\mathbf{G}_1 - \mathbf{G}_{N\times N}) \\ \vdots & \ddots & \vdots \\ \eta(\mathbf{G}_{N\times N} - \mathbf{G}_1) & \dots & \eta(\mathbf{G}_{N\times N} - \mathbf{G}_{N\times N}) \end{bmatrix}, \tag{3.20}$$

$$P = \begin{bmatrix} P(\mathbf{G}_1) \\ \vdots \\ P(\mathbf{G}_{N\times N}) \end{bmatrix}. \tag{3.21}$$

By truncating the system of equations with N components, we can solve the system given in (3.16) for each Bloch vector in the irreducible area of the first Brillouin zone.

3.3.2 Extended Plane Wave Expansion

From (3.16) we define the following vector, [29, 30]

$$\Phi_i = \Sigma \Gamma_i P. \tag{3.22}$$

With this definition it is possible to reformulate the eigenvalue problem (3.16) as the system of equations

$$\Phi_i = \Sigma \Gamma_i P, \tag{3.23}$$

$$\omega^2 \Omega P = \sum_{i=1}^{3} \Gamma_i \Phi_i. \tag{3.24}$$

In order to obtain an eigenvalue problem for $\mathbf{k}(\omega)$, we write $\mathbf{k} = k\alpha$, where α is a unit vector. Then (3.18) can be written as

$$\Gamma_i = \Gamma_i^0 + k\alpha_i I, \tag{3.25}$$

where I is the identity matrix, and

$$\Gamma_i^0 = \begin{bmatrix} G_i & 0 & \dots & 0 \\ 0 & G_i & \dots & 0 \\ \vdots & \vdots & \ddots & \vdots \\ 0 & \dots & \dots & G_i \end{bmatrix}, \quad \alpha_i = \begin{bmatrix} \alpha_i & 0 & \dots & 0 \\ 0 & \alpha_i & \dots & 0 \\ \vdots & \vdots & \ddots & \vdots \\ 0 & \dots & \dots & \alpha_i \end{bmatrix}. \tag{3.26}$$

Then, (3.16) can be written as

$$\begin{bmatrix} \omega^2 \Omega - \sum_{i=1}^{3} \Gamma_i^0 \Sigma \Gamma_i^0 & 0 \\ -\sum_{i=1}^{3} \Sigma \Gamma_i^0 & I \end{bmatrix} \begin{bmatrix} P \\ \Phi' \end{bmatrix} = k \begin{bmatrix} \sum_{i=1}^{3} \Gamma_i^0 \Sigma \alpha_i & I \\ \sum_{i=1}^{3} \Sigma \alpha_i & 0 \end{bmatrix} \begin{bmatrix} P \\ \Phi' \end{bmatrix}, \tag{3.27}$$

where $\Phi' = \sum_{i=1}^{3} \alpha_i \Phi_i$.

Equation (3.27) represents a generalized eigenvalue problem with eigenvalues k, possibly complex numbers, for each frequency. Complex band structures on the incidence direction α can be obtained by solving the eigenvalue equation for a discrete number of frequencies and then sorting them by looking for the continuity of k. In contrast with the $\omega(\mathbf{k})$ method, in this formulation the periodicity is not relevant and $k(\omega)$ does not belong to the first Brillouin zone, therefore, in addition to the solutions in the irreducible Brillouin zone, other solutions can be obtained out of it.

Figure 3.3 shows the results obtained by using both the PWE and the EPWE for a 2D SC made of rigid cylinders embedded in air arranged in a square array with lattice constant a and filling fraction $ff = 50\%$. (a) and (b) show the band structures for the ΓX and ΓM directions respectively. The left panel in both Fig. 3.3a, b shows the real part of the complex dispersion relation, obtained by using both PWE and EPWE. The right panel shows the imaginary part of the dispersion relation obtained by using the EPWE. We observe that modes inside the BG present complex wave vectors. In Fig. 3.3 one can observe that the imaginary part of the wave number for the frequencies inside the BG grows with values of frequency closer to the centre of the BG and disappears at the edges of the BG in the ΓX as well as in the ΓM directions, i.e., the rate of decay is larger for frequencies closer to the centre of the BG, as it was predicted in Sec. 3.2. We notice here that imaginary part of the wave vector connects propagating bands conserving the overall number of modes for a given frequency.

3.3.3 Structure Factor

For some particular scatterers, the structure factor can be calculated analytically. For example, for cylindrical scatterers with circular cross-sections of radius r, the structure factor is [12]

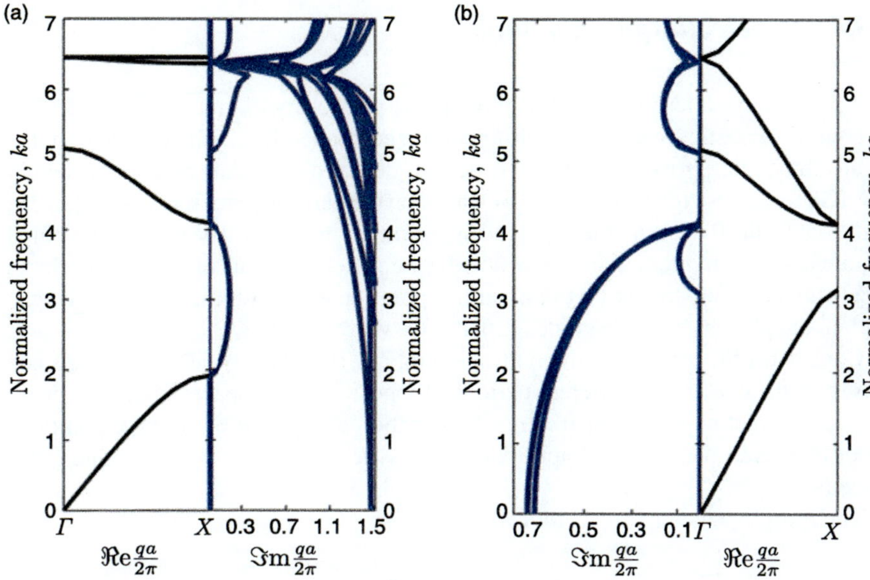

Fig. 3.3 Band Structure for a SC made of rigid cylinders arranged in a square array with lattice constant a and filling fraction $ff = 50\%$. **a** Left panel: Band structures calculated using PWE. Right panel: Complex band structures for the ΓX direction calculated using EPWE. **b** Left panel: Complex band structures for the ΓM direction calculated using EPWE. Right panel: Band structures calculated using PWE

$$F(\mathbf{G}) = \frac{1}{A_{uc}} \int\limits_{A_{cyl}} e^{-\imath \mathbf{G}\mathbf{r}} \mathbf{dr} = \frac{2ff}{Gr} J_1(Gr), \tag{3.28}$$

where A_{uc} is the area of the unit cell, A_{cyl} is the area of the considered cylinder and J_1 is the Bessel function of the first kind of order 1.

For example, for the case of square-rod scatterers of side l and angle of rotation θ, the structure factor is [32, 33]

$$F(\mathbf{G}, \theta) = \frac{1}{A_{uc}} \int\limits_{A_{cyl}} e^{-\imath \mathbf{G}\mathbf{r}} \mathbf{dr} = f \operatorname{sinc}\left(\frac{\mathcal{G}_x l}{2}\right) \operatorname{sinc}\left(\frac{\mathcal{G}_y l}{2}\right), \tag{3.29}$$

where,

$$\begin{bmatrix} \mathcal{G}_x \\ \mathcal{G}_y \end{bmatrix} = \begin{bmatrix} \cos(\theta) & \sin(\theta) \\ -\sin(\theta) & \cos(\theta) \end{bmatrix} \begin{bmatrix} G_x \\ G_y \end{bmatrix}. \tag{3.30}$$

Other analytical expressions for the structure factor can be found in the literature [34]. In the cases where the geometry of the scatterer is too complicated, numerical tools can be used to calculate it [35].

3.3.4 Supercell Approximation

One particularly interesting aspect of SC is the possibility of creating point defects to confine acoustic waves by exciting localized modes [36, 37]. Because of the locally breaking periodicity of the structure, defect modes can be created within the BG. To analyse the propagation of waves inside periodic structures with defects, traditionally the PWE with the supercell approximation is used. The supercell method requires as low an interaction as possible between defects. This results in a periodic arrangement of supercells that contain the point defect. With this method we can explain the physics of wave guides [26–28] or wave traps [38, 39].

Consider a SC with primitive lattice vectors $\mathbf{a}_i$ $(i = 1, 2, 3)$. The supercell is a cluster of $n_1 \times n_2 \times n_3$ scatterers periodically positioned in space. Then, the primitive lattice vectors in the supercell approximation are $\mathbf{a}'_i = n_i \mathbf{a}_i$, and the complete set of lattices in the supercell approximation is $\{R'|R' = l_i \mathbf{a}'_i\}$, where n_i and l_i are integers.

The primitive reciprocal vectors are then

$$\mathbf{b}'_i = 2\pi \frac{\varepsilon_{ijk} \mathbf{a}'_j \times \mathbf{a}'_k}{\mathbf{a}'_1 \cdot (\mathbf{a}'_2 \times \mathbf{a}'_3)}, \tag{3.31}$$

where ε_{ijk} is the three-dimensional Levi-Civita completely anti-symmetric symbol. The complete set of reciprocal lattice vectors in the supercell is $\{\mathbf{G}|\mathbf{G}_i = N_i \mathbf{b}'_i\}$ where N_i are integers.

If the supercell contains N_p point defects at the sites labelled by (l_s, m_s) in the periodic system, with $s = 1, ..., N_p$, then the Fourier coefficients of the expansion of the physical parameters involved in the problem satisfy the following equation

$$\beta(\vec{G}) = \begin{cases} \beta_A(N - N_p)ff + \beta_B(1 - (N - N_p)ff) & \text{if } \vec{G} = \vec{0}, \\ (\beta_A - \beta_B)\, F(\vec{G}) & \text{if } \vec{G} \neq \vec{0}. \end{cases} \tag{3.32}$$

The structure factor of such a supercell with N_p point defects is

$$F(\mathbf{G}) = \left[\sum_{i=-(n_1-1)/2}^{(n_1-1)/2} \sum_{j=-(n_2-1)/2}^{(n_2-1)/2} e^{\imath(ia|\mathbf{G}_1|+ja|\mathbf{G}_2|)} + \ldots \right.$$
$$\left. \ldots - \sum_{s=1}^{N_p} e^{\imath(l_s a|\mathbf{G}_1|+m_s a|\mathbf{G}_2|)} \right] P(\mathbf{G}), \tag{3.33}$$

where $P(\mathbf{G})$ is the structure factor of one of the elements of the supercell as described previously.

3.4 Multiple Scattering Theory

Multiple Scattering Theory (MST) has a long history. In 1913, Záviška [40] described the method for the scattering of finite arrays in 2D acoustic fields, that was applied in 1914 to the case of normal incidence on an infinite row of cylinders by von Ignatowsky [41]. After that, extensions of the work to the oblique incidence have been done [42, 43].

Multiple scattering can be understood as an interaction of waves with two or more obstacles [44, 45]. The classical multiple scattering problem of sound waves by rigid cylinders is presented briefly in this Section. MST solves the problem considering that the field scattered from one obstacle induces further scattered fields from all the other obstacles, which induces further scatterers fields from all the other obstacles, and so on. This characterizes the MST as a self-consistent method that is valid for situations where the scatterers have been placed either randomly or periodically.

In this Section we will show the formalism of MST for the case of cylindrical waves radiating N cylindrical scatterers, then restricting our study to the 2D case. More details about 3D multiple scattering, or other kind of radiating waves, can be found in the literature [46, 47].

3.4.1 Cylindrical Waves

In this case, an acoustic source transmitting monochromatic waves is placed at point $\vec{r}_s$, some distance from the system of scatterers. For simplicity without compromising generality, the acoustic source can be approximated as a line source located at origin, i.e., $\vec{r}_s = \vec{0}$. The Helmholtz equation with such a source is

$$\left(\vec{\nabla}^2 + k^2\right) p(\vec{r}) = -4\pi\delta^2(\vec{r}), \tag{3.34}$$

where $k = \omega/c$ and δ^2 is the 2-dimensional delta-function. In cylindrical coordinates, the solution is

$$p(\vec{r}) = \imath\pi H_0(kr), \tag{3.35}$$

where H_0 is the zero$-th$ order Hankel function of the first kind. The solution represents a line source located at origin.

We consider N straight cylinders located at $\vec{r}_i = (r_i, \theta_i)$ of radius a_i with $i = 1, 2, \ldots, N$ to form either a regular lattice or a random array perpendicular to the $x - y$ plane. The cylinders are parallel to the $z-$axis, then since the boundary conditions and the geometry do not change with z, the problem can be reduced to two uncoupled problems for the scalar Helmholtz equation. The final wave reaches a receiver located at $\vec{r}_r$ and it is formed by the sum of the direct wave from the source and the scattered waves from all the cylinders. The problem consist of solving the previous self-consistent method, satisfying some boundary conditions on the surfaces of the cylinders and a radiation condition at infinity. Such a scattering problem can be exactly formulated in cylindrical coordinates.

Solutions for the scattering of waves by cylindrical objects can be expressed in terms of two important functions: $H_n(kr)e^{in\theta}$ and $J_n(kr)e^{in\theta}$, Hankel and Bessel function of n-order and first kind respectively. Both are solutions of the Helmholtz equation for the planar problem. The first one satisfies the radiation condition at infinity and is singular at the origin. The second one is regular at the origin. Thus the first one could be used to represent outgoing cylindrical waves whereas the second one could express regular cylindrical waves.

Considering the presence of the N cylinders placed at $\vec{r}_i$, the scattered wave from the j-th cylinder can be expressed as

$$p_{\mathrm{sc}}^j = \sum_{n=-\infty}^{\infty} \imath\pi A_{jn} H_n(kr_j)e^{\imath n\theta_j}, \tag{3.36}$$

where H_n is the n-th order Hankel function of the first kind, A_{jn} are the coefficients to be determined, and θ_j is the azimuthal angle of the vector r_j.

In order to separate the governing equations into modes, we can express the total incident wave as:

$$p_{\mathrm{inc}}^i = \sum_{n=-\infty}^{\infty} B_{in} J_n(kr_i)e^{\imath n\theta_i}. \tag{3.37}$$

For determining the matrix relation (T-matrix) between B_{in} and A_{jn}, we need to write down the relation between p_{sc}^j and p_{inc}, for each $j \neq i$, in the coordinate system centered at the i-th scatterer. We use the Graf's addition theorem for $H_m(kr)e^{\imath m\theta}$ [48], for representing the p_{sc}^j in the form

$$P_{\mathrm{sc}}^j = \sum_{n=-\infty}^{\infty} C_{jin} J_n(kr_i)e^{\imath n\theta_i}, \tag{3.38}$$

where

$$C_{jin} = \sum_{l=-\infty}^{\infty} \imath\pi A_{jl} H_{l-n}(kr_{ij})e^{\imath(l-n)\theta_{ij}}. \tag{3.39}$$

On the other hand, using again the Graf's theorem, the incident pressure can be expressed as,

$$P_{\mathrm{inc}} = \sum_{l=-\infty}^{\infty} S_{il} J_l(kr_i)e^{\imath l\theta_i}, \tag{3.40}$$

where

$$S_{il} = \imath\pi H_{-l}(kr_i)e^{-il\theta_i}. \tag{3.41}$$

At this stage, the S_{il} is known, but both B_{in} and A_{jl} are unknown. Boundary conditions will provide another equation relating them with each other. In the previous Section we used the Neumann's boundary conditions as we were interested in rigid scatterers. Now, in this Section we will consider the general boundary condition, this means the continuity of both the pressure and the normal velocity across the interface between a scatterer and the surrounding medium (later, in Sect. 3.4.2, we will describe different boundary conditions).

Thus, the boundary conditions for the j-th cylinder are [46]

$$p_{\text{ext}}\Big|_{\partial\Omega_j} = p_{\text{int}}\Big|_{\partial\Omega_j}, \tag{3.42}$$

$$\frac{1}{\rho}\frac{\partial p_{\text{ext}}}{\partial n}\Big|_{\partial\Omega_j} = \frac{1}{\rho_j}\frac{\partial p_{\text{int}}}{\partial n}\Big|_{\partial\Omega_j}, \tag{3.43}$$

where $\partial\Omega_j$ is the boundary of the j-th scatterer, ρ is the density of the surrounding medium and ρ_j is the density of the j-th scatterer.

In order to apply the previous boundary conditions, we represent the pressure inside the i-th cylinder in the form of the Fourier series as follows

$$P^i_{\text{int}} = \sum_{n=-\infty}^{\infty} D_{in} J_n(k_{1i}r_i)e^{\imath n\theta_i}, \tag{3.44}$$

where k_{1i} is the wave number inside the i-th cylinder. Using the boundary conditions and the expressions for the incident, scattered and interior wave we can obtain the following relation,

$$B_{in} = \imath\pi\Gamma_{in}A_{in}, \tag{3.45}$$

where

$$\Gamma_{in} = \frac{H_n(ka_i)J'_n(ka_i/h_i) - g_ih_iH'_n(ka_i)J_n(ka_i/h)}{g_ih_iJ'_n(ka_i)J_n(ka_i/h_i) - J_n(ka_i)J'_n(ka_i/h_i)}. \tag{3.46}$$

Here, $g_i = \rho^i_1/\rho$ is the density ratio, $h_i = k/k^i_1 = c^i_1/c_0$ is the sound speed ratio for the i-th cylinder, and the primes indicate the derivatives of the Bessel and Hankel functions. Then, with the previous expression, it is possible to relate the coefficients B_{in} to the A_{in}. To do this, it is necessary to define the following values

$$\begin{aligned} T_{in} &= S_{in}/\imath\pi, & (3.47)\\ G_{ijln} &= H_{l-n}(kr_{ij})e^{\imath(l-n)\theta_{r_{ij}}}; \quad i \neq j. & (3.48)\end{aligned}$$

Then,

$$\Gamma_{in}A_{in} - \sum_{j=1,j\neq i}^{N}\sum_{l=-\infty}^{\infty} G_{ijln}A_{jl} = T_{in}, \tag{3.49}$$

is an infinite system of equations that can be used to determine A_{in} if it is properly truncated. Then, if the indexes n and l take values from $-M$ to M, both systems are reduced to $2M+1$ equations that can be expressed in matrix formulation as follows:

$$(\Gamma - G)A = T, \tag{3.50}$$

then the vector of coefficients A can be obtained as

$$A = (\Gamma - G)^{-1}T, \tag{3.51}$$

where the index -1 indicates the inverse of the matrix. The matrices Γ, G and the vectors A and T are represented by:

$$\Gamma = \begin{bmatrix} \Gamma_1 & 0 & \dots & 0 \\ 0 & \Gamma_2 & \dots & 0 \\ \vdots & \vdots & \ddots & \vdots \\ 0 & \dots & \dots & \Gamma_N \end{bmatrix}, \quad \text{and} \quad \Gamma_i = \begin{bmatrix} \Gamma_{i,-M} & 0 & \dots & 0 \\ 0 & \Gamma_{i,-M+1} & \dots & 0 \\ \vdots & \vdots & \ddots & \vdots \\ 0 & \dots & \dots & \Gamma_{i,M} \end{bmatrix}, \tag{3.52}$$

where N is the number of cylinders of the structure,

$$G = \begin{bmatrix}
0 & \dots & 0 & G^{1,2}_{-M,-M} & \dots & G^{1,2}_{M,-M} & \dots\dots & G^{1,N}_{-M,-M} & \dots & G^{1,N}_{M,-M} \\
\vdots & \ddots & \dots & \vdots & \ddots & \vdots & \dots\dots & G^{1,N}_{-M,-M} & \ddots & G^{1,N}_{M,-M} \\
0 & \dots & 0 & G^{1,2}_{-M,M} & \dots & G^{1,2}_{M,M} & \dots\dots & G^{1,N}_{-M,M} & \dots & G^{1,N}_{M,M} \\
G^{2,1}_{-M,-M} & \dots & G^{2,1}_{M,-M} & 0 & \dots & 0 & \dots\dots & G^{2,N}_{-M,-M} & \dots & G^{2,N}_{M,-M} \\
\vdots & \ddots & \vdots & \vdots & \ddots & \dots & \dots\dots & G^{2,N}_{-M,-M} & \ddots & G^{2,N}_{M,-M} \\
G^{2,1}_{-M,M} & \dots & G^{2,1}_{M,M} & 0 & \dots & 0 & \dots\dots & G^{2,N}_{-M,M} & \dots & G^{2,N}_{M,M} \\
\vdots & \vdots & \vdots & \vdots & \vdots & \vdots & \vdots\ \vdots & \vdots & \vdots & \vdots \\
\vdots & \vdots & \vdots & \vdots & \vdots & \vdots & \vdots\ \vdots & \vdots & \vdots & \vdots \\
G^{N,1}_{-M,-M} & \dots & G^{N,1}_{M,-M} & \dots & \dots & \dots & \dots\dots & 0 & \dots & 0 \\
\vdots & \ddots & \vdots & \dots & \dots & \dots & \dots\dots & \vdots & \ddots & \vdots \\
G^{N,1}_{-M,M} & \dots & G^{N,1}_{M,M} & \dots & \dots & \dots & \dots\dots & 0 & \dots & 0
\end{bmatrix},$$

and

$$A = \begin{bmatrix} A_{1,-M} \\ \vdots \\ A_{1,M} \\ A_{2,-M} \\ \vdots \\ A_{2,M} \\ \vdots \\ \vdots \\ A_{N,-M} \\ \vdots \\ A_{N,M} \end{bmatrix}, \quad T = \begin{bmatrix} T_{1,-M} \\ \vdots \\ T_{1,M} \\ T_{2,-M} \\ \vdots \\ T_{2,M} \\ \vdots \\ \vdots \\ T_{N,-M} \\ \vdots \\ T_{N,M} \end{bmatrix}, \tag{3.53}$$

where the values of the components of the previous matrices are given by Eqs. (3.46–3.48). We note that the size of the matrices Γ and G is $N(2M+1) \times N(2M+1)$ while the vectors T and A present a length equal to $N(2M+1)$. A good estimation for this truncation is $l = n = M = \text{floor}\left(kR_{\max} + 4.05(kR_{\max})^{1/3}\right) + 10$, with $R_{\max} = \max(a^i)$. Some works also propose the limit $l = M = n = \text{floor}\,(2.5kR_{\max})$.

Once the infinite system of equations to obtain the coefficients A_{in} is solved, the total wave at any point is

$$p(\vec{r}) = \imath\pi H_0(kr) + \sum_{i=1}^{N} \sum_{n=-\infty}^{\infty} \imath\pi A_{in} H_n(kr_i) e^{\imath n\theta_i}. \tag{3.54}$$

We note that the above derivation is valid for any finite configuration of the cylinders.

Figure 3.4a, b show the pressure distribution, $|p|$, produced by the acoustic scattering by a square array of cylinders for the two main directions of symmetry (0 and 45 degrees), with size $5a \times 5a$ and filling fraction $ff = 50\%$. Figure 3.4c also represents the acoustic spectra for the two main directions of symmetry measured at point $(x, y) = (11a, 0a)$ from the source location. We can observe the ranges of inhibition in both directions (pseudogaps shown in Fig. 3.3). The transmission outside these ranges can vary significantly as the number of scatterers or the shape of the array changes. The oscillatory behaviour for frequencies below $ka = 3$ is caused by external shape of the crystal. They may appear or not, depending on the arrangement of the array. But the inhibition behaviour remains quantitatively the same for both directions. Such a stable inhibition range is a clear indicator for the stop band.

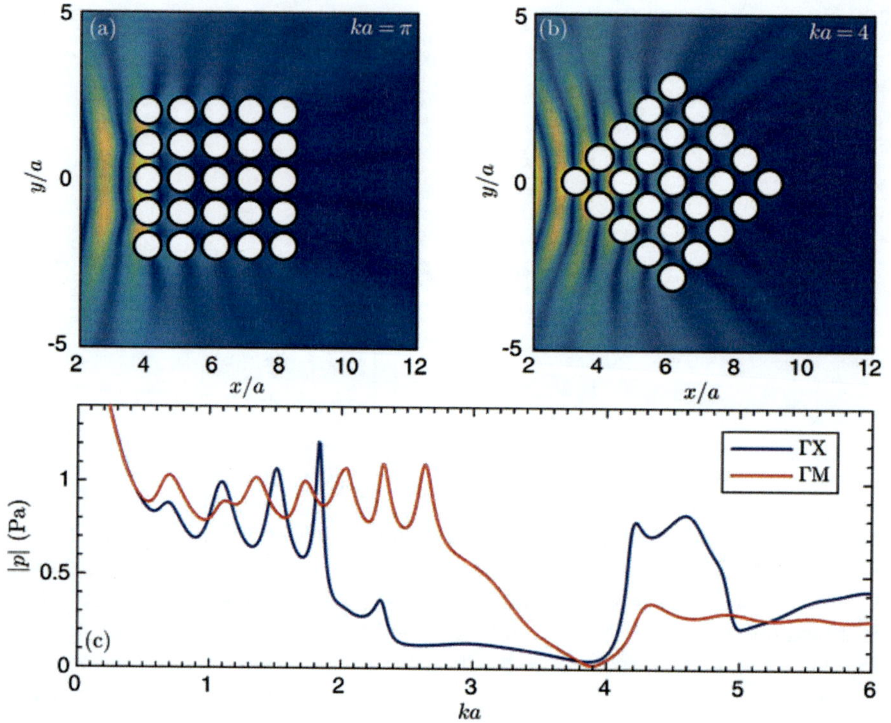

Fig. 3.4 Pressure maps and spectra calculated using MST for a cylindrical wave impinging a $5a \times 5a$ square array of rigid cylinders with a filling fraction $ff = 50\%$. **a** map for a frequency $ka = \pi$ inside the pseudogap at ΓX direction. **b** map for a frequency $ka = 4$ inside the pseudogap at ΓM direction. **c** Blue line (Red line) represents the spectrum for the ΓX (ΓM) direction measured at the point $(x/a, y/a) = (11, 0)$ from the source location

3.4.2 *Boundary Conditions*

3.4.2.1 Boundary Conditions for Rigid Scatterers

Equation (3.46) represents the results for the continuity of pressure and velocity in the boundaries of the scatterer. Considering the limiting case, $g_i \to \infty$ and $h_i \to \infty$, that means, considering large contrast between both the density and the sound speed of the host and the scatterer media, we can obtain the boundary condition for the case of rigid scatterers [46, 49]:

$$B_{in} = \imath \pi \Gamma_{in} A_{in}, \tag{3.55}$$

where

$$\Gamma_{in} = -\frac{H_n'(ka_i)}{J_n'(ka_i)}. \tag{3.56}$$

3.4.2.2 Viscous and Thermal Layers

The viscous and thermal dissipation phenomena occurring in the vicinity of the cylinders can be taken into account by introducing an equivalent surface admittance [50, 51]. This can be done by following the approach proposed by Tournat et al. [52] in the framework of the study of random cylinder networks. The expression of the admittance β in the case of a plane wave incident on a plane surface is, in the Cartesian coordinate system

$$\beta(\delta_v, \delta_h, \omega) = \frac{1-\imath}{2}k\left[\left(1-\frac{k_N^2}{k^2}\right)\delta_v + (\gamma-1)\delta_h\right]. \tag{3.57}$$

The viscous and the thermal boundary layers, defined respectively as

$$\delta_v = \sqrt{\frac{2\nu_v}{\omega}} \quad \text{and} \quad \delta_h = \sqrt{\frac{2\nu_h}{\omega}}, \tag{3.58}$$

depend on the coefficients of the cinematic viscosity ν_v and thermal diffusivity $\nu_h$of the fluid surrounding the cylinders, as well as on the pulsation ω. The wave number k_N represents the normal component of the wave vector k. A two-dimensional plane wave incident on a cylindrical scatterer can be represented as a series involving Bessel functions in polar coordinates (r,θ), as it was previously done. In this case, for each cylindrical component n of the pressure field $p_{i,n}$, the corresponding equivalent surface admittance is written as

$$\beta_n(\delta_v, \delta_h, \omega) = \frac{1-\imath}{2}k\left[\left(1-\frac{n^2}{k^2 a_i}\right)\delta_v + (\gamma-1)\delta_h\right], \tag{3.59}$$

where a_i is the radius of the i-th cylinder. Modifying the boundary condition at the boundary of the cylinders by a new condition involving the cylindrical components β_n of the equivalent surface admittance, we obtain

$$-\imath\beta_n k p_{in} = \frac{\partial p_{in}}{\partial r}, \tag{3.60}$$

and, as a consequence, we can obtain the relation

$$\Gamma_{in} = -\frac{H_n'(ka_i) + \imath\beta_n H_n(ka_i)}{J_n'(ka_i) + \imath\beta_n J_n(ka_i)} \tag{3.61}$$

between the incident and the scattered wave amplitudes.

3.4.2.3 Porous or Micro-Perforated Layers

Some times it is interesting to analyse the case of a rigid core covered with absorbent material, as it has been done in several works in the literature [53–56]. A MST procedure can be developed to study the scattering problem of scalar waves by periodic arrays of absorbent scatterers, if the boundary conditions are well defined. In this case, the scatterer presents two different kinds of boundary conditions. The core surface could be treated as rigid, i.e., a Neumann Boundary condition in applied. However, on the absorbing material-host medium interface, one should consider the continuity of the pressure and the velocity. Thus, the boundary conditions on the rigid wall, Γ, inside i-th scatterer is:

$$\left.\frac{\partial p^i_{int}}{\partial n}\right|_{\Gamma_i} = 0, \tag{3.62}$$

and the boundary conditions in the exterior interface of the scatterer are

$$\left.p^i_{\text{ext}}\right|_{\partial\Omega_j} = \left.p^i_{\text{int}}\right|_{\partial\Omega_j}, \tag{3.63}$$

$$\frac{Z_c(\omega)k_c(\omega)}{k_0}\left.\frac{\partial p_{\text{ext}}}{\partial n}\right|_{\partial\Omega_i} = \left.\frac{\partial p_{\text{int}}}{\partial n}\right|_{\partial\Omega_j}, \tag{3.64}$$

where $\partial\Omega_j$ is the boundary of the i-th scatterer, k_0 is the wave number in the host medium, $k_c(\omega)$ and $Z_c(\omega)$ are the propagation constant and the impedance of the absorbing material of the scatterer i.

The wave on the interior of the absorbing material of i-th cylinder is

$$p^i_{\text{int}}(\vec{r},\vec{r}_i) = \sum_{n=-\infty}^{\infty} A^i_n\Big(X^i_n H^{(1)}_n\left[k_c(\omega)|\vec{r}-\vec{r}_i|\right] + \cdots$$
$$+ Y^i_n J_n\left[k_c(\omega)|\vec{r}-\vec{r}_i|\right]\Big)e^{\imath n\phi_{\vec{r}-\vec{r}_i}}. \tag{3.65}$$

By applying the boundary condition given by (3.62), we can obtain a simple relation between coefficients X^i_n and Y^i_n:

$$Y^i_n = X^i_n T^i_n, \tag{3.66}$$

$$T^i_n = -\frac{H'_n(k_c(\omega)r^i_{in})}{J'_n(k_c(\omega)r^i_{in})}. \tag{3.67}$$

where the prime superscript represents the derivative with respect to the normal of the surface and r^i_{in} represents the radius of the rigid core of the i-th cylinder.

Then, the exterior wave outside the i-th cylinder is written as

$$p_{\text{ext}}(\vec{r}, \vec{r}_i) = \sum_{n=-\infty}^{\infty} B_n^i J_n \left[k|\vec{r} - \vec{r}_i|\right] e^{\imath n \phi_{\vec{r}-\vec{r}_i}} + \cdots$$
$$+ \sum_{n=-\infty}^{\infty} \imath \pi A_n^i H_n^{(1)} \left[k|\vec{r} - \vec{r}_i|\right] e^{\imath n \phi_{\vec{r}-\vec{r}_i}}. \quad (3.68)$$

Therefore, applying the boundary conditions at the $\partial\Omega$ interfaces, we get:

$$B_n^i = \imath \pi Z_n^i A_n^i, \quad (3.69)$$

where,

$$Z_n^i = -\frac{f(\omega) H_n'(kr_{\text{out}}^i) - H_n(kr_{\text{out}}^i)}{f(\omega) J_n'(kr_{\text{out}}^i) - J_n(kr_{\text{out}}^i)}, \quad (3.70)$$

$$f(\omega) = \frac{Z(\omega)k(\omega)}{k} \frac{H_n k_c(\omega r_{\text{out}}^i) + T_n J_n(k_c(\omega) r_{\text{out}}^i)}{H_n' k_c(\omega r_{\text{out}}^i) + T_n^i J_n'(k_c(\omega) r_{\text{out}}^i)}, \quad (3.71)$$

where r_{out}^i represents the radius of the outer porous core of the i-th cylinder. It is worth noting here, that $Z(\omega)$ represents the frequency dependent and possibly complex impedance of the layer around the rigid core.

3.5 Research on Sonic Crystals

Depending on the ratio between the wavelength of the incident wave, λ, and the lattice constant of the crystals, a, the basic mechanisms describing the effect of the crystal on the wave can be best interpreted in terms of two main regimes: the refraction [57] and the diffraction regimes [19]. In the long wavelength regime, i.e., $\lambda/a \gg 1$, crystals can be considered as homogeneous materials with effective properties [58, 59] therefore one can design refractive [57] or gradient index (GRIN) [60] lenses to control waves. In this direction, metamaterial acoustic GRIN lenses have recently been designed by using unit cells based on cross-shape scatterers [61] or coiling-up space, [62] providing a high transmission efficiency and small size. On the other hand, the case $\lambda \simeq a$ corresponds to the diffractive regime, where the crystal is strongly dispersive. Yang et al. [63] reported the first three dimension (3D) phononic crystal showing the focusing of ultrasonic waves in this regime. Since then, several phononic lenses have been designed by using the curvature properties of the isofrequency contours, making use of the all angle negative refraction [64].

In this Section we present a brief review of the research on sonic crystals developed in the last years. Instead of using chronological order, we will organize the presentation in terms of the frequency, i.e., λ/a ratio; starting by the long wavelength regime and proceeding to the diffraction regime where the bandgaps are found.

3.5.1 Long Wavelength Regime

3.5.1.1 Effective Parameters

Compact formulas for the effective density, ρ_{eff}, and sound speed, c_{eff}, that take into account the inner structure of the periodic medium have been obtained [65, 66]. In the limit case of low filling fractions, the homogenized parameters can be analytically obtained, and they are

$$\rho_{eff} = \rho_0 \frac{1 + ff}{1 - ff}, \tag{3.72}$$

$$c_{eff} = c_0 \frac{1}{\sqrt{1 + ff}}, \tag{3.73}$$

where ρ_0 and c_0 are the density and the sound speed of the surrounding fluid. These expressions recover the formula derived by Berryman [67] for the effective mass density as well as the effective sound speed reported in [57]. Both results have been recently corroborated by using a multiple scattering approach applied to an infinite system [59]. Expressions similar to that in (3.73) were also obtained for the case of cylinders of finite density and sound velocity. Also note that relationships in (3.73) have appeared in studying water waves, where homogenization was performed in the framework of the coherent-potential-approximation [46]. Another issue of great interest is to determine the cluster's minimum size in which the refractive effects dominate over the diffractive. It has been shown that the homogenization approach is valid at wavelength λ_c as low as a fourth of the cluster's diameter. Finally, the homogenization is valid, as shown in [65], up to a frequency equivalent to $\lambda \approx 4a$.

3.5.1.2 Gradient Refractive Index (GRIN) Systems

The properties of sonic crystals in the low-frequency limit have been studied by several groups [57, 60, 68–72] for its possible use as refractive devices. In the homogenization limit each layer of the GRIN lens can be treated as an effective medium. Then, previous Eqs. (3.73), or those shown in [66], can be used to calculate each layer's effective properties in order to create an index profile, as for example a hyperbolic secant profile that has been proved to reduce the aberration of the focal spot. One interesting result is that the symmetry matching between the source and the lens results in fundamental interest for lensing applications [72]. Axisymmetric gradient index (GRIN) lenses, as those shown in Fig. 3.5, made of rigid toroidal scatterers embedded in air consider this symmetry matching with radially symmetric sources. The sound amplification obtained in the focal spot of the reported lens (8.24 dB experimentally) shows the efficiency of the axisymmetric lenses with respect to the previous Cartesian acoustic GRIN lenses.

Fig. 3.5 Axisymmetric GRIN lens made of rigid toroidal scatterers used in [72]

3.5.1.3 Locally Resonant Sonic Crystals: Acoustic Metamaterials

The works of Liu et al. [73] and Fang et al. [74] described periodic arrays of locally resonant building blocks that exhibited attenuation bands with wavelength much larger than the periodicity of the structure. The origin of this phenomenon has been explained by means of the localized resonances associated with each scatterer: the hybridization of the flat resonance band and the propagating band of the periodic medium produces the stop band. Thus, the frequency and the width of the stop band can be tuned by varying the size and geometry of the resonators. These results were found before in the field of acoustics by the pioneering works of Bradley [75] and Sugimoto [76], however the breakthrough achieved by Liu et al. and Fang et al. was to define such structured systems as homogeneous materials, showing that, they can have negative effective properties, giving rise to acoustic materials not found in nature, i.e., the acoustic metamaterials.

In acoustic metamaterials, the sound speed is proportional to $\sqrt{\kappa_{eff}/\rho_{eff}}$, where κ_{eff} and ρ_{eff} are the effective bulk modulus and the mass density of the system, respectively. Sonic crystals made of rigid scatterers can be considered as an acoustic metamaterial showing real and positive effective properties, as we have previously seen in the previous Section. However, some interesting differences can appear in locally resonant sonic crystals with negative values of the effective parameters [73]. Depending on the values of these effective parameters several effects can be observed [77, 78]. If κ_{eff} and ρ_{eff} are both positive, the Poynting vector, $\vec{S}$, has the same direction as $\vec{k}$ and Snell's law is normally accomplished.

However, if κ_{eff} and ρ_{eff} are both negative, $\vec{S}$ and $\vec{k}$ have opposite directions. Physically, the negativity of both κ_{eff} and ρ_{eff} means that the medium displays an anomalous response at some frequencies such that it expands upon compression (negative bulk modulus) and moves to the left when being pushed to the right (negative density) at the same time. These double negative materials present unique properties, such as negative refractive index and subwavelength focusing [79].

However, if only one of these quantities is negative, the effective sound speed is complex valued, as is the wavenumber. Thus, when the real part of the expression for the Poynting vector is negative and sufficiently large, a frequency range where

$\text{Re}(\vec{k}\vec{S}) < 0$ is observed. A direct consequence of such behaviour is the exponential wave attenuation at these frequencies. It has been previously shown that low frequency attenuation bands can be induced by an effective bulk modulus that becomes negative near the resonance frequencies, giving rise to exponential decay of modes [74]. This properties has been exploited in the literature to create tunable acoustic filters by using locally resonant sonic crystals [54, 80–84].

This combination of material properties, not found before, have dramatically changed the conception of the acoustic materials. At this stage it is fair to say that metamaterials have recently revolutionized the field of wave control, particularly in acoustics showing extraordinary functionalities giving rise to innovations in material design. In many cases they can replace traditional treatments in practical situations because of their better performances in targeted and tunable frequency ranges with strongly reduced dimensions. Acoustic metamaterials themselves represent a scientific breakthrough with respect to the conventional treatments for noise and vibration problems.

3.5.2 Diffraction Regime

3.5.2.1 Band Gap

The periodicity of sonic crystals is introduced in the solution of the wave equation by means of Bloch's theorem. This solution leads to the phenomenon of bandgaps, frequency ranges where waves do not propagate through the crystal [16, 29]. Traditionally, wave propagation inside such systems was analysed by means of the band structures. Plane wave expansion transforms the wave equation into an eigenvalue problem that can be solved for each Bloch vector, k, in the irreducible first Brillouin zone; and so obtaining the eigenfrequencies $\omega(\vec{k})$ that constitute the band structures [12]. In the case of SCs, it has been proven that eigenfrequencies for an arbitrary crystal structure and an arbitrary filling fraction are real values for the propagating bands. Inside the bandgaps, the eigenvalue are complex [30, 31] and this allows us to defined the bandgap as ranges of frequencies where all Bloch waves must be evanescent [29]. A great number of applications based on SCs are explained by the existence of BGs: acoustic filters, acoustic barriers or waveguides.

Several works have been published, exploring the optimal unit cell designs that produce the widest bandgaps in terms of frequency. The problem has been treated in a variety of settings and using several techniques. For example, unit cells have been optimized in one-dimension [85] and in two-dimensions (2D) using gradient-based [86–88] as well as non-gradient-based [89, 90] techniques. The attempts to achieve as wide the bandgap as possible has also been performed outside the scope of the unit cell dispersion problem [91]. In all these optimization studies the focus has been primarily on crystals of infinite extends and consisting of two or more solid (or solid and fluid) phases with the exception of a few investigations that considered thin-plate single-phase models [87, 88]. For combined out-of-plane and in-plane waves in 2D infinite-thickness phononic crystals formed from silicon and a square lattice of circular voids, it has been shown that the bandgap size normalized with respect to the mid-gap frequency cannot exceed 40% [92]. Specialized optimization algorithms have been used in pursuit of the best unit cell solid-void distribution for the 2D plain-strain problem considering high-symmetry square lattices. The cases of (1) out-of-plane, (2) in-plane and (3) combined out-of- plane and in-plane elastic wave propagation have been analysed [35].

3.5.2.2 Point Defects

One of the most important properties of the periodic structures is the emergence of localized modes within the BG when a point defect is introduced [39]. The localized modes of a single point defect are characterized by evanescent behaviour [29, 30, 93]. Localized modes of N-point defects are mainly characterized by three properties: splitting of frequencies, the symmetry of the vibrational patterns and evanescent behaviour inside the crystal. EPWE in addition to PWE both with supercell approximation has been used to analyse the whole properties of the localized modes in a SC with a double-point defect [93]. Within the imaginary complex band structure, the localized modes present different values for the imaginary part of k meaning that each mode has a different decay rate inside the crystal. This property has been observed experimentally by fitting the exponential decay for each localized mode inside the crystal. The symmetry of the pressure profile in a double-point defect has also been analysed by means of multiple scattering calculations and experimental data [93].

3.5.2.3 Waveguiding

Linear point defects in sonic crystals have been used to create waveguides at frequencies inside the bandgap, where the walls of these linear point defects act as mirrors, creating the waveguide. It has been previously shown [94] that waveguiding effects due to periodicity go beyond rigid-wall waveguides. Due to the linear defect in the sonic-crystal waveguide, interesting physics appear in the bandgap. Different kinds of waveguides have been analysed in the literature. Interestingly, the properties of the guided modes in a linear waveguide, have been analysed by using the complex band structures [95, 96], showing that the attenuation is governed by a decay rate related to the imaginary part of the complex band structures obtained using EPWE. These results have raised more understanding of the deaf-modes [97]: modes with the real part of the complex band structure represented by antisymmetric modes, and imaginary parts represented by excitable symmetries. Therefore, in contrast to the classical interpretation of the attenuation in a deaf band, and based on geometric arguments, in [97] it was found that the attenuation is governed by a decay rate related to the imaginary part of the complex band structures.

Fig. 3.6 **a** Waveguide used in [95, 96] to analyse the stubbed waveguides. **b** Waveguide used in [33] for bending, splitting and spatial filtering waves

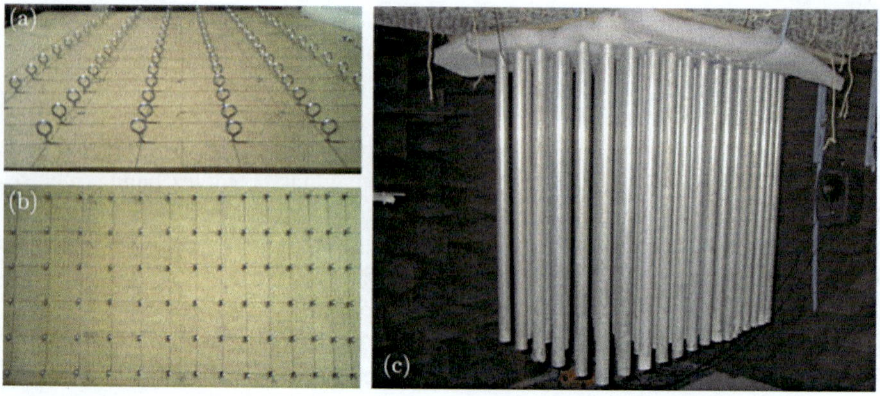

Fig. 3.7 Chirped crystal used in [101]. **a** and **b** holding structure. **c** Chirped sonic crystal

The simplicity of the SC in which only the longitudinal polarization can be excited has been used to observe without loss of generality the level repulsion between symmetric and antisymmetric bands in an stubbed waveguide. The level repulsion between symmetric and antisymmetric bands in antisymmetric sonic crystal stubbed waveguides have been interpreted as the presence of an evanescent mode connecting both bands [95]. These evanescent modes explain both the attenuation produced in this range of frequencies and the transfer of symmetry from one band to the other. Mode conversion in phononics have been also analysed in the literature by using similar systems [98–100].

Extensive simulations and experimental results have been performed in the literature in order to show tunable waveguides made of arrays made of rigid square-rod scatterers embedded in air. By rotating some of the square-rod scatterers of an array one can easily produce a point defect, and even multiple point or line defects [33]. Examples of the several properties of such a waveguides are the bending and splitting or the spatial filtering [33].

Fig. 3.8 **a–c** Different views of the sonic crystal acoustic barrier prototype

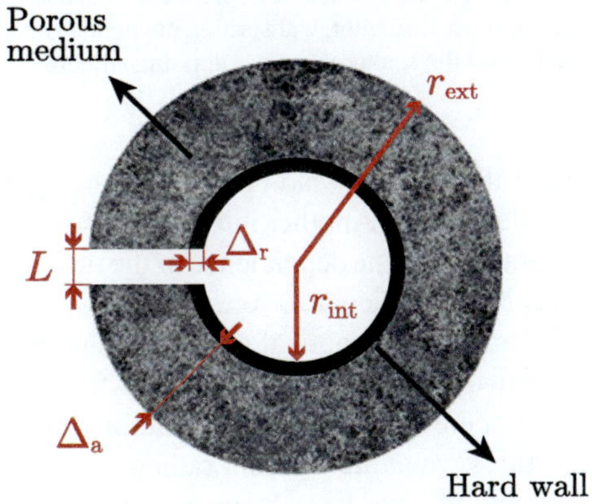

Fig. 3.9 The transverse view of the multi-physical phenomena scatterer. r_{int} and r_{ext} are the inner and exterior radii respectively, L the aperture, Δ_r the rigid wall thickness and Δ_a the thickness of the absorbing cover

3.5.2.4 Chirped Structures

Chirped (sometimes called graded or adiabatic tapered) crystals have been introduced in optics [102] and acoustics [103–105] for different purposes, such as opening wide full bandgaps in tandem structures [103] or waveguiding of beams. Chirped or graded materials are widely used in the wave physics community due to the opportunities they offer to manipulate the wave propagation. These artificial materials are emerging as promising tools for potential applications in several branches of research and technology. Several applications for focusing, trapping, bending waves, opening of

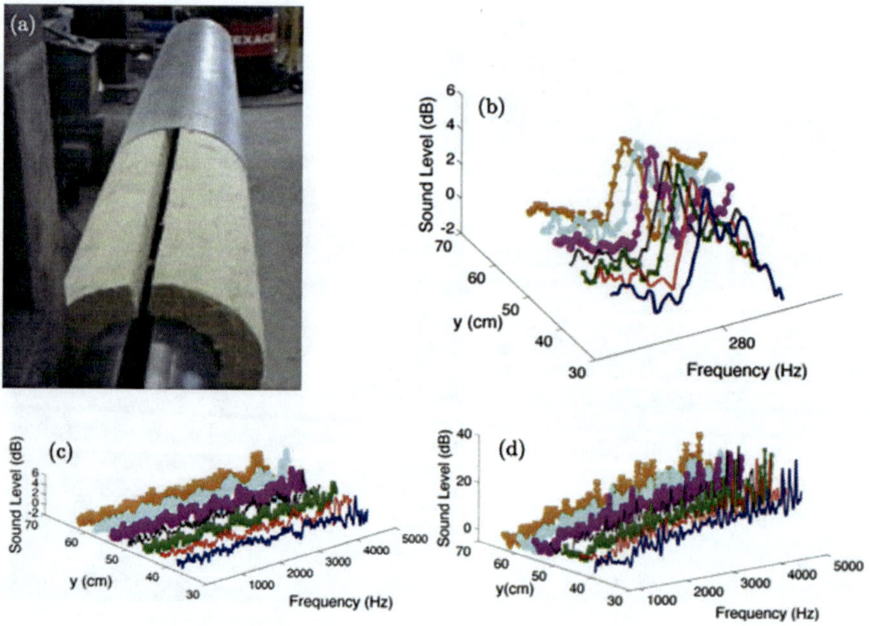

Fig. 3.10 Characterization of the several parts of the multi-physical phenomena scatterer. **a** Picture of each part of the scatterer: Perforated cover, absorbing cover and inner resonator. IL measurements of the **b** inner resonator, **c** absorbing cover and **d** perforated plate. The measurements were done at 25 cm from the scatterer at several points all along the symmetry axis (y-axis)

wide full bandgaps and controlling the spatial dispersion beams in reflection have been developed. Another interesting effect reported recently is the so-called rainbow trapping effect, the dependence of the turning point position on the colour of radiation. It has been predicted for one-dimensionally modulated chirped photonic structures [106] and tapered optical and plasmonic waveguides [107, 108]. Rainbow trapping and wave enhancement are two different physical effects (the latter occurs even for monochromatic radiation), although they may occur simultaneously in chirped structures when the incident radiation is broadband (Figs. 3.6 and 3.7).

Sonic crystals have been used to show that chirped structures can be used to enhance the acoustic amplitude due to the progressive decrease of the group velocity along the propagation direction [101, 109]. Wave reflection from a bandgap in a chirped structure is peculiar. The dispersion curves $\omega(\vec{k})$ close to the band-edges develop nearly horizontal segments, which corresponds to small or zero group velocity of the wave, since $v_g = \partial\omega/\partial k$. The occurrence of the controlled sound enhancement requires that the crystal at the entrance plane be within the transparency range for the incoming wave, whose frequency is above the first bandgap. The wave entering into the crystal is gradually slowing down, as the *local* bandgaps are approaching the wave frequency in the course of propagation. Finally, at a particular depth corresponding to the band-edge, the wave, literally speaking, stops, turns around, and starts propagating back. In other words, it experiences a *soft* reflection.

3.5.2.5 Beamforming and Focusing

The influence of the spatial periodicity on the spectral properties of such systems is represented by the dispersion relation, however it has come out that the spatial periodicity can affect not only this temporal dispersion, but also the spatial one making possible the control of the diffraction inside the periodic structures. Due to that, one can observe different types of behaviour depending on the spatial dispersion relation, i.e., on the curvature of the iso-frequency contours[110]. The so-called self-collimation effect, due to flat iso-frequency contours, consists in the propagation of a beam in the periodic system without apparent diffraction keeping its original width. This phenomenon has been experimentally demonstrated to date at different frequency ranges for both, electromagnetic waves [111, 112] and acoustic waves [113, 114] in photonic and sonic crystals, respectively. Conversely, when the curvature of the iso-frequency contour is negative, one can observe focusing due to the all-angle negative refraction phenomenon, which has been also observed in both electromagnetics [64, 115] and acoustics [63, 116, 117].

The previous spatial effects, the self-collimation and all-angle negative refraction, occur for connected (continuous) iso-frequency contours, however, there are frequencies in which the iso-frequency lines are discontinuous in the Brillouin zone, making it possible the existence of angles in which no iso-lines exist and, as a consequence, there is not propagation of waves. These ranges of angles are called angular bandgaps [110, 113].

3.6 Technological Application: Sonic Crystal Acoustic Barrier

In the last years, the possibility to manipulate the sound by means of SCs motivated the idea of using these periodic acoustic media as attenuation devices as, for example, an alternative to acoustic barriers[24, 54, 56, 81–84, 118–121] However, from the acoustical point of view, the mere existence of the BGs is not sufficient to use SCs as acoustic barriers because both the size and position of these BGs depend on several factors such as the angle of incidence of the wave or the arrangement of the scatterers. To avoid these problems, some strategies have been developed in the last few years. First, some authors have studied new and more efficient arrangements of scatterers out of the classical crystalline ones, such as quasi-crystals, [122] quasi-ordered structures [123, 124] or quasi-fractal [119] arrangements. Another strategy is the use of scatterers with additional properties motivated by the work of Liu et al. [73]. The use of local properties or scatterers can reduce the angular dependence of the attenuation achieved by the periodic arrangement and increases both the magnitude of the attenuation and the range of the attenuated frequencies. In this sense, recent works have shown the possibility to design this kind of structures with multi-physical phenomena scatterers to be applied in the audible ranges of frequencies as efficient acoustic barriers [54].

In this Section we show the realization and the acoustical and structural characterization of a prototype of sonic crystal acoustic barrier. As shown in Fig. 3.8. We have acoustically characterized this sonic crystal acoustic barrier in a laboratory approved for this purpose for its use as traffic noise reducing devices. To do that, we have followed the European Standards EN 1793:1997 relative to the intrinsic characteristics of sound absorption (Part 1) and to the intrinsic characteristics of airborne sound insulation (Part 2). The structural characterization has been done in a wind tunnel evaluating the loads produced by wind in these structures comparing the results with the ones obtained for a traditional acoustic barrier. In the following we briefly describe the acoustic standardization and the mechanical characterization of the prototype.

3.6.1 Design Process

The design and the experimental characterization of the building blocks of the prototype are discussed here. Three different mechanisms have been considered to attenuate sound in the desired range of frequencies: resonances, absorption and scattering. Resonances in the low frequency range, absorption in the medium-high frequency range and scattering, making use of the periodicity of the structure, i.e., using the Bragg interference, in the medium to high frequency range.

3.6.1.1 Design of the Scatterers

The scatterer used in the prototype is based on the concept of split-ring resonator (SRR) which is well-known in optics [125, 126]. Figure 3.9 shows the transversal view of the scatterer, that is basically a two dimensional split ring resonator covered by an absorbing material. One can observe three parts. The inner one, which is a resonant cavity, the rigid wall and the absorbing cover. The rigid wall is made of iron and plays acoustical and structural roles. The absorbing cover is a sheet of porous material (mineral-wool) with density $100\,\text{kg/m}^3$.

The resonant behaviour of the scatterer is influenced by the aperture L of the scatterer, the interior radius r_{int} and the thickness of the aperture, $\Delta_\text{a} + \Delta_\text{r}$. [54] Two kinds of scatterers with different cross-section areas of the resonant cavity have been designed in order to cover a wide band in the low range of frequencies (210 Hz 300 Hz). The cover of porous material has a thickness $\Delta_\text{a} = 4$ cm and it is selected in order to attenuate both the medium and the high ranges frequencies. Finally, taking into account the outdoor application of the prototype, we have used a perforated plate to protect the mineral-wool from the environment.

Prior to acoustic standardization process we have characterized the acoustical response of a scatterer in our anechoic chamber. We used the Insertion Loss (IL), defined as the difference between the sound level recorded without and with the sample at the same point. In order to observe that the scatterer has the same acoustic response all along its symmetry axis (defined in this work in the vertical direction along the y axis), we measure the IL in several heights. IL measurements give the dependence of the attenuation properties of the scatterer on the frequency. The three parts of the scatterers used in the prototype, the perforated plate, the absorbent cover and the inner resonator, can be seen in Fig. 3.10a. We have analysed the acoustical response of each part of the scatterer separately in order to observe the range of frequencies in which they are effective.

3.6.1.2 Resonators

The iron cylinders are slotted along its entire length with an aperture, L, equal to 2 cm. This forms the inner resonator. Figure 3.10b shows the IL for the inner resonator. The analysed scatterer has an external diameter of 0.247 m, which produces a resonant peak 210 Hz. One can observe that the resonance frequency is practically constant all along the scatterer. Notice that each scatterer attenuates around 6 dB due to resonance.

3.6.1.3 Absorbent

The absorbent cover consists of a sheet of rock-wool with a thickness of 4 cm. Porous materials work in the range of medium to high frequencies. Figure 3.11c shows IL measurements for the absorbent cover. One can observe that the IL increases with the frequency and that the attenuation starts 900 Hz. The average attenuation is similar for different heights all along the scatterer. Notice the difference of the IL values between the measurements of the absorbent covering and the resonators, see Fig. 3.10b.

3.6.1.4 Perforated Cover

The exterior cover of the scatterer is a perforated plate of 1 mm thickness and perforations of 5 mm of diameter. The role of the perforated cover is to protect the absorbent from the elements. Thus, it should be transparent to acoustic waves. One can see that it is transparent for the range of frequencies between 100 5000 Hz, which corresponds to traffic noise.

3.6.1.5 Periodic Structure

The prototype consists of a two dimensional periodic array of multi-physics phenomena scatterers arranged following a triangular periodicity. The SCs are characterized by two parameters: the lattice constant a, separation between scatterers, and the filling fraction, ff, the volume occupied by the scatterers with respect to the total volume occupied by the SC. For this periodicity, the main directions of symmetry are ΓX (0°) and ΓJ (30°). In this system, the inner part of the scatterer has a triple function.

On one hand, it serves as structural support for the scatterers. On the other hand, the inner part acts as a resonator producing attenuation peaks in the low frequency range, as we have seen in the previous Section. However, the exterior part of the inner core (rigid resonator) contributes to the increase of the scattering inside the structure that leads to the phenomenon of the BG. The prototype has a lattice constant $a = 0.28$ m. For this lattice constant, the central frequency of the BG (Bragg's frequency) at ΓX direction is $f_{BG} = 600$ Hz.

3.6.2 *Acoustic Standardization*

3.6.2.1 Sound Absorption: EN-1793-1:1997

According to the standard EN 1793-1:1997, we have evaluated the sound absorption coefficient α_s in order to calculate the evaluation index of acoustic absorption DL_α. The value of this index is used to classify the barrier with regard to its acoustic absorption characteristics. In our case, $DL_\alpha = 8$ dB, that correspond to the A3 category. This result shows that a non-continuous acoustic barrier formed by multi-physics phenomena cylinders can compete, from the acoustical point of view, with traditional acoustic barriers formed by continuous systems.

3.6.2.2 Airborne Sound Insulation: EN 1793-2:1997

The test corresponding to the intrinsic characteristics of the barrier relative to the airborne sound insulation has been done following the recommendation of the Standard EN 1793-2:1998. To do that, the evaluation index of the airborne sound insulation DL_R (dB) is calculated according to the standard EN-ISO 10140:2011. The measurements take into account the sound level for each third octave band of the normalized traffic noise spectrum, given by the standard EN-1793-3 1997. The value of this index allows classifying the capability of airborne sound insulation of the checked barrier. In this case, $DL_R = 20$ dB that correspond to the category B2.

3.6.3 *Wind Tunnel Analysis: Structural Load*

Traditional acoustic barriers basically consist of a solid system positioned between the sound source and the receiver. Apart from the acoustical isolation, one of the main problems of these systems is the high transmission of the mechanical stresses produced by several systems of loads supported by the structure to the ground. For example, the transmitted loads due to the wind are important, increasing with the height of the barrier, and as a consequence the volume of foundations and the product becomes technically and economically expensive. The prototype presented in this

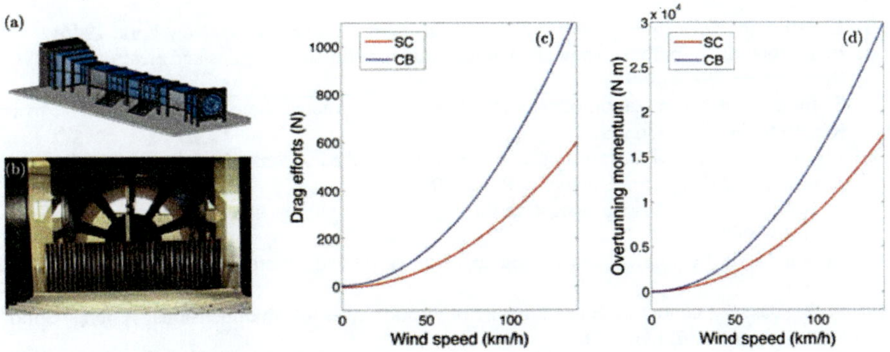

Fig. 3.11 Wind tunnel characterization. **a** Scheme of the wind tunnel. **b** Picture of the scaled sonic crystal acoustic barrier inside the tunnel. **c** and **d** show the dependence of the drag efforts an the overtunning momentum on the wind speed respectively

work is a sonic crystal acoustic barrier which is formed by separated inclusions, and this fact allows the wind to pass through, decreasing the load that is transmitted to the ground. To estimate the values of these loads in our barrier and to compare with those corresponding to a classical one, we have carried out some experiments in a wind tunnel with dimensions 2.14 m high, 1.8 m width and 12 m long (see Fig. 3.11a). As a consequence two models in a scale 1:5 for both a classical barrier and a sonic crystal acoustic barrier have been designed. Figure 3.11b shows the scaled sonic crystal acoustic barrier inside the wind tunnel.

The results obtained in the laboratory are shown in Fig. 3.11c–d. We have measured the drag load and the overtunning momentum. Taking into account these results one can conclude that the sonic crystal acoustic barrier produces loads considerably smaller than the classical barrier, with this reduction on average around 42% in the case of the drag loads in the wind direction and 37% in the case of the overturning momentum.

Acknowledgements Author would like to acknowledge A. Cebrecos, N. Jiménez, L.M. García-Raffi, V. J. Sánchez-Morcillo, R. Picó, J.-P. Groby, V. Tournat, C. Lagarrigue, S. Castiñeira-Ibáñez, J. V. Sánchez-Pérez and K. Staliunas for the collaboration and discussions on the topics described in this chapter.

References

1. I. Newton, *The Mathematical Principles of Natural Phylosophy*, vol. II (Daniel Adee, New York, 1846)
2. L. Kelvin, Pop. Lect. **III**, 185 (1881)
3. J.W. Strutt (Lord. Rayleigh), 'On the maintenance of vibrations by forces of double frequency, and on the propagation of waves through a medium endowed with a periodic structure, in *Philosophical Magazine*, vol. 24, no. 147, pp. 145–159 (1887)

4. L. Brillouin, *Wave Propagation in Periodic Structures* (Dover Books, New York, 1946)
5. E. Yablonovitch, Inhibited spontaneous emission in solid-state physics and electronics. Phys. Rev. Lett. **58**, 2059 (1987)
6. S. John, Strong localization of photons in certain disordered dielectric superlattices. Phys. Rev. Lett. **58**(23), 2486 (1987)
7. J.D. Joannopoulus, S.G. Johnson, J.N. Winn, R.D. Meade, *Photonic Crystals. Molding the Flow of Light* (Princeton University Press, 2008)
8. A. Ruffa, Acoustic wave propagation through periodic bubbly liquids. J. Acoust. Soc. Am. **91**, 1–6 (1992)
9. M. Sigalas, E. Economou, Elastic and acoustic wave band structure. J. Sound Vib. **158**, 377 (1992)
10. M. Sigalas, E. Economou, Band structure of elastic waves in two dimensional systems. Solid State Commun. **86**, 141 (1993)
11. M. Kushwaha, P. Halevi, L. Dobrzynski, B. Djafari-Rouhani, Acoustic band structure of periodic elastic composites. Phys. Rev. Lett. **71**(13), 2022–2025 (1993)
12. M.S. Kushwaha, P. Halevi, G. Martinez, L. Dobrzynski, B. Djafari-Rouhani, Theory of acoustic band structure of periodic elastic composites. Phys. Rev. B **49**(4), 2313–2322 (1994)
13. M. Sigalas, E. Economou, M. Kafesaki, Spectral gaps for electromagnetic and scalar weaves: possible explanation for certain differences. Phys. Rev. B **50**, 3393 (1994)
14. M. Sigalas, M.S. Kushwaha, E.N. Economou, M. Kafesaki, I.E. Psarobas, W. Steurer, Classical vibrational modes in phononic lattices: theory and experiment. Z. Kristallogr **220**, 765–809 (2005)
15. E. Economou, M. Sigalas, Classical wave propagation in periodic structures: cermet versus network topology. Phys. Rev. B **48**(18), 13434 (1993)
16. R. Martínez-Sala, J. Sancho, J.V. Sánchez, V. Gómez, J. Llinares, F. Meseguer, Sound attenuation by sculpture. Nature **378**, 241 (1995)
17. M. Kushwaha, Stop-bands for periodic metallic rods: sculptures that can filter the noise. Appl. Phys. Lett. **70**, 3218 (1997)
18. J.O. Vasseur, P.A. Deymier, Propagation of acoustic waves in periodic and random two-dimensional composite media. J. Matter. Res. **12**, 2207 (1997)
19. J.V. Sánchez-Pérez, D. Caballero, R. Martínez-Sala, C. Rubio, J. Sánchez-Dehesa, F. Meseguer, J. Llinares, F. Gálvez, Sound attenuation by a two-dimensional array of rigid cylinders. Phys. Rev. Lett. **80**(24), 5325–5328 (1998)
20. W.M. Robertson, J.F. RudyIII, Measurement of acoustic stop bands in two-dimensional periodic scattering arrays. J. Acoust. Soc. Am. **104**(2), 694–699 (1998)
21. M.S. Kushwaha, P. Halevi, Stop bands for cubic arrays of spherical balloons. J. Acoust. Soc. Am. **101**, 619 (1997)
22. C. Rubio, D. Caballero, J. Sánchez-Pérez, R. Martinez-Sala, J. Sánchez-Dehesa, F. Meseguer, F. Cervera, The existence of full gaps and deaf bands in two-dimensional sonic crystals. J. Lightwave Technol. **17**(11), 2202–2207 (1999)
23. M. Shen, W. Cao, Acoustic band-gap engineering using finite-size layered structures of multiple periodicity. Appl. Phys. Lett. **75**, 3713–3715 (2001)
24. J. Sánchez-Pérez, C. Rubio, R. Martínez-Sala, R. Sánchez-Grandia, V. Gómez, Acoustic barriers based on periodic arrays of scatterers. Appl. Phys. Lett. **81**, 5240 (2002)
25. A. Khelif, B. Djafari-Rouhani, J.O. Vasseur, P.A. Deymier, P. Lambin, L. Dobrzynski, Transmittivity through straight and stublike waveguides in a two-dimensional phononic crystal. Phys. Rev. B **65**, 174308 (2002)
26. A. Khelif, A. Choujaa, B. Djafari-Rouhani, M. Wilm, S. Ballandras, V. Laude, Trapping and guiding of acoustic waves by defect modes in a full-band-gap ultrasonic crystal. Phys. Rev. B **68**, 214301 (2003)
27. A. Khelif, M. Wilm, V. Laude, S. Ballandras, B. Djafari-Rouhani, Guided elastic waves along a rod defect of a two-dimensional phononic crystal. Phys. Rev. E **69**, 067601 (2004)
28. J.O. Vasseur, P.A. Deymier, B. Djafari-Rouhani, Y. Pennec, A.-C. Hladky-Hennion, Absolute forbidden bands and waveguiding in two-dimensional phononic crystal plates. Phys. Rev. B **77**, 085415 (2008)

29. V. Romero-García, J. Sánchez-Pérez, S. Castiñeira-Ibáñez, L. García-Raffi, Evidences of evanescent bloch waves in phononic crystals. Appl. Phys. Lett. **96**, 124102 (2010)
30. V. Romero-García, J. Sánchez-Pérez, L. García-Raffi, Evanescent modes in sonic crystals: Complex dispersion relation and supercell approximation. J. Appl. Phys. **108**, 044907 (2010)
31. V. Laude, Y. Achaoui, S. Benchabane, A. Khelif, Evanescent bloch waves and the complex band structure of phononic crystals. Phys. Rev. B **80**, 092301 (2009)
32. C. Goffaux, J.P. Vigneron, Theoretical study of a tunable phononic band gap system. Phys. Rev. B **64**, 075118 (2001)
33. V. Romero-García, C. Lagarrigue, J.-P. Groby, O. Richoux, V. Tournat., Tunability of band gaps and waveguides in periodic arrays of square-rod scatterers: Theory and experimental realization. J. Phys. D: Appl. Phys. **46**, 305108 (2013)
34. R. Wang, X.-H. Wang, B.-Y. Gu, G.-Z. Yang, Effects of shapes and orientations of scatterers and lattice symmetries on the photonic band gap in two-dimensional photonic crystals. J. Appl. Phys. **90**, 4307 (2001)
35. O.R. Bilal, M.I. Hussein, Ultrawide phononic band gap for combined in-plane and out-of-plane waves. Phys. Rev. E **84**, 065701(R) (2011)
36. L. Wu, L. Chen, C. Liu, Experimental investigation of the acoustic pressure in cavity of a two dimensional sonic crystal. Physica B **404**, 1766 (2009)
37. L.-Y. Wu, L.-W. Chen, C.-M. Liu, Acoustic pressure in cavity of variously sized two-dimensional sonic crystal with various filling fraction. Phys. Lett. A **373**, 1189–1195 (2009)
38. M. Sigalas, Elastic wave band gaps and defect states in two-dimensional composites. J. Acoust. Soc. Am. **101**, 1256 (1997)
39. M. Sigalas, Defect states of acoustic waves in a twodimensional lattice of solid cylinders. J. Appl. Phys. **84**, 3026 (1998)
40. F. Zaviska, Uber die beugung elektromagnetischer wellen an parallelen, unendlich langen kreiszylindern. Ann. Phys. **40**, 1023 (1913)
41. W. von Ignatowsky, Zur theorie der gitter. Ann. Phys. **44**, 369 (1914)
42. V. Twersky, Multiple scattering of radiation by an arbitrary configuration of parallel cylinders. J. Acoust. Soc. Am. **24**(1), 42 (1951)
43. V. Twersky, On scattering of waves by the infinite gratting of circular cylinders. IRE Trans. Antennas Propag. **10**, 737 (1962)
44. P. Martin, *Multiple Scattering. Interaction of Time-Harmonic Waves with N Obstcles* (Cambirdge University Press, UK, 2006)
45. C. Linton, P. McIver, *Handbook of Mathematical Techniques for Wave/Structure Interactions* (CRC Press, 2001)
46. Y.-Y. Chen, Z. Ye, Theoretical analysis of acoustic stop bands in two-dimensional periodic scattering arrays. Phys. Rev. E **64**, 036616 (2001)
47. R. Sainidou, N. Stefanou, I. Psarobas, A. Modinos, A layer-multiple-scattering method for phononic crystals and heterostructures of such. Comput. Phys. Commun. **166**, 197–240 (2005)
48. M. Abramowitz, I.A. Stegun, *Handbook of Mathematical Functions with Formulas, Graphs, and Mathematical Tables* (Dover, Ninth Dover Printing, Tenth GPO Printing ed., New York, 1964)
49. L. Sanchis, F. Cervera, J. Sánchez-Dehesa, J.V. Sánchez-Pérez, C. Rubio, R. Martínez-Sala, Reflectance properties of two-dimensional sonic band gap crystals. Jour. Ac. Soc. Am. **109**, 2598–2605 (2001)
50. A. Duclos, *Diffusion Multiple En Fluide Visco-Thermique, Cas Du Cristal Phononique à Deux Dimensions*. Ph.D. thesis, Laboratoire d'Acoustique de l'Université du Mans (2007)
51. A. Duclos, Transmission of acoustic wave s through 2D phononic crystal: visco-thermal and multiple scattering effects. Eur. Phys. J. Appl. Phys. **45**(11302) (2009)
52. V. Tournat, V. Pagneux, D. Lafarge, L. Jaouen, Multiple scattering of acoustic waves and porous absorbing media. Phys. Rev. E **70**(026609) (2004)
53. O. Umnova, K. Attenborough, C.M. Linton, Effects of porous covering on sound attenuation by periodic arrays of cylinders. J. Acoust. Soc. Am. **119**(1) (2006)

54. V. Romero-García, J.V. Sánchez-Pérez, L.M. García-Raffi, Tunable wideband bandstop acoustic filter based on two-dimensionalmultiphysical phenomena periodic systems. J. Appl. Phys. **110**, 149041–149049 (2011)
55. J. Sánchez-Dehesa, V. García-Chocano, D. Torrent, F. Cervera, S. Cabrera, F. Simon, Noise control by sonic crystal barriers made of recycled materials. J. Acous. Soc. Am. **129**(1173) (2011)
56. V. García-Chocano, S. Cabrera, J. Sánchez-Dehesa, Broadband sound absorption by lattices of microperforated cylindical shells. Appl. Phys. Lett. **101**(18), 184101 (2012)
57. F. Cervera, L. Sanchis, J.V. Sánchez-Pérez, R. Martínez-Sala, C. Rubio, F. Meseguer, Refractive acoustic devices for airborne sound. Phys. Rev Lett. **88**(2), 023902–4 (2002)
58. P. Sheng, *Introduction to Wave Scattering, Localization, and Mesoscopic Phenomena* (Academic Press, New York, 1995)
59. J. Mei, Z. Liu, W. Wen, P. Sheng, Effective mass density of fluid-solid composites. Phys. Rev. Lett. **96**(024301) (2006)
60. S.-C. Lin, T. Huang, J.-H. Sun, T.-T. Wu, Gradient-index phononic crystals. Phys. Rev. B, **79**(094302) (2009)
61. L. Zigoneanu, B.-I. Popa, S.A. Cummer, Design and measurements of a broadband two-dimensional acoustic lens. Phys. Rev. B **84**, 024305 (2011)
62. Y. Li, B. Liang, X. Tao, X. feng Zhu, X. ye Zou, J. Chun Cheng, Acoustic focusing by coiling up space. Appl. Phys. Lett. **101**(23), 233508 (2012)
63. S. Yang, J.H. Page, Z. Liu, M.L. Cowan, C. Chan, P. Sheng, Focusing of sound in a 3D phononic crystal. Phys. Rev. Lett. **93**(2) (2004)
64. C. Luo, S.G. Johnson, J.D. Joannopoulos, J.B. Pendry, All-angle negative refraction without negative effective index. Physics **65**, 201104(R) (2002)
65. D. Torrent, A. Håkansson, F. Cervera, J. Sánchez-Dehesa, Homogenization of two-dimensional cluster of rigid rods in air. Phys. Rev. Lett. **96**, 204302 (2006)
66. D. Torrent, J. Sánchez-Dehesa, Effective parameters of clusters of cylinders embedded in a nonviscous fluid or gas. Phys. Rev. B **74**, 224305 (2006)
67. J.G. Barryman, Long-wavelength propagation in composite elastic media i. spherical inclusions. J. Acoust. Soc. Am. **68**(1809) (1980)
68. L. Sanchis, A. Hakansson, F. Cervera, J. Sánchez-Dehesa, Acoustic interferometers based on two-dimensional arrays of rigid cylinders in air. Phys. Rev. B **67**, 035422 (2003)
69. T.P. Martin, M. Nicholas, G.J. Orris, L.-W. Cai, D. Torrent, J. Sánchez-Dehesa, Sonic gradient index lens for aqueous applications. Appl. Phys. Lett. **97**(11), 113503 (2010)
70. Y. Li, B. Liang, X. Tao, X.-F. Zhu, X.-Y. Zou, J.-C. Cheng, Acoustic focusing by coiling up space. Appl. Phys. Lett. **101**(23), 233508 (2012)
71. A. Climente, D. Torrent, J. Sánchez-Dehesa, Sound focusing by gradient index sonic lenses. Appl. Phys. Lett. **97**(10), 104103 (2010)
72. V. Romero-García, A. Cebrecos, R. Picó, V. Sánchez-Morcillo, L. García-Raffi, J. Sánchez-Pérez, Wave focusing using symmetry matching in axisymmetric acoustic gradient index lenses. Appl. Phys. Lett. **103**(26), 264106 (2013)
73. Z. Liu, X. Zhang, Y. Mao, Y. Zhu, Z. Yang, C. Chan, P. Sheng, Locally resonant sonic materials. Science **289**, 1734 (2000)
74. N. Fang, D. Xi, J. Xu, M. Ambati, W. Srituravanich, C. Sun, X. Zhang, Ultrasonic metamaterials with negative modulus. Nat. Mater. **5**, 452–456 (2006)
75. C.E. Bradley, Acoustic bloch wave propagation in a periodic waveguide, Technical Report, Technical Report of Applied Research Laboratories, Report No. ARL-TR-91-19 (July), The University of Texas at Austin (1991)
76. N. Sugimoto, T. Horioka, Dispersion characteristics of sound waves in a tunnel with an array of helmholtz resonators. J. Acoust. Soc. Am. **97**(3), 1446–1459 (1995)
77. S.H. Lee, C.M. Park, Y.M. Seo, Z.G. Wang, C.K. Kim, Composite acoustic medium with simultaneously negative density and modulus. Phys. Rev. Lett. **104**(054301) (2010)
78. Y.M. Seo, J.J. Park, S.H. Lee, C.M. Park, C.K. Kim, S.H. Lee., Acoustic metamaterial exhibiting four different sign combinations of density and modulus. J. Appl. Phys. **111**(023504) (2012)

79. S. Guenneau, A. Movchan, G. Pétursson, S.A. Ramakrishna, Acoustic metamaterial for sound focusing and confinement. New J. Phys. **9**, 399 (2007)
80. X. Hu, C.T. Chan, Two-dimensional sonic crystals with Helmholtz resonators. Phys. Rev. E **71**, 055601(R) (2005)
81. V. Romero-García, A. Krynkin, L. García-Raffi, O. Umnova, J. Sánchez-Pérez, Multi-resonant scatterers in sonic crystals: locally multi-resonant acoustic metamaterial. J. Soun Vib. **332**(1), 184 (2013)
82. C. Lagarrigue, J.P. Groby, V. Tournat, Sustainable sonic crystal made of resonating bamboo rods. J. Acoust. Soc. Am. **133**(1), 247 (2013)
83. A. Krynkin, O. Umnova, A.Y.B. Chong, S. Taherzadeh, K. Attenborough, Predictions and measurements of sound transmission through a periodic array of elastic shells in air. J. Acoust. Soc. Am. **128**(6), 3496–3506 (2010)
84. F. Koussa, J. Defrance, P. Jean, P. Blanc-Benon, Acoustical efficiency of a sonic crystal assisted noise barrier. Acta Acustica United Acustica **99**(3), 399–409 (2013)
85. M.I. Hussein, K. Hamza, G.M. Hulbert, R.A. Scott, K. Saitou, Multiobjective evolutionary optimization of periodic layered materials for desired wave dispersion characteristics. Struct. Multidiscip. O. **60**(2006), 31, 60 (2006)
86. O. Sigmund, J. Søndergaard Jensen, Systematic design of phononic band–GAP materials and structures by topology optimization, in *Philosophical Transactions of the Royal Society of London A: Mathematical, Physical and Engineering Sciences*, vol. 361, no. 1806, pp. 1001–1019 (2003)
87. A.R. Diaz, A.G. Haddow, L. Ma, Design of band-gap grid structures. Struct. Multidiscip. O. **29**(418) (2005)
88. S. Halkjaer, O. Sigmund, J.S. Jensen, Maximizing band gaps in plate structures. Struct. Multidiscip. O. **32**(263) (2006)
89. G.A. Gazonas, D.S. Weile, R. Wildman, A. Mohan, Genetic algorithm optimization of phononic bandgap structures. Int. J. Solids Struct. **43**(18), 5851–5866 (2006)
90. M.I. Hussein, K. Hamza, G.M. Hulbert, K. Saitou, Optimal synthesis of 2D phononic crystals for broadband frequency Isolation. Waves Random Compl. Media **17**(4), 491–510 (2007)
91. S. Halkjær, O. Sigmund, J.S. Jensen, Maximizing band gaps in plate structures. Struct. Multidiscip. Optim. **32**(4), 263–275 (2006)
92. M. Maldovan, E.L. Thomas, Simultaneous localization of photons and phonons in two-dimensional periodic structures. Appl. Phys. Lett. **88**(25) (2006)
93. V. Romero-García, J.V.S. Pérez, L.M. García-Raffi, Propagating and evanescent properties of duble-point defect in sonic crystals. New J. Phys. **12**, 1 (2010)
94. T. Miyashita, Sonic crystals and sonic wave-guides. Meas. Sci. Technol. **16**, R47 (2005)
95. V. Romero-García, J.O. Vasseur, A. Hladky-Hennion, L.M. García-Raffi, J.V.S. Pérez, Level repulsion and evanescent waves in sonic crystals. Phys. Rev. B **84**, 2123021 (2011)
96. V. Romero-García, J. Vaseur, L.M. García-Raffi, A. Hladky-Hennion, Evanescent modes and level repulsion states in sonic crystal stubbed waveguides. New J. Phys. **1**, 14 (2012)
97. V. Romero-García, J.V.S. Pérez, L.M. García-Raffi, Evanescent waves and deaf bands in sonic crystals. AIP Adv. **1** (2011)
98. J. Chen, B. Bonello, Z. Hou, Plate-mode waves in phononic crystal thin slabs: mode conversion. Phys. Rev. E **78**, 036609 (2008)
99. M. Bavencoffe, *Propagation d'ondes de Lamb dans un guide d'ondes à interface périodique*. Ph.D. thesis, University of Le Havre (2009)
100. Y. Achaoui, A. Khelif, S. Benchabane, V. Laude, Polarization state and level repulsion in two-dimensional phononic crystals and waveguides in the presence of material anisotropy. J. Phys. D: Appl. Phys. **43**, 185401 (2010)
101. V. Romero-García, R. Picó, A. Cebrecos, V. Sánchez-Morcillo, K. Staliunas, Enhancement of sound in chirped sonic crystals. Appl. Phys. Lett. **102**, 091906 (2013)
102. E. Cassan, C.K.-V.D. Caer, D. Marris-Morini, L. Vivien, Short-wavelength light propagation in graded photonic crystals. J. Lightwave Tech. **29**, 1937–1943 (2011)

103. M. Kushwaha, B. Djafari-Rouhani, L-Dobrynski, J. Vasseur, Sonic stop-bands for cubic arrays of rigid cylinders in air. Eur. Phys. J. B **3**, 155–161 (1998)
104. I.E. Psarobas, M.M. Sigalas, Elastic band gaps in a FCC lattice of mercury spheres in aluminum. Phys. Rev. B **66**, x (2002)
105. L. Wu, L. Chen, An acoustic bending waveguide designed by graded sonic crystals. J. Appl. Phys. **110**, 114507 (2011)
106. Y. Shen, J. Fu, G. Yu, Rainbow trapping in one-dimensional chirped photonic crystals composed of alternating dielectric slabs. Phys. Lett. A **375**, 3801–3803 (2011)
107. M. Stockman, Nanofocusing of optical energy in tapered plasmonicwaveguides. Phys. Rev. Lett. **93**, 137404 (2004)
108. V.N. Smolyaninova, I.I. Smolyaninov, A.V. Kildishev, V.M. Shalaev, Experimental observation of the trapped rainbow. Appl. Phys. Lett. **96**, 211121 (2010)
109. A. Cebrecos, R. Picó, V.J. Sánchez-Morcillo, K. Staliunas, V. Romero-García, L.M. García-Raffi, Enhancement of sound by soft reflections in exponentially chirped crystals. AIP Adv. **4**(124402) (2014)
110. V. Sánchez-Morcillo, K. Staliunas, V. Espinosa, I. Pérez-Arjona, J. Redondo, E. Soliveres, Propagation of sound beams behind sonic crystals. Phys. Rev. B **80**, 134303 (2009)
111. P.T. Rakich, M.S. Dahlem, S. Tandon, M. Ibanescu, M. Soljai, G.S. Petrich, J.D. Joannopoulos, L.A. Kolodziesjski, E.P. Ippen, Achieving centimeter-scale super-collimation in ultra large area photonic crystals. Nat. Mater. **5**, 93 (2006)
112. Z. Lu, S. Shi, A. Murakowski, G.J. Schneider, C.A. Schuetz, D.W. Prather, Experimental demonstration of self-collimation inside a three-dimensional photonic crystal. Phys. Rev. Lett. **96**, 173902 (2006)
113. I. Pérez-Arjona, V.J. Sánchez-Morcillo, J. Redondo, V. Espinosa, K. Staliunas, Theoretical prediction of the nondiffractive propagation of sonic waves through periodic acoustic media. Phys. Rev. B **75**(014304), 014304 (2007)
114. V. Espinosa, V.J. Sánchez-Morcillo, K. Staliunas, I. Pérez-Arjona, J. Redondo, Subdiffractive propagation of ultrasound in sonic crystals. Phys. Rev. B **76**, 140302(R) (2007)
115. C. Luo, S.G. Johnson, J.D. Joannopoulos, J.B. Pendry, Subwavelength imaging in photonic crystals. Phys. Rev B **68**, 045115 (2003)
116. M. Ke, Z. Liu, C. Qiu, W. Wang, J. Shi, W. Wen, P. Sheng, Negative-refraction imaging with two-dimensional phononic crystals. Phys. Rev. B **72**, 064306 (2005)
117. L. Feng, X. Liu, Y. Chen, Z. Huang, Y. Mao, Y. Chen, J. Zi, Y. Zhu, Negative refraction of acoustic waves in two-dimensional sonic crystals. Phys. Rev. B **72**, 033108 (2005)
118. R. Martńez-Sala, C. Rubio, L. García-Raffi, J. Sánchez-Pérez, E. Sánchez-Pérez, J. Llinares, Control of noise by trees arranged like sonic crystals. Jour. Sound Vib. **291**, 100 (2006)
119. S. Castiñeira-Ibáñez, V. Romero-García, J.V. Sánchez-Pérez, L.M. García-Raffi, Overlapping of acoustic bandgaps using fractal geometries. EPL **92**, 24007 (2010)
120. S. Castiñeira-Ibañez, C. Rubio, J. Sánchez-Pérez, Environmental noise control during its transmission phase to protect buildings. design model for acoustic barriers based on arrays of isolated scatterers. Build. Environ. **93**, 179 (2015)
121. A. Lardeau, J. Groby, V. Romero-García, Broadband transmission loss using the overlap of resonances in 3D sonic crystals. Crystals **6**(5), 51 (2015)
122. Y. Lai, X. Zhang, Z.-Q. Zhang, Large sonic band gaps in 12-fold quasicrystals. J. Appl. Phys. **91**(9), 6191–6193 (2002)
123. V. Romero-García, J.V. Sánchez-Pérez, L.M. García-Raffi, J. Herrero, S. García-Nieto, X. Blasco, Hole distribution in phononic crystals: design and optimization. J. Acoust. Soc. Am. **125**(6), 3774–3783 (2009)
124. J.M. Herrero, S. García-Nieto, X. Blasco, V. Romero-García, J.V. Sánchez-Pérez, L. García-Raffi, Optimization of sonic crystal attenuation properties by ev-moga multiobjective evolutionary algorithm. Struct. Multidisc. Optim. **39**, 203–215 203–215 203 (2009)
125. J.B. Pendry, A.J. Holden, W.J. Stewart, I. Youngs, Magnetism from conductors, and enhanced non-linear phenomena. IEEE Trans. Microw. Theory Technol. **47**, 2975 (1999)
126. A. Movchan, Guenneau, Split-ring resonators and localized modes. Phys. Rev. B **70**, 125116 (2004)

Chapter 4
The Transfer Matrix Method in Acoustics

Modelling One-Dimensional Acoustic Systems, Phononic Crystals and Acoustic Metamaterials

Noé Jiménez, Jean-Philippe Groby, and Vicent Romero-García

Abstract The transfer matrix method is a simple but powerful analytical tool used to model acoustic wave propagation in a wide range of one-dimensional problems. In this chapter, we present the method and summarize the most common building blocks encountered in one-dimensional acoustic systems. These include layers of fluids and porous media, ducts and waveguides of different geometries where thermoviscous losses can be accounted for, locally reacting elements such as Helmholtz or quarter-wavelength resonators, viscoelastic plates and membranes, micro-perforated panels or vibrating walls. Several examples are provided, including a multi-layered porous structure for room acoustics, the transmission problem of a double-leaf wall for building acoustics, and the analysis of the dispersion relations of acoustic waves in periodic media and metamaterials using locally resonant elements. Various one-dimensional wave-motion phenomena can be studied using the generalized framework provided by the transfer matrix method such as reflection, transmission, absorption, attenuation and dispersion, as illustrated in the examples.

4.1 Introduction

Many acoustic problems of practical interest can be reduced to one-dimensional ones under the hypothesis of plane-wave propagation. Analytical solutions can thus

N. Jiménez (✉)
Instituto de Instrumentación para Imagen Molecular, Universitat Politècnica de València, CSIC, Camino de Vera S/N, València, Spain
e-mail: nojigon@upv.es

J.-P. Groby · V. Romero-García
Laboratoire d'Acoustique de l'Université du Mans, UMR CNRS 6613, Le Mans Université, Avenue Olivier Messiaen, 72085 Le Mans, France
e-mail: jean-philippe.groby@univ-lemans.fr

V. Romero-García
e-mail: vicente.romero@univ-lemans.fr

N. Jiménez et al. (eds.), *Acoustic Waves in Periodic Structures, Metamaterials, and Porous Media*, Topics in Applied Physics 143,
https://doi.org/10.1007/978-3-030-84300-7_4

be obtained under this hypothesis even for complex problems. Among the different analytical techniques such as modal expansions [1], plane wave expansion methods (PWE) (See Chap. 1) or multiple scattering theory (MST) (See Chap. 2), the transfer matrix method (TMM) [2] has been widely applied to study wave propagation in one-dimensional systems such as porous materials [3, 4], duct acoustics and mufflers with [5] and without flow [6], stratified solids [7] and multilayer elastic and acoustic materials for noise control [8–13], fluid mechanics [14], piezoelectric transducers [15] or acoustic holograms [16], among others. TMM is also at the heart of standard procedures to measure the properties and performance of acoustic materials [17]. The method is also widely applied in electromagnetics [18], e.g., to study multilayer optical structures [19, 20]. It can be applied in Cartesian and in other coordinate system, such as cylindrical coordinates for radially-symmetric multilayer structures [21].

The transfer matrix method has recently been used to described wave propagation in periodic structures and phononic crystals as well as to study acoustic metamaterials. Wave dispersion and acoustic properties such as reflection and transmission in multilayer phononic crystals can effectively be described by a transfer matrix approach [22]. The method can be used to derive the effective parameters of resonant structures [23] and even of hyperbolic metamaterials [24]. Thermoviscous and viscoelastic losses can also be easily included. In this way, this method has been applied to design and analyse efficient or perfect metamaterial absorbers based on quarter-wavelength resonators [25], Helmholtz resonators [26–28] or membranes and plates [29, 30].

The transfer matrix method results in fast calculations to describe complex acoustic structures, in fact, it is one of its most important advantages. It can thus be easily combined with optimization techniques that would be prohibitive using other simulation methods such as finite-difference in time-domain (FDTD) or Finite Element Methods (FEM) due to extreme computational resources and simulation times they would require.

In this chapter, we present the method and summarize the most common building blocks. We will describe the modelling of layers of fluids and porous media, ducts of different geometries possibly accounting for thermoviscous losses, locally resonant elements such as Helmholtz and quarter-wavelength resonators, viscoelastic membranes and plates, micro-perforated plates and vibrating walls. The last section, several examples will be given comprising a multilayer porous structure for room acoustics, a double-leaf wall for building acoustics, the dispersion of acoustic waves in periodic multilayer media, and the design of acoustic metamaterials using different kinds of resonators.

4.2 The Transfer Matrix Method

We start deriving the basic relations between the acoustic magnitudes evaluated at the boundaries of a layer of homogeneous acoustic material, as shown in Fig. 4.1.

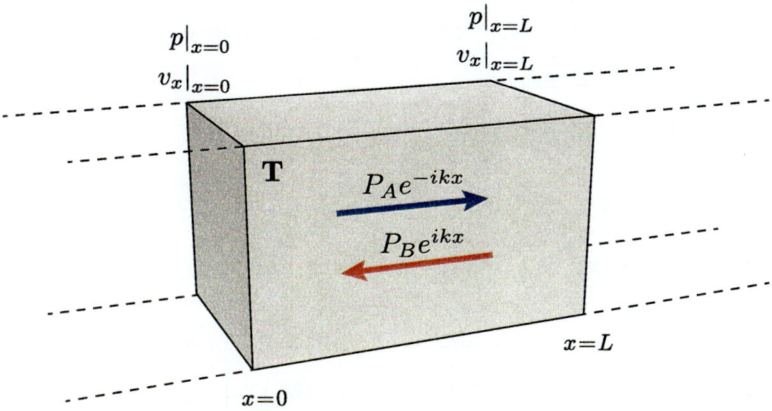

Fig. 4.1 Scheme of the acoustic material layer characterized by a transfer matrix **T**. Propagation inside the layer is modeled by counter propagating waves

Assuming that only longitudinal plane waves propagate in the layer and a temporal harmonic dependence of the type $e^{i\omega t}$, the total field inside the material is written as the superposition of two waves propagating in opposite directions as

$$p(x) = P_A e^{-ikx} + P_B e^{ikx}, \tag{4.1}$$

$$v_x(x) = \frac{P_A}{Z} e^{-ikx} - \frac{P_B}{Z} e^{ikx}, \tag{4.2}$$

where $Z = \rho c$ is the characteristic acoustic impedance, $k = \omega/c$ is the wavenumber at the angular frequency $\omega = 2\pi f$, with ρ the density and c the sound speed of the material, and the amplitudes of the two waves are given by P_A and P_B.

To evaluate these amplitudes we define the pressure and velocity at both sides of the slab. First, at $x = 0$ we obtain

$$p(x)\big|_{x=0} = P_A + P_B, \tag{4.3}$$

$$Z v_x(x)\big|_{x=0} = P_A - P_B, \tag{4.4}$$

while at $x = L$ we get

$$p(x)\big|_{x=L} = (P_A + P_B)\cos(kL) - i(P_A - P_B)\sin(kL), \tag{4.5}$$

$$v_x(x)\big|_{x=L} = \frac{P_A - P_B}{Z}\cos(kL) - i\frac{P_A - P_B}{Z}\sin(kL). \tag{4.6}$$

Then, we can relate the acoustic magnitudes at both boundaries by combining (4.3)–(4.4) with (4.5)–(4.6) via

$$p(x)\big|_{x=L} = \cos(kL)p(x)\big|_{x=0} - iZ\sin(kL)v_x(x)\big|_{x=0}, \tag{4.7}$$

$$v_x(x)\big|_{x=L} = \cos(kL)v_x(x)\big|_{x=0} - i\frac{1}{Z}\sin(kL)p(x)\big|_{x=0}. \tag{4.8}$$

Equations (4.7)–(4.8) can be expressed in a matrix form as

$$\begin{bmatrix} p \\ v_x \end{bmatrix}_{x=L} = \begin{bmatrix} \cos(kL) & -iZ\sin(kL) \\ \cos(kL) & -i\frac{1}{Z}\sin(kL) \end{bmatrix} \begin{bmatrix} p \\ v_x \end{bmatrix}_{x=0}. \tag{4.9}$$

After inversion, we retrieve the basic transfer matrix formulation of a layer of acoustic material, given by

$$\begin{bmatrix} p \\ v_x \end{bmatrix}_{x=0} = \begin{bmatrix} \cos(kL) & iZ\sin(kL) \\ i\frac{1}{Z}\sin(kL) & \cos(kL) \end{bmatrix} \begin{bmatrix} p \\ v_x \end{bmatrix}_{x=L}. \tag{4.10}$$

In this way, the acoustic magnitudes at both sides of the 1D fluid layer are related by a 2×2 matrix which only depends on the impedance and wavenumber in the material. It is interesting to note that additional elements can be introduced into the system in a simple and modular way. This allows to model complex materials and structures using a simple theoretical framework, as we will see below.

4.2.1 Total Transfer Matrix

For a given material, we can define a total transfer matrix, $\mathbf{T}$, that relates the sound pressure, p, and normal acoustic particle velocity, v_x, at the beginning, $x = 0$, and at the end of a structure, $x = L$ as

$$\begin{bmatrix} p \\ v_x \end{bmatrix}_{x=0} = \mathbf{T} \begin{bmatrix} p \\ v_x \end{bmatrix}_{x=L}, \tag{4.11}$$

therefore, the total transfer matrix $\mathbf{T}$ is a 2×2 matrix as

$$\begin{bmatrix} p \\ v_x \end{bmatrix}_{x=0} = \begin{bmatrix} T_{11} & T_{12} \\ T_{21} & T_{22} \end{bmatrix} \begin{bmatrix} p \\ v_x \end{bmatrix}_{x=L}. \tag{4.12}$$

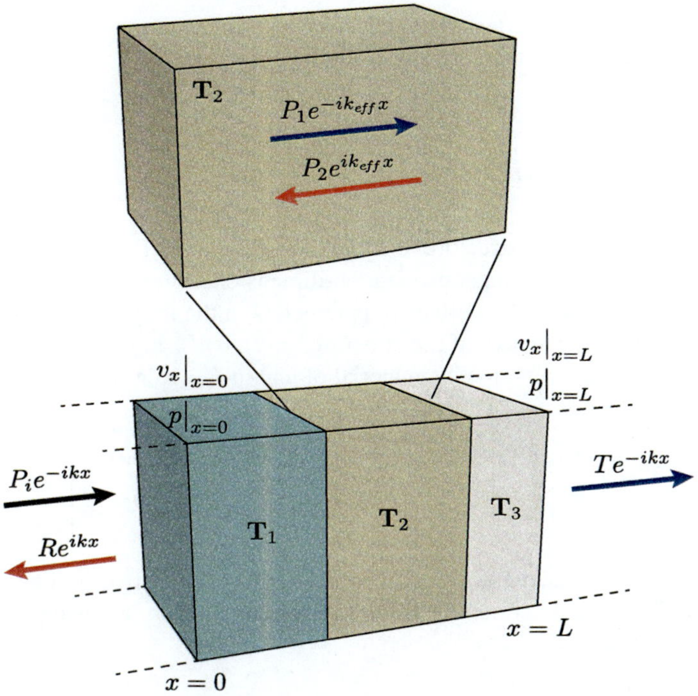

Fig. 4.2 Scheme of an acoustic structure composed by several layers of effective properties. The system is characterized by a total transfer matrix, $\mathbf{T}$, equal to the matrix product of the elements, i.e., $\mathbf{T} = \mathbf{T}_1\mathbf{T}_2\mathbf{T}_3$ in this example

When dealing with a complex structure, e.g., an arrangement of several elements as depicted in Fig. 4.2, the total transfer matrix $\mathbf{T}$ is given by the product of the transfer matrices of the N layers or elements of the system as

$$\mathbf{T} = \prod_{n=1}^{N} \mathbf{T}_n \,. \tag{4.13}$$

The continuity of pressure and normal particle velocity at each interface of the system are intrinsically satisfied. The transfer matrix of each layer or element, $\mathbf{T}_n$, is calculated according to its nature. Several basic examples will be given in Sect. 4.3 including thermoviscous fluids, fluid-saturated rigid-frame porous layers, or locally resonant elements such as elastic membranes or Helmholtz resonators.

The total transfer matrix $\mathbf{T}$ offers abundant information about the system such as:

1. *The effective parameters:* The effective impedance and wavenumber, in addition to the effective density and bulk modulus of the system in the long wavelength regime.

2. *The scattering of the system:* The reflection and transmission coefficients and, if looses are considered, the absorption coefficient.

4.2.2 Effective Parameters

In the case of a symmetric structure, the total transfer matrix **T** is symmetric and the system can be modelled as an equivalent on-dimensional fluid-like layer with complex and frequency-dependent effective parameters in the long wavelength regime. Thus, the total transfer matrix of the structure, given by (4.11), can be identified to the propagation matrix an effective material of length L, given by (4.10), as

$$\mathbf{T} = \begin{bmatrix} T_{11} & T_{12} \\ T_{21} & T_{22} \end{bmatrix} = \begin{bmatrix} \cos(k_{\text{eff}} L) & i Z_{\text{eff}} \sin(k_{\text{eff}} L) \\ i \dfrac{1}{Z_{\text{eff}}} \sin(k_{\text{eff}} L) & \cos(k_{\text{eff}} L) \end{bmatrix}, \tag{4.14}$$

where $k_{\text{eff}}(\omega)$ is the effective wavenumber and $Z_{\text{eff}}(\omega)$ is the effective characteristic impedance of the whole structure. Both effective parameters are usually complex and frequency dependent.

4.2.2.1 Effective Wavenumber

By relating the elements of (4.12) with those in (4.14) we can obtain the expression for the effective wavenumber as a function of the coefficients of the total transfer matrix

$$k_{\text{eff}} = \frac{1}{L} \cos^{-1}\left(\frac{T_{11} + T_{22}}{2}\right) + \frac{n\pi}{L}, \quad n \in \mathbb{Z}. \tag{4.15}$$

It is important to note that due to the trigonometric inversion the wavenumber is warped around $-\pi < k_{\text{eff}} L < \pi$. This is very useful for periodic structures as we will see in the examples in Sect. 4.4: obtaining the wavenumber $k_{\text{eff}}(\omega)$ of a unit cell provides the dispersion relation in the irreducible Brillouin zone. It is effectively important to note that $k_{\text{eff}}(\omega)$ matches the solution of the eigenvalue problem solved for recovering the dispersion relation of any symmetric system and thus is valid whatever the frequency range considered.

4.2.2.2 Effective Characteristic Impedance

In the same way, the characteristic acoustic impedance is identified from the total transfer matrix coefficients as

$$Z_{\text{eff}} = \sqrt{\frac{T_{12}}{T_{21}}}. \tag{4.16}$$

In this case, no trigonometric inversion is used and the retrieved impedance values are measured in [Rayls] units. Nevertheless, this identification is only valid in the long wavelength regime contrary to the recovery of $k_{\text{eff}}(\omega)$ and special attention should thus be paid on the frequency validity regime.

4.2.2.3 Effective Density and Bulk Modulus

The dynamic mass-density, ρ_{eff}, and bulk modulus, K_{eff}, of the slab of effective material can be obtained using

$$K_{\text{eff}} = Z_{\text{eff}} \frac{\omega}{k_{\text{eff}}}, \quad \text{and} \quad \rho_{\text{eff}} = Z_{\text{eff}} \frac{k_{\text{eff}}}{\omega}, \tag{4.17}$$

where Z_{eff} is given by (4.16) and k_{eff} by (4.15) with proper unwrap of the wavenumber.

4.2.3 The Scattering Matrix

The scattering matrix, **S**, relates the amplitudes of the incoming waves to those of the outgoing waves. The total pressure at both sides of the structure, at $x = 0$ and $x = L$, is given by

$$p(x) = \begin{cases} Ae^{-ikx} + Be^{ikx} & \text{for} \quad x < 0, \\ Ce^{-ikx} + De^{ikx} & \text{for} \quad x > L, \end{cases} \tag{4.18}$$

as shown in Fig. 4.3 for a structure of length L. Thus, the relation between the amplitudes of both waves is given by the **S**-matrix as

$$\begin{bmatrix} C \\ B \end{bmatrix} = \mathbf{S} \begin{bmatrix} A \\ D \end{bmatrix} = \begin{bmatrix} T^- & R^+ \\ R^- & T^+ \end{bmatrix} \begin{bmatrix} A \\ D \end{bmatrix}, \tag{4.19}$$

where the elements of the **S**-matrix give directly the transmission (T^- and T^+) and reflection (R^- and R^+) coefficients for a system excited from each side of the structure, i.e., the superscripts $(+, -)$ denote the direction of incidence: the positive

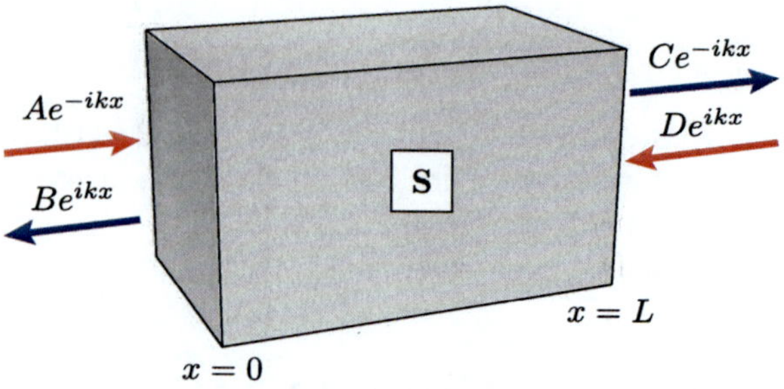

Fig. 4.3 Incoming and outgoing waves that define the scattering of the structure. The scattering matrix, **S**-matrix, completely describe the transmitted and reflected waves

and negative x-axis respectively. The **S**-matrix is widely used in wave physics to characterize and interpret the wave scattering. In the case of acoustics, the scattering matrix completely describe the transmitted and reflected waves, and when looses are included, the absorption of the system.

4.2.4 Reflection, Transmission and Absorption Coefficients

The system is usually geometrically bounded and therefore its scattering properties can be evaluated. Depending on the boundary conditions at $x = 0$ and $x = L$, we can distinguish between:

(i) *Reflection problem:* the material is located against an impervious wall. The acoustic impedance is that of the surrounding medium $Z_0 = \rho_0 c_0$ at the beginning ($x = 0$), while the rigid boundary condition is applied, i.e., $v_x = 0$, at the end of the material ($x = L$),
(ii) *Transmission problem:* the material is surrounded by a fluid on both sides. Then, the acoustic impedance is that of the surrounding media $Z_0 = \rho_0 c_0$ at $x = 0$ and $Z_L = \rho_L c_L$ at $x = L$.

For example, a layer of porous material obviously presents different reflection (and absorption) coefficient if it is located against a rigid-impervious wall (reflection problem) or if it is surrounded by air on both sides (transmission problem).

4.2.4.1 Transmission Problem

The transmission problem implies the waves impinging the structure on one side can propagate through it and be transmitted to the other side. First, (4.11) are written as

$$p(x)\big|_{x=0} = T_{11}\, p(x)\big|_{x=L} + T_{12}\, v_x(x)\big|_{x=L}\,, \tag{4.20}$$

$$v_x(x)\big|_{x=0} = T_{21}\, p(x)\big|_{x=L} + T_{22}\, v_x(x)\big|_{x=L}\,. \tag{4.21}$$

Then, assuming an incident pressure wave whose amplitude is the unity, the pressure and velocity at both sides of the structure can be defined as a function of the reflection coefficients R^+ and R^-, and the corresponding transmission coefficients, T^+ and T^-, as

$$p(x)\big|_{x=0} = 1 + R^-, \qquad p(x)\big|_{x=L} = T^- e^{ikL}, \tag{4.22}$$

$$v_x(x)\big|_{x=0} = \frac{1 - R^-}{Z_0}, \qquad v_x(x)\big|_{x=L} = \frac{T^- e^{ikL}}{Z_L}, \tag{4.23}$$

for an incident plane wave coming from $-\infty$ and propagating in the $+x$ direction, and

$$p(x)\big|_{x=0} = T^+ e^{ikL}, \qquad p(x)\big|_{x=L} = R^+ + 1, \tag{4.24}$$

$$v_x(x)\big|_{x=0} = -\frac{T^+ e^{ikL}}{Z_0}, \qquad v_x(x)\big|_{x=L} = \frac{R^+ - 1}{Z_L}, \tag{4.25}$$

for an incident plane wave coming from $+\infty$ and propagating in the $-x$ direction, where $Z_0 = \rho_0 c_0$ and $Z_L = \rho_L c_L$ are the impedances of the media on either side of the structure, i.e., $x < 0$ and $x > L$, respectively.

Combining (4.22)–(4.25) with (4.20)–(4.21), we obtain the following relations:

$$T^- = \frac{1 + R^-}{T_{11} + T_{12}/Z_L}, \tag{4.26}$$

$$T^- = \frac{1 - R^-}{T_{21} Z_0 + T_{22} Z_0 / Z_L}, \tag{4.27}$$

$$T^+ = T_{11}\left(1 + R^+\right) + \frac{T_{12}}{Z_L}\left(R^+ - 1\right), \tag{4.28}$$

$$T^+ = -T_{21} Z_0 \left(1 + R^+\right) - T_{22} \frac{Z_0}{Z_L}\left(R^+ - 1\right). \tag{4.29}$$

In the following, we will obtain the relation between the elements of the **S**-matrix and those of the **T**-matrix in specific configurations:

(i) *Non-reciprocal systems*
Structures in which the condition $T^- \neq T^+$ is fulfilled.

(ii) *Reciprocal systems*
Structures in which the condition $T^- = T^+$ is fulfilled.

(iii) *Reciprocal and symmetric systems*
Structures in which the conditions $T^+ = T^-$ and $R^- = R^+$ are fulfilled.

4.2.4.2 Non-reciprocal Systems

In a general case, we can combine (4.26)–(4.29) and get

$$T^- = \frac{2e^{ikL}}{T_{11} + T_{12}/Z_L + T_{21}Z_0 + T_{22}Z_0/Z_L}, \tag{4.30}$$

$$R^- = \frac{T_{11} + T_{12}/Z_L - T_{21}Z_0 - T_{22}Z_0/Z_L}{T_{11} + T_{12}/Z_L + T_{21}Z_0 + T_{22}Z_0/Z_L}, \tag{4.31}$$

$$T^+ = \frac{Z_0}{Z_L}\frac{2e^{ikL}\left(T_{11}T_{22} - T_{12}T_{21}\right)}{T_{11} + T_{12}/Z_L + T_{21}Z_0 + T_{22}Z_0/Z_L}, \tag{4.32}$$

$$R^+ = \frac{-T_{11} + T_{12}/Z_L - T_{21}Z_0 + T_{22}Z_0/Z_L}{T_{11} + T_{12}/Z_L + T_{21}Z_0 + T_{22}Z_0/Z_L}. \tag{4.33}$$

Equations (4.30)–(4.33) give the relation between the **T**-matrix and the **S**-matrix in the general form. However, some simplifications can be done.

4.2.4.3 Reciprocal Systems

The reciprocal behaviour of the system implies that the determinant of transfer matrix is the unity, i.e., $T_{11}T_{22} - T_{12}T_{21} = 1$. This property is satisfied by linear and time-invariant systems where the transmission does not depend on the direction of incident wave. These conditions are satisfied by most of the acoustic materials. Note this term appears in (4.32) and directly implies the transmission coefficients are identical whatever the direction of excitation. Therefore, in reciprocal systems

$$T^- = T^+ = T. \tag{4.34}$$

All the systems considered in this chapter are reciprocal. Note that in this case, the **S**-matrix, (4.19), possesses two eigenvalues given by

$$\lambda_{1,2} = T \pm \sqrt{R^+R^-}, \tag{4.35}$$

while the eigenvectors corresponding to λ_1 and λ_2 are

$$\vec{v}_1 = \left[\sqrt{R^+R^-}, R^+\right], \quad \vec{v}_2 = \left[R^-, -\sqrt{R^+R^-}\right], \tag{4.36}$$

respectively. The poles and zeros of the eigenvalues as well as the eigenvectors of the **S**-matrix in the complex-frequency plane provide rich information, as they are identified with the resonances of the system. A further description of the eigenvalues of the scattering matrix and its implications for perfect absorption are given in Chap. 5.

In addition, in most cases in acoustics the structure is surrounded by the same media $Z_0 = Z_L$, as occurs in metamaterials surrounded by air. Under this additional condition, the reflection and transmission coefficients given by (4.30)–(4.33) can be simplified to

$$T = \frac{2e^{ikL}}{T_{11} + T_{12}/Z_0 + T_{21}Z_0 + T_{22}}, \tag{4.37}$$

$$R^- = \frac{T_{11} + T_{12}/Z_0 - T_{21}Z_0 - T_{22}}{T_{11} + T_{12}/Z_0 + T_{21}Z_0 + T_{22}}, \tag{4.38}$$

$$R^+ = \frac{-T_{11} + T_{12}/Z_0 - T_{21}Z_0 + T_{22}}{T_{11} + T_{12}/Z_0 + T_{21}Z_0 + T_{22}}. \tag{4.39}$$

When the materials that constitute the acoustic structure present intrinsic losses, e.g., thermoviscous or viscoelastic ones, a portion of the energy is neither reflected nor transmitted: it is absorbed by the structure and irreversibly transformed into heat. The amount of absorbed energy, dissipated by intrinsic losses, with respect to the total energy is characterized by the absorption coefficient, $\alpha(\omega)$. For asymmetric systems $R^+ \neq R^-$ and, therefore, the absorption depends on the direction of propagation. For the positive x-axis incident wave, the absorption is given by

$$\alpha^- = 1 - \left|R^-\right|^2 - |T|^2, \tag{4.40}$$

while for the negative x-axis incident waves,

$$\alpha^+ = 1 - \left|R^+\right|^2 - |T|^2. \tag{4.41}$$

4.2.4.4 Symmetric Systems

For symmetric systems such as structures presenting a mirror symmetry with respect to $x = L/2$, the transfer matrix coefficients fulfil

$$T_{11} = T_{22}. \tag{4.42}$$

As a consequence, the reflection coefficients from both sides are the same, i.e., $R^+ = R^- = R$. The absorption coefficients from both sides of the system are thus identical and correspond to

$$\alpha^+ = \alpha^- = \alpha = 1 - |R|^2 - |T|^2. \tag{4.43}$$

4.2.4.5 Reflection Problems

When the system is rigidly backed, a rigid boundary condition $v_x|_{x=L} = 0$ is applied in (4.11) and, obviously, no waves are transmitted through the system, i.e. $T = 0$ in (4.19). Therefore, we obtain the following relations

$$p(x)\big|_{x=0} = T_{11}\, p(x)\big|_{x=L}, \tag{4.44}$$

$$v_x(x)\big|_{x=0} = T_{21}\, p(x)\big|_{x=L}. \tag{4.45}$$

The reflection coefficient is related to the pressure and velocity at $x = 0$ as

$$p(x)\big|_{x=0} = 1 + R^-, \tag{4.46}$$

$$v_x(x)\big|_{x=0} = \frac{1 - R^-}{Z_0}. \tag{4.47}$$

Combining (4.44)–(4.45) with (4.46)–(4.47), the reflection coefficient for a rigidly-backed system becomes:

$$R = \frac{T_{11} - T_{21} Z_0}{T_{11} + T_{21} Z_0}. \tag{4.48}$$

The absorption coefficient of the rigidly-backed system is

$$\alpha = 1 - |R|^2. \tag{4.49}$$

Finally, note that, if a rigid boundary condition is set at the beginning of the structure, $x = 0$, as $v_x|_{x=0} = 0$, and a wave impinges the structure from the opposite direction, i.e., a wave travelling in the $-x$ direction, the reflection coefficient is given by $R = (T_{22} - T_{21} Z_L)/(T_{22} + T_{21} Z_L)$.

4.2.4.6 Specific Acoustic Impedance

Note that the *characteristic* acoustic impedance, $Z_{\text{eff}} = \rho_{\text{eff}} c_{\text{eff}}$, is generally different from the *specific* acoustic impedance, $Z = p/v_x$. The characteristic impedance is a property of the material itself and does not depend on the boundary conditions at $x = 0$ and $x = L$. In the opposite, the effects of the boundary conditions are implicit for the acoustic impedance.[1] One important magnitude is the *specific* acoustic impedance of the system at the input of the system, or normal acoustic specific impedance, given by,

[1] Note that $Z = Z_{\text{eff}}$, i.e., $p/v_x = \rho_{\text{eff}} c_{\text{eff}}$, only stands for plane waves travelling in an infinite medium.

$$Z_{\text{in}} = \left. \frac{p}{v_x} \right|_{x=0} . \tag{4.50}$$

For transmission problems, it can be calculated directly from the reflection and transmission coefficients as

$$Z_{\text{in}} = Z_0 \left(\frac{1+R}{1-R} \right) \left(\frac{1+T}{1-T} \right) . \tag{4.51}$$

For rigidly-backed problems this expression reduces to

$$Z_{\text{in}} = Z_0 \left(\frac{1+R}{1-R} \right) . \tag{4.52}$$

This quantity is very useful to analyse the impedance matching of a given structure with the surrounding media.

4.3 Review of the Usual Transfer Matrices

A given system can be subdivided in N elements of respective transfer matrices $\mathbf{T}_n$, $n = 1, ..., N$ to evaluate its full transfer matrix $\mathbf{T}$. Depending on the nature of each element, its individual transfer matrix is calculated in different way. In the following, we review the most usual transfer matrices used to solve 1D problems in acoustics.

4.3.1 Particle Velocity Verses Flow Formulation

First of all, we shall differentiate the problem of wave propagation through layers of fluids or fluid-like materials of undefined section, with the problem of wave propagation through fluids confined in ducts of finite cross-sectional area, as shown in Fig. 4.4.

4.3.1.1 Particle Velocity Formulation

For layers of undefined section, Fig. 4.4a, the problem can be described by using the *particle-velocity formulation*, as given in the previous section. For example, this is the case when considering multilayer porous material absorbers.

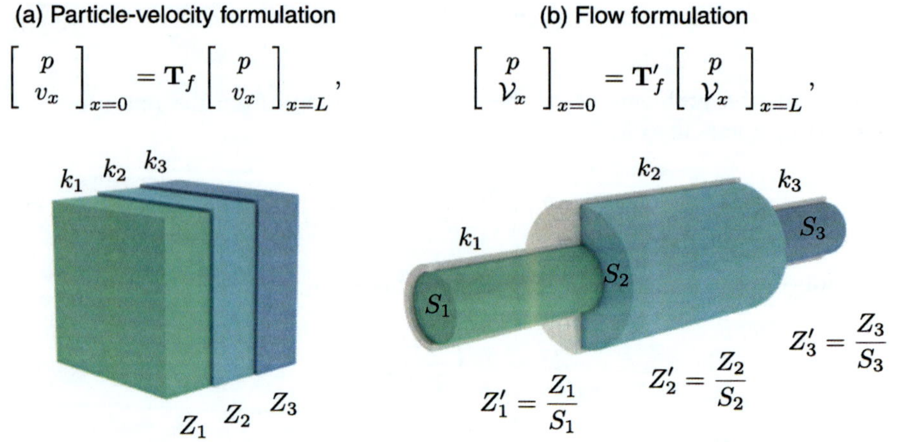

Fig. 4.4 **a** Formulation of the problem for layers of fluid-like materials of undefined section, i.e., particle-velocity formulation. **b** Formulation of the problem for waveguides of different cross-sectional areas, i.e., flow formulation

4.3.1.2 Flow Formulation

However, when considering waveguides of different cross-section, Fig. 4.4b, as it is usual when describing mufflers, metamaterials or similar structures, it is convenient to formulate the problem using the *flow*.

4.3.2 Fluid Layers: Particle Velocity Formulation

The transmission matrix $\mathbf{T}_f$ of a fluid layer of length L takes the form

$$\mathbf{T}_f = \begin{bmatrix} \cos(k_f L) & i Z_f \sin(k_f L) \\ i \dfrac{1}{Z_f} \sin(k_f L) & \cos(k_f L) \end{bmatrix}, \tag{4.53}$$

where $k_f = \omega/\sqrt{K_f/\rho_f}$ and $Z_f = \sqrt{K_f \rho_f}$ are wavenumber and the characteristic impedance in the fluid, respectively, where K_f and ρ_f are the effective bulk modulus and mass density of the fluid. Note that, in general, when losses are accounted for, K_f and ρ_f, and thus k_f and Z_f are complex and frequency dependent variables.

The intrinsic losses can be neglected in the case of sound confined in a cavity, e.g., an air cavity between two walls, if the size of the latter is much larger than the thicknesses of the viscous and thermal boundary layers of the fluid. The wavenumber in the fluid thus reduces to $k_f = \omega/c_0$, and the acoustic impedance becomes $Z_f =$

$\rho_0 c_0$, where the sound speed is $c_0 = \sqrt{K_0/\rho_0}$, with ρ_0 and K_0 respectively the density and bulk modulus of the lossless fluid.

However, the intrinsic losses of the fluid should be accounted for in highly viscous fluids, when sound wave is propagating over a long path, i.e., when the attenuation along the propagation path L is not negligible, and at high frequencies.

4.3.2.1 Thermoviscous Fluid Layers

Before considering confined sound propagation, note that, even in the absence of boundaries, the propagation of acoustic waves in fluid with a high viscosity is characterized by a complex wavenumber and impedance, which take the form [31]

$$k_f(\omega) = \frac{\omega}{c_0} + i\frac{b\omega^2}{2\rho_0 c_0^3}, \quad Z_f(\omega) = \rho_0 \frac{\omega}{k_f}, \tag{4.54}$$

where $b = \eta + (4/3)\mu + \kappa\left(C_V^{-1} - C_P^{-1}\right)$ is the diffusion coefficient, η and μ are the shear and bulk viscosities, κ the thermal conductivity of the thermoviscous fluid and C_V and C_P are the heat capacity at constant volume and pressure, respectively. Note the losses in the thermoviscous fluid show a quadratic dependence on frequency. Thus, the imaginary part of the complex wavenumber can be neglected for relatively low frequencies.

4.3.3 Ducts: Flow Formulation

When the fluid is confined in ducts or cavities, i.e., waveguides of different cross-sectional area, reflections are produced at the discontinuities. In this case, it is convenient to formulate the transfer matrix considering the flux,

$$\mathcal{V}_x = S v_x, \tag{4.55}$$

across a cross-sectional area S instead of using the particle velocity, v_x. The transfer matrix for the flow formulation, $\mathbf{T}'_f$, relates the pressure and flow at the inlet and at the outlet of the waveguide of length L as

$$\begin{bmatrix} p \\ \mathcal{V}_x \end{bmatrix}_{x=0} = \mathbf{T}'_f \begin{bmatrix} p \\ \mathcal{V}_x \end{bmatrix}_{x=L}. \tag{4.56}$$

By substituting (4.55) in (4.7)–(4.8) we can obtain the transfer matrix of a waveguide or duct for the flow formulation as

$$\mathbf{T}'_f = \begin{bmatrix} \cos(k_f L) & i Z'_f \sin(k_f L) \\ i \dfrac{1}{Z'_f} \sin(k_f L) & \cos(k_f L) \end{bmatrix}, \tag{4.57}$$

where $Z'_f = Z_f/S$ is the normalized impedance. Note this matrix is equivalent to the one given by (4.53): the impedances appearing in the transfer matrix are divided by S, while the wavenumbers are not modified. In addition, transfer matrices of varying cross-section waveguides, e.g., conical, are also available in the literature [32–34].

On the one hand, to calculate the effective wavenumber for the flow problem (4.15) holds. However, to calculate the effective impedance for a flow formulation, instead of (4.16), the following equation must be used

$$Z_{\text{eff}} = S\sqrt{\frac{T_{12}}{T_{21}}}. \tag{4.58}$$

On the other hand, the transmission and reflection coefficients for the flow formulation can be obtained in a straightforward manner: one might use normalized impedances instead of impedances. For example, the scattering for the flow formulation corresponding to the non-reciprocal case can be obtained by changing Z by Z/S in (4.30)–(4.33).

In addition, if the fluid is confined in a narrow duct or cavity, strong losses are observed when the transversal dimension of the cavity is of the same order of the thermal and/or viscous boundary layers. The thermal and viscous layers are given by

$$\delta_{\text{thermal}} = \sqrt{\frac{2\kappa}{\omega C_P}}, \quad \delta_{\text{viscous}} = \sqrt{\frac{2\eta}{\omega \rho_0}}, \tag{4.59}$$

therefore, $\delta_{\text{viscous}} = \sqrt{\text{Pr}}\,\delta_{\text{thermal}}$, where $\text{Pr} = C_P\eta/\kappa$ is the Prandtl number. One approach to include the thermoviscous loses in the TMM is to model the fluid inside the duct as an equivalent fluid with effective parameters, therefore $\rho_f(\omega),\ K_f(\omega) \in \mathbb{C}$, and they depend on the geometry and dimensions of the duct.

4.3.3.1 Slits

When considering the propagation of acoustic waves in narrow slits, as shown in Fig. 4.5, the thermoviscous losses should be accounted for if the length of the thermal and/or viscous boundary layers are of the same order of the slit height. Instead of solving the full Navier-Stokes equations with non-slip boundary conditions, the thermoviscous losses can be modelled by using effective complex and frequency dependent parameters accounting for both attenuation and dispersion.

Thus, assuming that only plane waves propagate inside a slit, the complex effective parameters expressed as [35]:

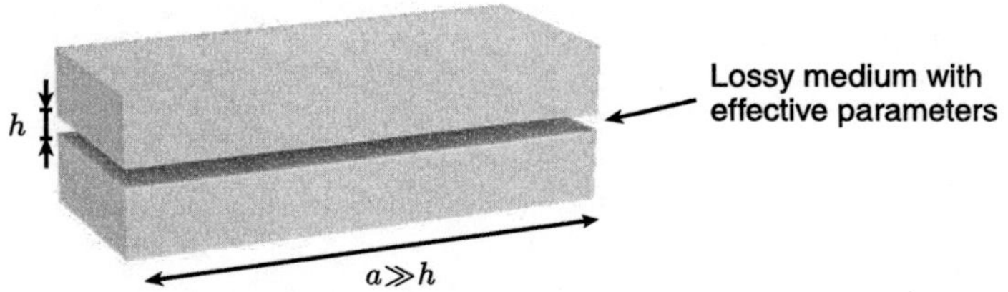

Fig. 4.5 Slit of height h. The system is modelled as a fluid-like medium with effective properties: complex and frequency dependent effective parameters

$$\rho_f(\omega) = \rho_0 \left[1 - \frac{\tanh \frac{h}{2} G_\rho(\omega)}{\frac{h}{2} G_\rho(\omega)} \right]^{-1}, \tag{4.60}$$

$$K_f(\omega) = K_0 \left[1 + (\gamma - 1) \frac{\tanh \frac{h}{2} G_K(\omega)}{\frac{h}{2} G_K(\omega)} \right]^{-1}, \tag{4.61}$$

where h is the width of the slit.

The functions $G_\rho(\omega)$ and $G_K(\omega)$ are

$$G_\rho(\omega) = \sqrt{\frac{i\omega\rho_0}{\eta}}, \tag{4.62}$$

$$G_K(\omega) = \sqrt{\frac{i\omega\mathrm{Pr}\rho_0}{\eta}}, \tag{4.63}$$

and $\gamma = C_P/C_V$ is the ratio of specific heats of the fluid, $K_0 = \gamma P_0$ is the adiabatic bulk modulus with P_0 the static pressure, η the dynamic viscosity and ρ_0 the density. The normalized acoustic impedance, for a 2D problem, is given by $Z'_f = \sqrt{K_f \rho_f}/ah$, where a is the width of the slit that must fulfil $a \gg h$, as shown in Fig. 4.5. In the case of slits of small a, its effective parameters must be calculated as a rectangular waveguide (see Sect. 4.3.3.3 below).

4.3.3.2 Cylindrical Cross-Section Ducts

In the same way, the propagation of acoustic waves in a narrow cylindrical duct, as shown in Fig. 4.6, can be described via a complex and frequency dependent density and bulk modulus given by [35].

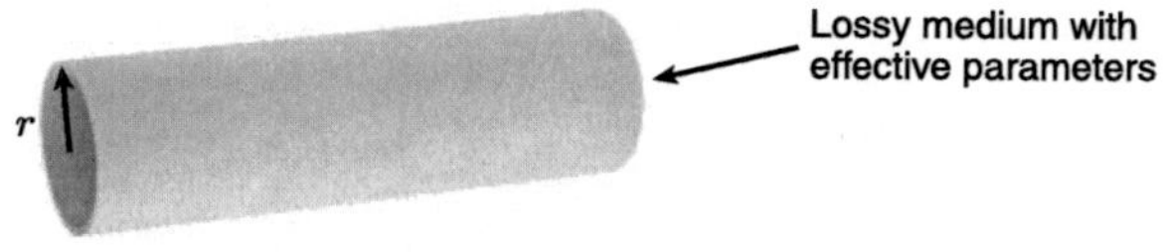

Fig. 4.6 Cylindrical cross-sectional duct of radius r

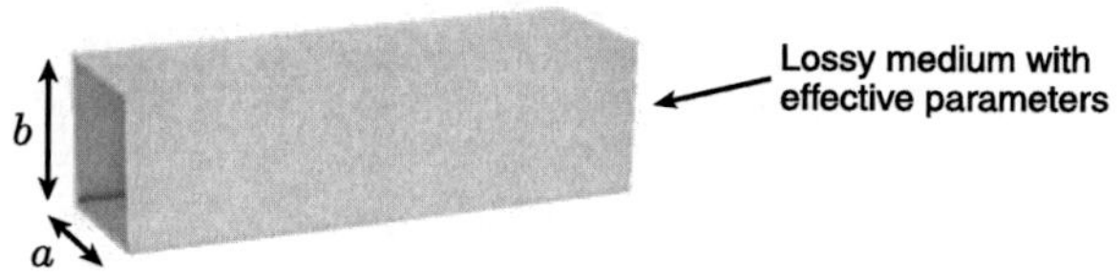

Fig. 4.7 Rectangular cross-section duct of sides a and b

$$\rho_f(\omega) = \rho_0 \left[1 - \frac{2}{rG_\rho(\omega)} \frac{J_1(rG_\rho(\omega))}{J_0(rG_\rho(\omega))}\right]^{-1}, \tag{4.64}$$

$$K_f(\omega) = K_0 \left[1 + \frac{2(\gamma-1)}{rG_K(\omega)} \frac{J_1(rG_K(\omega))}{J_0(rG_K(\omega))}\right]^{-1}, \tag{4.65}$$

where r is the radius of the cylindrical duct, J_n is the Bessel function of the first kind and order n, and the functions $G_\rho(\omega)$ and $G_K(\omega)$ are given by (4.62)–(4.63). In this case, the normalized acoustic impedance is given by $Z'_f = \sqrt{K_f\rho_f}/\pi r^2$.

It is important to note that TMM calculations should be restricted for frequencies lower that the cut-off frequency of the cylindrical duct, i.e., in the range given by $\omega < 1.84c_0/r$. For higher frequencies the duct can exhibit high-order modes and the plane-wave approximation does not hold.

4.3.3.3 Rectangular Cross-Section Ducts

Finally, again assuming that only plane waves propagate inside a rectangular cross-sectional duct, as depicted Fig. 4.7, the propagation can be modelled with effective density and bulk modulus given by [35].

$$\rho_f(\omega) = \rho_0 \frac{(a/2)^2(b/2)^2}{4G_\rho^2(\omega) \sum_{m\in\mathbb{N}} \sum_{n\in\mathbb{N}} \left[\alpha_m^2\beta_n^2\left(\alpha_m^2 + \beta_n^2 - G_\rho^2(\omega)\right)\right]^{-1}}, \tag{4.66}$$

$$K_f(\omega) = K_0 \frac{1}{\gamma + \frac{4(\gamma-1)G_K^2(\omega)}{(a/2)^2(b/2)^2} \sum_{m\in\mathbb{N}} \sum_{n\in\mathbb{N}} \left[\alpha_m^2\beta_n^2\left(\alpha_m^2 + \beta_n^2 - G_K^2(\omega)\right)\right]^{-1}}, \tag{4.67}$$

where a and b are the dimensions of the rectangular duct, $\alpha_m = (2m+1)\pi/a$ and $\beta_n = (2n+1)\pi/b$. The acoustic impedance is $Z'_f = \sqrt{K_f\rho_f}/ab$.

The frequency range where the plane-wave assumption is valid for a rectangular duct is given by $\omega < \pi c_0/\max(a,b)$. As occur in the cylindrical duct, at higher frequencies the waveguide can exhibit high-order modes and TMM calculations will be inaccurate.

4.3.4 Porous Media Layers

Porous materials are biphasic materials composed of a solid frame filled with a fluid, generally air. Typically, the pore size of foams or fibrous materials is much smaller than the characteristic wavelength of the sound waves. Thus, the complex processes occurring during the propagation of acoustic waves at the micro-scale can be modelled at the macro-scale via an effective medium with effective properties. Although elastic waves can propagate in the solid frame, the solid skeleton can be considered motionless in most practical cases in acoustics because of its high impedance contrast with the light saturating fluid, e.g. the air. In this way, the waves emerging from the interaction of elastic waves in the solid matrix and acoustic waves in the saturating fluid, i.e., the Biot waves, can be neglected. Under this assumption, a layer of porous material can be modelled as an equivalent fluid with effective density and bulk modulus. This is generally accurate for thick porous layers because flexural modes appear at very high frequencies and are then strongly attenuated.

The transfer matrix of a porous layer, $\mathbf{T}_p$, is thus written as

$$\mathbf{T}_p = \begin{bmatrix} \cos(k_p L_p) & i Z_p \sin(k_p L_p) \\ i \frac{1}{Z_p} \sin(k_p L_p) & \cos(k_p L_p) \end{bmatrix}, \tag{4.68}$$

where Z_p and k_p are the characteristic effective impedance and wavenumber of the porous material. These effective properties are complex and frequency dependent and several models have been developed providing comprehensive expressions. These models usually consider parameters of the structural properties of the porous frame, in addition to the thermoviscous properties of the saturating fluid (Fig. 4.8).

Accurate models with many input parameters are required to account for most of the physics concerning the thermoviscous processes. However, simpler models relying on a reduced number of input parameters are available, also providing a lower accuracy and possessing a thinner frequency range of validity. Models with a reduced number of parameters are generally sufficient (and practical in most cases), because the input parameters are usually difficult to estimate and the samples are not exactly identical due to the random nature of the porous structure. In the following, we will review several models relying on one, five and six input parameters.

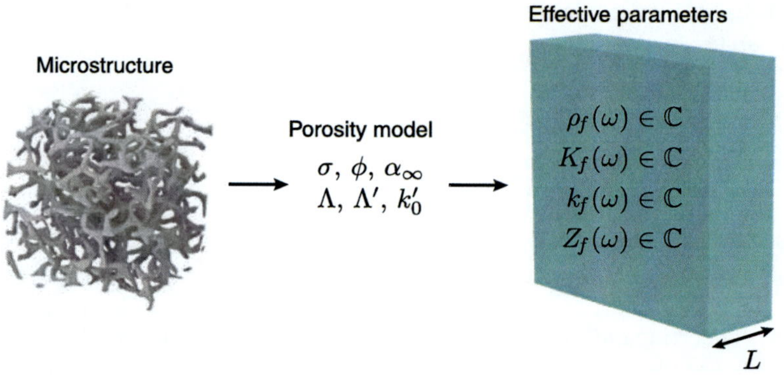

Fig. 4.8 The complex structure of a porous material in the micro-scale can be described by several input parameters. Then, complex and frequency dependent expressions for the effective parameters can be obtained

4.3.4.1 Empirical Models

One of the simplest models of porous materials is the one-parameter model proposed by Delany-Bazley (D&B) [36]. This unique parameter is the flow resistivity σ. This model was derived by fitting the measured characteristic impedance and wavenumber of a wide variety of porous materials with porosity close to the unit and flow resistivity covering the range $1 < \sigma < 50$ kNs/m^4 with a power law of $\rho_0 f/\sigma$. The model validity is thus $0.01 < \rho_0 f/\sigma < 1$.

The complex and frequency dependent effective wavenumber and characteristic impedance are given by

$$Z_p(\omega) = Z_0 \left[1 + a_1^z \left(\frac{\rho_0 \omega}{2\pi\sigma} \right)^{b_1^z} - i a_2^z \left(\frac{\rho_0 \omega}{2\pi\sigma} \right)^{b_2^z} \right], \tag{4.69}$$

$$k_p(\omega) = k_0 \left[1 + a_1^k \left(\frac{\rho_0 \omega}{2\pi\sigma} \right)^{b_1^k} - i a_2^k \left(\frac{\rho_0 \omega}{2\pi\sigma} \right)^{b_2^k} \right], \tag{4.70}$$

where the coefficients a_1^z, a_2^z, b_1^z, b_2^z, a_1^k, a_2^k, b_1^k, b_2^k are given in Table 4.1. It is worth noticing here that the behaviour of some porous materials does not exactly follow this model.

This approach has been extended by several authors. Table 4.1 summarizes alternative empirical models where the coefficients were fitted for specific types of porous and fibrous materials: Dunn and Davern [38] obtained the coefficients for polyurethane foams of low flow resistivity, Qunli [39] for porous plastic open-cell foams, Kirby and Cummings[42] for fibrous materials A glass, E glass, basalt wool and steel wool, Muehleisen et al. [41] for highly porous, rigid and open cell carbon foams, (reticulated vitreous carbon), and finally Garai and Pompoli [43] for polyester fibrous materials.

Table 4.1 Coefficients of empirical models to calculate the effective parameters of porous materials using [*Use (4.71)–(4.72)]. Courtesy of Rodolfo Venegas

Model		a_1^z	a_2^z	b_1^z	b_2^z	a_1^k	a_2^k	b_1^k	b_2^k
D&B [36]		0.0571	−0.087	−0.754	−0.732	0.0978	−0.189	−0.7	−0.595
Mechel [37]	$X < 0.025$	0.0688	−0.196	−0.707	−0.549	0.135	−0.396	−0.646	−0.458
	$X \geq 0.025$	0.0235	−0.0875	−0.887	−0.77	0.102	−0.179	−0.705	−0.674
Dunn&Davern [38]		0.114	−0.099	−0.369	−0.758	0.136	−0.168	−0.491	−0.715
Qunli [39]		0.212	−0.105	−0.607	−0.607	0.163	−0.188	−0.592	−0.544
Miki* [40]		0.07	−0.107	−0.632	−0.632	0.109	−0.16	−0.618	−0.618
Muehleisen et al. [41]		0.2283	−0.1803	−0.4884	−0.8155	0.0898	−0.2049	−0.8453	−−0.7215
K&C A glass [42]		0.0924	−0.1457	−0.7177	−0.5951	0.1443	−0.2251	−0.7088	−0.5827
K&C E glass [42]		0.0954	−0.1689	−0.6687	−0.5707	0.2010	−0.2202	−0.5829	−0.5850
K&C Basalt wool [42]		0.0599	−0.1376	−0.7664	− 0.6276	0.1281	−0.2178	−0.6746	−0.6051
K&C Steel wool [42]		0.0877	−0.0876	−0.5557	− 0.7609	0.1328	−0.1540	−0.5571	−0.7093
G&P [43]		0.078	−0.074	−0.623	−0.660	0.121	−0.159	−0.530	−0.571

In particular, a further refinement of the D&B model was carried out by Miki [40], using same data as those used to derive the D&B model, to correct non-physical behaviour of the effective properties in the low frequency regime. The equations for this model are

$$Z_p(\omega) = Z_0 \left[1 + a_1^z \left(\frac{\omega}{2\pi\sigma} \right)^{b_1^z} - i a_2^z \left(\frac{\omega}{2\pi\sigma} \right)^{b_2^z} \right], \tag{4.71}$$

$$k_p(\omega) = k_0 \left[1 + a_1^k \left(\frac{\omega}{2\pi\sigma} \right)^{b_1^k} - i a_2^k \left(\frac{\omega}{2\pi\sigma} \right)^{b_2^k} \right], \tag{4.72}$$

and the range of validity is then enlarged to $0.01 < f/\sigma < 1$. The model proposed by Miki should be used instead of the original proposed by Delany and Bazley.

4.3.4.2 Semi-empirical Models

Another widely used class of equivalent models for porous materials is the semi-empirical one. These models attempt to link high and low frequency asymptotic exact behaviours of porous material by simple functions and rely on the separation of viscous (in the density) and thermal (in the bulk modulus) losses. Among several, the Johnson-Champoux-Allard and the Johnson-Champoux-Allard-Lafarge models are the most commonly encountered.

4.3.4.3 The Johnson-Champoux-Allard Model

The Johnson-Champoux-Allard (JCA) model [44, 45] involves 5-parameters and provides the expressions of the dynamic effective density and bulk modulus of a porous material saturated by a fluid of density ρ_0 and bulk modulus K_0 considering a rigid frame. The porous material is characterized by its porosity, ϕ, its tortuosity, α_∞, its flow resistivity, σ, and the thermal and viscous characteristic lengths, Λ' and Λ respectively.

The dynamic effective density and bulk modulus given by the JCA model are

$$\rho_p(\omega) = \rho_0 \frac{\alpha_\infty}{\phi} \left[1 - i G_1(\omega) \sqrt{1 + i G_2(\omega)} \right], \tag{4.73}$$

$$K_p(\omega) = K_0 \frac{\phi^{-1}}{\gamma - (\gamma - 1) \left[1 - i G_1'(\omega) \sqrt{1 + i G_2'(\omega)} \right]^{-1}}, \tag{4.74}$$

where the adiabatic bulk modulus is given by $K_0 = \gamma P_0$. A description of the physical meaning and the measurement procedure for the input parameters can be found in the literature [2, 46].

The functions $G_1(\omega)$, $G_2(\omega)$, $G_1'(\omega)$, $G_2'(\omega)$ are given by

$$G_1(\omega) = \frac{\sigma\phi}{\alpha_\infty \rho_0 \omega}, \tag{4.75}$$

$$G_2(\omega) = \frac{4\alpha_\infty^2 \rho_0 \eta \omega}{\sigma^2 \phi^2 \Lambda^2}, \tag{4.76}$$

$$G_1'(\omega) = \frac{8\eta}{\rho_0 \mathrm{Pr}\, \Lambda'^2 \omega}, \tag{4.77}$$

$$G_2'(\omega) = \frac{\rho_0 \mathrm{Pr}\, \Lambda'^2 \omega}{16\eta}. \tag{4.78}$$

Using these expressions, both the effective wavenumber and the characteristic acoustic impedance of the porous material can be obtained by using

$$k_p = \frac{\omega}{c_p} = \omega\sqrt{\frac{\rho_p}{K_p}}, \tag{4.79}$$

where c_p is the effective sound speed in the porous material and

$$Z_p = \sqrt{\rho_p K_p}. \tag{4.80}$$

4.3.4.4 The Johnson-Champoux-Allard-Lafarge Model

The JCA model was further be extended by Lafarge [47] to accurately describe the thermal effects in the low frequency regime. The extended model, namely the Johnson-Champoux-Allard-Lafarge (JCAL) model, involves a new parameter, the static thermal permeability, k_0'. Only the bulk modulus is modified when compared to the JCA model and $G_1'(\omega)$ and $G_2'(\omega)$ read as

$$G_1'(\omega) = \frac{\phi\eta}{\rho_0 \mathrm{Pr} k_0' \omega}, \tag{4.81}$$

$$G_2'(\omega) = \frac{4\mathrm{Pr}\,\rho_0 k_0'^2 \omega}{\eta \phi^2 \Lambda'^2}. \tag{4.82}$$

Please note that the JCAL model is usually written in terms of the static viscous permeability $k_0 = \eta/\sigma$ which is an intrinsic parameter of the material. In this case $G_1(\omega)$ and $G_2(\omega)$ become

$$G_1(\omega) = \frac{\phi\eta}{\alpha_\infty \rho_0 k_0 \omega}, \tag{4.83}$$

$$G_2(\omega) = \frac{4\alpha_\infty^2 \rho_0 k_0^2 \omega}{\eta\phi^2 \Lambda^2}. \tag{4.84}$$

4.3.5 Locally Resonant Elements

The previously described elements only considered continuity of pressure and flux along the main propagation direction. However, some elements can be added to the transmission line that rely on pressure drop or flux continuity in other directions and are usually locally resonant elements.

Let us assume an element of lateral dimension Δx much smaller than the wavelength along the main waveguide. This element can be considered as a punctual resonator. The upstream and downstream pressure, p_u and p_d, and flux, $\mathcal{V}_d$ and $\mathcal{V}_u$, are first introduced allowing to define the transfer matrix of this infinitesimal element.

$$\begin{bmatrix} p_d \\ \mathcal{V}_d \end{bmatrix}_x = \mathbf{T}' \begin{bmatrix} p_u \\ \mathcal{V}_u \end{bmatrix}_{x+\Delta x}, \tag{4.85}$$

The pressure drop

$$\Delta p = p_d - p_u, \tag{4.86}$$

and flux continuity

$$\Delta \mathcal{V} = \mathcal{V}_d - \mathcal{V}_u, \tag{4.87}$$

can subsequently be defined as a function of the resonator.

4.3.5.1 Side-Branch (Parallel) Elements

On the one hand, locally resonant elements can load the main waveguide, as shown in Fig. 4.9. This loading element can be an open or closed duct, a Helmholtz resonator, or a dead-end cavity of any complex shape. In this case, the pressure is constant along the element, i.e., $\Delta p = 0$ or $p_d = p_u$, while the flux continuity implies $\Delta \mathcal{V} = \mathcal{V}_d - \mathcal{V}_u = p_u/Z_r'$, where Z_r' is the resonator impedance. Please note that the continuity of pressure has already been accounted for in the last expression.

The equations relating pressures and velocities at both sides of the infinitesimal element are then

$$p_d = p_u, \tag{4.88}$$

$$v_d = v_u + p_u/Z_r, \tag{4.89}$$

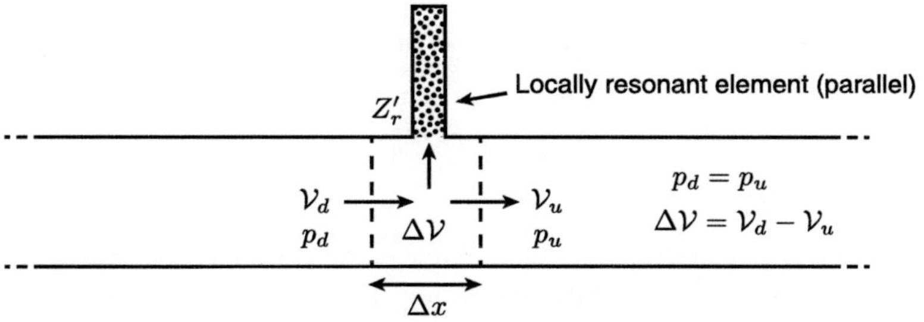

Fig. 4.9 Parallel configuration with a locally resonant element loaded in a waveguide. The resonator introduces a *flow drop* while pressure is continuous

which can be re-written in matrix form to give the transmission matrix for "parallel" connected elements, $\mathbf{T}'$, as

$$\mathbf{T}' = \begin{bmatrix} 1 & 0 \\ \dfrac{1}{Z'_r} & 1 \end{bmatrix}. \tag{4.90}$$

4.3.5.2 In-Line (Series) Elements

On the other hand, locally resonant elements can be in series in a main waveguide, as shown in Fig. 4.10. For example, this series elements can be membranes or elastic plates located in a waveguide.

This time the flux is continuous across the infinitesimal element, i.e., $\Delta\mathcal{V} = 0$ or $\mathcal{V}_d = \mathcal{V}_u$, but the punctual resonator induces a pressure drop $\Delta p = p_d - p_u = Z'_r \mathcal{V}_u$, where Z'_r is the resonator impedance. Please note that the flux continuity has already been applied. The equations relating pressures and velocities at both sides are

$$p_d = p_u + Z'_r v_u, \tag{4.91}$$

$$\mathcal{V}_d = \mathcal{V}_u, \tag{4.92}$$

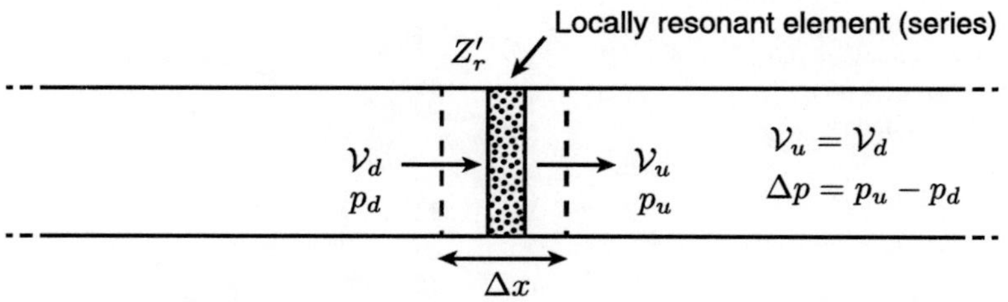

Fig. 4.10 Series configuration with a locally resonant element located in-line in a waveguide. The resonator introduces a *pressure drop* while flow is continuous

an can then be written in matrix form to obtain the transfer matrix of "series" elements $\mathbf{T}'$ as

$$\mathbf{T}' = \begin{bmatrix} 1 & Z_r' \\ 0 & 1 \end{bmatrix}. \tag{4.93}$$

In the following subsections we will review particular impedance expressions of different resonators that are commonly arranged in-series or in-parallel to waveguides.

4.3.6 Side Resonating Ducts

When a waveguide is loaded by a secondary duct of length l, as shown for example in Fig. 4.11 for a sealed secondary duct, the pressure and flux at $z = 0$ and $z = l$ can be related by a transfer matrix of the form

$$\begin{bmatrix} p \\ \mathcal{V}_z \end{bmatrix}_{z=0} = \begin{bmatrix} \cos(k_f l) & i Z_f' \sin(k_f l) \\ i \dfrac{1}{Z_f'} \sin(k_f l) & \cos(k_f l) \end{bmatrix} \begin{bmatrix} p \\ \mathcal{V}_z \end{bmatrix}_{z=l}. \tag{4.94}$$

where $Z_f' = Z_f / S_f$ and k_f are the characteristic impedance and wavenumber, and S_f the cross-section area of the secondary duct. Note that thermoviscous losses in this duct can be accounted for using an appropriate complex and frequency dependent wavenumber and impedance, that can be calculated accordingly to (4.60)–(4.67) as a function of the cross-sectional geometry. Depending on the boundary condition at $z = l$, the input resonator impedance at $z = 0$ can be evaluated via $Z_r' = p(0)/\mathcal{V}_z(0)$.

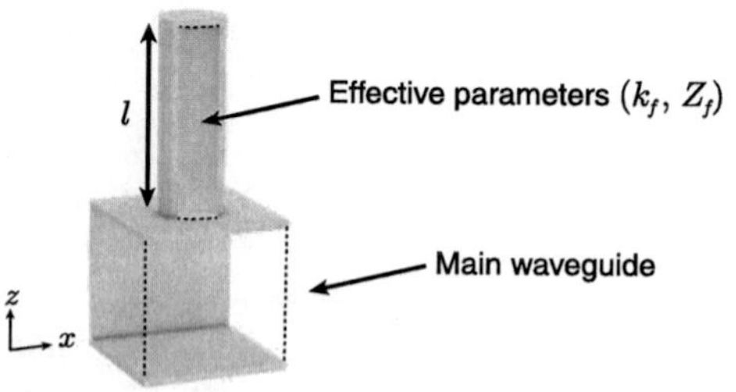

Fig. 4.11 Quarter-wavelength resonator (QWR), of length l, loaded on a main waveguide. Resonator is modelled using the complex and frequency dependent wavenumber and impedance that depend on the resonator geometry

4.3.6.1 Closed Secondary Duct: The Quarter-Wavelength Resonator (QWR)

When the secondary duct is sealed, i.e., a rigid boundary condition is applied at $z = l$, i.e., $\mathcal{V}_z(l) = 0$, the system of equations given by (4.94) reduces to

$$p(0) = \cos(k_f l)\, p(l) \quad \text{and} \quad \mathcal{V}_z(0) = i\frac{1}{Z_f}\sin(k_f l)\, p(l). \tag{4.95}$$

The input impedance is then

$$Z'_r = -iZ'_f \cot\left(k_f l\right) = -i\frac{Z_f}{S_f}\cot\left(k_f l\right), \tag{4.96}$$

where l is the QWR length. The first resonance is observed when $Z'_r = 0$, implying $\cot(k_f l) = 0$, i.e., $l \approx \lambda/4$ where λ is the wavelength (note k_f can be complex).

4.3.6.2 "End" Correction

Note that the well-known "end" correction must be incorporated to this impedance to account for the radiation of the resonator in the main waveguide. This radiation is modelled by adding a subwavelength, i.e. $k_0 \Delta l \ll 1$, fluid transfer matrix of length Δl, i.e., the end correction, with wavenumber and impedance k_0 and $Z'_0 = Z_0/S$ of the main waveguide of section S to (4.94):

$$\begin{bmatrix} p \\ \mathcal{V}_z \end{bmatrix}_0 = \begin{bmatrix} \cos(k_0 \Delta l) & iZ'_0 \sin(k_0 \Delta l) \\ i\dfrac{1}{Z'_0}\sin(k_0 \Delta l) & \cos(k_0 \Delta l) \end{bmatrix} \begin{bmatrix} \cos(k_f l) & iZ'_f \sin(k_f l) \\ i\dfrac{1}{Z'_f}\sin(k_f l) & \cos(k_f l) \end{bmatrix} \begin{bmatrix} p \\ \mathcal{V}_z \end{bmatrix}_l.$$

Taylor expanding the first matrix and applying the rigid boundary condition at $z = l$, (4.96) becomes

$$Z'_r = -i\frac{Z_f}{S_f}\cot\left(k_f l\right) - i\omega\rho_0\frac{\Delta l}{S}, \tag{4.97}$$

which is the input impedance of a QWR with "end" correction. The "end" corrections Δl are given in the Appendix 4.6.2.

4.3.6.3 Open Secondary Duct: The Half-Wavelength Resonator (HWR)

When the secondary duct is open, pressure-release boundary condition can be applied in (4.94) at $z = l$, i.e., $p(l) = 0$. This leads to the system of equations

$$p(0) = i Z_f \sin(k_f l)\, p(l), \tag{4.98}$$

$$\mathcal{V}_z(0) = \cos(k_f l)\, p(l). \tag{4.99}$$

The input impedance is then

$$Z'_r = i Z'_f \tan\left(k_f l\right) = i \frac{Z_f}{S_f} \tan\left(k_f l\right), \tag{4.100}$$

where l is the resonator length. The resonance is no longer achieved when the length of the duct is a quarter of the wavelength, but rather half of it, i.e. the system is a half-wavelength resonator (HWR). However, the present model for HWR is less accurate than the one for QWR because the radiation to the outer domain must be inaccurate, i.e., the system can no longer be one-dimensional.

4.3.7 Helmholtz Resonators

Another type of commonly used side-branch resonator is the Helmholtz one that comprises a neck coupled to a cavity, as sketched in Fig. 4.12. The input impedance of the Helmholtz resonator (HR) can also be derived by relating the pressure and flux at the beginning, $z = 0$, and at the end, $z = l = l_n + l_c$, of the system. The ducts corresponding to the neck and the cavity are characterized by their length l_n and l_c and their sections S_n and S_c, respectively. The transfer matrix problem is written as

$$\begin{bmatrix} p \\ \mathcal{V}_z \end{bmatrix}_0 = \begin{bmatrix} \cos(k_n l_n) & i Z'_n \sin(k_n l_n) \\ i \dfrac{1}{Z'_n} \sin(k_n l_n) & \cos(k_n l_n) \end{bmatrix} \begin{bmatrix} \cos(k_c l_c) & i Z'_c \sin(k_c l_c) \\ i \dfrac{1}{Z'_c} \sin(k_c l_c) & \cos(k_c l_c) \end{bmatrix} \begin{bmatrix} p \\ \mathcal{V}_z \end{bmatrix}_l .$$

Note that specific complex and frequency dependent effective parameters must be used for each element if thermoviscous losses are accounted for because the neck and the cavity are of different cross-sections: k_n and $Z'_n = Z_n / S_n$ for the neck, and k_c and

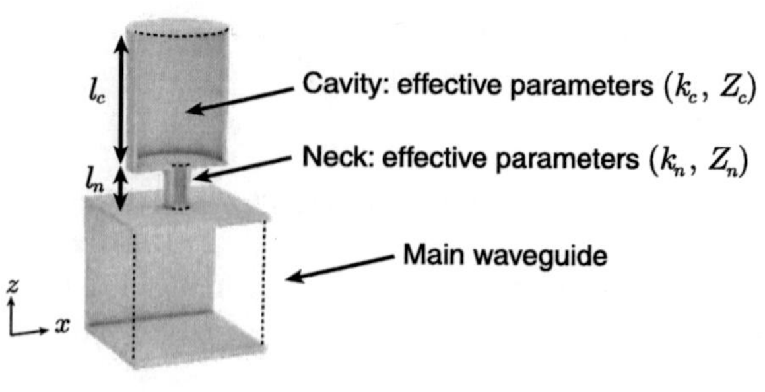

Fig. 4.12 Helmholtz resonator loaded on a main waveguide. The resonator is modelled using the complex and frequency dependent parameters that depend on its geometry

$Z'_c = Z_c/S_c$ for the cavity. Theses effective parameters are calculated accordingly to (4.60)–(4.67) as a function on the geometry of the ducts.

Applying rigid boundary condition at $z = l = l_n + l_c$, and manipulating the previous transfer matrix, the impedance of the HR can be written as $Z'_r = p(0)/\mathcal{V}_z(0)$

$$Z'_r = -iZ'_n \frac{Z'_c/Z'_n - \tan k_n l_n \tan k_c l_c}{Z'_c/Z'_n \tan k_n l_n + \tan k_c l_c}. \tag{4.101}$$

Taylor expanding the latter expression at low frequencies, i.e., $k_n l_n \ll 1$ and $k_c l_c \ll 1$, leads to

$$Z'_r = -iZ'_n \frac{Z'_c/Z'_n - k_n l_n k_c l_c}{Z'_c/Z'_n k_n l_n + k_c l_c}. \tag{4.102}$$

If losses are not considered, $k_n = k_c = k_0$ and $Z_n = Z_c = Z_0$, where $k_0 = \omega/c_0$ and $Z_0 = \rho_0 c_0$. The first resonance of the HR is then observed when $\mathrm{Im}(Z'_r) = 0$, leading to

$$\omega_R = c_0 \sqrt{\frac{S_n}{l_n l_c S_c}} = c_0 \sqrt{\frac{S_n}{l_n V_c}}, \tag{4.103}$$

which is the usual expression for the resonance frequency of a HR, where $V_c = S_c l_c$ is the volume of the cavity.

However, when thermoviscous losses are included such a compact expression cannot be derived. In this case the resonant frequency can be estimated numerically by looking for the frequency at which $\mathrm{Im}(Z'_r) = 0$ with the corresponding complex k_n, k_c, Z_n and Z_c. Usually, the so-calculated HR resonance frequency is moderately reduced as compared with (4.103).

It is worth noting here that (4.101) is not exact because corrections due to the radiation at the discontinuities must be included. Using a transfer matrix approach, see e.g., [48, Suppl. mat.], we can model the pressure radiation between the different elements. Thus, we can express the system as

$$\begin{bmatrix} p \\ \mathcal{V}_z \end{bmatrix}_0 = \begin{bmatrix} \cos(k_n \Delta l_2) & iZ'_n \sin(k_n \Delta l_2) \\ i\frac{1}{Z'_n}\sin(k_n \Delta l_2) & \cos(k_n \Delta l_2) \end{bmatrix} \begin{bmatrix} \cos(k_n l_n) & iZ'_n \sin(k_n l_n) \\ i\frac{1}{Z'_n}\sin(k_n l_n) & \cos(k_n l_n) \end{bmatrix} \times \ldots$$
$$\begin{bmatrix} \cos(k_n \Delta l_1) & iZ'_n \sin(k_n \Delta l_1) \\ i\frac{1}{Z'_n}\sin(k_n \Delta l_1) & \cos(k_n \Delta l_1) \end{bmatrix} \begin{bmatrix} \cos(k_c l_c) & iZ'_c \sin(k_c l_c) \\ i\frac{1}{Z'_c}\sin(k_c l_c) & \cos(k_c l_c) \end{bmatrix} \begin{bmatrix} p \\ \mathcal{V}_z \end{bmatrix}_l.$$

The first length correction, Δl_1, is due to pressure radiation at the discontinuity from the cavity to the neck of the HR [49], while the second length correction, Δl_2, comes from the radiation at the discontinuity from the neck to the principal waveguide [50]. Please note that rigorously, this second correction matrix should be written in terms of the principal waveguide parameters, k_0 and Z_0'. However, replacing this parameters by those of the neck provides more accurate results. After application of the rigid boundary condition at $x = l$, Taylor expansion ($k_n \Delta l_1 \gg 1$ and $k_n \Delta l_2 \gg 1$) of the length correction matrices, and rearrangement, the impedance of the HR accounting for the "end" corrections becomes

$$Z_r' = -iZ_n' \frac{\cos k_n l_n \cos k_c l_c - \dfrac{k_n \Delta l Z_n'}{Z_c'} \cos k_n l_n \sin k_c l_c - \dfrac{Z_n'}{Z_c'} \sin k_n l_n \sin k_c l_c}{\sin k_n l_n \cos k_c l_c - \dfrac{k_n \Delta l Z_n'}{Z_c'} \sin k_n l_n \sin k_c l_c + \dfrac{Z_n'}{Z_c'} \cos k_n l_n \sin k_c l_c},$$

where the correction length $\Delta l = \Delta l_1 + \Delta l_2$ is the addition of the two correction lengths. The specific values for the correction lengths are given in the Appendix 4.6.1.

4.3.8 *Rigid Micro-perforated Plates*

When a rigid thin panel is densely perforated with holes whose dimensions are much smaller than the wavelength, it can be modelled as a local impedance that accounts for the wave propagation in the small ducts including the thermoviscous effects. Maa's model [51] consider a local impedance as

$$Z_r = -\frac{i\omega\rho_0 h}{\phi}\left[1 - \frac{2}{\sigma\sqrt{i}} \frac{J_1\left(\sigma\sqrt{i}\right)}{J_0\left(\sigma\sqrt{i}\right)}\right]^{-1} - i0.85\omega\frac{\rho_0 d}{\phi} + \frac{h\sqrt{2}\sigma\eta}{d\phi}, \tag{4.104}$$

where $\sigma = d\sqrt{\omega\rho_0/4\eta}$, d is the perforation diameter, h is the thickness of the plate, and $\phi = \pi d^2/4a^2$ is the surface porosity in case of circular perforation arranged in a square lattice of side a, as depicted in Fig. 4.13. The first term on right-hand-side of (4.104) accounts for the hole impedance, the second term models the reactive radiation at both appendicular parts of the perforation, while the purely resistive term models the friction at these appendicular parts. Note that the provided impedance is Z_r and not Z_r' and holds for a perforated plate of infinite lateral extends. An approximation of this model is given by

$$Z_r = -\frac{i\omega\rho_0 h}{\phi}\sigma_i + \frac{32\eta h}{\phi d^2}\sigma_r, \tag{4.105}$$

where the coefficients σ_i and σ_r are given by

$$\sigma_i = 1 + \left[1 + \frac{\sigma^2}{2}\right]^{-\frac{1}{2}} + 0.85\frac{d}{h}, \quad \sigma_r = \sqrt{1 + \frac{\sigma^2}{32}} + \frac{4\sqrt{2}}{32}\sigma\frac{d}{h}. \tag{4.106}$$

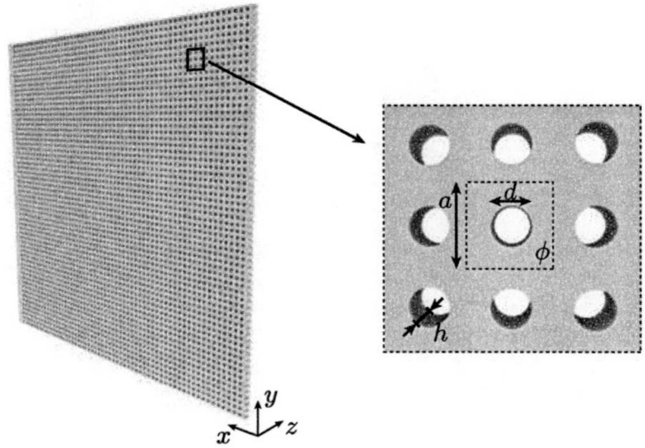

Fig. 4.13 Micro-perforated panel and the geometrical parameters used

This approximation is valid for $1 < \sigma < 10$. Further details, extended and alternative models can be found in [52].

4.3.9 *Elastic Plates*

Another type of often encountered locally resonant elements are elastic plates. The acoustic impedance of the plate relies on the pressure drop and flux continuity across the plate. Thus, different plate geometries and boundary conditions leads to different expressions for the acoustic impedance.

4.3.9.1 Circular Elastic Plate

Let us first consider a clamped circular plate of radius r and section $S = \pi r^2$, as the one depicted Fig. 4.14a. Note the clamped condition implies that the transverse displacement and its first normal derivative vanish at the plate boundary. Assuming that only the axisymmetric modes can be excited, the associated acoustic impedance can be derived analytically and reads as [53]

$$Z_r' = -\frac{i\omega\rho h}{S}\frac{J_0(k_m r)I_1(k_m r) + J_1(k_m r)I_0(k_m r)}{J_2(k_m r)I_1(k_m r) - J_1(k_m r)I_2(k_m r)}, \tag{4.107}$$

where I_n is the modified Bessel's function of the first kind of order n, h is the thickness of the plate, and k_m and ρ are respectively the wavenumber and density of the plate, given by

$$k_m^2 = \omega\sqrt{\frac{\rho h}{D}}, \quad \text{and} \quad D = \frac{Eh^3}{12(1-\nu^2)}, \tag{4.108}$$

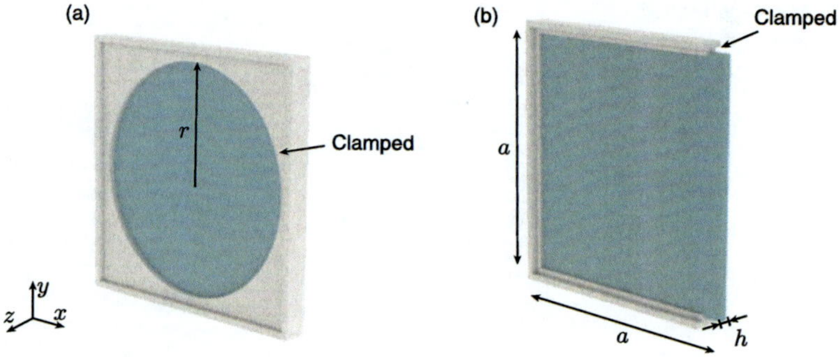

Fig. 4.14 Elastic plate resonators and the geometrical parameters used. (a) Circular elastic and (b) square clamped plates

with D is the bending stiffness (or flexural rigidity), and E and ν, the Young's modulus and Poisson's ratio of the material plate, respectively.

4.3.9.2 Approximation

Taylor expanding (4.107) at low frequencies, i.e., $k_m r \ll 1$, provides a lumped impedance model, which reads as

$$Z_r' = \frac{1}{i\omega C_p} + i\omega M_p, \tag{4.109}$$

where the compliance and the acoustic mass are [54] respectively:

$$C_p = \frac{\pi r^6}{196.51 D}, \quad M_p = 1.8830\,\frac{\rho h}{\pi r^2}. \tag{4.110}$$

The first resonance frequency is then given by the following formulae

$$f_0 = 0.4694 \frac{h}{r^2} \sqrt{\frac{E}{\rho(1-\nu^2)}}. \tag{4.111}$$

Note that this model only accounts for the first resonance and, therefore, (4.109) is only valid for $f \lessapprox f_0$.

4.3.9.3 Squared Elastic Plate

The impedance of a squared clamped elastic plate, as shown in Fig. 4.14b, can be obtained assuming the system obeys the flexural wave equation which is valid for thin elastic plates. Using separation of variables, the full expression for the impedance of a square cross-sectional elastic plate of side a is given by [55]

$$Z_r' = \left[i\omega \int_0^a \int_0^a \left(\sum_{m=1}^{\infty} \sum_{n=1}^{\infty} \frac{\int_0^a \int_0^a X_m Y_n dx dy}{D(I_1 I_2 + 2I_3 I_4 + I_5 I_6) - \rho h \omega^2 I_2 I_6} X_m Y_n \right) dx dy \right]^{-1} \tag{4.112}$$

where $X_m(x)$ and $Y_n(y)$ are the eigenfunctions describing the shape of the n-th and m-th eigenmode of the clamped plate. They both have the same form and are given by

$$X_m(x) = G\left(\frac{\lambda_m x}{a}\right) - \frac{G(\lambda_m)}{H(\lambda_m)} H\left(\frac{\lambda_m x}{a}\right), \tag{4.113}$$

$$Y_n(y) = G\left(\frac{\lambda_n y}{a}\right) - \frac{G(\lambda_n)}{H(\lambda_n)} H\left(\frac{\lambda_n y}{a}\right). \tag{4.114}$$

The functions $G(u)$ and $H(u)$ satisfy the clamped boundary conditions and are given by

$$G(u) = \cosh(u) - \cos(u), \quad H(u) = \sinh(u) - \sin(u), \tag{4.115}$$

while λ_m and λ_n satisfy

$$\cosh(\lambda)\cos(\lambda) = 1. \tag{4.116}$$

In practice, λ_m and λ_n can be found using a root-finding algorithm such as Muller's method [56].

Finally, the integrals, I_i, are given by

$$I_1 = \int_0^a X_m(x) \frac{\partial^4 X_m(x)}{\partial x^4} dx, \qquad I_2 = \int_0^a Y_n(y)^2 dy, \tag{4.117}$$

$$I_3 = \int_0^a X_m(x) \frac{\partial^2 X_m(y)}{\partial x^2} dx, \qquad I_4 = \int_0^a Y_n(y) \frac{\partial^2 Y_n(y)}{\partial x^2} dy, \tag{4.118}$$

$$I_5 = \int_0^a Y_n(y) \frac{\partial^4 Y_n(y)}{\partial x^4} dy, \qquad I_6 = \int_0^a X_m(x)^2 dx. \tag{4.119}$$

While the derivatives can be calculated analytically, the integrals must be calculated numerically, e.g., using Simpson's method.

The resonance frequencies of the n-th and m-th modes of the square clamped plate are given by

$$\omega_{m,n} = \sqrt{\frac{D(I_1 I_2 + 2I_3 I_4 + I_5 I_6)}{\rho h I_2 I_6}}. \tag{4.120}$$

Finally, note that in (4.112) the term $\int_0^a\int_0^a X_m Y_n dxdy$ vanish for $m, n = 2, 4, \ldots$, so that the even modes do not contribute to impedance: for these modes, the pressure drop at both sides of the plate is zero.

4.3.9.4 Approximation

The impedance of a square clamped plate given by (4.112) can be approximated for frequencies around and below the first resonance frequency of the plate. The lumped impedance model for the square clamped plate can be written as

$$Z_r' = \frac{1}{i\omega C_p} + i\omega M_p, \tag{4.121}$$

where the compliance and the acoustic mass are [57], respectively

$$C_p = 3.73 \times 10^{-4}\,\frac{a^6}{D}, \quad M_p = 2.06\,\frac{\rho h}{a^2}, \tag{4.122}$$

with $D = Eh^3/12(1-\nu^2)$ the bending stiffness, and E, ν, ρ and h the Young's modulus, Poisson's ratio, density and thickness of the square plate, respectively. The first resonance frequency ω_0 of the clamped elastic square plate satisfies $\mathrm{Im}(Z_r') = 0$ and is thus given by $\omega_0 = \sqrt{1/M_p C_p}$.

4.3.10 Membranes

A clamped membrane is a thin elastic plate clamped at the boundaries under a large pre-stretch tension. As shown in Fig. 4.15a, b, the membrane behaviour arises when the tension T is much higher than the bending stiffness of the elastic plate, i.e., $T \gg D$, which is easily achieved by thin elastic plates, $h \ll \sqrt[3]{12(1-\nu^2)T/E}$, of soft materials, $\nu \to 0.5$. As in elastic plates, different geometries led to different expressions for the acoustic impedance of membranes.

4.3.10.1 Circular Membrane

Considering only the axisymmetric modes of a circular membrane of radius r and area $S = \pi r^2$, and thickness h, as depicted in Fig. 4.15a, an analytical expression can be obtained for the acoustic impedance of the clamped circular clamped membrane as

$$Z_r' = -\frac{i\omega\rho h}{S}\frac{J_0(k_m r)}{J_2(k_m r)}, \tag{4.123}$$

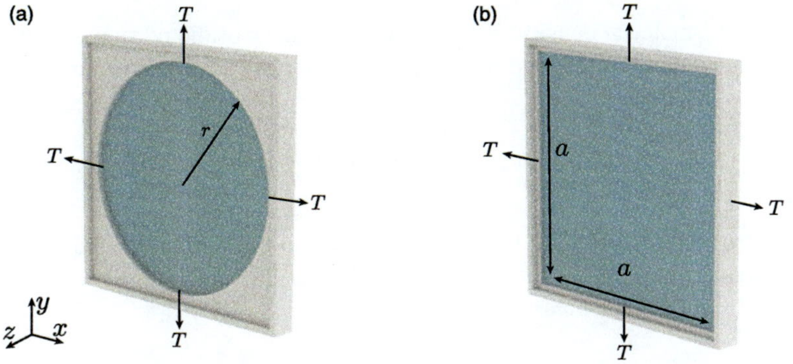

Fig. 4.15 Elastic membrane resonators and the geometrical parameters used. **a** Circular and **b** square membranes

where ρ is the material density and $k_m = \omega\sqrt{\rho/T}$ is the wavenumber in the membrane, with T the tension applied to the membrane.

4.3.10.2 Approximation

The Taylor expansion of (4.123) at the low frequencies ($k_m r \ll 1$) provides the lumped impedance model, which takes the form

$$Z_r' = \frac{i\omega\rho h}{S}\left[1 - \frac{f_0}{f}\right], \tag{4.124}$$

where the first resonance of the plate f_0 is given by the following approximation

$$f_0 = 0.38274\frac{1}{r}\sqrt{\frac{T}{\rho}}. \tag{4.125}$$

Note this model only accounts for the first resonance and, therefore, is only valid for $0 \leq f \lessapprox f_0$.

4.3.10.3 Squared Membrane

In the case of a square clamped membrane of side a and thickness h, as shown in Fig. 4.15b, the acoustic impedance can be written as [57]:

$$Z_r' = \left[i\omega a^2 \sum_{m=1}^{\infty}\sum_{n=1}^{\infty} \frac{4\left[1-(-1)^m\right]^2\left[1-(-1)^n\right]^2}{\rho h\left(\omega_{m,n}^2 - \omega^2\right)m^2 n^2 \pi^4}\right]^{-1}, \tag{4.126}$$

where ρ is the material density, m and n are the indices of the normal modes of the membrane and their corresponding resonance frequency $\omega_{m,n}$ is given by

$$\omega_{m,n} = \frac{\pi}{a}\sqrt{\frac{T}{\rho}\left(m^2 + n^2\right)}, \tag{4.127}$$

with the tension T. Note that the even modes do not contribute to the impedance because the term $[1-(-1)^m]^2\,[1-(-1)^n]^2$ in (4.126) vanish for $n, m = 2, 4, \ldots$, i.e., for these modes the pressure drop is zero.

4.3.10.4 Approximation

The impedance of a square clamped membrane given by (4.126) can be approximated for frequencies around or below the first resonance by

$$Z_r' = \frac{1}{i\omega C_m} + i\omega M_m, \tag{4.128}$$

where the acoustic compliance and acoustic mass are given by

$$C_m = 0.035\frac{a^6}{T}, \quad M_m = 1.44\frac{\rho h}{a^2}. \tag{4.129}$$

These coefficients are calculated from the full analytic model. Note the impedance of the membrane is independent of the material elasticity. These formulae are valid for thin clamped square membranes where the pre-stretched tension dominates over the bending stiffness, i.e., $T \gg Eh^3/12(1-\nu^2)$. The first resonance frequency of the clamped square membrane, ω_0, again satisfies $\mathrm{Im}(Z_r') = 0$ and is thus given by $\omega_0 = \sqrt{1/M_p C_p}$.

4.3.11 *Infinite Elastic Vibrating Wall*

A thin and unbounded elastic material of thickness h can be included in a TMM formulation as a series element given by (4.93), under the assumption that $k_l h \ll 1$, where k_l is the wavenumber associated with the longitudinal waves in the elastic material. The acoustic impedance of this thin, unbounded and elastic plate as shown in Fig. 4.16, can be written as

$$Z_r = Z_\mathrm{w} = \frac{D}{i\omega}\left(k_0^4 \sin^4\theta - k_\mathrm{w}^4\right), \tag{4.130}$$

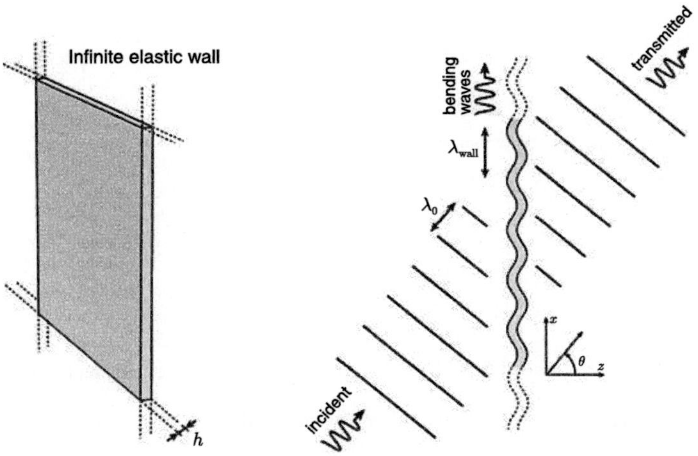

Fig. 4.16 (Left) Geometry of the unbounded elastic wall and geometrical parameters used. (Right) Scheme of the induced flexural waves at the coincidence frequency for a given angle of incidence

where θ is the incidence angle and the dispersion relation $k_\mathrm{w} = \omega/c_\mathrm{w}$ gives the wavenumber in the wall. The phase speed of the bending waves is $c_\mathrm{w} = (\omega^2 D/m')^{1/4}$, $k_0 = \omega/c_0$ is the wavenumber of the surrounding medium with sound speed c_0, $m' = \rho h$ is the surface mass density, i.e., the mass per unit area, $D = Eh^3/12(1-\nu^2)$ is the bending stiffness of the wall and E, ν, and ρ are the Young's modulus, Poisson's ratio and density of the elastic material, respectively. Please note that the provided impedance corresponds to Z_r and not to Z_r' and holds for a thin elastic plate of infinite lateral extents.

Equation (4.130) exhibits a resonance at $k_0^4 \sin^4\theta = k_\mathrm{w}^4$, i.e., when the wavenumber of the bending waves in the wall matches the transverse component of the wavenumber in the surrounding medium, as shown in Fig. 4.16. This occurs when $\lambda_0/\sin\theta = \lambda_\mathrm{w}$, i.e., at the *coincidence frequency* given by

$$f_c' = \frac{c_0^2}{2\pi \sin\theta}\sqrt{\frac{m'}{D}}. \tag{4.131}$$

The lowest coincidence frequency occurs at grazing angles when $\theta = \pi/2$. We can thus define this frequency as the *critical frequency* as

$$f_c = \frac{c_0^2}{2\pi}\sqrt{\frac{m'}{D}} = f_c' \sin\theta. \tag{4.132}$$

This critical frequency only depends on the properties of the wall and the surrounding fluid. The acoustic impedance of the thin wall can then be written as a function of this critical frequency as

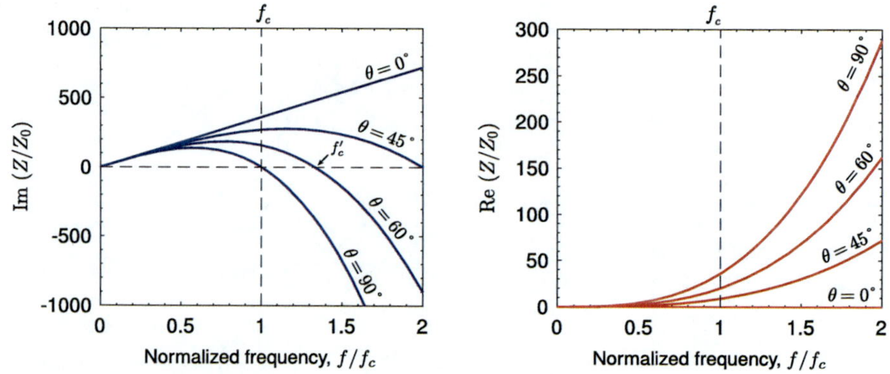

Fig. 4.17 (Left) Imaginary part of the impedance of the elastic wall as a function of the frequency normalized to the critical frequency. (Right) Corresponding real part

$$Z_w = im'\omega\left[1 - \left(\frac{f}{f_c}\right)^2 (1 + i\eta_w)\sin^4\theta\right]. \tag{4.133}$$

Note that a complex bending stiffness has been introduced to account for the viscoelasticity of the solid material with $D' = D(1 + i\omega\eta_w)$, where η_w is the loss factor. For normal incidence or for frequencies $f \ll f_c$, the impedance reduces to $Z_w \approx im'\omega$, i.e., only the mass of the wall contributes to the acoustic impedance.

As an example, the impedance of a $h = 1.5$-cm thick vibrating gypsum wall with $\rho = 850$ kg/m^3, $E = 4.1$ GPa, $\nu = 0.3$ and $\eta_w = 0.1$ Pa/s, is given in Fig. 4.17. First, we can observe that the imaginary part of the impedance under normal incidence, $\theta = 0$ does not cross at zero. However, the imaginary part of the impedance vanishes for oblique incidence at some specific frequencies, i.e., at the coincidence frequencies of the wall, and thus the wall resonates. The real part of the impedance accounts for the losses of the element. At normal incidence, no loss is noticed, (4.133). At oblique incidence, the losses are activated, and they present a quadratic dependence on frequency.

4.4 Examples of Application

4.4.1 Absorption of Multilayered Porous Structure

One of the simplest configurations where the TMM has been widely used is in the modelling of absorption properties of multilayer rigidly-backed porous materials. These structures are widely used in room acoustics, as well as in automotive and aerospace applications. In most of these situations, the goal is to produce structures

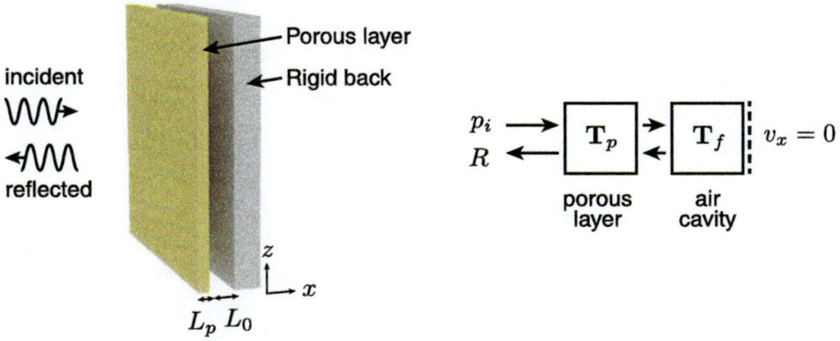

Fig. 4.18 Geometry of the configuration, composed of a porous layer with a rigidly-backed air cavity (plenum)

with high absorption coefficients. In this case, we consider that the porous material is rigidly backed, i.e., there is no transmission.

In this example we present a simple multilayer structure composed of a layer of porous material and a layer of air, i.e., an air plenum, as Fig. 4.18 shows. Note the extension of the proposed example to a higher number of layers is straightforward [12, 13].

The total transfer matrix of the system is composed of two transfer matrices, each one corresponding to the propagation in one media, that is written as

$$\mathbf{T} = \mathbf{T}_p \mathbf{T}_f, \tag{4.134}$$

where $\mathbf{T}_p$ is the transfer matrix of the porous layer of length L_p, given by (4.68), and $\mathbf{T}_f$ is the transfer matrix of the air gap of length L_0 given by (4.53). In this example, we consider a JCA model for the layer of porous material, with parameters corresponding to a mineral wool material: $\sigma = 20.6 \times 10^3$ Ns/m^4, $\alpha_\infty = 1.01$, $\phi_p = 0.98$ $\Lambda = 85$ µm, $\Lambda' = 2\Lambda$, while for the gap and the exterior media air at room temperature is considered with parameters $P_0 = 101325$ Pa, $\gamma_0 = 1.4$, $\rho_0 = 1.213$ kg/m^3, Pr $= 0.71$ $\eta_0 = 1.839 \times 10^{-5}$ Pa·s, $K_0 = \gamma_0 P_0$ Pa.

4.4.1.1 Normal Incidence

Once the full transfer matrix, $\mathbf{T}$, is obtained, the reflection and absorption coefficients are calculated using (4.48), (4.49). The total length of the multilayer structure was $L = 6$ cm. Two configurations are shown, with and without the air cavity. First, Fig. 4.19 shows the absorption of a structure calculated using a porous layer of length $L_p = 6$ cm (blue curve). In this case, the cavity of air was not included ($L_0 = 0$). Second, we show the absorption of a layer of porous material of $L_p = 3$ cm and an air cavity of $L_0 = 3$ cm (red curve). The absorbing features of the multilayer structure can be easily modelled by the TMM using a compact and fast calculation,

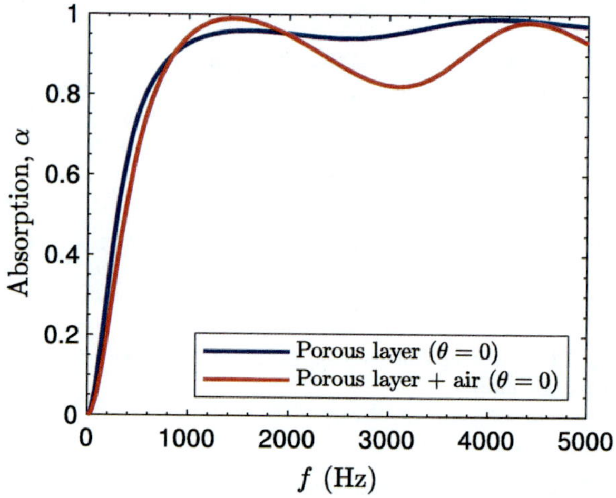

Fig. 4.19 Absorption at normal incidence angle of the rigidly-backed porous layer (black) and layer of porous material with a rigidly-backed air cavity (red)

allowing the fast optimization of the layer properties to, i.e., maximize the absorption [13]. Adding more layers with different parameters is straightforward by adding their corresponding transfer matrix terms.

4.4.1.2 Oblique Incidence

In a real situation, acoustic waves impinge the structure in more than one incidence angle. Thus, normal incidence absorption can be misleading as, in general, the absorption properties depend on the angle of incidence. We consider a plane wave with an angle of incidence θ defined with respect to the outward normal to the structure, see Fig. 4.20. Due to continuity of the transversal component of the wavevector along the interfaces, the wavenumber in the transverse direction x is the same in all media, leading to

$$k_{0,x} = k_0 \sin(\theta), \qquad k_{0,z} = \sqrt{k_0^2 - k_0^2 \sin^2(\theta)}, \tag{4.135}$$

$$k_{p,x} = k_0 \sin(\theta), \qquad k_{p,z} = \sqrt{k_p^2 - k_p^2 \sin^2(\theta)}, \tag{4.136}$$

where $k_{0,x}$ and $k_{p,x}$ are the transversal components (along x direction), and $k_{0,z}$ and $k_{p,z}$ are the normal component (along z) of the wavenumbers in the air and in the porous layer respectively, as shown in Fig. 4.20. Note that, due to symmetry considerations it is sufficient to consider only 2 components (k_z and k_z) to describe the problem: in the case of an oblique incidence with a component in the direction y, a simple rotation of the reference system can be applied to reduce the transversal wavenumber to one component.

The transfer matrix of the porous layer, $\mathbf{T}_p$, under oblique incidence is given by

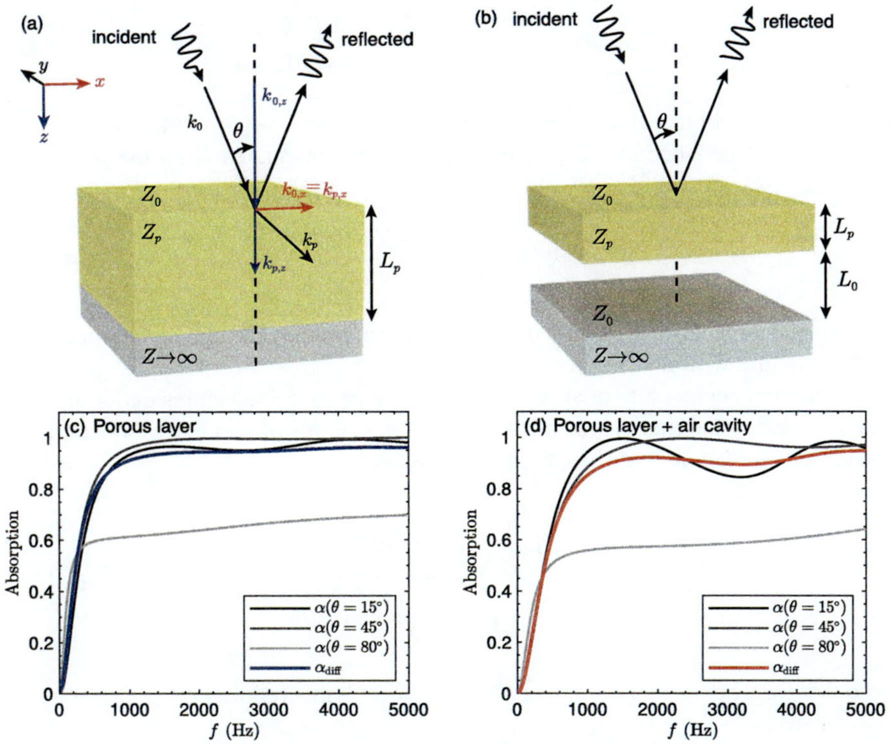

Fig. 4.20 **a** Geometry of the configuration, composed of a porous layer with rigidly-backed air cavity, **b** scheme of the transfer matrices used. **c** Absorption at of the rigidly-backed porous layer at several incidence angles (grey) and random incidence absorption (blue). **d** Absorption of layer of porous material with a air plenum and its corresponding random incidence absorption (red)

$$\mathbf{T}_p = \begin{bmatrix} \cos(k_{p,z}L_p) & iZ_p \dfrac{k_p}{k_{p,z}} \sin(k_{p,z}L_p) \\ \dfrac{i}{Z_p}\dfrac{k_{p,z}}{k_p} \sin(k_{p,z}L_p) & \cos(k_{p,z}L_p) \end{bmatrix}, \tag{4.137}$$

while the transfer matrix of the air layer, $\mathbf{T}_0$, is written as

$$\mathbf{T}_0 = \begin{bmatrix} \cos(k_{0,z}L_0) & iZ_0 \dfrac{k_0}{k_{0,z}} \sin(k_{0,z}L_0) \\ \dfrac{i}{Z_0}\dfrac{k_{0,z}}{k_0} \sin(k_{0,z}L_0) & \cos(k_{0,z}L_0) \end{bmatrix}. \tag{4.138}$$

Finally, the reflection coefficient of the rigidly-backed structure under oblique incidence is obtained by setting $v_z|_{z=L} = 0$. After some algebra, it can be written as a function of the elements of the total transfer matrix given in (4.13) as

$$R(\omega,\theta)=\frac{T_{11}(\omega)\cos(\theta)-Z_0T_{21}(\omega)}{T_{11}(\omega)\cos(\theta)+Z_0T_{21}(\omega)}. \tag{4.139}$$

Then, the absorption is calculated as usual as $\alpha(\omega,\theta)=1-|R(\omega,\theta)|^2$. In addition, the specific impedance at the first interface can be calculated from the reflection coefficient as

$$Z_s(\omega,\theta)=\frac{Z_0}{\cos(\theta)}\frac{(1+R(\omega,\theta))}{(1-R(\omega,\theta))}. \tag{4.140}$$

The specific impedance at the first interface is useful to study the impedance matching of the structure with the exterior medium.

A common approach to quantify the performance of the structures under oblique incidence is to integrate the absorption for all the angles of incidence to obtain the absorption coefficient in diffuse field, α_{diff}, as [46]

$$\alpha_{\text{diff}}(\omega)=\frac{\displaystyle\int_{\theta=0}^{\theta_m}\alpha(\omega,\theta)\sin(\theta)\cos(\theta)d\theta}{\displaystyle\int_{\theta=0}^{\theta_m}\cos(\theta)\sin(\theta)d\theta}. \tag{4.141}$$

Note that for a hemispherical integration, $\theta_m=\pi/2$, this reduces to

$$\alpha_{\text{diff}}(\omega)=\int_{\theta=0}^{\pi/2}\alpha(\omega,\theta)\sin(2\theta)d\theta, \tag{4.142}$$

known as the Paris' formula. This coefficient is closely related to the measurement of sound absorption under random incidence in a reverberant chamber following the standardized acoustic test ISO 354:2003 [58]. Note the experimental test should differ from the calculations due to many factors, including a non-diffuse field in the reverberant chamber or the absorption at the boundaries of the finite-dimension material [59]. Figure 4.20c–d show the predicted absorption for various angles of incidence for both configurations. The absorption coefficient in diffuse field is also shown, where it can be observed that under random incidence the absorption curve is smoothed.

4.4.2 Noise Transmission Through a Double Wall

Other interesting problem that has been widely solved using the TMM in the literature is the transmission of acoustic waves travelling through multiple elastic walls. This concerns the classical problem of noise transmission in building acoustics for sound-proofing, as well as sound transmission in automotive and aerospace applications [60]. In the following, we present the solution of a classical example of a double-

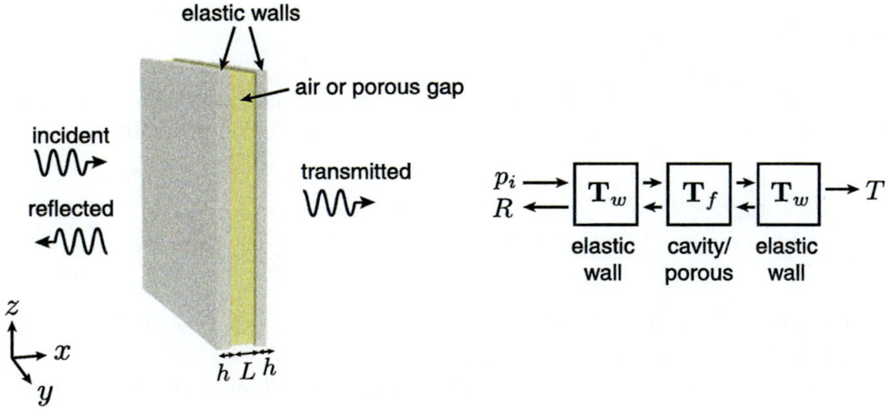

Fig. 4.21 Geometry of the configuration, composed of a double elastic wall separated by a cavity and scheme of the transfer matrices used

wall sound proofing system composed of two layers of gypsum board separated a distance L, with an air cavity between the walls, as shown in Fig. 4.21.

In this case, the total transfer matrix is given by

$$\mathbf{T} = \mathbf{T}_w^{[1]} \mathbf{T}_f \mathbf{T}_w^{[2]}, \tag{4.143}$$

where $\mathbf{T}_w^{[1,2]}$ are the transfer matrices of the gypsum boards given by (4.133), and $\mathbf{T}_f$ is the transfer matrix of the air cavity given by (4.53). The total transfer matrix under oblique incidence then

$$\mathbf{T} = \begin{bmatrix} 1 & Z_1 \\ 0 & 1 \end{bmatrix} \begin{bmatrix} \cos(k_{0,z}L) & iZ_0 \frac{k_0}{k_{0,z}} \sin(k_{0,z}L) \\ \frac{i}{Z_0} \frac{k_{0,z}}{k_0} \sin(k_{0,z}L) & \cos(k_{0,z}L) \end{bmatrix} \begin{bmatrix} 1 & Z_2 \\ 0 & 1 \end{bmatrix}, \tag{4.144}$$

where the perpendicular and transverse wavenumbers are given by

$$k_{0,z} = \sqrt{k_0^2 - k_0^2 \sin^2(\theta)} \quad \text{and} \quad k_{0,x} = k_0 \sin(\theta), \tag{4.145}$$

and Z_1 and Z_2 are the impedances of the infinite elastic walls given by (4.133). Once the total transfer matrix is calculated, we can obtain the transmission coefficient T using (4.37). Then, we can represent the transmission loss in logarithmic scale as

$$TL = -10 \log_{10} |T|^2, \tag{4.146}$$

which is useful to quantify the soundproofing performance of structures.

4.4.2.1 Normal Incidence

For normal incidence, $\theta = 0$, then $k_{0,z} = k_0$. We can calculate the total transfer matrix from (4.144), and then the transmission coefficient using (4.37) to obtain

$$T = \frac{1}{\left(1 + \frac{Z_1 + Z_2}{2Z_0}\right)\cos(k_0 L) + i\left(1 + \frac{Z_1 + Z_2}{2Z_0} + \frac{Z_1 Z_2}{2Z_0^2}\right)\sin(k_0 L)}. \quad (4.147)$$

For normal incidence (4.133) leads to $Z_1 \approx i\omega m'_1$ and $Z_2 \approx i\omega m'_2$, i.e., only the mass of each wall contributes to their impedance.

We show an example for normal incidence using a double-wall composed of two identical gypsum panels of $h = 1.5$ cm thickness, separated by a distance $L = 5$ cm with a density of $\rho = 850$ kg/m^3, Young modulus of $E = 4.1$ GPa, Poison ratio of $\nu = 0.3$ and losses modelled by $\eta = 0.05$. Air is assumed to fill the cavity. Figure 4.22 shows that (4.147) gives the same solution as the numerical evaluation of the transfer matrix (4.144). Moreover, approximate expressions for each regime (dashed lines), can be obtained, as well as the transition frequencies.

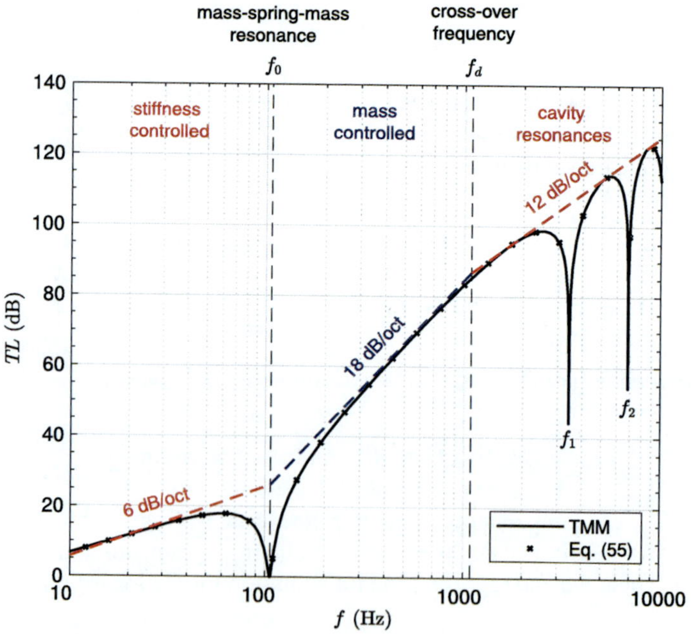

Fig. 4.22 Transmission loss (TL) at normal incidence angle using an air cavity (black) and using (4.147). Approximate solution are marked in dashed lines

4.4.2.2 Low Frequency Approximation

For low frequencies, $k_0 L \ll 1$, we can approximate $\sin(k_0 L) \approx k_0 L$ and $\cos(k_0 L) \approx 1 - k_0^2 L^2/2$. Then, neglecting high order terms we obtain

$$TL \approx 20 \log_{10} \left(\frac{\omega(m_1 + m_2)}{2Z_0} \right), \quad \text{for} \quad f < f_0. \tag{4.148}$$

This frequency regime corresponds to the well-known *mass law*: the double-wall system behaves as a single wall with total surface mass $(m'_1 + m'_2)$. Thus, the transmission decreases with the square of the frequency, i.e., the transmission loss increases 6 dB/octave.

This expression is valid for frequencies up to the double-wall resonance frequency, f_0, given by

$$f_0 = \frac{1}{2\pi} \sqrt{\frac{K_0}{L} \left(\frac{1}{m'_1} + \frac{1}{m'_2} \right)}. \tag{4.149}$$

At this particular frequency the double-wall resonates as a mass-spring-mass system, where walls contribute to both masses (m'_1 and m'_2) and the bulk modulus of the inner cavity ($K_0 = \rho_0 c_0^2$) contributes to the stiffness of the equivalent spring. At this resonance frequency a peak of transmission is observed, producing a deep dip in the transmission loss.

4.4.2.3 Medium Frequency Approximation

For intermediate frequencies up to $k_0 L \approx 1$, the transmission loss of the system can be simplified to

$$TL \approx 20 \log_{10} \left(\frac{\omega^3 m'_1 m'_2 L}{2\rho_0^2 c_0^3} \right), \quad \text{for} \quad f_0 < f < f_d. \tag{4.150}$$

In this regime the transmission increases with the sixth power of the frequency, i.e., the transmission loss increases with 18 dB/octave. Here, the transmission loss depends on the product of the masses and the separation of the panels, i.e., the transmission loss is the addition of the insulation of both walls plus an additional term as $TL \approx TL_1 + TL_2 + 20\log_{10}(2k_0 L)$, where $TL_1 = 20\log_{10}(\omega m'_1/2Z_0)$ and $TL_2 = 20\log_{10}(\omega m'_2/2Z_0)$ are the transmission losses of both independent walls, respectively. Note the soundproofing performance of the double-wall in this regime is greatly improved if compared with the performance of a panel with equivalent mass.

This expression is valid for frequencies up to $k_0 L \approx 1$, leading to a cut-off frequency of

$$f_d = \frac{c_0}{2\pi L}. \tag{4.151}$$

Above this frequency the resonances of the inner air cavity start to dominate.

4.4.2.4 High Frequency Regime

Above f_d, the cavity between the two walls shows several resonances given by $k_0 L = n\pi$, where $n = 1, 2, 3, ...$ Thus, the resonance frequencies are

$$f_n = n\frac{c_0}{2L}. \tag{4.152}$$

At each resonance frequency a peak of transmission is produced, leading to a dip in the transmission loss. The overall TL in this regime, neglecting the dips given by the resonances, can be approximated by

$$TL \approx 20\log_{10}\left(\frac{\omega^2 m_1' m_2'}{2Z_0^2}\right), \quad \text{for} \quad f_d < f. \tag{4.153}$$

In this regime the transmission decreases with the fourth power of the frequency, i.e., the transmission loss increases with a slope of 12 dB/octave. This is roughly equivalent to $TL \approx TL_1 + TL_2 + 6$ dB.

4.4.2.5 Oblique Incidence

As occurs with the reflection problem, the transmission at normal incidence does not describe a real situation where acoustic waves will impinge the walls at more than one incidence angle. In addition, for each oblique incidence angle there exist a frequency at which the corresponding wavelength of air matches the wavelength of the bending waves travelling trough the infinite wall. Then, the impedance of each wall should introduce a contribution modelling these phenomena, as shown in (4.133).

Under oblique incidence, the transmission coefficient is calculated as

$$T(\omega, \theta) = \frac{2}{T_{11} + T_{12}\cos(\theta)/Z_0 + T_{22} + T_{21}Z_0/\cos(\theta)}. \tag{4.154}$$

In addition, manipulating the total transfer matrix we can obtain the characteristic frequencies under oblique incidence as

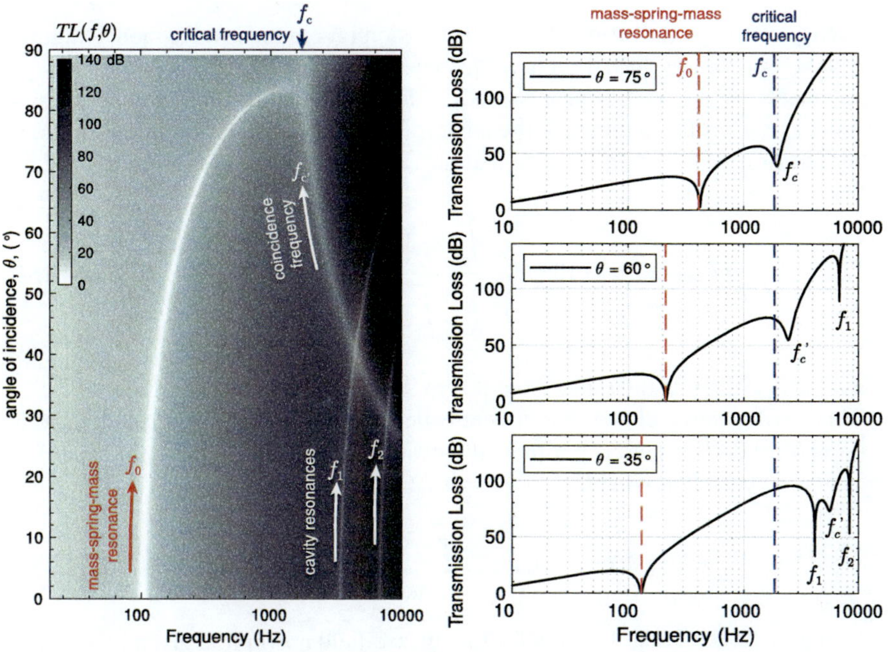

Fig. 4.23 (left) Transmission loss as a function of the frequency and the angle of incidence. The white dips mark the existence of resonance modes that led to transmitted energy. (right-bottom) Transmission loss near the normal, at $\theta = 35°$. The effect of the coincidence mode is visible at frequency f_c'. Cavity resonances are also clearly visible at f_1 and f_2. (Right-center) Transmission loss at $\theta = 60°$. The effect of the coincidence mode is visible at frequency f_c', cavity resonances have increased their frequency. (Right-top) Transmission loss near the grazing angle, at $\theta = 75°$. The effect of the coincidence mode dominates and its frequency almost match the critical frequency. The effect of cavity resonances in the transmission loss is negligible

$$\text{Mass-spring-mass resonance:} \qquad f_0(\theta) = \frac{1}{2\pi\cos(\theta)}\sqrt{\frac{K_0}{L}\left(\frac{1}{m_1'} + \frac{1}{m_2'}\right)},$$

$$\text{Coincidence frequency:} \qquad f_c'(\theta) = \frac{f_c}{\sin^2(\theta)},$$

$$\text{Cavity resonances:} \qquad f_n(\theta) = n\frac{c_0}{2L\cos(\theta)} \text{ with } n = 1, 2, 3, \ldots$$

where f_c is the critical frequency given by (4.132), i.e., the coincidence frequency for an incidence angle of $\theta = \pi/2$.

Figure 4.23 shows the transmission loss as a function of the incidence angle and frequency. We can observe that the overall transmission loss is reduced when the incident wavefront is tilted, mainly caused by the existence of a resonant transmission due to the coincidence effect. The mass-spring-mass resonance also shifts in frequency causing a reduction of the transmission loss in the low frequency regime. The cavity modes are shifted-up in frequency.

To quantify the overall transmission loss in realistic situation including random incidence, it is useful to integrate the angle-dependent transmission of the single to obtain the diffuse-field transmission loss as

$$TL_{\text{diff}}(\omega) = -10\log_{10}\frac{\displaystyle\int_{\theta=0}^{\theta_m} T(\omega,\theta)\cos(\theta)\sin(\theta)d\theta}{\displaystyle\int_{\theta=0}^{\theta_m}\cos(\theta)\sin(\theta)d\theta}. \tag{4.155}$$

where θ_m is usually in the range from 80° to 90°.

This coefficient is related (but not equivalent) to the standardized test defined in the ASTM E90-09 [61] to experimentally evaluate the insertion loss of a panel in a transmission chamber. Note that, if $\theta_m = \pi/2$, then

$$TL_{\text{diff}}(\omega) = -10\log_{10}\int_{\theta=0}^{\pi/2} T(\omega,\theta)\sin(2\theta)d\theta, \tag{4.156}$$

in analogy with the Paris' formula for the diffuse-field absorption given by (4.142).

4.4.2.6 Porous Layer in the Cavity

A common solution to increase the transmission loss at the cavity resonances is to add a layer of porous material in the cavity. Thus, we can modify the total transfer matrix as

$$\mathbf{T} = \mathbf{T}_w^{[1]}\mathbf{T}_f\mathbf{T}_p\mathbf{T}_f\mathbf{T}_w^{[2]}, \tag{4.157}$$

where $\mathbf{T}_p$ is the transfer matrix of the porous material given by (4.68) and $\mathbf{T}_f$ is the transfer matrix of the air gap between the porous material and the walls, given by (4.53). We consider a cavity of 5 cm, with a layer of porous material of thickness $L_p = 4$ cm, modelled using the JCA model and using a static air flow resistivity of $\sigma = 9958$ Ns/m^4, a high frequency limit of the tortuosity $\alpha_\infty = 1.03$, and open porosity of $\phi_p = 0.982$ a viscous characteristic length of $\Lambda = 203$ μm and a thermal characteristic length of $\Lambda' = 2\Lambda$ μm. Therefore, the air gap at each side of the porous layer is $L_0 = 0.5$ cm.

The transmission loss of the system is shown in Fig. 4.24. Note that, in this case, instead of obtaining an explicit analytical form like (4.147), it is more straightforward to directly evaluate numerically the transfer matrix of each element in (4.157). First, for normal incidence we can see that at the cavity resonances ($f_1, f_2, \ldots$) the transmission loss is greatly increased due to the damping of the purely acoustic modes propagating in the air between the layers. This is the principal benefit of

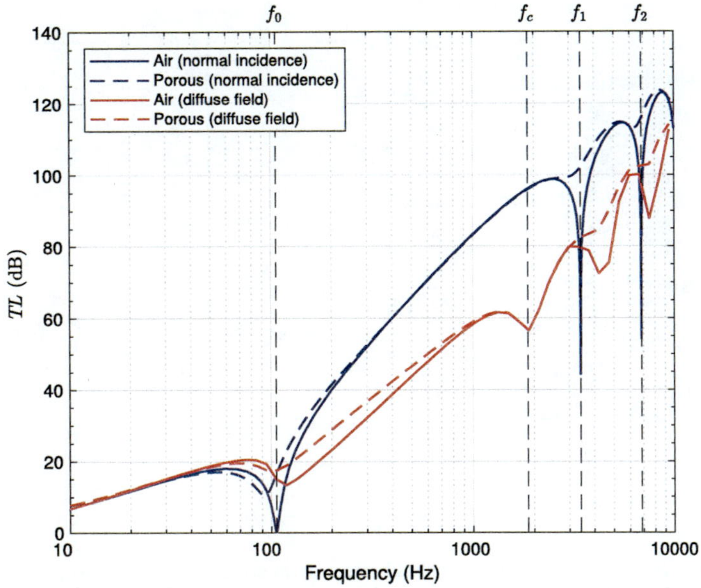

Fig. 4.24 Transmission loss of the double-wall system without (continuous) and with (dashed) the porous absorber. (Blue) normal incidence, (red) diffuse field

using a porous material inside the cavity. At the mass-spring-mass resonance (f_0) the transmission is not longer perfect but the value of the TL remains low.

On the other hand, for diffuse field calculations, we can see that the transmission loss is greatly reduced as compared with the normal incidence. This is mainly caused by the coincidence effect that sweeps an almost perfectly transmitted mode all over the spectra down to the critical frequency, as shown previously in Fig. 4.23. When the layer of porous material is introduced, its effect is clearly visible at high frequencies as it mainly attenuates the modes in the cavity. However, the contribution of the porous material to reduce the transmission of the coincidence effect is very low, and, therefore, the TL value at the critical frequency (f_c) remains low.

Note that, using the TMM the inclusion of more elements and layers to the system is straightforward. However, in a realistic situation in building acoustics other important phenomena should be considered: the transmission by the flanks, stubs and supporting systems, the finite size of the elastic walls and many other vibro-acoustic effects [60].

4.4.3 Phononic Crystals

We present the analysis of 1D periodic structures and metamaterials using the TMM. An acoustic periodic media is an arrangement of acoustic elements, unit cells, whose

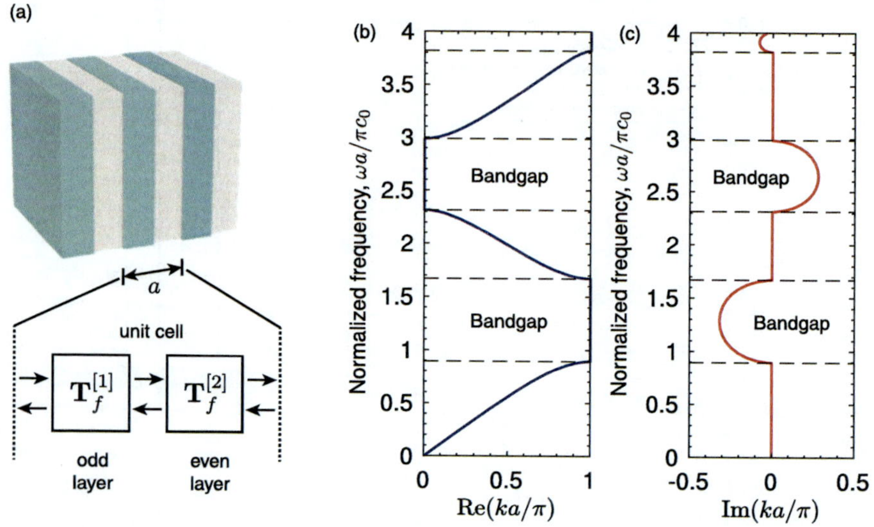

Fig. 4.25 **a** Geometry of a phononic crystal made of alternating layers of two materials. **b–c** Dispersion relations obtained for the phononic crystal, real and imaginary part of the wavenumber, respectively. Bandgap regions are shaded

structure is repeated in space. A simple example of such media is a periodic-multilayer structure composed of two fluid layers with alternating properties, as shown in Fig. 4.25a, also named a phononic crystal in analogy with photonic crystals in optics.

The total transfer matrix of the unit cell considered in this problem, $\mathbf{T}_{\text{u.c.}}$, is given by the product of the transfer matrices of each layer as

$$\mathbf{T}_{\text{u.c.}} = \mathbf{T}_f^{[1]}\mathbf{T}_f^{[2]}, \tag{4.158}$$

where $\mathbf{T}_f^{[1]}$ is the transfer matrix of the first fluid and $\mathbf{T}_f^{[2]}$ is the transfer matrix of the second fluid, both given by (4.53). Here, we consider a unit cell of total length a, where the length of each layer is a_1 and a_2 respectively. Due to the periodicity of the structure, it is sufficient to analyse a single unit cell to obtain the dispersion relations, as well as the transmission and reflection properties.

When analysing period structures it is important to note that their dispersion relations are defined for infinite period structures, while to obtain their reflection and transmission properties these structures must be bounded in space, i.e., the scattering properties depend on the number of unit cells considered.

4.4.3.1 Derivation of the Dispersion Relation of a 1D Multilayer System

We consider two fluids with different wavenumbers and impedances, given by $k_1 = \omega/c_1, k_2 = \omega/c_2$ and $Z_1 = \rho_1 c_1$ and $Z_2 = \rho_2 c_2$, where ρ_1 and ρ_2 are the density and c_1 and c_2 the sound speed of each fluid. The periodicity of the system is $a = a_1 + a_2$. Then, the transfer matrices of both fluids are given by (4.53), respectively

$$\mathbf{T}_f^{[1]} = \begin{bmatrix} T_{11}^{[1]} & T_{12}^{[1]} \\ T_{21}^{[1]} & T_{22}^{[1]} \end{bmatrix} = \begin{bmatrix} \cos(k_1 a_1) & iZ_1 \sin(k_1 a_1) \\ i\dfrac{1}{Z_1} \sin(k_1 a_1) & \cos(k_1 a_1) \end{bmatrix} \tag{4.159}$$

and

$$\mathbf{T}_f^{[2]} = \begin{bmatrix} T_{11}^{[2]} & T_{12}^{[2]} \\ T_{21}^{[2]} & T_{22}^{[2]} \end{bmatrix} = \begin{bmatrix} \cos(k_2 a_2) & iZ_2 \sin(k_2 a_2) \\ i\dfrac{1}{Z_2} \sin(k_2 a_2) & \cos(k_2 a_2) \end{bmatrix}. \tag{4.160}$$

In acoustics, for these periodic systems, the pressure and velocities accomplish the Bloch-Floquet theorem providing the following relation between the boundaries of the unit cell

$$\begin{bmatrix} p \\ v_x \end{bmatrix}_{x=0} = \mathbf{T}_{\text{u.c.}} \begin{bmatrix} p \\ v_x \end{bmatrix}_{x=a} = \begin{bmatrix} T_{11} & T_{12} \\ T_{21} & T_{22} \end{bmatrix} \begin{bmatrix} e^{-ika} p \\ e^{-ika} v_x \end{bmatrix}_{x=0}. \tag{4.161}$$

Rearranging this equation we obtain

$$\left(\begin{bmatrix} T_{11} & T_{12} \\ T_{21} & T_{22} \end{bmatrix} - \begin{bmatrix} e^{ika} & 0 \\ 0 & e^{ika} \end{bmatrix} \right) \begin{bmatrix} p \\ v_x \end{bmatrix}_{x=a} = 0. \tag{4.162}$$

Defining $\Lambda = e^{ika}$, this system only has solution if

$$\left\| \begin{bmatrix} T_{11} - \Lambda & T_{12} \\ T_{21} & T_{22} - \Lambda \end{bmatrix} \right\| = 0. \tag{4.163}$$

By using the condition of reciprocity: $T_{11}T_{22} - T_{12}T_{21} = 1$, we obtain the following dispersion relation

$$\cos(k_e a) = \frac{T_{11} + T_{22}}{2} = \frac{\text{Tr}(\mathbf{T}_{\text{u.c.}})}{2}. \tag{4.164}$$

Thus, by solving the matrix product between (4.159) and (4.160) we obtain the coefficients of the total transfer matrix as

$$T_{11} = \cos(k_1 a_1)\cos(k_2 a_2) - \frac{Z_1}{Z_2}\sin(k_1 a_1)\sin(k_2 a_2), \tag{4.165}$$

$$T_{12} = iZ_1\cos(k_2 a_2)\sin(k_1 a_1) + iZ_2\cos(k_1 a_1)\sin(k_2 a_2), \tag{4.166}$$

$$T_{21} = \frac{i}{Z_1}\sin(k_1 a_1)\cos(k_2 a_2) + \frac{i}{Z_2}\cos(k_1 a_1)\sin(k_2 a_2), \tag{4.167}$$

$$T_{22} = \cos(k_1 a_1)\cos(k_2 a_2) - \frac{Z_2}{Z_1}\sin(k_1 a_1)\sin(k_2 a_2). \tag{4.168}$$

On the one hand, the effective wavenumber can be calculated by the terms T_{11} and T_{22} using (4.15). Then, the effective wavenumber is given by

$$k_e = \frac{1}{L}\cos^{-1}\left[\cos(k_1 a_1)\cos(k_2 a_2) - \frac{Z_1^2 + Z_2^2}{2Z_1 Z_2}\sin(k_1 a_1)\sin(k_2 a_2)\right]. \tag{4.169}$$

On the other hand, the effective impedance, given by (4.16), can be obtained as $Z_e = \sqrt{T_{12}/T_{21}}$. Then we obtain

$$Z_e = \sqrt{\frac{Z_1^2 Z_2\cos(k_2 a_2)\sin(k_1 a_1) + Z_2^2 Z_1\cos(k_1 a_1)\sin(k_2 a_2)}{Z_2\cos(k_2 a_2)\sin(k_1 a_1) + Z_1\cos(k_1 a_1)\sin(k_2 a_2)}}. \tag{4.170}$$

Using these expressions the effective mass density, ρ_e, and bulk modulus, K_e, of the multilayer system are

$$\rho_e = \frac{Z_e k_e}{\omega} \quad \text{and} \quad K_e = \frac{Z_e \omega}{k_e}. \tag{4.171}$$

Note that, due to periodicity, the wavenumber is obtained only in the first Brillouin zone. However, for the calculations of the effective parameters ρ_e, K_e, a proper unwrap of the wavenumber is required.

Figure 4.25b–c show the dispersion relation of the system. First, we can see that the system is highly dispersive and the wavenumber does not depend linearly on frequency: the phase and group speeds strongly depend on frequency. Second, we can see that for certain frequency bands the wavenumber becomes imaginary. The waves that propagate in these frequency bands are evanescent, therefore, these frequency bands are called bandgaps, as explained in detail in Chaps. 1–3.

4.4.3.2 Transmission and Reflection of Bounded Phononic Crystals

The fact that waves propagating in bandgap frequencies are evanescent does not imply that energy cannot propagate in a phononic crystal. In fact, for bounded (finite) structures some amount of energy penetrates and, therefore, some acoustic transmission is expected.

For a finite phononic crystal of N unit cells the total transfer matrix is

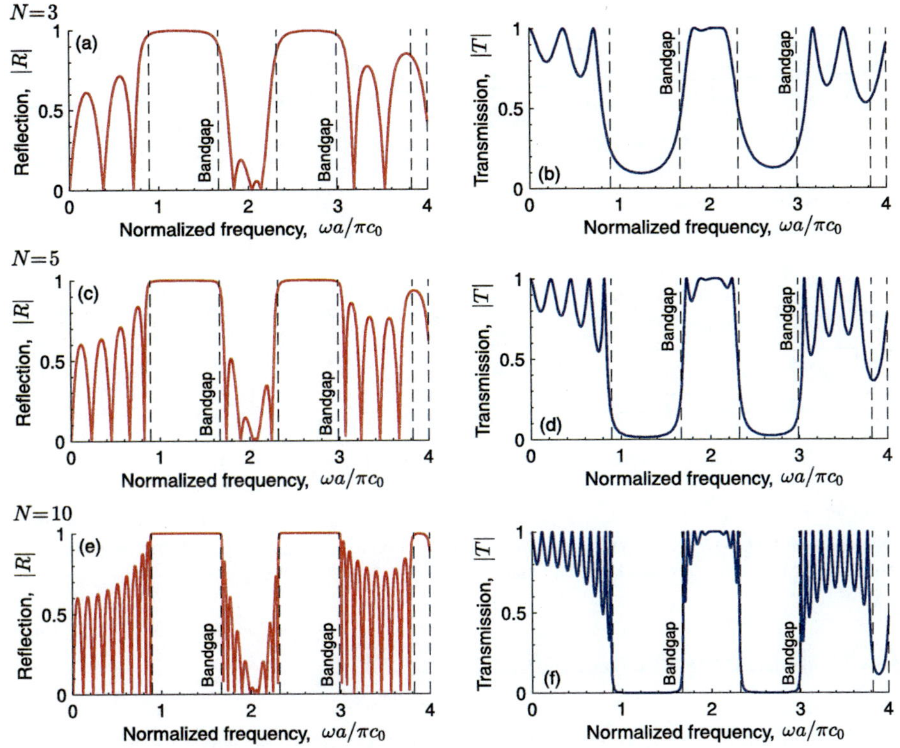

Fig. 4.26 **a, c, e** Reflection and **b, d, f** transmission coefficients of a phononic crystal of $N = 3, 5$ and 10 unit cells

$$\mathbf{T} = \mathbf{T}_{\text{u.c.}}^{N}, \tag{4.172}$$

where $\mathbf{T}_{\text{u.c.}}$ is the total transfer matrix given by (4.158).

Figure 4.26 shows the reflection and transmission coefficients of the finite phononic crystal using $N = 3, 5$ and 10 unit cells. We can see that, first, in the propagating bands there exist some peaks of transmission. In particular for frequencies $\omega a/\pi c_0 < 1$ there exist $N - 1$ peaks (neglecting the peak 0 Hz) of transmission corresponding to the Fabry-Pérot resonances of the bounded system. Obviously, at each peak of transmission the reflection vanishes as the considered system here is conservative. However, at the frequency bands corresponding to bandgap regions, the transmission drops. The energy is then strongly reflected by the structure by the constructive interference in the backward direction corresponding to the Bragg resonance.

As more unit cells are considered, the number of resonances increases accordingly: the filtering effect of the band-gap is more evident and the value of the transmission inside it decreases. However, in a finite structure without intrinsic looses some portion of the energy will always be transmitted.

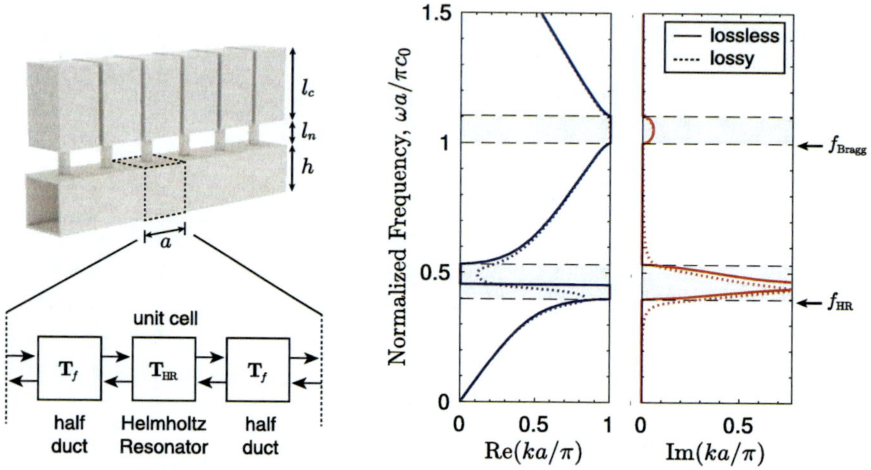

Fig. 4.27 Geometry of the negative bulk-modulus metamaterial composed of a waveguide loaded with an array of HR. Dispersion relations of the system, real and imaginary part of the wavenumber, respectively. Bandgaps are marked in shaded areas

4.4.4 Metamaterial Modelling Using TMM

The transfer matrix method has been applied to model acoustic waves propagating in locally resonant structures. Here, we show two examples, a negative bulk-modulus metamaterial made of a waveguide loaded with HR, and a negative mass-density metamaterial made of a waveguide with embedded elastic plates.

4.4.4.1 Negative Bulk Modulus Metamaterial: Waveguide Loaded with HRs

First, we show the model of a locally-resonant metamaterial composed of a waveguide loaded with HR, as shown in Fig. 4.27. The total transfer matrix of the unit cell is given by

$$\mathbf{T}_{\text{u.c.}} = \mathbf{T}_f \mathbf{T}_{\text{HR}} \mathbf{T}_f, \tag{4.173}$$

where $\mathbf{T}_f$ is the transfer matrix of the waveguide of length $a/2$ given by (4.53), and $\mathbf{T}_{\text{HR}}$ is the transfer matrix of the HR loaded in parallel, given by (4.90). Therefore, the transfer matrix of the unit cell is

$$\mathbf{T}_{\text{u.c.}}=\begin{bmatrix} \cos(k_f a') & iZ'_f \sin(k_f a') \\ i\dfrac{1}{Z'_f}\sin(k_f a') & \cos(k_f a') \end{bmatrix}\begin{bmatrix} 1 & 0 \\ \dfrac{1}{Z'_r} & 1 \end{bmatrix}\begin{bmatrix} \cos(k_f a') & iZ'_f \sin(k_f a') \\ i\dfrac{1}{Z'_f}\sin(k_f a') & \cos(k_f a') \end{bmatrix},$$

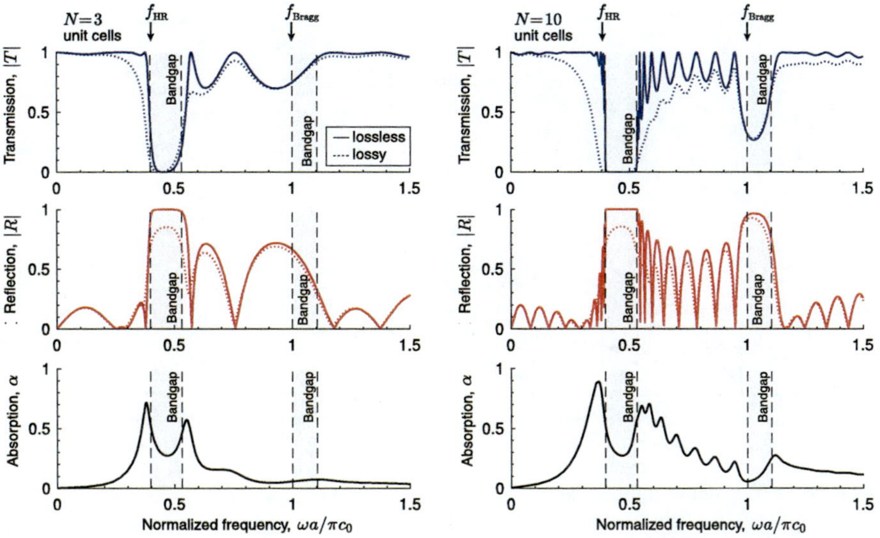

Fig. 4.28 Scattering of a negative bulk-modulus metamaterial composed of (left column) $N = 3$ unit cells and (right column) $N = 10$ unit cells. (Top row) transmission, (center row) reflection and (bottom row) absorption

where $a' = a/2$. For this example, we use a squared cross-section waveguide loaded by an array of HR, where the neck and cavity of the HR are also rectangular. Substituting (4.3.7) in (4.90), and evaluating (4.173), we can obtain the elements $T_{i,j}$ of the transfer matrix. Then, we can calculate the dispersion relations of the system by using (4.15).

Figure 4.27 shows the real and imaginary part of the wavenumber. We can see that strong dispersion is generated when the HRs are loaded in the waveguide. Bandgaps are also observed. In the example shown here, the first bandgap appears just above the resonance frequency of the HR. We can see that just in the limits of the bandgap strong dispersion is produced. Moreover, at $\omega a/\pi c_0 = 1$ a second bandgap is generated. This bandgap corresponds to the Bragg's resonance and is produced by the multiple scattering of the waves due to periodicity. Interestingly, by using the effective parameters of the duct we can include the thermoviscous losses. Obtaining an expression for the wavenumber including thermoviscous losses is straightforward but the expressions will be too much complex. Instead, in practice we simply evaluate the total transfer matrix numerically, as shown in Fig. 4.27. Note that, when introducing thermoviscous losses the dispersion relations are modified: in general, the wavenumber becomes complex in propagating bands and the strongly-dispersive flat propagating bands are smoothed.

Once the metamaterial is bounded we can obtain the reflection, transmission and absorption properties. To calculate the transmission, the reflection and the absorption of a structure with N unit cells we first evaluate the total transfer matrix as

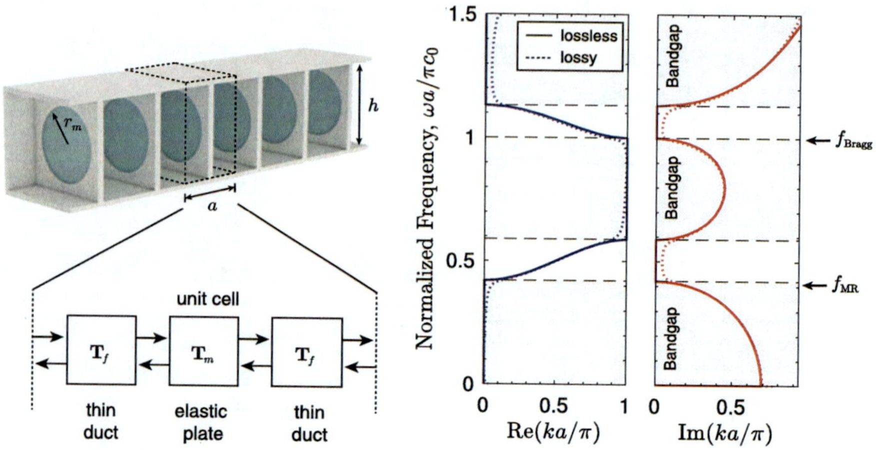

Fig. 4.29 (left) Scheme of the negative mass-density metamaterial with thin circular elastic plates. Dispersion relation of the system, (center) real and (right) imaginary part of the wavenumber. Bandgaps are marked in shaded areas

$$\mathbf{T} = \mathbf{T}_{\text{u.c.}}^{N}, \tag{4.174}$$

where $\mathbf{T}_{\text{u.c.}}$ is the total transfer matrix given by (4.173). Then, (4.37)–(4.43) are used. Figure 4.28 shows the scattering properties of the metamaterial using $N = 3$ and $N = 10$ unit cells. We can see that waves with frequencies in the range of the bandgap cannot easily propagate through the material. In the rest of the cases the structure is almost transparent to sound, specially at the Fabry-Pérot resonances of the structure. Note that the bandgap generated by the resonance introduces far more attenuation than the one caused by Bragg scattering. In addition, by adding more unit cells to the structure this effect becomes more evident. Finally, when losses are taken into account, absorption is produced mainly at the resonances of the structure. Absorption will be studied in detail in Chap. 5.

Finally, note that TMM also allows the calculation of non-periodic metamaterials based on this topology e.g., waveguides loaded by resonators of different geometry. In this case, the **T**-matrix of each unit must be calculated and, using (4.13), the total transfer matrix is obtained as the product of all matrices [28].

4.4.4.2 Negative Mass-Density Metamaterial: Waveguide with Thin Plates

In the same way, we can design a metamaterial composed of a waveguide with an embedded array of thin elastic clamped plates, as shown in Fig. 4.29. These materials present negative mass-density effective parameters.

The total transfer matrix of the unit cell is given by

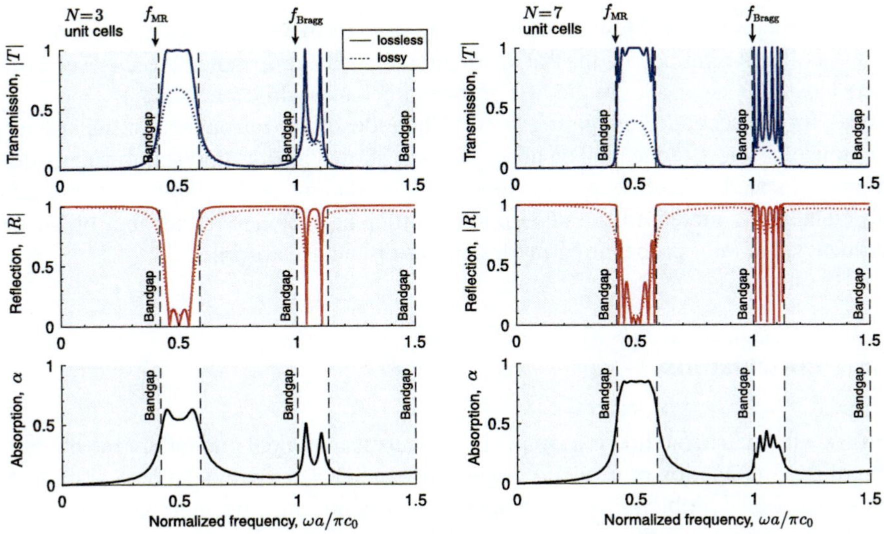

Fig. 4.30 Scattering of a negative mass-density metamaterial composed of (left column) $N = 3$ unit cells and (right column) $N = 7$ unit cells. (Top row) transmission, (center row) reflection and (bottom row) absorption

$$\mathbf{T}_{\text{u.c.}} = \mathbf{T}_f \mathbf{T}_{\text{PR}} \mathbf{T}_f, \tag{4.175}$$

where $\mathbf{T}_f$ is the transfer matrix of the waveguide of length $a/2$ given by (4.53), and $\mathbf{T}_{\text{PR}}$ is the transfer matrix of the plate resonator embedded in series, given by (4.93). Therefore, the transfer matrix of the unit cell is

$$\mathbf{T}_{\text{u.c.}}=\begin{bmatrix} \cos(k_f a') & i Z'_f \sin(k_f a') \\ i \dfrac{1}{Z'_f} \sin(k_f a') & \cos(k_f a') \end{bmatrix} \begin{bmatrix} 1 & Z'_r \\ 0 & 1 \end{bmatrix} \begin{bmatrix} \cos(k_f a') & i Z'_f \sin(k_f a') \\ i \dfrac{1}{Z'_f} \sin(k_f a') & \cos(k_f a') \end{bmatrix},$$

where $a' = a/2$. For this example, we use a squared cross-section waveguide loaded by an array of circular thin plates. Substituting (4.123) in (4.93), and evaluating (4.175), we can obtain the elements $T_{i,j}$ of the transfer matrix. Then, we can calculate the dispersion relations of the system by using (4.15).

The typical dispersion relation of this kind of system is shown in Fig. 4.29. We can see that, contrary to the previous case, waves in the low-frequency regime cannot propagate. However, above the resonance of the plates we observe that the wavenumber becomes real. In addition, a secondary bandgap is observed below Bragg frequency. Note when looses are introduced waves are damped in the propagation band and bandgap limits smooth.

The scattering properties of a system using $N = 3$ and $N = 7$ membranes are shown in Fig. 4.30. First, we can see that even for a system with few resonating elements the transmission of waves for frequencies below the resonance frequency of the membrane is very low, as shown in Fig. 4.30 (top). Only waves corresponding to the collective modes of the set of membranes can propagate. We can see that there exist $M = N$ modes in the first propagating band, while we found $M = N - 1$ modes for the second propagating band. When looses are introduced in the system the transmission is reduced. The absorption, as shown in Fig. 4.30 (bottom), results in peaks located at each resonance mode of the system. As the number of resonating elements is increased the absorption coefficient becomes higher and broad in frequency, but only propagating modes can be efficiently absorbed.

4.5 Conclusions

The transfer matrix method is a simple but powerful analytical method for the prediction of the propagation of acoustic waves that can be used to model a broad range of one-dimensional problems. In this chapter, we have presented the method in a general way, and summarized the most common building blocks to solve one-dimensional systems in Acoustics. We have included layers of fluid and porous media, ducts and waveguides of different geometries including thermoviscous losses, locally reacting elements as Helmholtz or quarter-wavelength resonators, viscoelastic plates and membranes, or micro-perforated panels.

Through the presented examples, we have reviewed the basic applications of the method to evaluate the scattering properties of acoustic structures of particular interest. These include multi-layered porous absorbers for room acoustics, the transmission problem in double-leaf wall for building acoustics, and the analysis of the dispersion of acoustic waves in periodic media and metamaterials using locally resonant elements. Using the generalized framework provided by the transfer matrix method, many one-dimensional wave-motion phenomena as reflection, transmission, absorption, attenuation and dispersion can be studied, as illustrated in the examples of this chapter. One of the most interesting features of the TMM is that it provides fast calculations of one-dimensional structures, allowing fast and robust optimization procedures. In summary, the TMM allows the modelling of complex materials and structures using a simple theoretical framework.

Acknowledgements The authors would like to thank Julien Leng for his useful comments. N.J. acknowledges financial support from Spanish Ministry through grant "Juan de la Cierva - Incorporación" program (IJC2018-037897-I). We acknowledge support from the Metaudible (ANR-ANR-13-BS09-003) and Metaroom (ANR-18-CE08-0021) projects.

4.6 Appendix—End Corrections

Radiation corrections must be applied when there exist discontinuities in the waveguides due to a change of section. The radiation correction of a waveguide segment due to cross-section changes, $\mathbf{M}_{\Delta l}$, is modelled by a *in-series* transfer matrix as

$$\mathbf{M}_{\Delta l} = \begin{bmatrix} 1 & Z_{\Delta l} \\ 0 & 1 \end{bmatrix}, \tag{4.176}$$

where $Z_{\Delta l} = -i\omega\rho\Delta l$ is the characteristic radiation impedance and Δl is the end-correction length that depends on the geometry of the discontinuity. Below we summarize some examples.

4.6.1 Change of Section in a Waveguide

When a discontinuity is caused by a change of section in a waveguide, as those produced between the ducts that conform the neck and the cavity of a Helmholtz resonator, the length correction, Δl, is approximated by [49],

$$\Delta l = 0.82\left[1 - 1.35\frac{r_n}{r_c} + 0.31\left(\frac{r_n}{r_c}\right)^3\right] r_n. \tag{4.177}$$

where r_n is the radius of the narrower waveguide, e.g., the neck, and r_c is the radius of the wider waveguide, e.g., the cavity of a cylindrical Helmholtz resonator.

4.6.2 Side Branch

Another kind of discontinuity arises when a duct is loaded in parallel to a principal waveguide, e.g., as the one shown in Fig. 4.11. The length of the end correction Δl is given by [50]

$$\Delta l = 0.82\left[1 - 0.235\frac{r_n}{r_s} - 1.32\left(\frac{r_n}{r_t}\right)^2 + 1.54\left(\frac{r_n}{r_t}\right)^3 - 0.86\left(\frac{r_n}{r_t}\right)^4\right] r_n, \tag{4.178}$$

where r_n is the radius of the loaded waveguide and r_t is the radius of the main waveguide. This correction only depends on the radius of the waveguides, so it becomes important when the length of the duct is comparable to its radius.

4.6.3 Periodic Array of Slits

Another kind of end correction comes from the radiation from a series of slits to the free media, as occurs in slotted panels. The radiation correction for a periodic distribution of slits can be expressed as [62].

$$\Delta l = h\phi_t \sum_{n=1}^{\infty} \frac{\sin^2(n\pi\phi_t)}{(n\pi\phi_t)^3}. \tag{4.179}$$

where h is the height of the slit, $\phi_t = h/d$ and d is the distance between slits. For $0.1 \leq \phi_t \leq 0.7$ this expression reduces to

$$\Delta l_{\text{slit}} \approx -\sqrt{2}\ln\left[\sin\left(\pi\phi_t/2\right)\right]/\pi. \tag{4.180}$$

Note that, while these end corrections are good approximations, to accurately model the radiation all Bloch waves must be calculated, e.g., using modal expansion methods [27].

References

1. P.M. Morse, K.U. Ingard, *Theoretical Acoustics* (Princeton University Press, 1986)
2. J. Allard, N. Atalla, *Propagation of Sound in Porous Media: Modelling Sound Absorbing Materials 2e* (Wiley, 2009)
3. J.S. Bolton, R.J. Yun, J. Pope, D. Apfel, Development of a new sound transmission test for automotive sealant materials, in *SAE transactions*, pp. 2651–2658 (1997)
4. B.H. Song, J.S. Bolton, A transfer-matrix approach for estimating the characteristic impedance and wave numbers of limp and rigid porous materials. J. Acoust. Soc. Am. **107**(3), 1131–1152 (2000)
5. T. Lung, A. Doige, A time-averaging transient testing method for acoustic properties of piping systems and mufflers with flow. J. Acoust. Soc. Am. **73**(3), 867–876 (1983)
6. K. Peat, A transfer matrix for an absorption silencer element. Journal of Sound Vibration **146**, 353–360 (1991)
7. W.T. Thomson, Transmission of elastic waves through a stratified solid medium. J. Appl. Phys. **21**(2), 89–93 (1950)
8. W. Lauriks, P. Mees, J.F. Allard, The acoustic transmission through layered systems. J. Sound Vib. **155**(1), 125–132 (1992)
9. M. Munjal, Response of a multi-layered infinite plate to an oblique plane wave by means of transfer matrices. J. Sound Vib. **162**(2), 333–343 (1993)
10. B. Brouard, D. Lafarge, J.-F. Allard, A general method of modelling sound propagation in layered media. J. Sound Vib. **183**(1), 129–142 (1995)
11. J. Bolton, N.-M. Shiau, Y. Kang, Sound transmission through multi-panel structures lined with elastic porous materials. J. Sound Vib. **191**(3), 317–347 (1996)
12. N. Jiménez, V. Romero-García, J.-P. Groby, Perfect absorption of sound by rigidly-backed high-porous materials. Acta Acust. Acust. **104**(3), 396–409 (2018)
13. N. Jiménez, V. Romero-García, A. Cebrecos, R. Picó, V.J. Sánchez-Morcillo, L.M. García-Raffi, Broadband quasi perfect absorption using chirped multi-layer porous materials. AIP Adv. **6**(12), 121605 (2016)
14. D. Johnston, D.K. Longmore, J.E. Drew, A technique for the measurement of the transfer matrix characteristics of two-port hydraulic components, in *ASME International Congress & Exposition* (1994)
15. A. Arnau, *Piezoelectric Transducers and Applications* (Springer, 2004)
16. S. Jiménez-Gambín, N. Jiménez, J.M. Benlloch, F. Camarena, Holograms to focus arbitrary ultrasonic fields through the skull. Phys. Rev. Appl. **12**(1), 014016 (2019)
17. ASTM E2611-09, Standard test method for measurement of normal incidence sound transmission of acoustical materials based on the transfer matrix method (2009). ASTME2611

18. M. Born, E. Wolf, *Principles of Optics: Electromagnetic Theory of Propagation, Interference and Diffraction of Light* (Elsevier, 2013)
19. B. Harbecke, Coherent and incoherent reflection and transmission of multilayer structures. Appl. Phys. B **39**(3), 165–170 (1986)
20. D. Whittaker, I. Culshaw, Scattering-matrix treatment of patterned multilayer photonic structures. Phys. Rev. B **60**(4), 2610 (1999)
21. D. Torrent, J. Sánchez-Dehesa, Acoustic resonances in two-dimensional radial sonic crystal shells. New J. Phys. **12**(7), 073034 (2010)
22. P. King, T. Cox, Acoustic band gaps in periodically and quasiperiodically modulated waveguides. J. Appl. Phys. **102**(1), 014902 (2007)
23. X. Zhu, K. Li, P. Zhang, J. Zhu, J. Zhang, C. Tian, S. Liu, Implementation of dispersion-free slow acoustic wave propagation and phase engineering with helical-structured metamaterials. Nat. Commun. **7**, 11731 (2016)
24. J. Li, L. Fok, X. Yin, G. Bartal, X. Zhang, Experimental demonstration of an acoustic magnifying hyperlens. Nat. Mater. **8**(12), 931 (2009)
25. P. Leclaire, O. Umnova, T. Dupont, R. Panneton, Acoustical properties of air-saturated porous material with periodically distributed dead-end pores. J. Acoust. Soc. Am. **137**(4), 1772–1782 (2015)
26. N. Jiménez, W. Huang, V. Romero-García, V. Pagneux, J.-P. Groby, Ultra-thin metamaterial for perfect and quasi-omnidirectional sound absorption. Appl. Phys. Lett. **109**(12), 121902 (2016)
27. N. Jiménez, V. Romero-García, V. Pagneux, J.-P. Groby, Quasiperfect absorption by subwavelength acoustic panels in transmission using accumulation of resonances due to slow sound. Phys. Rev. B **95**, 014205 (2017)
28. N. Jiménez, V. Romero-García, V. Pagneux, J.-P. Groby, Rainbow-trapping absorbers: Broadband, perfect and asymmetric sound absorption by subwavelength panels for transmission problems. Sci. Rep. **7**(1), 13595 (2017)
29. V. Romero-García, G. Theocharis, O. Richoux, A. Merkel, V. Tournat, V. Pagneux, Perfect and broadband acoustic absorption by critically coupled sub-wavelength resonators. Sci. Rep. **6**, 19519 (2016)
30. A.A. Fernández-Marín, N. Jiménez, J.-P. Groby, J. Sánchez-Dehesa, V. Romero-García, Aerogel-based metasurfaces for perfect acoustic energy absorption. Appl. Phys. Lett. **115**, 061901 (2019)
31. A.D. Pierce, *Acoustics: An Introduction to Its Physical Principles and Aplications. 1989 Edition*. Acoustical Society of America, 3rd edn. (1990)
32. M.L. Munjal, *Acoustics of Ducts and Mufflers with Application to Exhaust and Ventilation System Design*. (Wiley, 1987)
33. M. Munjal, Muffler acoustics, in *Formulas of Acoustics*, pp. 689–740 (Springer, 2004)
34. Y. Kulik, Transfer matrix of conical waveguides with any geometric parameters for increased precision in computer modeling. J. Acoust. Soc. Am. **122**(5), EL179–EL184 (2007)
35. M.R. Stinson, The propagation of plane sound waves in narrow and wide circular tubes, and generalization to uniform tubes of arbitrary cross-sectional shape. J. Acoust. Soc. Am. **89**(2), 550–558 (1991)
36. M. Delany, E. Bazley, Acoustical properties of fibrous absorbent materials. Appl. Acoust. **3**(2), 105–116 (1970)
37. F. Mechel, Extension to low frequencies of the formulae of delany and bazley for a bsorbing materials. Acta Acust. Acust. **35**(3), 210–213 (1976)
38. I. Dunn, W. Davern, Calculation of acoustic impedance of multi-layer absorbers. Appl. Acoust. **19**(5), 321–334 (1986)
39. W. Qunli, Empirical relations between acoustical properties and flow resistivity of porous plastic open-cell foam. Appl. Acoust. **25**(3), 141–148 (1988)
40. Y. Miki, Acoustical properties of porous materials-modifications of delany-bazley models. J. Acoust. Soc. Jpn. (E) **11**(1), 19–24 (1990)
41. R.T. Muehleisen, C.W. Beamer IV., B.D. Tinianov, Measurements and empirical model of the acoustic properties of reticulated vitreous carbon. J. Acoust. Soc. Am. **117**(2), 536–544 (2005)

42. R. Kirby, A. Cummings, Prediction of the bulk acoustic properties of fibrous materials at low frequencies. Appl. Acoust. **56**(2), 101–125 (1999)
43. M. Garai, F. Pompoli, A simple empirical model of polyester fibre materials for acoustical applications. Appl. Acoust. **66**(12), 1383–1398 (2005)
44. D.L. Johnson, J. Koplik, R. Dashen, Theory of dynamic permeability and tortuosity in fluid-saturated porous media. J. Fluid Mech. **176**, 379–402 (1987)
45. Y. Champoux, J.-F. Allard, Dynamic tortuosity and bulk modulus in air-saturated porous media. J. Appl. Phys. **70**(4), 1975–1979 (1991)
46. T.J. Cox, P. D'antonio, *Acoustic Absorbers and Diffusers: Theory, Design and Application* (CRC Press, 2009)
47. D. Lafarge, P. Lemarinier, J.F. Allard, V. Tarnow, Dynamic compressibility of air in porous structures at audible frequencies. J. Acoust. Soc. Am. **102**(4), 1995–2006 (1997)
48. G. Theocharis, O. Richoux, V.R. García, A. Merkel, V. Tournat, Limits of slow sound propagation and transparency in lossy, locally resonant periodic structures. New J. Phys. **16**(9), 093017 (2014)
49. J. Kergomard, A. Garcia, Simple discontinuities in acoustic waveguides at low frequencies: critical analysis and formulae. J. Sound Vib. **114**(3), 465–479 (1987)
50. V. Dubos, J. Kergomard, A. Khettabi, J.-P. Dalmont, D. Keefe, C. Nederveen, Theory of sound propagation in a duct with a branched tube using modal decomposition. Acta Acust. Acust. **85**(2), 153–169 (1999)
51. D.-Y. Maa, Potential of microperforated panel absorber. J. Acoust. Soc. Am. **104**(5), 2861–2866 (1998)
52. P. Cobo, F. Simón, Multiple-layer microperforated panels as sound absorbers in buildings: A review. Buildings **9**(2), 53 (2019)
53. Z. Skvor, Vibrating systems and their equivalent circuits. Stud. Electr. Electron. Eng. **40**, 11–241 (1991)
54. F. Bongard, H. Lissek, J.R. Mosig, Acoustic transmission line metamaterial with negative/zero/positive refractive index. Phys. Rev. B **82**(9), 094306 (2010)
55. C.-C. Sung, J. Jan, The response of and sound power radiated by a clamped rectangular plate. J. Sound Vib. **207**(3), 301–317 (1997)
56. D.E. Muller, A method for solving algebraic equations using an automatic computer. Mathematical tables and other aids to computation **10**(56), 208–215 (1956)
57. T.-Y. Huang, C. Shen, Y. Jing, On the evaluation of effective density for plate-and membrane-type acoustic metamaterials without mass attached. J. Acoust. Soc. Am. **140**(2), 908–916 (2016)
58. ISO 354:2003, Acoustics-measurement of sound absorption in a reverberation room (2003)
59. Y. Makita, T. Hidaka, Comparison between reverberant and random incident sound absorption coefficients of a homogeneous and isotropic sound absorbing porous material-experimental examination of the validity of the revised cosθ law. Acta Acust. Acust. **66**(4), 214–220 (1988)
60. T.E. Vigran, *Building Acoustics* (CRC Press, 2014)
61. ASTM E90-09, Standard test method for laboratory measurement of airborne sound transmission loss of building partitions and elements (2016)
62. F.P. Mechel, *Formulas of Acoustics*, 2nd edn. (Springer Science & Business Media, 2008)

Part II
Wave Propagation in Absorbing Metamaterials and Porous Media

Chapter 5
Acoustic Metamaterial Absorbers

Jean-Philippe Groby, Noé Jiménez, and Vicent Romero-García

Abstract Porous and fibrous materials provide effective and broadband acoustic absorption at mid/high audible frequencies. However, these traditional treatments result in thick and heavy layers when designed for low frequency audible sound. To overcome these limitations, in the recent years metamaterials have been proposed as an alternative to design sub-wavelength and efficient absorbing structures. In the current chapter, we review the recent advances in metamaterial absorbers and their underlying physics based on the analysis of the scattering matrix. Particularly, we exploit the physical interpretation of the eigenvalues and eigenvectors of the scattering matrix in the complex frequency plane to design efficient absorbers. We start by the discussion of the reflection problem, where these eigenvalues and eigenvectors collapse to the reflection coefficient. We state the conditions to design efficient and perfect metamaterial absorbers. Then, we increase the complexity of the scattering considering the transmission problem, where additional conditions are needed for designing perfect absorbers. Through this chapter we discuss different acoustic metamaterials based on metaporous absorbers, arrays of Helmholtz resonators and air cavities, as well as membrane and elastic-plate resonators.

J.-P. Groby (✉) · V. Romero-García
Laboratoire d'Acoustique de l'Université du Mans, UMR CNRS 6613, Le Mans Université, Avenue Olivier Messiaen, 72085 Le Mans, France
e-mail: jean-philippe.groby@univ-lemans.fr

V. Romero-García
e-mail: vicente.romero@univ-lemans.fr

N. Jiménez
Instituto de Instrumentación para Imagen Molecular, Universitat Politècnica de València, CSIC, Camino de Vera S/N, València, Spain
e-mail: nojigon@upv.es

N. Jiménez et al. (eds.), *Acoustic Waves in Periodic Structures, Metamaterials, and Porous Media*, Topics in Applied Physics 143,
https://doi.org/10.1007/978-3-030-84300-7_5

5.1 Introduction

When acoustic waves encounter in their path a physical material, a scattering event is produced: part of the energy of the initial wave is scattered-back creating a reflected wave, and some portion of the energy is transmitted through the structure. In lossless systems, energy conservation grants that the energy of the incident wave is equal to the sum of the energies of the reflected and transmitted waves, i.e. the scattered energy. However, real materials present intrinsic losses and some fraction of the total energy is irreversibly transformed into heat. Thus, a part of the incident wave energy is absorbed by the material and, as a consequence, the scattering can be strongly controlled if the intrinsic losses are properly adjusted. This can be done by tuning either the structure of the material or its acoustic properties.

The sources of acoustic dissipation can be mainly divided in viscous losses and thermal conduction. These viscothermal losses are at the basis of the absorbing mechanisms of traditional porous and fibrous absorbers providing effective and broadband acoustic absorption in the mid/high audible frequencies [1, 2]. For a rigidly backed material, as observed in practical situations when a layer of porous material is placed in front of an impervious wall, the resonance frequencies of the material match the quarter-wavelength resonance frequencies given by

$$f_n = \frac{(2n-1)c_p}{4L}, \tag{5.1}$$

where c_p is the sound speed inside the porous material, L the total thickness of the material and $n = 1, 2, 3 \ldots$. In this way, traditional absorbing materials become thick and heavy for low frequency sound [3], which is not desirable for practical applications such as sound-proofing or, in general, for building acoustics applications.

To overcome these limitations, the existing solutions rely on the use of other mechanisms to shift this kind of resonance to the *deep sub-wavelength* regime, i.e., the wavelength associated to the new resonance frequency is much larger than the characteristic size of the material. Arrangements of locally-resonant building blocks, including Helmholtz resonators (HRs) when using air cavities, or the use of membranes and elastic plates, have been used to design acoustic *metamaterials*. Using sub-wavelength resonators, the density of states at low frequencies can be increased. These artificial structures have been manufactured providing exotic dynamical features around their resonance [4, 5]. These include negative mass density [6], negative bulk modulus [7] or simultaneous negative mass density and bulk modulus [8] materials.

In the recent years, metamaterials have been also exploited to design efficient acoustic absorbers in the sub-wavelength regime [9]. Once the problem of the increasing the density of states at low frequencies is solved by using locally resonant building blocks, the second problem to be solved is the impedance matching to the surrounding medium. The resonant building blocks, basically based on air cavities, membranes or elastic plates, present a huge degree of freedom to modify their structure and, in addition, to control the amount of losses in an unprecedented way. This allows to design systems with sub-wavelength resonances being impedance matched to the surrounding medium, i.e., the realization of perfect and sub-wavelength absorbing structures. In the present chapter, we will review the recent advances in metamaterial absorbers and their underlying absorbing mechanisms.

5.1.1 The Scattering Matrix

When a physical system finds itself in the propagation path of an acoustic wave, part of the waves is deflected from its original course leading to scattered acoustic waves as depicted Fig. 5.1 for a 1D example.

The scattering matrix, **S**, relates the amplitudes of the incoming waves to those of the outgoing waves of any physical system. For one-dimensional or planar problems, the scattered waves take the form of reflected waves, i.e. backward-propagating waves, and transmitted waves, i.e. forward-propagating waves. The total pressures at both sides of a structure, considering a harmonic temporal dependence of the type $e^{-i\omega t}$, are then given by

$$p(x) = Ae^{-ikx} + Be^{ikx}, \text{ for } x < 0, \tag{5.2}$$

$$p(x) = Ce^{-ikx} + De^{ikx}, \text{ for } x > L, \tag{5.3}$$

as described in Fig. 5.1. Thus, the relation between the amplitudes of both waves is given by the **S**-matrix as

$$\begin{bmatrix} A \\ D \end{bmatrix} = \mathbf{S} \begin{bmatrix} C \\ B \end{bmatrix} = \begin{bmatrix} T^- & R^+ \\ R^- & T^+ \end{bmatrix} \begin{bmatrix} C \\ B \end{bmatrix}, \tag{5.4}$$

where the elements of the **S**-matrix are directly the transmission (T^+ and T^-) and reflection coefficients (R^- and R^+) from each side of the structure. The asymmetry of the structure translates into the fact that R^+ and R^- are different, while its possible non-reciprocity translates into the fact that T^+ and T^- are different. All along the present chapter, only reciprocal systems will be considered, whose fact implies the

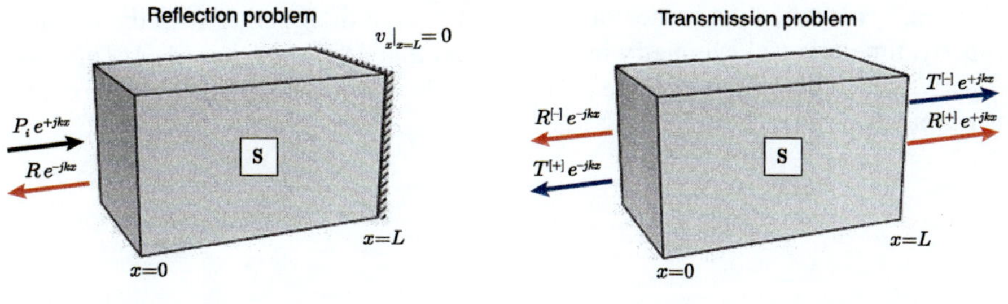

Fig. 5.1 Definition of incoming and outgoing waves for 1D reflection and transmission problems

transmitted fields are identical when the system is excited from one or the other side, i.e. $T = T^+ = T^-$. In this latter case, the **S**-matrix reduces to

$$\mathbf{S} = \begin{bmatrix} T & R^+ \\ R^- & T \end{bmatrix}. \tag{5.5}$$

The **S**-matrix is widely used in wave physics to characterize and interpret the wave scattering. In particular, the scattering matrix (5.5) possesses two eigenvalues given by

$$\lambda_1 = T + \sqrt{R^- R^+}, \quad \lambda_2 = T - \sqrt{R^- R^+}, \tag{5.6}$$

while the eigenvectors corresponding to λ_1 and λ_2 are

$$\vec{v}_1 = \left[\sqrt{R^- R^+}, R^-\right], \quad \vec{v}_2 = \left[R^+, -\sqrt{R^- R^+}\right], \tag{5.7}$$

respectively.

We can study some particular cases:

Symmetric systems
In case of symmetric physical systems both eigenvalues λ_1 and λ_2 reduce to the reflection coefficients of the associated symmetric and asymmetric problems:

$$\lambda_1 = R_{\text{sym}} = T + R \quad \text{and} \quad \lambda_2 = R_{\text{asym}} = T - R. \tag{5.8}$$

Reflection problems
For purely reflecting problems, where $T = 0$, both eigenvalues collapse into the reflection coefficient:

$$\lambda_1 = \lambda_2 = R. \tag{5.9}$$

In the lossless case, the scattering matrix is unitary, i.e. $\mathbf{S}^\dagger \mathbf{S} = \mathbf{S}\mathbf{S}^\dagger = \mathbf{I}$, where $\mathbf{S}^\dagger$ is the complex conjugate transpose of **S** and **I** is the identity matrix. In particular, this unitary property reflects the energy conservation relations, $|R^+|^2 + |T|^2 = 1$ and $|R^-|^2 + |T|^2 = 1$. The time-reversal symmetry leads in addition to the condition $\mathbf{S}^\star \mathbf{S} = \mathbf{S}\mathbf{S}^\star = \mathbf{I}$, where $\mathbf{S}^\star$ is the complex conjugate of **S**. Together with the unitary property, time-reversal symmetry induce the fact that $\lambda_i^\star = 1/\lambda_i$, $i = 1, 2$. This last property is the central point of the design of perfect absorbers by the complex frequency analysis.

5.1.2 Complex Frequency Plane Analysis

The poles and zeros of the eigenvalues as well as the eigenvectors of the **S**-matrix in the complex-frequency plane provide rich information, as they are identified with the resonances of the system. In this section, a lossless reflection problem, i.e., $T = 0$, will be considered for the clarity of the presentation and without loss of generality. The structure is subjected to leakage, because it is open to the surrounding medium. In addition, specular reflection is assumed. Therefore, the scattering matrix as well as the eigenvalue reduce to R. In the absence of losses, the unitary property provides $|R(\omega)|^2 = 1$, i.e., the structure is perfectly reflecting. The time-reversal symmetry provides

$$R^\star(\omega) = \frac{1}{R(\omega)}. \tag{5.10}$$

Introducing a complex frequency $\Omega = \omega + i\zeta$, this last property also implies

$$|R(\Omega^\star)| = \frac{1}{|R(\Omega)|} \tag{5.11}$$

and translates the fact that if the reflection coefficient possesses a pole associated to a resonance in the complex frequency plane at frequency Ω_p, it also possesses a zero at Ω_z such that

$$\Omega_p = \Omega_z^\star. \tag{5.12}$$

The location of the pole in the complex frequency plane Ω_p provides the quality factor of the associated resonance, which reflects the leakage of the structure

$$Q_{\text{leak}} = \frac{\omega_p}{2\zeta}. \tag{5.13}$$

In this way perfect absorption is associated with the zeros of the reflection coefficient of a resonant system. Therefore, the analysis of this coefficient in the complex frequency plane also provides useful insights for the design of absorbers [10, 11].

5.1.3 Perfect Absorption and Critical Coupling

The viscothermal losses induces a deformation of the complex frequency representation of the reflection coefficient. In case of very large quality factor, the reflection coefficient may be rewritten in the form

$$|R(\omega)|^2 \approx \frac{\left(Q_{\text{leak}}^{-1} - Q_{\text{diss}}^{-1}\right)^2}{\left(Q_{\text{leak}}^{-1} + Q_{\text{diss}}^{-1}\right)^2}, \tag{5.14}$$

where Q_{diss} represents the dissipation induced by the losses. It immediately follows that if

$$Q_{\text{diss}} = Q_{\text{leak}}, \tag{5.15}$$

the structure perfectly absorb the incoming energy. The losses induce a deformation of the complex frequency plane that enable to locate the zero of the reflection coefficient on the real frequency axis. In other words, *when the losses exactly compensate the leakage of the structure, perfect absorption is reached.* When this condition is fulfilled, the structure is *critically coupled* to the incoming wave. Therefore, tracking the poles and the zeros of the eigenvalues of the scattering matrix when modifying the amount of losses is a very powerful tool to design perfect absorbers in acoustics or electromagnetism [10–13].

5.2 Reflection Problems

The ability to perfectly absorb an incoming wave field by a sub-wavelength structure is advantageous for several applications in wave physics as energy conversion [14], time reversal technology [15], coherent perfect absorbers [16], or sound-proofing [17] among others. The solution to this challenge requires to solve a complex problem: reducing the geometric dimensions of the structure while increasing the density of states at low frequencies and finding the good conditions to match the impedance with the background medium.

Until now, efficient sound absorption has been mostly achieved by using porous materials or microperforated materials [2]. These materials attenuate sound waves through viscothermal losses arising from the interaction of the sound wave with the usually motionless skeleton. In these open and lossy systems, when the impedance matching condition is fulfilled, perfect absorption is achieved. Throughout this chapter, porous materials will refer to motionless skeleton materials, while poroelastic materials will refer to porous structures with deformable frames.

The present section is organized as follows. We start by describing the limits of absorption of regular acoustic porous materials. Then, metaporous and metaporoelastic materials, consisting on porous our poroelastic materials embedding rigid or resonant inclusions respectively, will be presented. These materials represent an efficient alternative to purely porous/poroelastic materials in the inertial regime, in the sense that they provide broadband perfect absorption of incident waves, the wavelengths of which are smaller than 10 times the thickness of the materials. We continue the discussion on the recent advances using metamaterials made of resonant air cavities as well as elastic plates or membranes for sub-wavelength perfect absorption.

5.2.1 Rigidly-Backed Porous Materials

Acoustic wave propagation in rigid-frame porous materials is usually modelled through semi-phenomenological fluid models [18, 19]. These equivalent-fluid models, presented in Chap. 4, rely on complex and frequency dependent equivalent density and compressibility, which respectively accounts for the viscous and thermal losses. Different regimes might be considered, but the viscous and inertial regimes are of particular interest because they are only related to the viscous losses. The Biot's frequency represents the transition frequency between these two regimes and is given by

$$f_{\text{Biot}} = \frac{\sigma\phi}{2\pi\rho_0\alpha_\infty}, \tag{5.16}$$

where σ is the flow resistivity, ϕ is the porosity, α_∞ is the tortuosity, and ρ_0 is the density of the saturating fluid. Below this frequency, the effective density of the porous material is mainly imaginary and the pressure field satisfies a diffusion equation, while above this frequency, the density is mainly real and the pressure field satisfies a Helmholtz equation with losses.

The lowest frequency absorption peak of a rigidly-backed porous layer corresponds to the so-called quarter-wavelength resonance. By using a first-order Taylor expansion around this frequency of the numerator of the reflection coefficient of a rigidly-backed layer of porous material, we end up with an optimal length which reads as [20]

$$L_{opt} = \frac{iZ_0}{\omega\rho_{eff}} + \frac{\pi c_{eff}}{2\omega}, \tag{5.17}$$

where Z_0 is the characteristic impedance of the air medium, and ρ_{eff} and c_{eff} are respectively the effective density and sound speed of the rigid-frame porous material. This relation imposes a purely complex value of the effective density and a purely real value of the effective sound speed. In other words, this can only be achieved for a frequency which is slightly above the viscous/inertial regime transition frequency. This results in two main limitations:

(i) Rigidly-backed porous materials can only efficiently absorb sound for wavelengths smaller than about 4 times their thickness,
(ii) Rigidly-backed porous materials can only perfectly absorb sound for frequencies higher than the so-called quarter wavelength resonance frequency when the latter corresponds to the viscous/inertial regime transition frequency [20]. Below this frequency, attenuation is too large, while above this frequency, attenuation is too poor.

Figure 5.2 illustrates this by using the Delaney–Blazley–Miki (DBM) model for the effective parameters of the porous material, given in Chap. 4. The absorption for a layer of porous material of different thicknesses is shown in Fig. 5.2b. As expected,

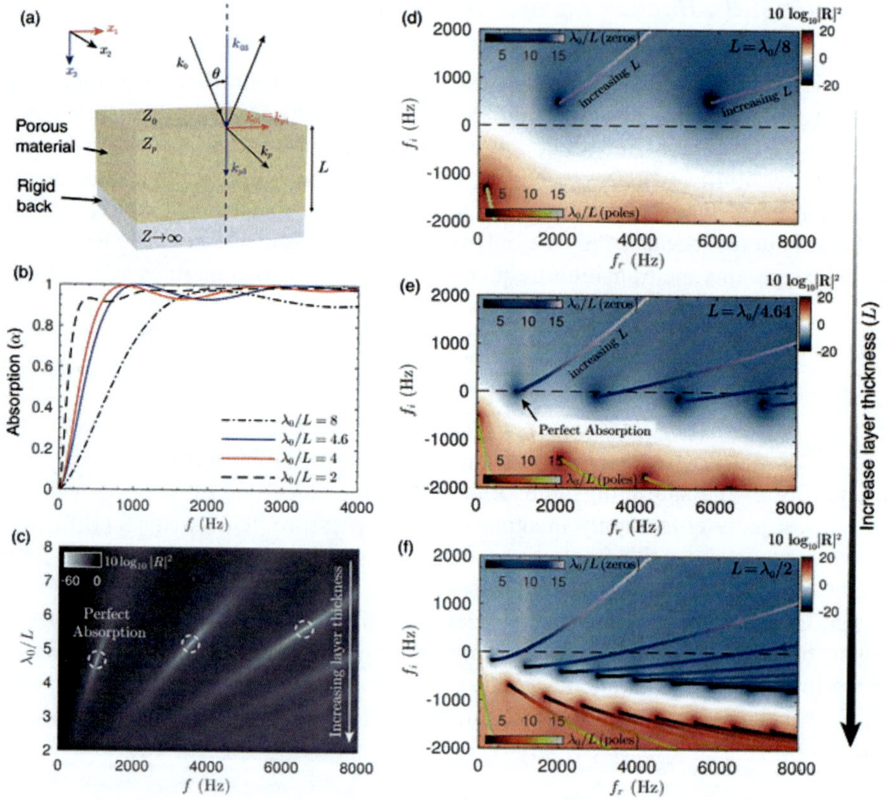

Fig. 5.2 **a** Scheme of a rigidly-backed layer of porous material. **b** Absorption for layer thickness $L = \lambda_0/8$, $\lambda_0/4.6$, $\lambda_0/4$, $\lambda_0/2$. **c** Map of the absorption showing perfect absorption peaks. **d–e** Complex frequency plane representation of the reflection coefficient where the location of the zeros and poles have been tracked as the thickness of the panel was varied. Adapted with permission from [20]

when the thickness of the layer is increased the material becomes more efficient in the low frequency regime as shown in Fig. 5.2c. This is due to the down-shift of the quarter wavelength resonance when the thickness of the layer is increased.

A complete picture of the absorption process can be observed using the complex frequency plane representation of the reflection coefficient. Note the reflection coefficient is the eigenvalue of the scattering matrix for the reflection problem. Figure 5.2d shows the complex frequency plane of the reflection coefficient for layer thinner that the optimal, $L = \lambda_0/8$, where $\lambda_0 = c_0/f_{\text{Biot}}$. Here, the zeros of the reflection coefficient (blueish dots) are located far from the real frequency axis, therefore, absorption is not efficient. Moreover, the zeros of the reflection coefficient are located in the opposite half-plane than their corresponding poles (reddish dots), therefore, the structure lacks of losses to produce efficient absorption at the resonances. We notice that each zero-pole pair is associated with a resonance, in this case each pair corresponds with the quarter-wavelength resonances given by (5.1).

When the thickness of the layer is increased, the resonance is shifted-down in frequency and, consequently, the location of the zeros of the reflection coefficient in the complex frequency plane changes. The trajectories of the zeros have been tracked, and, eventually, they cross the real frequency axis, as shown in Fig. 5.2e. This occurs, using the DBM model, for a layer of $L = \lambda_0/4.64 = L_{opt}$. Note using other equivalent-fluid models for the porous material this value can be slightly different [20]. Finally, if the layer is bigger than L_{opt} the zeros associated to the resonances of the layer are located into the same half-complex plane of their corresponding poles. In this case, the whole structure presents an excess of intrinsic losses and, even while the lowest absorption peak appears at lower frequency, the magnitude of the absorption is decreased and it becomes far from optimal.

Therefore, bulky and heavy structures are required to absorb sound at low frequency and very low frequency cannot be absorbed by homogeneous porous materials.

5.2.2 Metaporous Absorbers

An efficient way of designing broadband and thin sound absorbing materials is to combine porous materials with locally resonant elements in the same composite. Therefore, the viscothermal losses arising from porous materials will be used to attenuate sound, while the role of the locally resonant elements is twofold: on the one hand trapping the sound energy inside the composite at frequencies much lower than the quarter-wavelength resonance and, on the other hand, modifying/tuning the system attenuation. Figure 5.3 shows three possible topologies of metaporous absorbers: a sonic-crystal metaporous,[1] i.e., periodic array of rigid cylindrical inclusions embedded in a layer of porous material (Fig. 5.3a), a split-ring metaporous absorber, i.e., split ring resonators embedded in a layer of porous material (Fig. 5.3b) and a HR porous absorber, i.e., HR embedded in a layer of porous material (Fig. 5.3c).

Figure 5.4a shows the absorption coefficient of a porous layer without and with rigid inclusions. We notice that the first absorption peak for the case without inclusions is not unity due to the fact that this frequency lies in the inertial regime. However, once the array of rigid inclusions are included, an enhancement of the absorption of the structure at low frequency is clearly shown. This enhancement is due to the excitation of a trapped mode, which localize the acoustic energy between the inclusion set and the rigid backing, Fig. 5.4b. For a given filling fraction and position of the inclusions, the energy leakage of the structure can be exactly compensated by the intrinsic losses and a perfect absorption peak is reached for a frequency much lower than the so-called quarter wavelength frequency of the homogeneous porous layer [21, 22]. The Bragg interference arising from the interaction of the inclusions with

[1] Using only one layer of scatters it becomes a grating metaporous absorber.

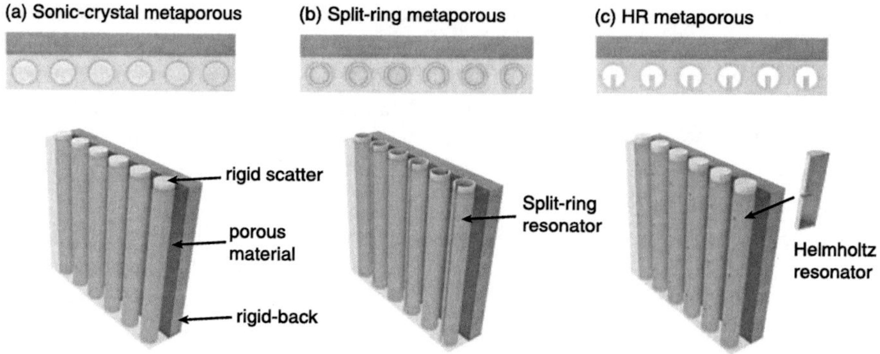

Fig. 5.3 Three topologies of metaporous absorbers. **a** Sonic-crystal metaporous: periodic rigid-cylindrical inclusions embedded in a layer of porous material. **b** Split ring resonators embedded in a layer of porous material. **c** HRs embedded in a layer of porous material

their image with respect to the rigid backing unfortunately leads to a large reflection of the structure and therefore a lower absorption, as depicted Fig. 5.4a. At higher frequencies, the absorption might be enhanced at the resonance frequencies of the porous plate, which are discretely excited thanks to the periodicity. These modified modes of the plate correspond to the Wood's anomaly in the presence of the porous plate, Fig. 5.4a. Note that the required filling fraction for perfect absorption is larger for three-dimensional inclusions than for two-dimensional ones. This implies that perfect absorption might be impossible to achieve for some porous materials by employing some three-dimensional inclusion shapes, typically sphere inclusions arranged on a cubic lattice.

The absorption can be further enhanced at low frequencies by embedding resonant inclusions, like split-ring resonators [23], HRs [24] or coiled structures [25]. At their resonance frequency, the acoustic energy is trapped in the resonant inclusions and attenuated by viscothermal losses. The resonant inclusions can be coupled with the rigid backing. Therefore, broadband absorption can be achieved by considering a supercell composed of various resonators with different orientations. It is worth noting here, that perfect absorption can only be achieved when the resonance of the inclusion lies in the inertial regime of the porous matrix, Fig. 5.4c–d. When it lies in the viscous regime, absorption peaks are usually noticed but cannot be unity either because the acoustic energy cannot travel to the resonators, as depicted Fig. 5.4e–f in the case of completely embedded HRs, or because the resonator can hardly resonate when filled with a porous material in the viscous regime. When a metaporous layer is flush-mounted in a partially lined duct, both the eigenvalues and the eigenvectors of the lower two Bloch modes coalesce at an exceptional point leading to a maximal sound attenuation because of a localized mode in the liner [26].

Whilst the effect of the Bragg's frequency may be erased by adjusting the resonance of resonant inclusions at this frequency, it can be more efficient to structure the rigid backing by adding quarter wavelength resonators or HRs [27, 28]. Therefore,

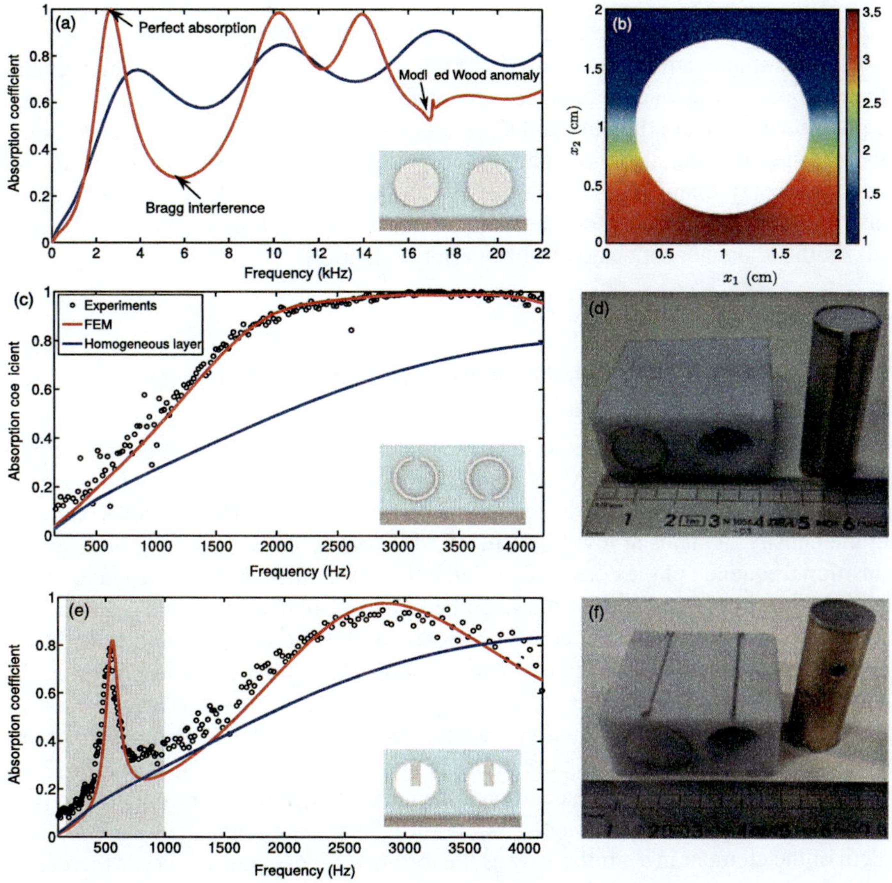

Fig. 5.4 **a** Absorption coefficient of a 2 cm-thick porous layer of Fireflex without inclusion (blue curve) and with 2 cm periodic rigid circular inclusions of radius 7.5 mm embedded in (red curve); **b** snapshot of the pressure field at the perfect absorption frequency (see arrows in (**a**)); **c** Absorption coefficient of a 2 cm-thick melamine layer without inclusion (blue curve), with 4.2 cm periodic supercell composed of two split-ring resonators the opening of which being in opposite directions embedded in (red curve), and experimental validation (**o**); **d** Sample picture; **e** Absorption coefficient of a 2 cm-thick melamine layer without inclusion (blue curve), with 2.1 cm periodic HRs embedded in (red curve), and experimental validation (**o**); **f** Sample picture

the embedded resonant inclusions leads to perfect absorption at low frequencies while the structured rigid backing avoids the Bragg interference associated to an absorption loss at higher frequencies in optimised structures. Such metaporous materials possess enhanced acoustic properties to drastically reduce treatment thickness.

Accounting for the possible motion of the skeleton paves the way to remove the limitation of the metaporous materials to the inertial regime. Beyond the apparent increasing of the stiffness of the poroelastic structure by the presence of the purely elastic inclusions, elastic resonators usually resonate at lower frequencies than acoustic ones and can therefore be efficiently designed for low frequency absorption purpose. For example, a preliminary study [29] has shown that the periodic embedment of viscoelastic shells in a poroelastic layer enables the enhancement of the poroelastic plate absorption coefficient thanks to the excitation of trapped modes, the volume mode of the shell, but also the higher order modes of the shell which occur at much lower frequency than the Biot's frequency. If optimally excited, elastic resonances may offer new possibilities for the design of sub-wavelength metaporoelastic materials, both for the acoustic and elastic energy mitigation.

5.2.3 Metamaterials Based on Resonant Air Cavities

As in the case of metaporous structures, the design of efficient sub-wavelength absorbers based on resonant cavities also requires avoiding the viscous regime. Using sub-wavelength resonators, with their associated viscothermal losses, allows increasing the density of states at low frequencies while down-shifting the viscous/inertial transition frequency of the constitutive elements. Once the resonators are made of air cavities with simple geometries, the viscothermal losses can be effectively accounted for in these constitutive elements by making use of the appropriate complex and frequency dependent effective density and compressibility, as described in Chap. 4. Usual designs are made of structured walls (also known as metasurfaces) comprising a periodic arrangement of quarter-wavelength resonators derived from diffraction gratings [3, 30] possibly making use of the Wood's anomaly [31–33], or of HRs [34–36]. The quarter-wavelength resonators or the neck of the HRs can be coiled in order to make them deep sub-wavelength [34, 37–40]. The coiling enhances the effective length of the element in a similar way as the tortuosity for porous materials therefore down-shifting the resonance frequency of the resonators, while keeping the resistivity of the structure relatively low in order to fulfil the impedance matching condition. Nevertheless, the effect of coiling on the losses is usually not accounted for in the models [34] but can be relevant if this coiling results in narrow channels.

Because of the sub-wavelength nature of these types of metasurface when the Wood's anomaly is not used, resonant surfaces made by periodically arranging HRs upon or slightly above an impervious substrate can be homogenized [41]. The method relies on the existence of a locally periodic boundary layer developed in the vicinity of the metasurface where strong near-field interactions of the resonators with each other and with the substrate take place. Effective boundary conditions which account for both the surface corrugation and the low-frequency resonances are derived allowing to critically couple the structure.

Another way of decreasing the resonance frequency of such a metasurfaces consists of exploiting the strong dispersion introduced by the resonators in order to induce slow sound regimes in the structure. Slow sound propagation is currently a growing topic in acoustics because of the direct analogy with electromagnetic induced transparency. This phenomenon appears when an opaque medium exhibits enhanced transmission in a narrow frequency window along with strong dispersion. This rapid

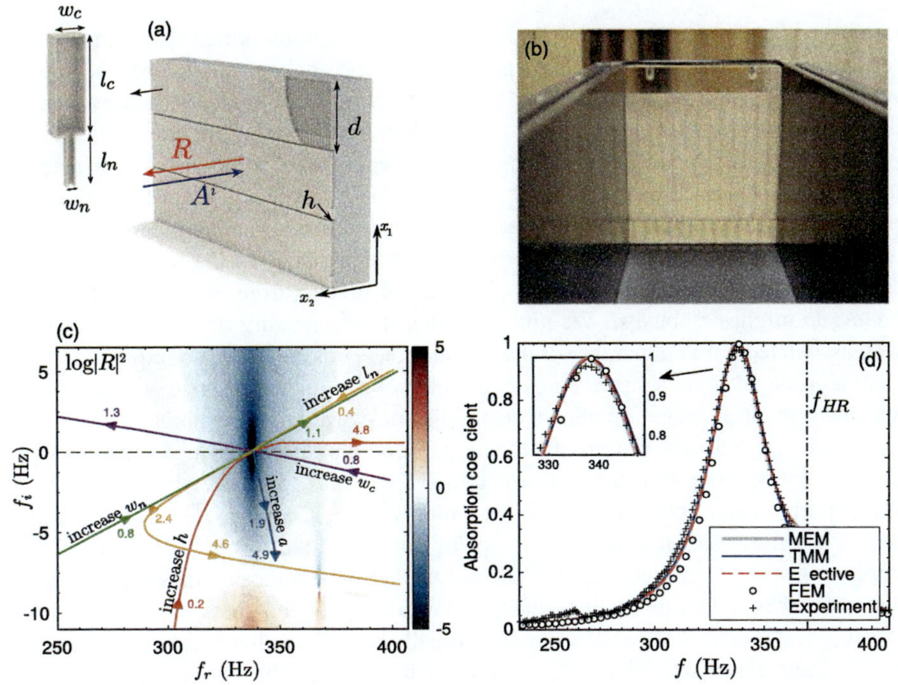

Fig. 5.5 **a** Sketch of the perfect absorber for reflection problem; **b** Picture of the perfect absorber sample; **c** Map of $\log|R|^2$ in the complex frequency plane; **d** Absorption coefficient calculated with different methods and measured. Adapted with permission from [36]

change in transmission leads to strong dispersion giving rise to slow phase or group velocity waves whose frequency is centred on a narrow transmission band [42]. In acoustics, most of the theoretical and experimental evidences of slow sound have been achieved by considering sound propagation in pipes with a series of detuned resonators (mostly HRs) separated by a sub-wavelength distance [43, 44], tuned or detuned resonators separated by half of the wavelength giving rise to a coupling between the resonators and the Bragg bandgap [45], in a sonic-crystal waveguide [46], in lined ducts [47], among others.

Only a few studies have been focusing on the dissipation (dispersion and attenuation) of slow sound propagation [48, 49], even if it has sometimes been noticed or discussed. Dissipation was considered as a side effect of an unexpected adverse reaction. The key point is to make use of slow sound propagation, which appears for a broadband frequency range below the bandgap associated with the local resonances, or in a narrow frequency band between the bandgaps associated with detuned resonator resonance (induced transparency band), together with the associated dissipation (attenuation and dispersion) to design a sound absorbing metamaterial.

The dispersion relation in a duct loaded by identical [50], detuned [12] or different [51] quarter-wavelength or Helmholtz [36] resonators was studied. From a description based on the locally reacting impedance condition mimicking the loading by the resonators, the effective parameters are derived. The analysis shows that the sound wave propagating in the loaded duct possesses the specific features of slow sound, in the induced transparency band, but also at low frequency. While the sound speed is necessarily dispersive in the induced transparency band, it possesses a plateau at low frequency with lower sound velocity than air. As expected from [52], the real part of the effective bulk modulus becomes negative inside the band gap due to the resonance of the loading resonators. Note that the presence of the losses induces a small amount of propagation in the band gap [49]. Therefore, the use of the term *band gap* might be abusive. We note here that the decreasing of sound speed is as well associated with an increase of the attenuation. Thus, very low sound speed wave cannot propagate over a large distance in acoustics due to the presence of losses. This specific type of propagation allows designing sub-wavelength resonators.

In case of rigidly-backed structures, these resonators can be slits loaded by other air cavities, e.g., quarter wavelength resonators [51] or HRs [36] as shown in Fig. 5.5a–b. When periodically arranged, they form a metasurface similar to dead-end porosity ones [53], the resonance of which appears for wavelength much larger than the impinging wave one. Note that the sub-wavelength resonance of these slits is only due to a drastic reduction of the sound speed inside the loaded ducts and not to coiling effect [34, 37]. These structures can be critically coupled to the exterior medium by tuning their geometry. When this condition is fulfilled, the perfect absorption is obtained because the attenuation of the structure exactly compensate its leakage and the zero of the scattering matrix (R in this case) exactly lies on the real frequency axis, see Fig. 5.5c. Several structures were designed by using this technique, the absorption band of some being broaden by using detuned sub-wavelength resonators in the unit cell and the absorption of others being deeply sub-wavelength [36], i.e. $L = \lambda/88$, see Fig. 5.5a–d. All these structures were validated experimentally. It was found in practice that the limiting parameters in low frequency is the manufacturing of the sample by stereo-lithography, the structure being not acoustically rigid any more when the walls become very thin.

It is also important to note that these structures are locally reacting ones and therefore can only be critically coupled for a single angle of incidence [54]. Effectively, increasing the number of resonant elements per unit cell will result in a single impedance value for each frequency.

Concerning broadband metasurfaces based on air cavities two basic phenomena must be included by the structure. The first is a mechanism to produce a large amount of resonance, while the second concerns a physical mechanism to control the losses for each resonance. The common approach is to design a structure with different set of resonators, each one producing a resonance at a particular frequency. To obtain broadband absorption the structure must be impedance matched with the exterior not only around the resonance frequency, but over a broad frequency range. This was achieved first by Wu et al. [3] using a set of QWRs with different lengths as those used in phase-grating sound diffusers. While each resonator presented a different resonance and they were efficiently coupled to the air, a porous screen placed in top of the structure provided viscothermal losses to produce broadband and quasi-perfect sound absorption. This structure was then improved by coiling the QWRs [30] to obtain optimal absorbing structures and it was demonstrated that the resonances must be distributed logarithmically in frequency to obtain optimal absorption. In addition, it was proved that causality imposes a physical limit between the thickness of an absorber and the frequency range of the absorbed spectrum [30] for any real acoustic structure. This important relation should be considered prior to any optimization procedure during the design of broadband absorbers of any kind.

Recent advances include the combination of slow-sound propagation due to QWRs and porous materials [55], QWRs and perforated sheets in honeycomb structures [56], split ring resonators [57], labyrinthine metasurfaces [58], coupled micro-slits [59], acoustic *black-hole* absorbers [60, 61] or bio-inspired metamaterials [62].

Finally, it is worth noting here that perfect absorbers based on metamaterials have been used to design sub-wavelength-thickness sound diffusers based on slow-sound [63, 64]. In this case, some of the unit cells corresponding to ideal absorbers are replaced by absorbing metamaterials leading to high-accuracy sound diffusers designed using number-theory sequences. For example, perfect sound absorption can be used to generate ternary-sequence sound diffusers using thin panels, and to contribute in the generation of uniform scattering patterns for broadband sound metadiffusers.

5.2.4 Metamaterials Based in Resonant Elastic Plates and Membranes

Other mechanism to produce sub-wavelength resonances is the use of elastic plates and membranes. Thin elastic plates and prestressed membranes present low bending stiffness that allows flexural waves to travel through their surface with a relatively slow phase speed.[2] Thus, when these materials are bounded, and they vibrate constrained by some boundary conditions, usually clamped ones, their resonance frequency can appear in the low frequency regime. In addition, at the resonance frequency, the impedance of the membrane can be efficiently coupled with the impedance of the air, giving rise to a broad range of efficient and sub-wavelength absorbers based on these resonating elements.

One of the first metamaterials using membranes for sound absorption was presented by Mei et al. [17] in 2012, where it was reported efficient absorption using membranes decorated with asymmetric rigid platelets. Decorated membranes were previously proposed in the past to design negative mass-density metamaterials [68]. At the resonance, the energy density in the membrane is increased several orders of magnitude with respect to the incident wave energy density. However, the platelets and the membrane develop flapping oscillations that are weakly coupled with radiation modes. Therefore, energy remains trapped in the structure and it is transformed into heat efficiently due to the intrinsic viscoelastic losses of the membranes. This

[2] See Chap. 4 for details about the modelling.

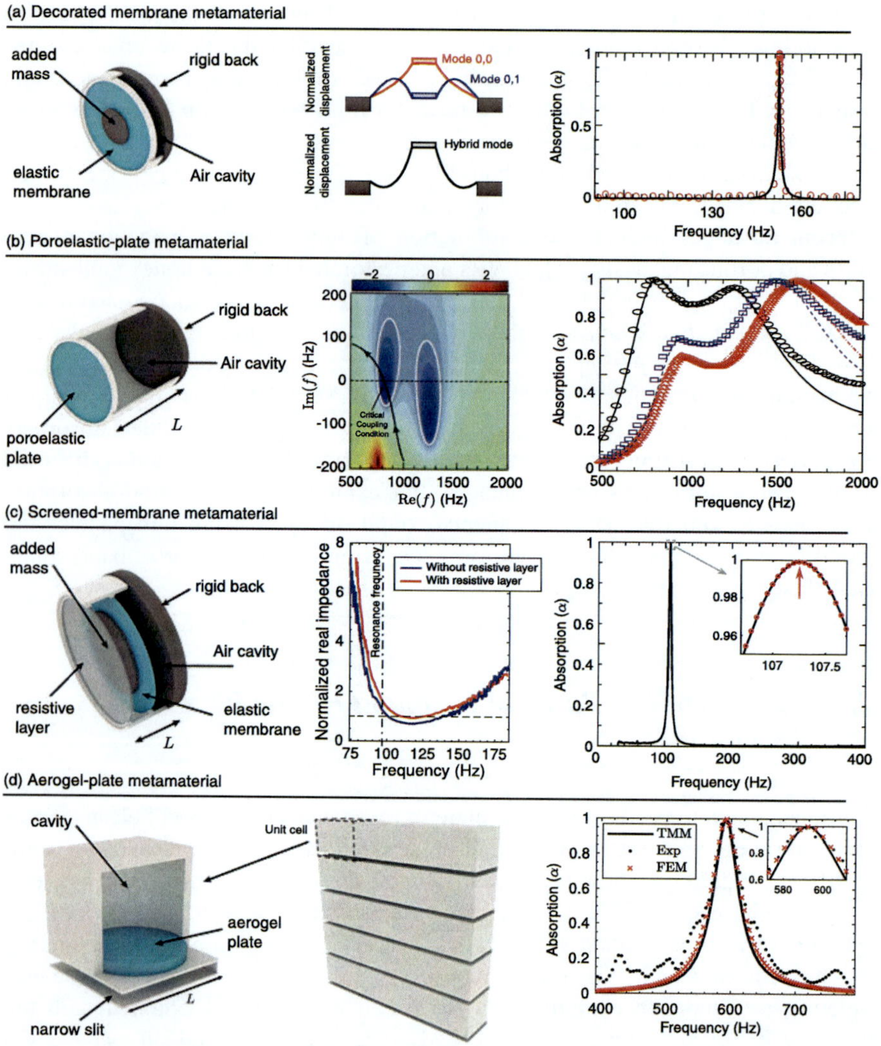

Fig. 5.6 **a** Decorated membrane metamaterial using a membrane with a rigid attached mass. The intrinsic losses of the system are fixed by the viscoelasticity of the membrane and the cavity, the mass and the tension of the membrane allow tunning the resonance frequency. Adapted with permission from [65]. **b** Poroelastic-plate metamaterial using a poroelastic plate backed by a tuned air cavity. The intrinsic losses of the system are fixed by the porous material, while the cavity is tuned to critically-couple the structure producing perfect absorption. Adapted with permission from [13]. **c** Screened-membrane metamaterial using a membrane with a rigid attached mass backed by a cavity. A resistive sheet is placed in front. The resonance frequency is fixed by the resonance of the decorated-membrane and the cavity, while the resistive screen allows the accurate control of the intrinsic losses of the system. This enables the critical-coupling and a peak of perfect absorption is observed. Adapted with permission from [66]. **d** Aerogel-plate metamaterial using an viscoelastic plate of aerogel material in a slow-sound configuration. The resonance frequency of the system is tuned using the cavity and the Aerogel properties, while the intrinsic losses of the system can be controlled by tuning the viscothermal processes in the narrow slit. Adapted with permission from [67]

approach was studied in more detail later by Yang et al. [69], while the vibro-acoustic behaviour of these resonators including attached masses of several geometries has also been studied for membranes [70] and for thin elastic plates [71].

Later, Ma et al. [65] reported efficient absorption by using decorated membranes using a similar set-up than the one shown in Fig. 5.6a. It consists, essentially, in an elastic membrane where a rigid platelet was glued to its surface. The added mass by the platelet contributed, first, to downshift the resonance frequency of the membrane. Then, when a cavity is placed behind the decorated membrane, a hybrid mode was observed [65, 69]. At this resonance, the system becomes impedance-matched to airborne sound at low frequencies and, therefore, a peak of absorption is observed.

Other approaches are the use of thin poroelastic plates to increase the intrinsic losses of the material and enhance the impedance-coupling of the clamped elastic plate with the air, as illustrated in Fig. 5.6b. Using poroelastic plates, Romero-García et al. [13] proposed efficient and broadband absorbers where the absorption of energy was demonstrated to be perfect. In this configuration, the intrinsic losses of the system are fixed by the viscothermal and viscoelastic processes in the poroelastic plate, while the resonance frequency was tuned controlling the length of the cavity. A similar approach has been also reported by Aurégan [66], where a deep-subwavelength metamaterial was designed using a decorated membrane and a cavity to fix the resonance frequency of the system (mainly controlling the imaginary part of the impedance of the system), and a resistive layer consisting in a thin metal-wire mesh was placed at the front to control the intrinsic losses of the system (the real part of the impedance of the system). Using this set-up, as shown in Fig. 5.6c, nearly perfect absorption was reported for a structure 200 times thinner than the wavelength at around 100 Hz.

Recently, exotic materials such as aerogels were used as resonating plates to design absorbing metamaterials [67], as illustrated in Fig. 5.6d. In this case, the structure included a narrow slit. In this way, the resonance frequency of the system can be tuned, mainly, by the resonance frequency of the resonant plate together with the length backing cavity. The use of the aerogel material introduced some viscoelastic losses. In addition, the introduction of the narrow slit adds an extra degree of freedom for the accurate tuning of the intrinsic losses of the system by introducing viscothermal losses. Then, by modifying the geometry both, resonance frequency and intrinsic losses can be adjusted to produce perfect sound absorption [67].

Finally, broadband absorption have been proposed using combination of decorated-membrane metamaterials [72]. Moreover, membrane metamaterials can also be used to design efficient absorbers for transmission problems, as those using degenerate resonators [73], but this will be explained in the next section.

5.3 Transmission Problems

In the previous section, we have reviewed the case of rigidly-backed materials. In a reflection problem both eigenvalues collapse to one single value, the reflection coefficient. Thus, to obtain perfect sound absorption, the control of single scattering event is sufficient, i.e., efficient absorbers can be produced using a single resonator. However, when transmission is allowed, the scattering matrix presents two different eigenvalues. To produce perfect absorption both eigenvalues must vanish, i.e., the structure must be critically coupled simultaneously for the two eigenvalues of the scattering matrix.

5.3.1 *Absorption and Fabry–Pérot Modes: Limits of Absorption by Single Resonators in the Transmission Problem*

Let us start considering a slab of homogeneous material of length L, with effective properties under normal incidence, as sketched in Fig. 5.7a. The reflection and transmission coefficients are linked to the effective parameters, i.e., the complex and frequency dependent effective bulk modulus, κ_{eff}, and effective density, ρ_{eff} as

$$R = \frac{i(\bar{Z}^2 - 1)\sin(k_{eff}L)}{2\bar{Z}\cos(k_{eff}L) - i(\bar{Z}^2 + 1)\sin(k_{eff}L)}, \tag{5.18}$$

$$T = \frac{2\bar{Z}}{2\bar{Z}\cos(k_{eff}L) - i(\bar{Z}^2 + 1)\sin(k_{eff}L)}, \tag{5.19}$$

with normalized effective impedance and effective wavenumber as

$$\bar{Z} = \frac{\rho_{eff}\kappa_{eff}}{\rho_0\kappa_0}, \quad k_{eff} = \omega\sqrt{\frac{\rho_{eff}}{\kappa_{eff}}}, \tag{5.20}$$

where ρ_0 and κ_0 are the density and bulk modulus of the surrounding medium, typically air for sound absorbing materials.

Finally, the absorption of the system is given by

$$\alpha = 1 - |T|^2 - |R|^2. \tag{5.21}$$

This material can be, e.g., a slotted rigid material, being the separation between slits much smaller than the wavelength of the impinging wave. Thus, when the slits are thick in comparison with the viscous and thermal boundary layers, the effective parameters of the material are the same as the surrounding medium and the intrinsic losses are almost negligible. A homogeneous structure like this presents symmetrical and anti-symmetrical resonance modes, as shown in Fig. 5.7b. These modes are in fact the Fabry–Pérot (FP) resonances of the slab of material. Each resonance, is associated with a zero and a pole which are visible in the complex-frequency plane representation of the eigenvalues of the scattering matrix shown in Fig. 5.7e–f. Note symmetrical resonances (marked in blue) appear in the λ_1 complex-frequency plane and anti-symmetrical resonances (marked in red) are represented by a zero-pole pair

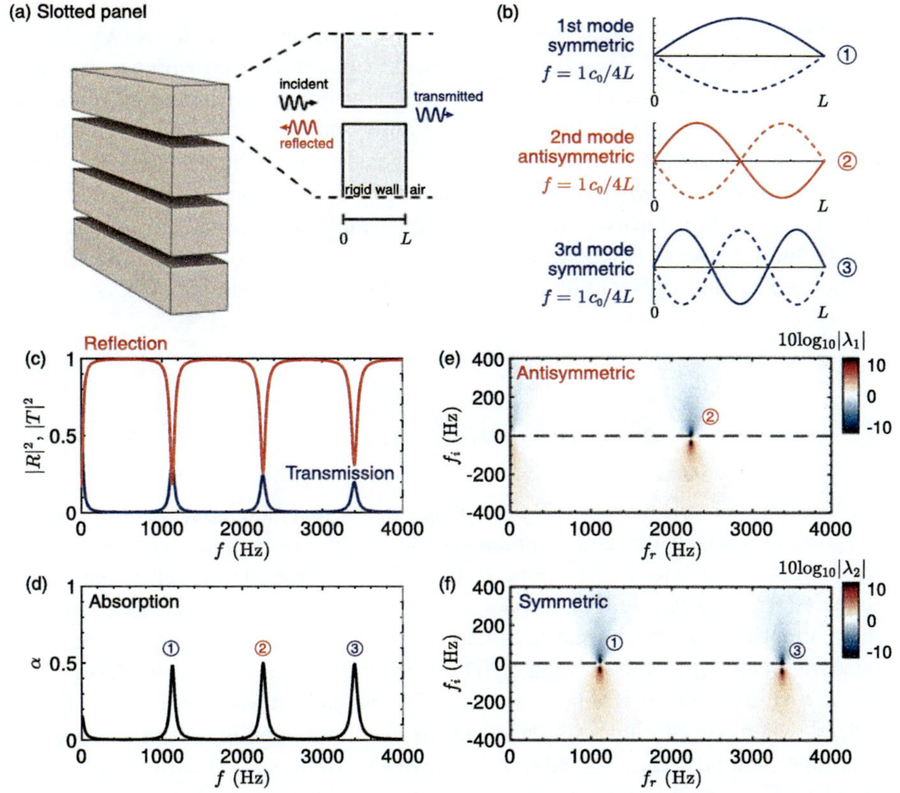

Fig. 5.7 **a** Example of a simple absorbing panel: a slotted rigid panel of length L. **b** Corresponding FP modes. **c** Reflection and Transmission and **d** absorption up to the third FP resonance frequency. **e–f** Eigenvalues of the scattering matrix represented in the complex frequency plane, corresponding to the anti-symmetric and symmetric problems

in the complex-frequency plane corresponding to the eigenvalue λ_2. At the resonance frequency associated with each FP mode, a peak of transmission is produced and reflection is reduced, as shown in Fig. 5.7c.

However, when losses are introduced in the system and, in this particular example, the slits have a thickness comparable with the viscous and thermal boundary layers, sound absorption is produced. In general, each peak of sound absorption corresponds to a different resonance of the material. However, symmetric and antisymmetric resonances appear at different frequencies in a homogeneous and symmetric material such as this slotted panel or in a layer of porous material. The position of the zeros in the complex frequency plane can be modified by tuning the intrinsic losses of the system. In the present example, the losses can be adjusted by tuning the thickness of the slits. Eventually, a zero of one eigenvalue associated with one resonance can be placed at the real frequency axis. In this situation, we obtain the higher possible absorption. In the present example we see that the maximum possible absorption is

$\alpha \leq 0.5$ because the resonances show a very high quality factor.[3] In this way, peaks of $\alpha \approx 0.5$ appear at each FP resonance, each peak being associated with a symmetric and antisymmetric mode. However, note that only one eigenvalue can vanish at the same frequency because symmetric and antisymmetric resonances appears at different and frequencies. In a single scattering event, as those obtained using a bubble immersed in a liquid, or when the quality factor of a resonance is very sharp, only 50% of the energy can be absorbed [74, 75]. However, in most practical situations the effect of overlapping between resonances is present. Thus, and in general, the maximum absorption of a slab of homogeneous and symmetric materials is $0.5 \leq \alpha_{\max} < 1$ [76, 77].

5.3.2 *Monopolar and Dipolar Resonances*

The preceding example used the FP resonances of a slotted panel, but the conclusions apply to any homogeneous and symmetric material. In this example, symmetric and antisymmetric resonances correspond to the FP resonances. However, when using metamaterials, sub-wavelength resonators are embedded in the structure. In this case, symmetric and antisymmetric modes arise in a different manner, and become dependent on the type of the resonator. Thus, it is interesting to distinguish between *monopolar* and *dipolar* resonances. Figure 5.8 shows examples of both types. On the one hand, when considering a single scattering event, monopolar resonances appear as a symmetric mode, as produced by a single resonator in parallel with a waveguide (Fig. 5.8a–c). On the other hand, when considering a single scattering event, dipolar resonances produce an antisymmetric mode, as those produced by a single resonator in series with a waveguide (Fig. 5.8d–e).

Finally, note that when considering a multiple scattering event, as those produced in a metamaterial with a high number of resonators, as shown in Fig. 5.9, some resonances will correspond to asymmetric modes (Fig. 5.9a–c), and other to symmetric ones (Fig. 5.9d–e). Those modes are in fact the collective resonance modes of the resonators that compose the metamaterial. On the one hand, in a structure composed of only identical monopolar resonators, as in Fig. 5.9a–c, the first resonance is a symmetric mode (not shown in the figure), the second resonance corresponds to an antisymmetric mode (shown in the figure), and so on. On the other hand, in a material using only dipolar resonators, as those in Fig. 5.9d–e, the first resonance corresponds to an antisymmetric mode (not shown in the figure), the second to a symmetric mode (shown in the figure), and so on. In general, there will exist, at least, the same number of collective resonances as the number of resonators existing in the structure.

[3]The quality factor of the resonances is associated to the impedance mismatch between the slit and the exterior medium due to the change of section. In the case of broadband resonances the absorption can reach $\alpha > 0.5$ due to overlapping between symmetric and antisymmetric modes.

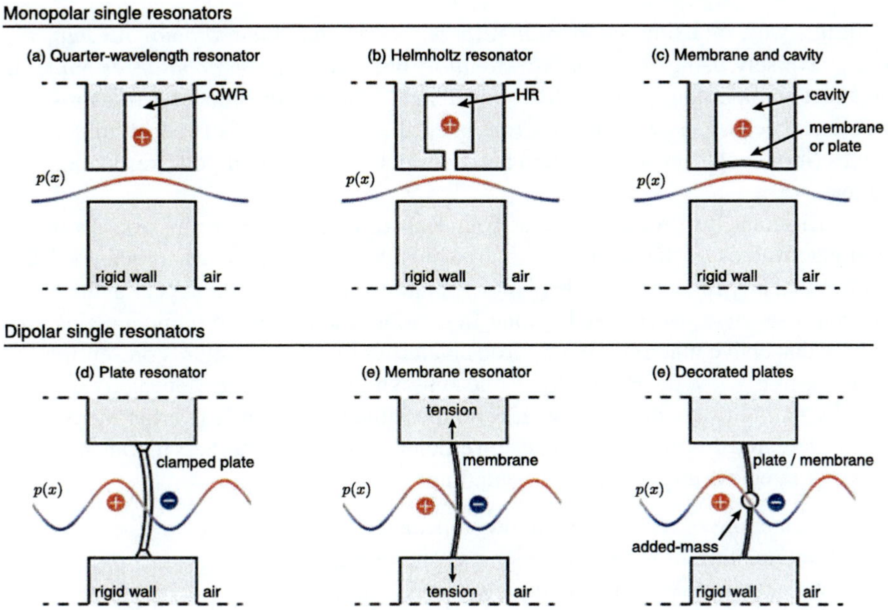

Fig. 5.8 Examples of single monopolar and dipolar resonances using simple resonators

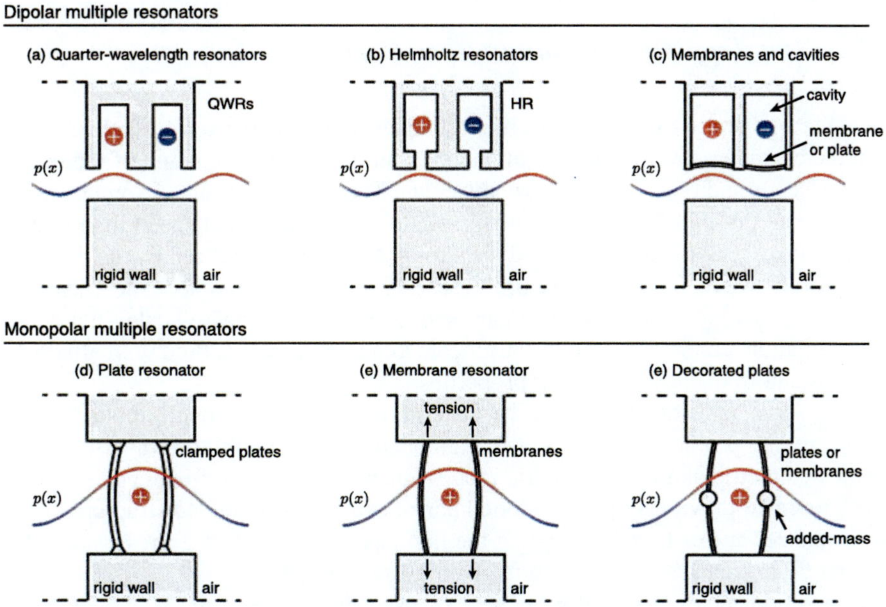

Fig. 5.9 Examples of single monopolar and dipolar resonances using multiple resonators

In this way, by using *monopolar resonators* one can obtain *dipolar resonances*, and, conversely, using only *dipolar resonators* one can obtain *monopolar resonances*. This fact is important to design sub-wavelength efficient absorbers for transmission problems because, as we will describe later, the overlapping between symmetrical modes (monopolar resonances) and anti-symmetrical modes (dipolar resonances) is critical.

In particular, to produce perfect sound absorption in a transmission problem, both eigenvalues of the scattering matrix must vanish at the same frequency. This implies that a symmetrical mode (a monopolar resonance) should appear at the same frequency as an anti-symmetrical one (a dipolar resonance). In homogeneous and weakly dispersive materials, e.g., porous materials or those structures presenting FP modes as in Fig. 5.7, the symmetrical and anti-symmetrical modes appear at different and alternating frequencies. Thus, it is not possible to efficiently overlap them.

To overcome this problem and design efficient or perfect absorbers in transmission several strategies can be followed, including:

(i) Overlapping of resonances in low impedance materials [78].
(ii) Accumulation of resonances using strong dispersion [38].
(iii) Symmetry breaking metamaterials [75, 79, 80].
(iv) Combination of monopolar and dipolar resonators [73, 81].

5.3.3 Overlapping of Resonances in Porous Layers

The micro-structure of porous media presents a representative scale much smaller than the characteristic wavelength of sound in air, the pore scale being of the order of the viscous and thermal boundary layers [1]. These materials allow sound waves to propagate through the structure with a slightly reduced sound speed than in air, mainly due to both the tortuosity and the strong attenuation produced by the thermal and viscous losses. The complex processes that experience the acoustic waves at the micro-scale in porous materials can be described at the macro-scale, in a good approximation, considering the porous media as a homogeneous fluid with effective complex and frequency dependent properties.

To observe acoustic absorption the material must be geometrically bounded. In a transmission problem, a porous layer cannot absorb efficiently the waves because (a) the symmetric and antisymmetric modes appear at staggered frequencies, as we have presented previously for the slotted panel, and (2) the bounded structure lacks of resonances at low frequency, being the first peak of absorption corresponding to the first FP resonance of the structure given by $f_1 = c_{eff}/2L$.

In addition, below its Biot's frequency, given by $f_{Biot} = \sigma\phi/2\pi\alpha_\infty\rho_0$, a bulk porous material is not impedance matched with the surrounding air. This important parameter indicates the transition between the inertial and the viscous regime of the porous material, and only depends on the material parameters. Absorption cannot be achieved efficiently below this frequency [20]. However, the resonances provided by

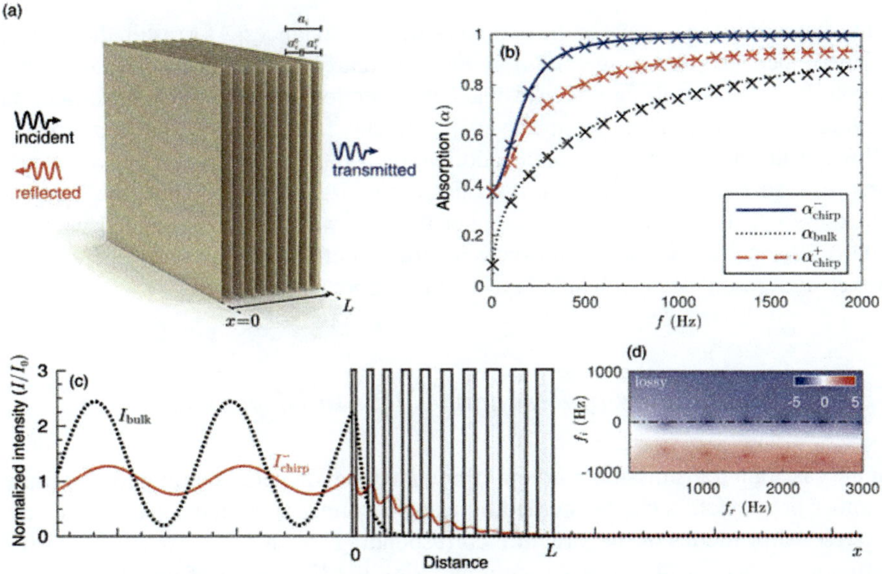

Fig. 5.10 **a** Chirped multilayer porous material, **b** absorption produced by the multilayer structure from both sides (blue and red) and absorption of a bulk structure using the same porous material (black). **c** Intensity as a function of the distance for the chirped multilayer structure and for a homogeneous porous layer. **d** Complex frequency plane of the sum of the eigenvalues of the scattering matrix. Adapted from [78] with permission

a porous layer present a very low quality factor because the characteristic effective parameters of porous materials are close to those of the air, and this fact can be exploited to design effective sound absorbers.

One way to modify these constraints is the use of materials with graded properties. An example of this mechanism is a chirped multilayer porous material [78], composed of alternating air cavities and porous layers, as shown in Fig. 5.10a. This structure, as occurs with a homogeneous porous material, behaves as a resonant FP cavity.

The smooth variation of the physical properties inside the structure allows characterizing the system using local effective parameters, that could be used to design the properties of the structured medium. In particular, using the chirped structure the effective parameters can be graded throughout the structure as a function of the depth. First, this allows obtaining an efficient impedance matching below the Biot's frequency of the material at the surface of the material, i.e., it provides efficient absorption at low frequencies. Second, it also allows progressively controlling the intrinsic losses of system by adding thicker porous layers.

Moreover, representing the eigenvalues of the scattering matrix in the complex frequency plane gives a clear picture of the existing low-quality-factor resonances, where each resonance is associated with a zero-pole pair. The complex frequency plane helps to fulfil the perfect absorption conditions: by tuning the intrinsic losses we can compensate the leakage of the system at the FP resonances. As a result, the control of the impedance matching condition and the control of the intrinsic losses of the porous layers allows designing broadband unidirectional quasi-perfect absorbers. In addition, as the resonances present a very low quality factor they overlap in frequency allowing the design of broadband unidirectional perfect absorbers. However, the first resonance of the structure corresponds to the first FP mode. Thus, the material thickness is only about two times smaller than the wavelength associated with the first resonance. To produce a sub-wavelength and efficient sound absorber, a sub-wavelength resonance mechanism must be included.

5.3.4 Accumulation of Resonances Due to Dispersion

In the case of a transmission problem, a bounded metamaterial composed of a finite number of resonators can be studied by using its effective parameters. In this way, the resonance modes of the structure corresponding to the collective modes of the different resonators [49], can be identified with the FP modes of the slab [7, 38]. Examples of such systems include arrays of monopolar resonators, e.g. as in arrays of QWRs [82], arrays of HRs [7, 38], arrays of HRs producing Bragg interference in addition to local resonances [83] or arrays of two concentrically placed QWRs producing Fano resonances [84].

When the system presents intrinsic losses, absorption is produced at the resonance frequencies. For symmetric systems, the scattering problem can be subdivided in its corresponding symmetric and antisymmetric problems, as sketched in Fig. 5.11b. Thus, by setting rigid, $\partial p/\partial x = 0$ (symmetric), and soft, $p = 0$ (anti-symmetric), boundary conditions at the symmetry plane of the system, the reflection coefficients of each sub-problem can be obtained as a combination of the original reflection and transmission coefficients of the global scattering problems as follows

$$R_{\text{sym}} = T + R = \frac{\bar{Z}\sin\left(k_{eff}L/2\right) - i\cos\left(k_{eff}L/2\right)}{\bar{Z}\sin\left(k_{eff}L/2\right) + i\cos\left(k_{eff}L/2\right)}, \tag{5.22}$$

$$R_{\text{asym}} = T - R = \frac{\bar{Z}\cos\left(k_{eff}L/2\right) + i\sin\left(k_{eff}L/2\right)}{i\sin\left(k_{eff}L/2\right) - \bar{Z}\cos\left(k_{eff}L/2\right)}, \tag{5.23}$$

for the symmetric and antisymmetric problems respectively. Then, the absorption coefficient of the full problem can be obtained from the absorption of each sub-problem as

$$\alpha = \frac{\alpha_{\text{sym}} + \alpha_{\text{asym}}}{2}, \quad \text{where} \quad \alpha_{\text{sym(asym)}} = 1 - |R_{\text{sym(asym)}}|^2. \tag{5.24}$$

It is worth noting that the reflection coefficient of each sub-problem in reflection represents the eigenvalues of the scattering matrix. Therefore, as perfect absorption

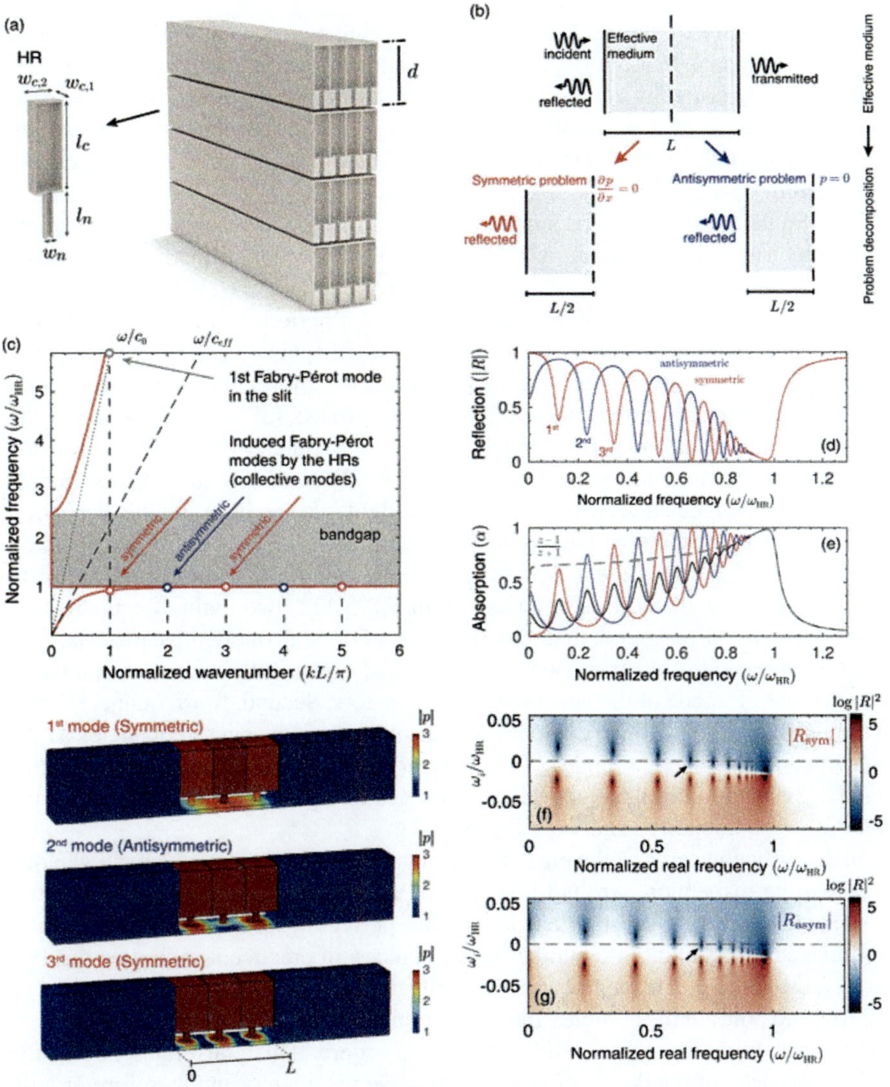

Fig. 5.11 **a** Example of a panel composed of only identical monopolar resonators (HRs). **b** Problem decomposition in symmetric and antisymmetric problems. **c** Dispersion relation of a strongly dispersive metamaterial showing the symmetric and antisymmetric modes. **d–e** Reflection and absorption of the symmetric problem (blue), antisymmetric problem (red) and total problem (black). **f–g** complex frequency plane representation of the eigenvalues of the scattering matrix corresponding to the symmetric and antisymmetric problems. Adapted from [38] with permission

requires both zeros of the eigenvalue of the scattering matrix must vanish at the same frequency, it means that both symmetric and antisymmetric sub-problems must be simultaneously critically coupled. In the absence of embedded resonators, the zero and pole structure of these two problems is staggered in frequency, corresponding each zero-pole pair to one FP resonance mode. Therefore, perfect absorption cannot be achieved.

When a periodic array of identical monopolar or dipolar resonators is included, the symmetric and antisymmetric resonances of the system still exist at staggered frequencies and, therefore, simultaneous critical coupling of the symmetric and antisymmetric modes is not possible. However, when a periodic array of resonators is loaded in the material as shown in Fig. 5.11a, strong dispersion can be generated and quasi perfect absorption can be obtained. Figure 5.11c shows the dispersion relation of a metamaterial composed of a periodic array of HRs [38]. The presence of the resonators induces a band gap around the resonance frequency of the HRs and strong dispersion in the medium (grey area in Fig. 5.11c). Figure 5.11c also shows the differences between the dispersion relation of free field (dashed line) and the dispersion relation once the resonators are loaded (red continuous line). The strong dispersion in the case of the loaded resonators induces slow-sound propagation conditions. In this way, once the material is bounded, as in Fig. 5.11a, two main effects are produced due to the slow-sound: first, the FP modes of the slab of effective material, i.e., the collective modes of the resonators, are shifted-down in frequency as compared with the first FP mode of the panel without resonators. Second, for frequencies close to the bandgap, the symmetric modes appear almost at the same frequencies as the antisymmetric resonance modes. This is clearly visible in the complex frequency plane representation of the eigenvalues of the scattering matrix where the zeros and poles are accumulated as shown in Fig. 5.11f–g.

In this way, the zeros of the eigenvalues are accumulated below the band gap due to the strong dispersion. Around this accumulation, symmetric and antisymmetric modes exist at very close frequencies. While perfect absorption can theoretically be achieved using an infinite number of resonances, in practice it cannot be reached because the number of resonators is necessarily finite. In other words, the total number of zeros and poles in the complex frequency domain equals the number of resonators and the accumulation point could not exist any more. This constitutes one of the main drawbacks of the description using effective parameters and therefore of bulk material description in practice. In this way, special attention should be paid to the physical modelling strategy that should account for the finite number of resonances. This can be done by the Transfer Matrix Method (see Chap. 4).

Accounting for the finite number of resonators, this problem was tacked and, while perfect absorption cannot be achieved, a quasi-perfect absorber structure can be designed using a slotted panel loaded by an array of HRs [38]. Due to the loading HRs, strong dispersion is observed in the interior of each slit and the cavity resonances accumulate below the band-gap frequency, being the symmetrical and antisymmetrical modes staggered but very close in frequency. In this frequency range, and by tuning the geometry, the system can be quasi-critically coupled with the exterior medium and therefore quasi-perfect absorption can be obtained. Using a finite number of resonators, it was demonstrated experimentally quasi-perfect absorption of $\alpha = 0.9$ for a sub-wavelength panel ten times thinner than the wavelength of the absorbed wavefront, $L = \lambda/10$. In this way, the absorption can exceed 50% even using only monopolar resonators due to this accumulation of resonances.

These results underline the necessity of breaking the symmetry of the system to achieve perfect absorption in transmission or the use of degenerate resonators with simultaneous symmetric (monopolar) and antisymmetric (dipolar) resonances, as we will see in the following sections.

5.3.5 Symmetry Breaking Materials

One approach to overcome the limitations of homogeneous absorbers in transmission problems consists in breaking the symmetry of the structure. This mechanism was first proposed by Merkel et al. [75] in waveguides and then extended to realistic metamaterial panels by Jiménez et al. [79].

5.3.5.1 Asymmetric Absorbers

The symmetry of the system can be broken by using detuned resonators as shown in Fig. 5.12a using HRs of different geometry.

Using such asymmetric system, a multiple scattering process occurs between both resonators. First, the inner resonator creates a drop of the transmission for frequencies just above its resonance frequency, f_1, acting it as a reflecting wall. Then, when loading a second resonator with slightly higher resonance frequency, f_2, and tuning its geometry to introduce the correct amount of losses, the system can be critically coupled to the exterior medium and perfect absorption is produced, as shown in Fig. 5.12b–c. This allows the impedance matching to the surrounding medium at this frequency. Therefore, both the reflection and the transmission vanish for waves impinging the structure in this direction and the structure becomes critically coupled.

The representation of the eigenvalues in the complex frequency plane, given by (5.6), is shown in Fig. 5.12e–f. Using detuned double-interacting resonators a zero for both eigenvalues can be located at the real axis at same frequency, producing perfect unidirectional sound absorption once the incident wave corresponds to the good eigenvector. In fact, as the system is asymmetric, the analysis of the scattering matrix eigenvalues is not sufficient to completely describes the acoustic behaviour of the designed structure. The eigenvectors of the system, given by (5.7), can also be represented in the complex frequency plane. The eigenvectors give us information about the direction in which the waves are absorbed: in this case only the waves impinging the panel in the $+x_2$ are perfectly absorbed: these panels present unidirectional absorption at a single frequency. In the case of a sub-wavelength asymmetric panel, perfect absorption at 300 Hz for a structure whose thickness is 40 times smaller than the working wavelength was reported [63].

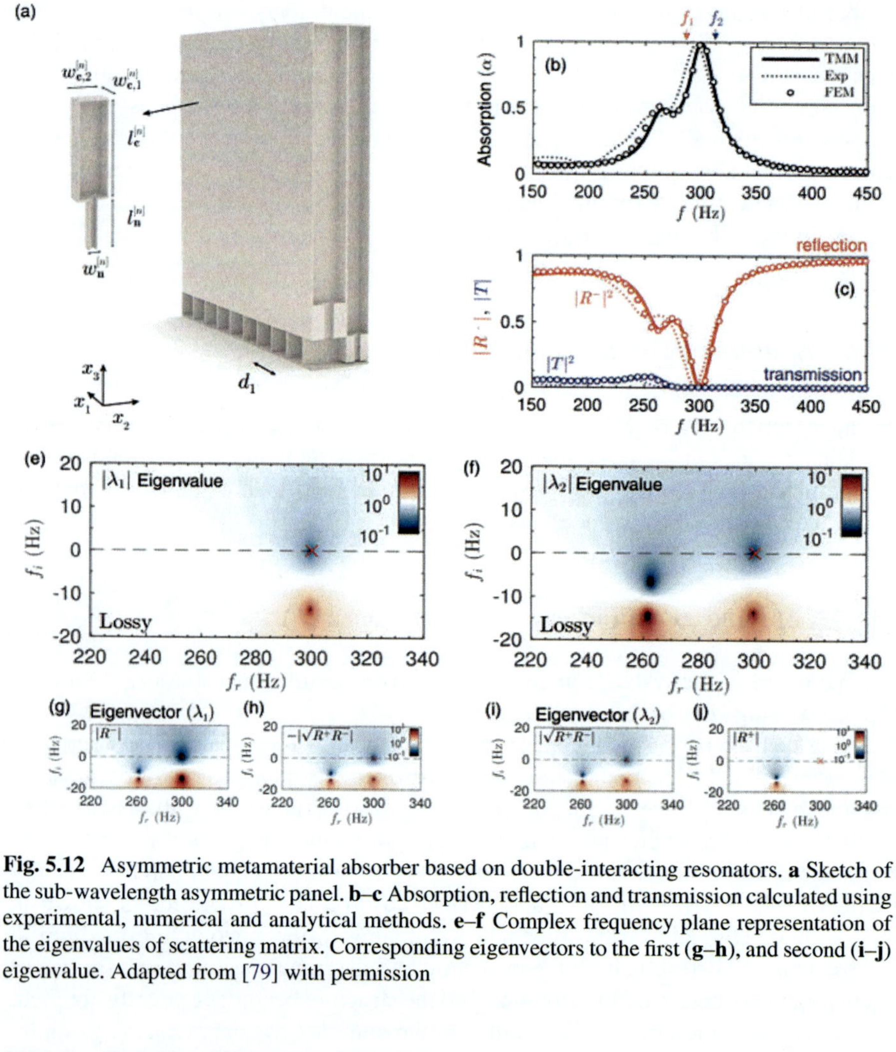

Fig. 5.12 Asymmetric metamaterial absorber based on double-interacting resonators. **a** Sketch of the sub-wavelength asymmetric panel. **b–c** Absorption, reflection and transmission calculated using experimental, numerical and analytical methods. **e–f** Complex frequency plane representation of the eigenvalues of scattering matrix. Corresponding eigenvectors to the first (**g–h**), and second (**i–j**) eigenvalue. Adapted from [79] with permission

5.3.5.2 Rainbow-Trapping Absorbers

The system discussed in the previous section can be used to create a cascade of critically-coupled resonators in order to obtain a perfect broadband unidirectional absorption. One can note that the second resonator also reduces the transmission for frequencies just above its resonance, f_2. Thus, we can repeat the process in cascade, by adding additional HRs to the waveguide, each of them with a higher resonance frequency than the preceding one. As illustrated in Fig. 5.13a, the panels are now composed of a periodic array of varying cross-section waveguides, each of them being loaded by HRs with graded dimensions. The low cut-off frequency of the absorption band is fixed by the resonance frequency of the deepest HR. Thus, by

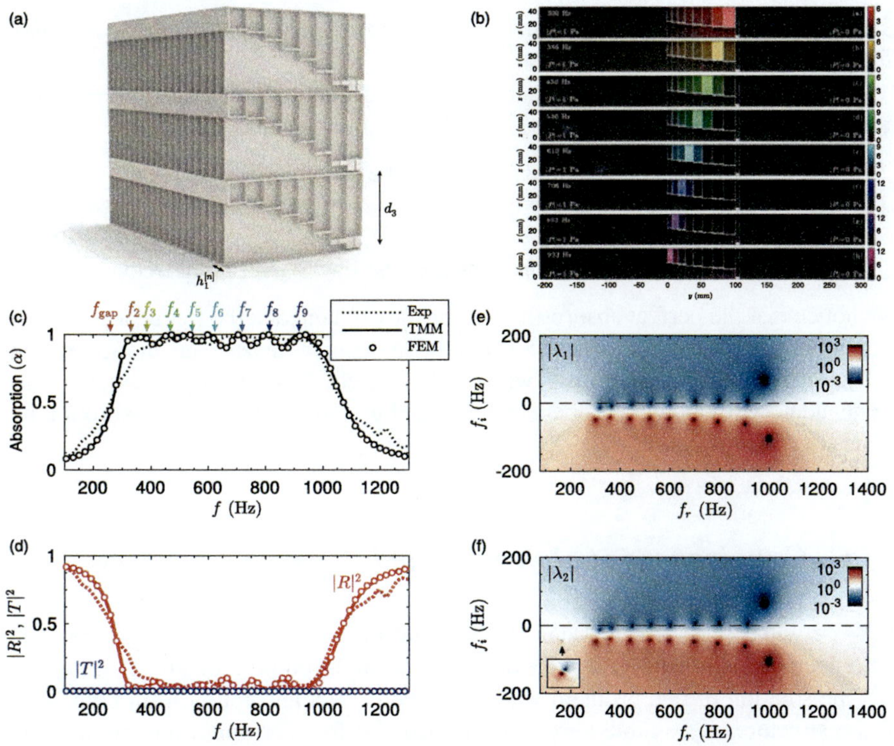

Fig. 5.13 Broadband metamaterial absorber based on rainbow-trapping phenomenon. **a** Sketch of the rainbow-trapping absorber. **b** Pressure distribution for different frequencies. **c** Absorption and **d** corresponding reflection and transmission calculated using experimental, numerical and analytical methods. **e–f** Complex frequency plane representation of the eigenvalues of of scattering matrix. Adapted from [79] with permission

progressively tuning their geometry, each resonator impedance-matches the system at a frequency slightly higher than the resonance frequency of the preceding one.

The varying cross-section waveguide introduces a progressive change in the opening surface that contributes to produce the impedance matching, leading low quality factor resonances. As a result, broadband absorption can be obtained as shown in Fig. 5.13c–d.

For each resonance frequency of the system, the zeros of both eigenvalues of the scattering matrix, λ_1 and λ_2, now lie on the real frequency axis and at the same frequency, see Fig. 5.13e–f, ensuring that perfect absorption is produced and the system is critically coupled. Using this frequency-cascade effect, perfect sound absorption over almost two frequency octaves ranging from 300 to 1000 Hz for a transparent panel composed of 9 resonators with a total thickness of 11 cm, i.e., 10 times smaller than the wavelength at 300 Hz was reported. Note, each frequency component becomes localised at a different location inside the structure as shown in Fig. 5.13b.

Because of this behaviour these structures have been denominated *Rainbow-trapping absorbers* [79].

In summary, to obtain broadband, perfect and unidirectional absorption in the transmission problem, three conditions must be simultaneously fulfilled:

(i) The zeros of both eigenvalues, $\lambda_{1,2}$, must be at same frequencies,
(ii) The zeros of the eigenvalues of the scattering matrix must be located on the real frequency axis,
(iii) The quality factor of the resonances must be low to overlap in frequency.

We notice that the perfect absorption will be only obtained at the frequencies where the zeros are in the real frequency axis. For the frequencies in between, the absorption will be very high but rigorously will not be perfect. The above three conditions are mandatory to obtain broadband, perfect and unidirectional sound absorption by panels in transmission.

5.3.6 *Combinations of Monopolar and Dipolar Resonances*

Now we will discuss the problem of perfect absorption in symmetric systems, i.e., the reflection coefficient is the same on both sizes of the point resonator. This implies that the solution of that problem will introduce perfect absorption from both sides of the structure. A possible approach to obtain perfect absorption in a symmetric transmission problem is to employ degenerate resonators as proposed by Yang et al. [73] using membrane resonators and, later, by Romero et al. [81] using Helmholtz resonators.

For membranes, two different structures were reported, both based in a resonator made of a combination of a monopolar and a dipolar resonance with identical resonance frequency, i.e., a degenerate resonator. In the first structure, a small decorated membrane with a coupled membrane resonator composed of two decorated membranes is located inside the waveguide as shown in Fig. 5.14a. The simple decorated-membrane resonator produces a dipolar resonance, while the coupled membrane system can develop a monopolar resonance when both membranes oscillate with opposite phases. This is in fact the second collective mode of the system, as shown previously in Fig. 5.9f. If the system is accurately tuned, both monopolar and dipolar resonances can be produced at the same frequency. Therefore, the monopolar resonance introduces a zero-pole pair in the corresponding eigenvalues of the scattering matrix, i.e., $\lambda_1 = R + T$, a symmetric mode; while the dipolar resonance does the same for the corresponding anti-symmetric mode, i.e., $\lambda_2 = R - T$. In these conditions, the intrinsic viscoelastic losses of the membranes are enough to critically-couple the system at this particular frequency, i.e., at the degenerate mode.

In the second structure, using the configuration shown in Fig. 5.14b an analogous situation can be obtained. Here, the dipolar resonance is obtained using a single membrane while the monopolar resonance is produced using a Helmholtz-membrane resonator, i.e., a membrane backed by a cavity located at the wall of the waveguide.

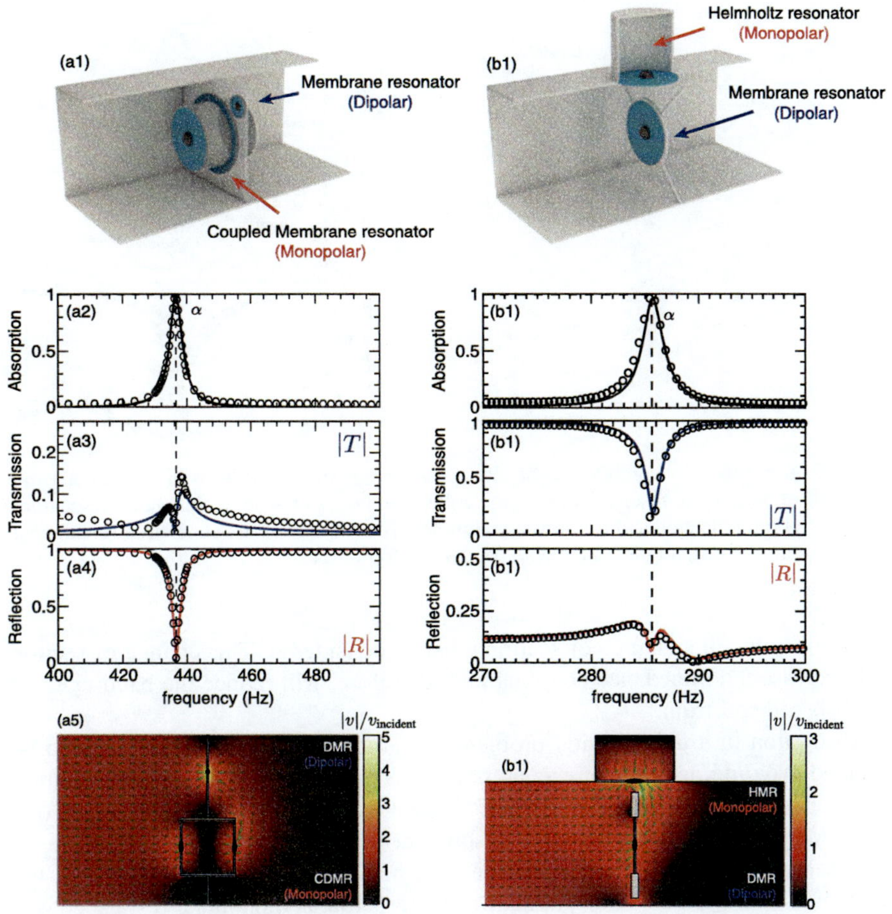

Fig. 5.14 Absorption by combination of monopolar and dipolar resonances. **a** Degenerate resonator based on coupled decorated membranes. The double membrane produces a monopolar resonance while the single membrane produces a dipolar resonance at the same frequency. **b** Degenerate resonator based on monopolar and dipolar membrane resonators. The HR produces a monopolar resonance while the single membrane produces a dipolar resonance at the same frequency. Adapted with permission from [73]

Therefore, a degenerate mode can be obtained if both, monopolar and dipolar resonances appear at the same frequency. Each mode introduces a zero-pole pair in the corresponding eigenvalues of the scattering matrix. Then, by the proper tuning of the intrinsic viscoelastic losses of the membranes the whole system can be critically-coupled with the exterior medium.

The second configuration offers a broader bandwidth absorption, but it is worth to mention that in this last configuration the Helmholtz-membrane resonator was located *outside* the waveguide. Therefore, to produce a metamaterial absorber panel

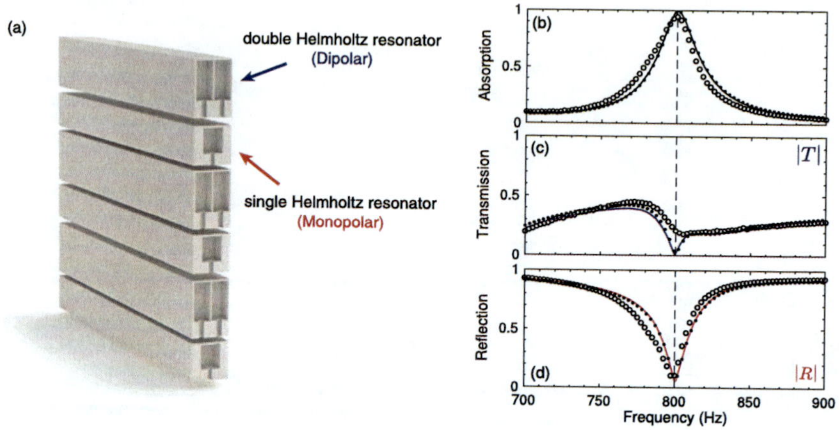

Fig. 5.15 Absorption by combination of monopolar and dipolar resonances in a panel. **a** Panel composed of degenerate resonators using the collective modes of several Helmholtz resonators. **b–d** Absorption, reflection and transmission. (circles) Experiment, (dots) simulation (continuous lines) theoretical absorption. Adapted with permission from [81]

using this unit cell a change of section must be considered. This change of section will decrease the impedance coupling and, therefore, will reduce the bandwidth of the absorbed spectrum.

Absorption in a transmission problem using degenerate resonators can also be obtained avoiding membrane resonators [81], as e.g. in mirror-symmetric metamaterials using Helmholtz resonators. Two slits were drilled into a panel, and each one was loaded by a different number of resonators with different geometries. A showcase material was manufactured, loading the first slit with one resonator, and the second slit with two resonators, as shown in Fig. 5.15a. A monopolar resonance was generated by the first slit as a result of the first resonance of the single resonator, while a dipolar resonance was produced by the second slit as a result of the second collective mode of the pair of resonators, i.e., the second Fabry–Pérot mode of the slit (see Fig. 5.9 previously). By tuning the geometrical parameters, perfect absorption can be observed, see Fig. 5.15b–d. Note that resonators were embedded in the panel to account for the impedance mismatching produced by the change of section.

Remark that the resonant scattering produced by this metascreen was analytically studied by using a hybrid model mixing the transfer-matrix and modal-decomposition methods to account for the Bloch waves. This was performed in order to consider the possible coupling between the slits, effect that was found to be particularly important for these absorbers. If no higher-order Bloch waves are accounted for, i.e., if only a plane wave is included outside the metascreen, the analytical model will not be in agreement with either the full-wave numerical simulation or the experimental results. The absorber was made of acoustically rigid materials, without any vibrating elements. Therefore, the life duration of this metascreen is thus expected to be longer than that of absorbers composed of viscoelastic membranes or plates. In addition, the heights of the slits of the proposed absorber are large enough for the system to be used in ventilation problems where air flow is of importance. Finally, note that the bandwidth can also be extended using a cascade of resonances as done in rainbow trapping absorbers, but in this case combining monopolar and dipolar resonances of two different slits.

5.4 Summary and Conclusions

In this chapter we have reviewed the vast literature of metamaterial absorbers trying to identify the key advances and to present the basic physical mechanisms to design efficient and sub-wavelength sound absorbers.

To obtain efficient absorption, i.e, perfect, all the eigenvalues of the scattering matrix must vanish at the single frequency. When this condition is fulfilled the system is critically coupled with the exterior medium and the structure does not produce any scattered wave, i.e., all energy is absorbed by it at this particular frequency.

To do that, we must pay attention to two aspects. First, the system must present a resonance to introduce its associated zero-pole pair into the eigenvalues of the scattering matrix. The nature of the resonance is, in principle, not relevant, but metamaterials allow deep-sub-wavelength resonances which are interesting to obtain absorbers of reduced dimensions. Second, once the resonance is introduced, the system must present the exact losses to compensate the energy leakage of the structure. In this sense, metamaterials present a huge degree of freedom to tune their intrinsic losses as commonly the geometry of the material for a given topology can be modified. For example, a metamaterial based on air cavities allows modifying its intrinsic losses by tuning the viscothermal process by narrowing or expanding the section of the air channels.

Another important aspect is to distinguish between reflection and transmission problems, being the former much easier to tackle than the later.

In reflection problems, both eigenvalues of the scattering matrix collapse to the reflection coefficient. Therefore, the critical-coupling condition reduces to the well-known impedance matching condition, which is easily fulfilled around a resonance, specially if the system allows tuning its intrinsic losses. For reflection problems efficient, and in some cases perfect, sound absorbers based on metamaterials have been reported. These included metaporous absorbers, i.e., porous materials with embedded resonating elements; metamaterials based on air cavities, i.e., those composed of QWRs, HRs, coiled and labyrinthine structures; and metamaterials based on membranes and thin elastic plates as those composed of decorated-membrane resonators, platelets or poroelastic membranes.

In transmission problems the scattering matrix presents two different eigenvalues. We have identified that perfect absorption is not possible in homogeneous and symmetric structures, as those presenting FP modes and weak dispersion. This is caused because the zeros of the eigenvalues of the scattering matrix, associated with their corresponding symmetric and antisymmetric modes, appear staggered in frequency.

One approach is to introduce strong dispersion in the system by using an array of monopolar or dipolar resonators. The strong dispersion accumulate the collective modes of the resonators around the same frequency. As the zeros of both eigenvalues of the scattering matrix overlap, efficient absorption is produced. Other approach is to break this symmetry using detuned resonators, allowing broadband perfect and unidirectional sound absorption as occurs in *rainbow-trapping absorbers*. Finally, a last approach is to design a metamaterial exhibiting degenerate resonances, i.e., a system able to produce a dipolar and a monopolar resonance at the same frequency. This is typically obtained by combining monopolar and dipolar resonators. In this situation the monopolar resonator introduces a resonance into the symmetric eigenvalue of the scattering matrix, while the dipolar resonator does the same into the antisymmetric one. As both resonances appear at the same frequency, by tuning the losses, perfect absorption can be obtained in transmission.

It is important to remark that causality imposes limits to the optimal absorption for any structure, metamaterial or not. In this sense, it has been demonstrated that there exist a constraint between the total thickness of the material and the range of frequencies being absorbed [30]. In essence, there is always a trade off between the bandwidth of the absorbed spectrum and the sub-wavelength dimension of a given structure. Remark that these limitations apply to the bandwidth, but not to the central frequency of the absorption. Causality-imposed limits explain why absorption becomes narrow when designing very thin structures for extremely-low frequency.

Finally, there are many other research lines, not detailed here, as nonlinear metamaterial absorbers, active metamaterials absorbers [85], sound diffusers [63, 64], multifunctional and reconfigurable absorbers. In selected industrial applications, as in automotive or aerospace industries, absorbing metamaterials interacting with flow is an active research line. The inclusion of additional physical mechanisms into the acoustic metamaterials as adsorption and desorption of air molecules in multi-scale porous materials [86], strong local nonlinearities, or efficient energy harvesting by critical-coupling will open new doors to go beyond traditional sound absorbers using thin, broadband and effective metamaterial absorbers.

Acknowledgements We acknowledge the contribution to the results reported here of V. Pagneux, G. Theocharis, O. Richoux, Y. Auregán, O. Dazel, C. Lagarrigue, V. Tournat, B. Brouard, B. Nennig, A.A. Fernández-Marín, J. Sánchez-Dehesa, A. Cebrecos, and those to contributed to this review. This article is based upon work from COST Action DENORMS CA15125, supported by COST (European Cooperation in Science and Technology). The authors gratefully acknowledge the ANR-RGC METARoom (ANR-18-CE08-0021) project and the project HYPERMETA funded under the program Étoiles Montantes of the Région Pays de la Loire. N.J. acknowledges financial support from Spanish Ministry of Science, Innovation and Universities through grant "Juan de la Cierva-Incorporación" IJC2018-037897-I.

References

1. J. Allard, N. Atalla, *Propagation of Sound in Porous Media: Modelling Sound Absorbing Materials 2e* (Wiley, New York, 2009)
2. T. Cox, P. d'Antonio, *Acoustic Absorbers and Diffusers: Theory, Design and Application* (CRC Press, Boca Raton, 2016)
3. T. Wu, T. Cox, Y. Lam, From a profiled diffuser to an optimized absorber. J. Acoust. Soc. Amer. **108**(2), 643–650 (2000)
4. S.A. Cummer, J. Christensen, A. Alù, Controlling sound with acoustic metamaterials. Nat. Rev. Mater. **1**, 16001 (2016)
5. G. Ma, P. Sheng, Acoustic metamaterials: from local resonances to broad horizons. Sci. Adv. **2**(2), e1501595 (2016)
6. Z. Liu, X. Zhang, Y. Mao, Y. Zhu, Z. Yang, C.T. Chan, P. Sheng, Locally resonant sonic materials. Science **289**(5485), 1734–1736 (2000)
7. N. Fang, D. Xi, J. Xu, M. Ambati, W. Srituravanich, C. Sun, X. Zhang, Ultrasonic metamaterials with negative modulus. Nat. Mater. **5**(6), 452 (2006)
8. S.H. Lee, C.M. Park, Y.M. Seo, Z.G. Wang, C.K. Kim, Composite acoustic medium with simultaneously negative density and modulus. Phys. Rev. Lett. **104**(5), 054301 (2010)
9. M. Yang, P. Sheng, Sound absorption structures: From porous media to acoustic metamaterials. Annu. Rev. Mater. Res. **47**, 83–114 (2017)
10. V. Romero-García, G. Theocharis, O. Richoux, V. Pagneux, Use of complex frequency plane to design broadband and sub-wavelength absorbers. J. Acoust. Soc. Amer. **139**(6), 3395–3403 (2016)
11. T.S. Luk, S. Campione, I. Kim, S. Feng, Y.C. Jun, S. Liu, J.B. Wright, I. Brener, P.B. Catrysse, S. Fan, M.B. Sinclair, Directional perfect absorption using deep subwavelength low-permittivity films. Phys. Rev. B **90**, 085411 (2014)
12. J.-P. Groby, R. Pommier, Y. Aurégan, Use of slow sound to design perfect and broadband passive sound absorbing materials. J. Acoust. Soc. Amer. **139**(4), 1660–1671 (2016)
13. V. Romero-García, G. Theocharis, O. Richoux, A. Merkel, V. Tournat, V. Pagneux, Perfect and broadband acoustic absorption by critically coupled sub-wavelength resonators. Sci. Rep. **6**, 19519 (2016)
14. M. Law, L.E. Greene, J.C. Johnson, R. Saykally, P. Yang, Nanowire dye-sensitized solar cells. Nat. Mater. **4**, 455–459 (2005)
15. P. Derode, A.P. Roux, M. Fink, Robust acoustic time reversal with high-order multiple scattering. Phys. Rev. Lett. **75**, 4206 (1995)
16. L. Chong, Y. Ge, H. Cao, A.D. Stone, Coherent perfect absorbers: time-reversed lasers. Phys. Rev. Lett. **105**, 053901 (2010)
17. J. Mei, G. Ma, M. Yang, Z. Yang, W. Wen, P. Sheng, Dark acoustic metamaterials as super absorbers for low-frequency sound. Nat. Commun. **3**, 756 (2012)
18. D. Johnson, J. Koplik, R. Dashen, Theory of dynamic permeability and tortuosity in fluid-saturated porous media. J. Fluid Mech. **176**, 379–402 (1987)
19. Y. Champoux, J.-F.J. Allard, Dynamic tortuosity and bulk modulus in air-saturated porous media. J. Appl. Phys. **70**, 1975–1979 (1991)
20. N. Jiménez, V. Romero-García, J.-P. Groby, Perfect absorption of sound by rigidly-backed high-porous materials. Acta Acust. Acust. **104**, 396–409 (2018)
21. J.-P. Groby, A. Duclos, O. Dazel, L. Boeckx, L. Kelders, Enhancing absorption coefficient of a backed rigid frame porous layer by embedding circular periodic inclusions. J. Acoust. Soc. Amer. **130**, 3771–3780 (2011)
22. J.-P. Groby, C. Lagarrigue, B. Brouard, O. Dazel, V. Tournat, B. Nennig, Using simple shape three-dimensional rigid inclusions to enhance porous layer absorption. J. Acoust. Soc. Amer. **136**(3), 1139–1148 (2014)
23. C. Lagarrigue, J. Groby, V. Tournat, O. Dazel, O. Umnova, Absorption of sound by porous layers with embedded periodic arrays of resonant inclusions. J. Acoust. Soc. Amer. **134**(6), 4670–4680 (2013)

24. J.-P. Groby, C. Lagarrigue, B. Brouard, O. Dazel, V. Tournat, B. Nennig, Enhancing the absorption properties of acoustic porous plates by periodically embedding helmholtz resonators. J. Acoust. Soc. Amer. **137**(1), 273–280 (2015)
25. Y. Zhou, D. Li, Y. Li, T. Hao, Perfect acoustic absorption by subwavelength metaporous composite. Appl. Phys. Lett. **115**(9), 093503 (2019)
26. L. Xiong, B. Nennig, Y. Aurégan, B. Wenping, Sound attenuation optimization using metaporous materials tuned on exceptional points. J. Acoust. Soc. Amer. **142**, 2288–2297 (2017)
27. J.-P. Groby, A. Duclos, O. Dazel, L. Boeckx, W. Lauriks, Absorption of a rigid frame porous layer with periodic circular inclusions backed by a periodic grating. J. Acoust. Soc. Amer. **129**(5), 3035–3046 (2011)
28. C. Lagarrigue, J.-P. Groby, O. Dazel, V. Tournat, Design of metaporous supercells by genetic algorithm for absorption optimization on a wide frequency band. Appl. Acoust. **102**, 49–54 (2016)
29. T. Weisser, J.-P. Groby, O. Dazel, F. Gaultier, E. Deckers, S. Futatsugi, L. Monteiro, Acoustic behavior of a rigidly backed poroelastic layer with periodic resonant inclusions by a multiple scattering approach. J. Acoust. Soc. Amer. **139**(2), 617–629 (2016)
30. M. Yang, S. Chen, C. Fu, P. Sheng, Optimal sound-absorbing structures. Mater. Horiz. **4**(4), 673–680 (2017)
31. R. Wood, A suspected case of the electrical resonance of minute metal particles for light-waves. A new type of absorption. Proc. Phys. Soc. Lond. **18**, 166–182 (1902)
32. J.-P. Groby, W. Lauriks, T. Vigran, Total absorption peak by use of a rigid frame porous layer backed with a rigid multi-irregularities grating. J. Acoust. Soc. Amer. **127**, 2865–2874 (2010)
33. J.-P. Groby, B. Brouard, O. Dazel, B. Nennig, L. Kelders, Enhancing rigid frame porous layer absorption with three-dimensional periodic irregularities. J. Acoust. Soc. Amer. **133**, 821–831 (2013)
34. Y. Li, B.M. Assouar, Acoustic metasurface-based perfect absorber with deep subwavelength thickness. Appl. Phys. Lett. **108**(6), 063502 (2016)
35. J. Li, W. Wang, Y. Xie, B.-I. Popa, S.A. Cummer, A sound absorbing metasurface with coupled resonators. Appl. Phys. Lett. **109**, 091908 (2016)
36. N. Jiménez, W. Huang, V. Romero-García, V. Pagneux, J.-P. Groby, Ultra-thin metamaterial for perfect and quasi-omnidirectional sound absorption. Appl. Phys. Lett. **109**(12), 121902 (2016)
37. X. Cai, Q. Guo, G. Hu, J. Yang, Ultrathin low-frequency sound absorbing panels based on coplanar spiral tubes or coplanar helmholtz resonators. Appl. Phys. Lett. **105**, 121901 (2014)
38. N. Jiménez, V. Romero-García, V. Pagneux, J.-P. Groby, Quasiperfect absorption by subwavelength acoustic panels in transmission using accumulation of resonances due to slow sound. Phys. Rev. B **95**(1), 014205 (2017)
39. C. Chen, Z. Du, G. Hu, J. Yang, A low-frequency sound absorbing material with subwavelength thickness. Appl. Phys. Lett. **110**(22), 221903 (2017)
40. Y. Wang, H. Zhao, H. Yang, J. Zhong, J. Wen, A space-coiled acoustic metamaterial with tunable low-frequency sound absorption. EPL (Europhysics Letters) **120**(5), 54001 (2018)
41. L. Schwan, O. Umnova, C. Boutin, J.-P. Groby, Nonlocal boundary conditions for corrugated acoustic metasurface with strong near-field interactions. J. Appl. Phys. **123**, 091712 (2018)
42. L. Hau, S. Harris, Z. Dutton, C. Behroozi, Light speed reduction to 17 metres per second in an ultracold atomic gas. Nature **397**, 594–598 (1999)
43. A. Santillán, S.I. Bozhevolnyi, Acoustic transparency and slow sound using detuned acoustic resonators. Phys. Rev. B **84**(6), 064304 (2011)
44. A. Santillan, S. Bozhevolnyi, Demonstration of slow sound propagation and acoustic transparency with a serie of detuned resonators. Phys. Rev. B **89**, 184301 (2014)
45. G. Yu, X. Wang, Acoustical "transparency" induced by local resonance in bragg bandgaps. J. Appl. Phys. **115**, 044913 (2014)
46. A. Cicek, O. Kaya, M. Yilmaz, B. Ulug, Slow sound propagation in a sonic crystla linear wavegauide. J. Appl. Phys. **111**, 013522 (2012)
47. Y. Aurégan, V. Pagneux, Slow sound in lined flow ducts. J. Acoust. Soc. Amer. **138**, 605 (2015)

48. W. Robertson, C. Baker, C. Brad Bennet, Slow group velocity propagation of sound via defect coupling in a one-dimensional acoustic band gap array. Amer. J. Phys. **72**, 255–257 (2003)
49. G. Theocharis, O. Richoux, V.R. García, A. Merkel, V. Tournat, Limits of slow sound propagation and transparency in lossy, locally resonant periodic structures. J. Phys. **16**(9), 093017 (2014)
50. J.-P. Groby, W. Huang, A. Lardeau, Y. Aurégan, The use of slow sound to design simple sound absorbing materials. J. Appl. Phys. **117**, 124903 (2015)
51. X. Jiang, B. Liang, R.-Q. Li, X.-Y. Zou, L.-L. Yin, J.-C. Cheng, Ultra-broadband absorption by acoustic metamaterials. Appl. Phys. Lett. **105**(24), 243505 (2014)
52. N. Fang, D. Xi, J. Xu, M. Ambati, W. Srituravanich, C. Sun, X. Zhang, Ultrasonic metamaterials with negative modulus. Nat. Mater. **5**, 452–456 (2006)
53. P. Leclaire, O. Umnova, T. Dupont, R. Panneton, Acoustical properties of air-saturated porous material with periodically distributed dead-end pores. J. Acoust. Soc. Amer. **137**, 1772–1782 (2015)
54. N. Jiménez, J.-P. Groby, V. Pagneux, V. Romero-García, Iridescent perfect absorption in critically-coupled acoustic metamaterials using the transfer matrix method. Appl. Sci. **7**(6), 618 (2017)
55. J. Yang, J.S. Lee, Y.Y. Kim, Multiple slow waves in metaporous layers for broadband sound absorption. J. Phys. D: Appl. Phys. **50**(1), 015301 (2016)
56. X. Peng, J. Ji, Y. Jing, Composite honeycomb metasurface panel for broadband sound absorption. J. Acoust. Soc. Amer. **144**(4), EL255–EL261 (2018)
57. H. Long, Y. Cheng, J. Tao, X. Liu, Perfect absorption of low-frequency sound waves by critically coupled subwavelength resonant system. Appl. Phys. Lett. **110**(2), 023502 (2017)
58. L. Liu, H. Chang, C. Zhang, X. Hu, Single-channel labyrinthine metasurfaces as perfect sound absorbers with tunable bandwidth. Appl. Phys. Lett. **111**(8), 083503 (2017)
59. H. Zhao, Y. Wang, J. Wen, Y.W. Lam, O. Umnova, A slim subwavelength absorber based on coupled microslits. Appl. Acoust. **142**, 11–17 (2018)
60. A. Climente, D. Torrent, J. Sánchez-Dehesa, Omnidirectional broadband acoustic absorber based on metamaterials. Appl. Phys. Lett. **100**(14), 144103 (2012)
61. A. Elliott, R. Venegas, J. Groby, O. Umnova, Omnidirectional acoustic absorber with a porous core and a metamaterial matching layer. J. Appl. Phys. **115**(20), 204902 (2014)
62. W. Huang, L. Schwan, V. Romero-García, J.-M. Génevaux, J.-P. Groby, 3d-printed sound absorbing metafluid inspired by cereal straws. Sci. Rep. **9**(1), 8496 (2019)
63. N. Jiménez, T.J. Cox, V. Romero-García, J.-P. Groby, Metadiffusers: Deep-subwavelength sound diffusers. Sci. Rep. **7**(1), 5389 (2017)
64. E. Ballestero, N. Jiménez, J.-P. Groby, S. Dance, H. Aygun, V. Romero-García, Experimental validation of deep-subwavelength diffusion by acoustic metadiffusers. Appl. Phys. Lett. **115**(8), 081901 (2019)
65. G. Ma, M. Yang, S. Xiao, Z. Yang, P. Sheng, Acoustic metasurface with hybrid resonances. Nat. Mater. **13**(9), 873 (2014)
66. Y. Aurégan, Ultra-thin low frequency perfect sound absorber with high ratio of active area. Appl. Phys. Lett. **113**(20), 201904 (2018)
67. A.A. Fernández-Marín, N. Jiménez, J.-P. Groby, J. Sánchez-Dehesa, V. Romero-García, Aerogel-based metasurfaces for perfect acoustic energy absorption. Appl. Phys. Lett. **115**(6), 061901 (2019)
68. Z. Yang, J. Mei, M. Yang, N. Chan, P. Sheng, Membrane-type acoustic metamaterial with negative dynamic mass. Phys. Rev. Lett. **101**(20), 204301 (2008)
69. M. Yang, Y. Li, C. Meng, C. Fu, J. Mei, Z. Yang, P. Sheng, Sound absorption by subwavelength membrane structures: a geometric perspective. Comptes Rendus Mécanique **343**(12), 635–644 (2015)
70. Y. Chen, G. Huang, X. Zhou, G. Hu, C.-T. Sun, Analytical coupled vibroacoustic modeling of membrane-type acoustic metamaterials: membrane model. J. Acoust. Soc. Amer. **136**(3), 969–979 (2014)

71. Y. Chen, G. Huang, X. Zhou, G. Hu, C.-T. Sun, Analytical coupled vibroacoustic modeling of membrane-type acoustic metamaterials: plate model. J. Acoust. Soc. Amer. **136**(6), 2926–2934 (2014)
72. C.R. Liu, J.H. Wu, K. Lu, Z.T. Zhao, Z. Huang, Acoustical siphon effect for reducing the thickness in membrane-type metamaterials with low-frequency broadband absorption. Appl. Acoust. **148**, 1–8 (2019)
73. M. Yang, C. Meng, C. Fu, Y. Li, Z. Yang, P. Sheng, Subwavelength total acoustic absorption with degenerate resonators. Appl. Phys. Lett. **107**(10), 104104 (2015)
74. V. Leroy, A. Strybulevych, M. Lanoy, F. Lemoult, A. Tourin, J.H. Page, Superabsorption of acoustic waves with bubble metascreens. Phys. Rev. B **91**(2), 020301 (2015)
75. A. Merkel, G. Theocharis, O. Richoux, V. Romero-García, V. Pagneux, Control of acoustic absorption in one-dimensional scattering by resonant scatterers. Appl. Phys. Lett. **107**(24), 244102 (2015)
76. Y. Chong, L. Ge, H. Cao, A.D. Stone, Coherent perfect absorbers: time-reversed lasers. Phys. Rev. Lett. **105**(5), 053901 (2010)
77. P. Wei, C. Croënne, S. Tak Chu, J. Li, Symmetrical and anti-symmetrical coherent perfect absorption for acoustic waves. Appl. Phys. Lett. **104**(12), 121902 (2014)
78. N. Jiménez, V. Romero-García, A. Cebrecos, R. Picó, V.J. Sánchez-Morcillo, L.M. García-Raffi, Broadband quasi perfect absorption using chirped multi-layer porous materials. AIP Adv. **6**(12), 121605 (2016)
79. N. Jiménez, V. Romero-García, V. Pagneux, J.-P. Groby, Rainbow-trapping absorbers: broadband, perfect and asymmetric sound absorption by subwavelength panels for transmission problems. Sci. Rep. **7**(1), 13595 (2017)
80. C. Fu, X. Zhang, M. Yang, S. Xiao, Z. Yang, Hybrid membrane resonators for multiple frequency asymmetric absorption and reflection in large waveguide. Appl. Phys. Lett.**110**(2), 021901 (2017)
81. V. Romero-García, N. Jiménez, J.-P. Groby, A. Merkel, V. Tournat, G. Theocharis, O. Richoux, V. Pagneux, Perfect absorption in mirror-symmetric acoustic metascreens. Phys. Rev. Appl. **14**(5), 054055 (2020)
82. C. Bradley, Time harmonic acoustic bloch wave propagation in periodic waveguides. part i. theory. J. Acoust. Soc. Amer. **96**(3), 1844–1853 (1994)
83. N. Sugimoto, T. Horioka, Dispersion characteristics of sound waves in a tunnel with an array of helmholtz resonators. J. Acoust. Soc. Amer. **97**(3), 1446–1459 (1995)
84. M. Amin, A. Elayouch, M. Farhat, M. Addouche, A. Khelif, H. Bağcı, Acoustically induced transparency using fano resonant periodic arrays. J. Appl. Phys. **118**(16), 164901 (2015)
85. H. Lissek, R. Boulandet, R. Fleury, Electroacoustic absorbers: bridging the gap between shunt loudspeakers and active sound absorption. J. Acoust. Soc. Amer. **129**(5), 2968–2978 (2011)
86. R. Venegas, C. Boutin, O. Umnova, Acoustics of multiscale sorptive porous materials. Phys. Fluids **29**(8), 082006 (2017)

Chapter 6
Acoustic Wave Propagation in Viscothermal Fluids

An Electromagnetic Analogy

Denis Lafarge

Abstract First, we recall the Navier-Stokes-Fourier model linearized equations, which govern the propagation of small amplitude, long wavelength waves in viscothermal fluids; we specify how these equations are derived from several thermodynamic simplifications, and examine some of their solutions. Then, we analyze the general pattern of macroscopic nonlocal equations of propagation of small amplitude electromagnetic waves in effective homogeneous media, taking into account both the temporal and spatial dispersion. We argue that we lack a whole thermodynamics to fully precise all intervening quantities; proceeding by analogy, we then suggest that for the general acoustics of a homogeneous fluid, an analogous general pattern of nonlocal equations of propagation would arise, if we had sufficient thermodynamics. These ideas are finally implemented to obtain, within the available Navier-Stokes-Fourier's model, a nonlocal description of compressional waves.

6.1 Introduction

In the many works carried out on acoustic phononic crystals and metamaterials, viscous and thermal losses are usually neglected. This is often unrealistic. In this chapter, we first recall the classical theory of small amplitude wave propagation in a viscothermal fluid, and next reexamine it in a new light: that of a deep *electromagnetic analogy*. Indeed, we show that to some extent it can be put in the form of "nonlocal acoustic" equations allowing for both temporal and spatial dispersion. This is closely similar to the general nonlocal Maxwell equations that describe electromagnetic wave propagation in dispersive materials with temporal and spatial dispersion. Working within the usual (Navier-Stokes-Fourier) near–equilibrium thermodynamic framework, only a reduced version of this analogy, concerned with compressional motions

The original version of this chapter was revised: The belated corrections have been corrected. The correction to this chapter is available at https://doi.org/10.1007/978-3-030-84300-7_13.

D. Lafarge (✉)
Laboratoire d'Acoustique de l'Université du Mans, UMR CNRS 6613, Le Mans, France
e-mail: denis.lafarge@univ-lemans.fr

N. Jiménez et al. (eds.), *Acoustic Waves in Periodic Structures, Metamaterials, and Porous Media*, Topics in Applied Physics 143,
https://doi.org/10.1007/978-3-030-84300-7_6

only, will be accessible here. As outlined in the Appendix, we believe that the analogy would undertake its more complete nondegenerate form, (including Frenkel's solid-type of shearing at short times, see Sect. 6.6.2), if the thermodynamic framework could be sufficiently extended. This extension, however, is at present time a distant perspective.

In the next chapter the reduced analogy describing compressional waves in the homogeneous fluid, then is generalized to describe macroscopic compressional waves along a symmetry axis in macroscopically homogeneous unbounded metamaterials constituted of the viscothermal fluid permeating a rigid[1] porous structure having arbitrary microgeometry. It expresses in a nonlocal dynamic homogenization attributing to the medium, from microstructure, an equivalent nonlocal density and bulk modulus, playing the role of nonlocal electric and (inverse) magnetic susceptibilities. The developed nonlocal homogenization description is new and, we believe, will predict the right metamaterial properties when generalized further to describe inhomogeneous materials, and in particular, materials having finite dimensions.[2]

Finally, in Appendix of next chapter we recall how the developed nonlocal description simplifies in a classic local one, in some simple microgeometries. It is more convenient not to derive this simplified local description from the general nonlocal one, but to calculate afresh the relevant quantities.

Usually, this limit is deduced from a long-wavelength condition by blindly applying the method of two-scale asymptotic homogenization[3]; in reality, this "deduction" is not only illusory, (in it, is tacitly introduced a prerequisite of the local theory, namely, that the microgeometry is "simple" i.e. does not involve very different characteristic pore sizes), it is also fundamentally faulty. Here, we deduce the same limit in a better physical way by explicitly granting *as a simplification of the real problem*, the total absence of spatial dispersion.[4] As this total absence is the exact, (but unphysical), expression of the leading order terms of the method of two-scale asymptotic homogenization, (considered here,in the case of rigid porous structure) it should be clear, a priori, that the latter method cannot be an entirely consistent method of homogenization, (despite its large literature, practical usefulness, and original introduction by mathematicians).[5]

---

[1] If deformable, the macroscopic description is expected to manifest the full nonlocal pattern of equations expressed in Appendix: see the concluding considerations.

[2] In the context of a macroscopic nonlocal description any material having finite dimensions is to be viewed as an inhomogeneous material.

[3] And retaining only the leading order, see e.g. [3] (Appendix A) and [2] and [4] in next chapter.

[4] To take an image, this simplification transforms the full Kirchhoff's theory of the propagation of sound waves in cylindrical circular tubes, in Zwikker and Kosten's, classical but incomplete treatment, only slightly inexact at long wavelengths.

[5] In particular, we can anticipate that the addition of all higher order terms appearing in sequence in this abstract homogenization process, will always prove to yield results departing significantly from the exact ones.

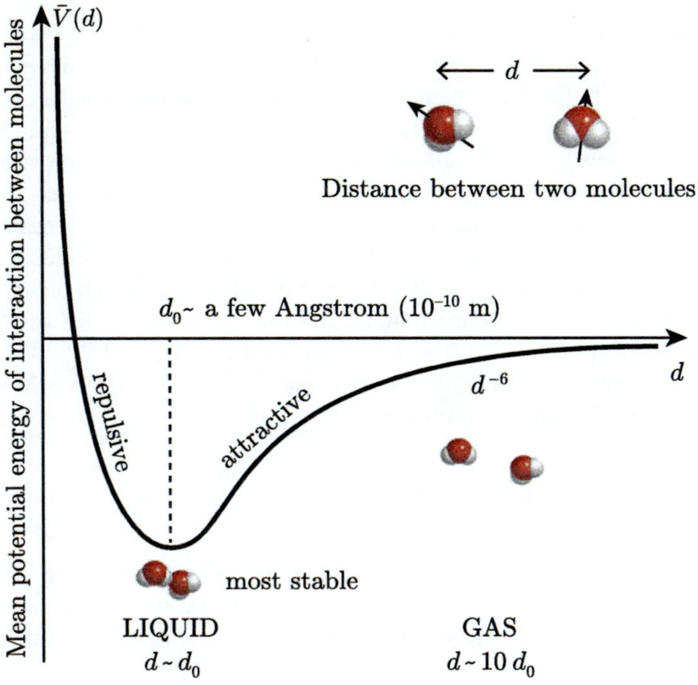

Fig. 6.1 Typical pair-interaction potential

6.2 Molecular Constitution of Fluids

We start by recalling basic facts about the molecular constitution of fluids. Virtually everything we are sensitive to, on earth and in the sky, is made up of countless molecules that respond to electromagnetic interactions. Molecules are usually small electrical dipoles, (either permanent or induced): a small separation of negative and positive charges is present, which produces a dipole moment equal to the amount of separated charges times the distance. The dipole averaged pair-interaction energy potentials $\bar{V}(d)$, have a simple Boscovich form, shown in Fig. 6.1, in function of the distance d between two of them.[6]

A fluid, liquid or gas, comprises an incredibly large number of molecules in any "physically infinitesimal" volume. A number so large (recall Avogadro's number: $N_A \cong 6.10^{23}$) that it corresponds to a so-called "thermodynamic limit".

In a liquid (solid) such as water (ice), the molecules are bound at distances d on the order of a few Ångström. Typically, $d_0 \sim 3.1 \cdot 10^{-10}$ m, for the distance where the pair-interaction potential $\bar{V}(d)$ is close to its minimum. In a gas such as air,

[6] In the averaging over the relative orientations of two dipoles, a thermalization is performed, so that a configuration having pair-interaction energy V is considered to have a probability proportional to $e^{-V/kT}$ to occur.

the molecules are widely distanced on the order of, say, mean distances $\overline{d} \sim 10\,d_0$. The molecules then are mostly non-interacting, in uniform inertial motion except for binary collisions. The mean free path, (mean distance travelled by a molecule between two collisions), is on the order of the microscopic distance $\ell = \eta/\rho_0 c_0$, (as can be estimated by elementary kinetic theory considerations), where η is first viscosity, ρ_0 ambient density, and c_0 is (adiabatic) speed of sound. This distance is very small, ($\sim 4.5 \cdot 10^{-8}$ m in air), only one order of magnitude higher than the mean distance $\overline{d}$ between molecules, ($\ell \sim 10\overline{d} \sim 100 d_0$). In a liquid, as the molecules are bound, a mean free path distance is not well-defined, but can be considered on the order of the mean distance ($\overline{d} \sim d_0$) between molecules, (in water we find $5.9 \cdot 10^{-10}$ m for the above microscopic distance ℓ, which is roughly two-times d_0).

The thermal expansion coefficient β_0 of a fluid, (liquid or gas), in thermodynamic equilibrium at temperature T and pressure P, is defined as

$$\beta_0 = \frac{1}{v}\left(\frac{\partial v}{\partial T}\right)_P = -\frac{1}{\rho}\left(\frac{\partial \rho}{\partial T}\right)_P, \tag{6.1}$$

where $v = 1/\rho$ is the specific volume, (volume per unit of mass). The ideal gas is the limit where the molecules are considered point mass, possessing mass but no significant volume. Simple kinetic theory considerations lead to the ideal gas law

$$PV = NkT,$$

where V is the volume, N is the number of particles in it, and k is Boltzmann's constant. From this law and the definition (6.1), we see that the thermal expansion coefficient of an ideal gas is given by

$$\beta_0 = 1/T_0, \tag{6.2}$$

with T_0 the ambient temperature. In a real gas, there are deviations from (6.2) to account for the nonzero tail of the actual potential at large distances; but as the potential goes to zero very rapidly, (see the exponent -6 in Fig. 6.1), the above is only slightly modified:

$$\beta_0 \cong 1/T_0. \tag{6.3}$$

A physical expression of these molecular characteristics is that a gas is easily compressible and also significantly expands when heated.

In a liquid, because the molecules are bound, trapped at the minimum distance d_0 allowed by intermolecular forces, but not occupying fixed mean positions, (on the contrary, undergoing permanent diffusive wandering from one location to the other), the coefficient of thermal expansion is drastically lower, in order of magnitude:

$$\beta_0 \lll 1/T_0. \tag{6.4}$$

Indeed, when some heat quantity is given to the fluid so that the equilibrium temperature increases, the molecules augment their characteristic diffusive velocities, wandering more rapidly from one location to the other. In so doing, however, they almost remain trapped at the minimum distance d_0. The thermal expansion coefficient is non zero, $\beta_0 \neq 0$, because there always subsists a small thermal expansion on the order of that found in solids.

In what follows, we limit ourselves to considering simple liquids and gases whose composition is invariable, i.e. so-called trivariate fluids whose thermodynamic state is specified by two independent thermodynamic variables, (such as pressure, temperature, density, etc.). A brief recap on the thermodynamics of trivariate fluids is given in Sect. 6.5. There is the following important thermodynamic identity, between fluid parameters, (see below (6.20.1)):

$$\gamma - 1 = \frac{T_0 \beta_0^2 c_0^2}{c_P}, \tag{6.5}$$

with γ and c_P, the heat capacity ratio (or adiabatic exponent) and specific heat coefficient at constant pressure. In view of this general identity, the above molecular characteristics of the liquid and gas phases, can be seen to have the following important consequences.

For a liquid, as β_0 is very small, the deviation (6.5) is almost zero, (it is a quadratic effect on β_0), and γ turns out to be nearly indistinguishable from 1. Therefore, the isothermal bulk modulus K_0, is nearly indistinguishable from the adiabatic bulk modulus[7] $K_a = \gamma K_0$, i.e.: $K_a \cong K_0$. At variance, in a gas, β_0 is not to be considered a small quantity, and it turns out that the factor $\gamma - 1$ is generally of order 1. Simple kinetic theory considerations give: $\gamma - 1 = 2/n_d$, where n_d is the number of excited degrees of freedom of a typical molecule. For example, for air, 99% composed of diatomic molecules (N_2 and O_2) whose vibrational movements are only slightly excited, $n_d \cong 5$, as there are 3 translational and only 2 rotational degrees of freedom to consider. This gives $\gamma - 1 \cong 0.4$, and in normal conditions the adiabatic bulk modulus of air, $K_a = \gamma P_0$, is 40% greater than the isothermal bulk modulus, $K_0 = P_0$, where P_0 is atmospheric pressure.

To finish, we recall that a special notation χ_0 is used for the inverse of the adiabatic bulk modulus, the adiabatic compressibility:

$$1/K_a \equiv \chi_0. \tag{6.6}$$

For the case of propagation in the fluid with losses disregarded, no heat enters or leaves an element of fluid during its alternate cooling and warming due to the passage of sound waves, so that the pressure-density cycle is adiabatic, and the following relation occurs between the fluid ambient density ρ_0, the fluid adiabatic compressibility χ_0, and the (adiabatic) speed of sound c_0

[7] See below (6.19.1) and (6.19.3) for the general definition of the adiabatic and isothermal bulk moduli K_a and K_0.

$$\rho_0 \chi_0 c_0^2 = 1. \tag{6.7}$$

This relation reminds us that observed for the propagation of electromagnetic waves in a lossless non-dispersive medium equivalent to vacuum:

$$\epsilon_0 \mu_0 c^2 = 1. \tag{6.8}$$

In the presence of losses, we will see that the density and compressibility eventually become operators $\hat{\rho}$ and $\hat{\chi}$. They will play the role assigned in electromagnetism to the electrical susceptibility $\hat{\epsilon}$ and magnetic susceptibility $\hat{\mu}$ operators, respectively.[8]

6.3 Wave Propagation in a Viscothermal Fluid: *Use of Near-Equilibrium Thermodynamics Because of Scale Separation*

Let us now precise notions such as, “fluid particle”, “macroscopic level” and “scale separation”. A “fluid particle” refers to the macroscopic collection of molecules contained in a “physically infinitesimal” volume. Despite its small size, it is a “macroscopic” object in the sense that it contains an enormous number – “thermodynamic” – of molecules. To have an order of magnitude in mind, we typically have 10^9 molecules in the smallest experimentally resolvable volumes [3]. A given volume $\ell^3 = dV$ containing a “thermodynamic” number of molecules will be considered an infinitesimal macroscopic volume, or “microscopic” volume, when its size ℓ is very much smaller than any macroscopic characteristic distance L, such as sample size, characteristic structure size, or wavelength λ.

As an example of a microscopic size ℓ in a homogeneous fluid, we may think of the mean free path, $\ell \sim \eta/\rho_0 c_0$. We do not aim at a description of the underlying motion of interacting molecules in such volumes. With numbers $N > 10^9$ this would be not only impossible but also useless. Instead, we aim at a macroscopic description statistical in nature, feasible because of there being a “thermodynamic-limit”, very large number of particles. In fact, through the statistical Law of Large Numbers, the physical state of each infinitesimal “fluid particle” is specified by remarkably few macroscopic variables.

This is related to the very large “scale separation” between the typical macroscopic and microscopic scales, as shown in Fig. 6.2.

Because of the assumed “scale separation”, the fluid in the infinitesimal volume $\ell^3 = dV$ is very nearly in a state of *local thermodynamic equilibrium*. Instead of

[8] More precisely and as discussed in the text and in Appendix, this affinity will be fully expressed only when the thermodynamic framework is sufficiently extended, with density and compressibility kernels becoming second order and fourth order tensors. In the main text they will remain scalars because we will restrict to long-wavelengths compressional waves only described within classical near-equilibrium framework.

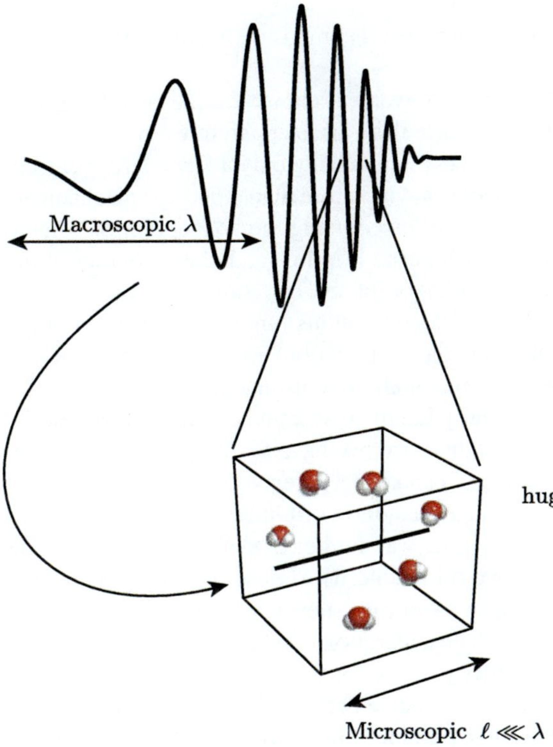

Fig. 6.2 Scale separation

doing a microscopic description, we can make a macroscopic description in which the state of "macroscopic motion" of each fluid particle is summed up in terms of only five numbers: the mean or macroscopic velocity, and two equilibrium, (for a trivariate fluid), thermodynamic variables.

Indeed, our wide "scale separation" condition implies in particular that any typical relaxation time associated with irreversible processes in the fluid, will be very short compared to the typical periods of the acoustic vibrations. As a result, we will be justified in discarding the effects of the so-called molecular relaxation. Consider for example the case of air. It is principally made of polyatomic – diatomic – molecules. Their energy will be distributed between the translational (external) degrees of freedom and the rotational and vibrational (internal) degrees of freedom. Obviously the sound wave disturbs to some extent the equilibrium distribution of energy between external and internal degrees of freedom. The concomitant departures from local thermodynamic equilibrium have tendency to return to equilibrium after certain relaxation times. Here, assuming that the typical periods of motion are sufficiently large compared to the molecular relaxation times, we have that the distribution of energy between external and internal degrees of freedom always remains that of thermodynamic equilibrium. In practice, this limit is well-satisfied for audible sounds.

Pressure and temperature is a possible convenient choice for the two thermodynamic variables. These thermodynamic variables – as is apparent from the fact that they become meaningless for the case of a system having only a few degrees of freedom – are macroscopic statistical variables having more than a purely "mechanical" meaning. Their averaged nature can be conceived using the conception of volume average, referred here to H.A. Lorentz who used it in his landmark program of the "theory of electrons" [4], (see Sect. 6.11.2), or, better, the conception of ensemble average, referred to J.W. Gibbs, who introduced it in his fundamental studies of the theory of gases and the statistical mechanics foundations of thermodynamics [1]: the average is performed over the ensemble of all systems that are "macroscopically identical". This notion of macroscopically identical systems can be clarified in the case of thermodynamic equilibrium. In this case, we have to think as equiprobable, all configurations that a set of molecules can take in a given volume and which share a given total energy. We will imagine that the notion of Gibbs ensemble could also in principle be generalized in an appropriate way in the case of deviations from the thermodynamic equilibrium, which would enable us to describe in a generalized Gibbs sense the propagation of waves of arbitrarily short wavelengths or arbitrarily high frequencies,[9] not respecting any scale-separation condition.

As for the macroscopic mean velocity $\boldsymbol{v}$, we have the same two conceptions. With volume averaging, we can define $\boldsymbol{v}$ as the velocity of the center of mass of the collection of molecules present in the fluid particle volume. With ensemble averaging, we will assume possible to define a collection of different molecular realizations of the same fluid medium and wave propagation phenomenon, that are equivalent from a macroscopic point of view. We can define $\boldsymbol{v}$ at a given position, as, say, the velocity of the closest molecule, averaged over all realizations. (Of course, these naive definitions would have to be refined as they are, both, not consistent quantum mechanically, but this is not our focus here). At long wavelengths, we will assume that ergodicity properties will ensure the equivalence of the two points of view.

6.4 Governing Navier-Stokes-Fourier Equations for Small Amplitude Wave Propagation

We are interested here in the first place in describing, at the macroscopic thermodynamic level, small amplitude long-wavelength wave propagation in a homogeneous fluid, (see e.g. the situation depicted on (Fig. 6.3) with $\lambda \gg \ell$, ℓ the mean free path). At each spatial position $\boldsymbol{x}$ and instant of time t, the state of the fluid, nearly in equilibrium, is for the essential described by the velocity $\boldsymbol{v} = \boldsymbol{v}(t, \boldsymbol{x})$, pressure $P(t, \boldsymbol{x})$ and temperature $T(t, \boldsymbol{x})$. But as the fluid is not, exactly, in a local state of pure thermo-

[9] We will later need thinking in this strongly out-of-equilibrium, hypothetical Gibbs averaging sense, when discussing hypothetical Frenkel-type of motions in gases.

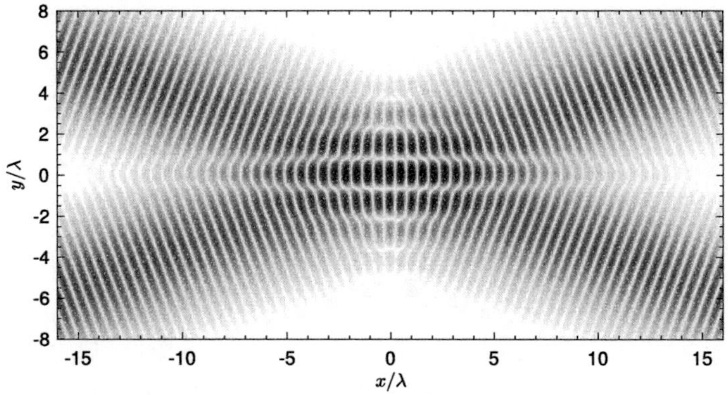

Fig. 6.3 Wave propagation example: two intersecting Gaussian beams

dynamic equilibrium, this description is not complete. To account for the dissipative processes induced by the nonequilibrium, there are additional variables, such as the dissipative viscous stresses σ'_{vij} and heat flux $\boldsymbol{q}$, that will be necessary to introduce, and relate, by constitutive laws, to the variations of the former variables. All variables are "total variables", I, sum of an equilibrium variable, I_0, representing the value of this variable in the ambient thermodynamic state, (zero for the velocity and additional variables), and a wave variable, I_w, representing an excess value associated to the wave:

$$I \begin{pmatrix} \text{total} \\ \text{variable} \end{pmatrix} = I_0 \begin{pmatrix} \text{equilibrium} \\ \text{variable} \end{pmatrix} + I_w \begin{pmatrix} \text{wave} \\ \text{variable} \end{pmatrix}. \tag{6.9}$$

We now recall how are derived the basic Navier-Stokes-Fourier viscothermal linearized model equations of motion, governing the small "wave variables", $\boldsymbol{v}$, p, τ and b, namely: $\boldsymbol{v}$ = part "w" of velocity, its equilibrium part being zero, p = part "w" of pressure, or "excess pressure", the equilibrium part being P_0, τ = part "w" of temperature, or "excess temperature", the equilibrium part being T_0, and $b = \rho'/\rho_0$, the condensation, equal to excess density ρ' – part "w" of density – over ambient density ρ_0[10]

[10] This variable b, as we already have p and τ in the set of variables, is redundant thermodynamically, see (6.12). It is introduced here mainly because it will later facilitate the discussion of the electromagnetic analogy. In the context of the electromagnetic analogy, however, b will in fact not be viewed as the quantity $b = \rho'/\rho_0$. It will either be viewed as the purely kinematic quantity directly defined with (6.10), (see Sect. 6.12.1 and the definitions (6.98.1) and (6.98.2), equivalent to (6.10)); more precisely, in the full analogy we introduce a symmetric tensor b_{ij} playing the role of antisymmetric magnetic tensor field, obeying $\partial b_{ij}/\partial t = -(\partial_i v_j + \partial_j v_i)/2$, and representing the opposite strain, with $b \equiv \sum_i b_{ii}$, (more details are in Appendix).

$$\frac{\partial b}{\partial t} + \partial \cdot \boldsymbol{v} = 0, \tag{6.10}$$

$$\begin{aligned}\rho_0 \frac{\partial \boldsymbol{v}}{\partial t} &= -\partial p - \eta \partial \times \partial \times \boldsymbol{v} + \left(\frac{4\eta}{3} + \zeta\right) \partial \left(\partial \cdot \boldsymbol{v}\right), \\ &= -\partial p + \eta \partial^2 \boldsymbol{v} - \left(\frac{\eta}{3} + \zeta\right) \partial \left(\frac{\partial b}{\partial t}\right),\end{aligned} \tag{6.11}$$

$$\gamma \chi_0 p = b + \beta_0 \tau, \tag{6.12}$$

$$\rho_0 c_P \frac{\partial \tau}{\partial t} = \beta_0 T_0 \frac{\partial p}{\partial t} + \kappa \partial^2 \tau. \tag{6.13}$$

The parameters η and ζ, first and second viscosities, and κ, thermal conduction coefficient, are constitutive constants of the fluid. Their appearance is the expression of the mentioned additional irreversible-thermodynamic variables, (σ'_{vij} and $\boldsymbol{q}$), related by constitutive laws to the former variables, (here $\boldsymbol{v}$ and τ, resp.). To arrive at these linearized model equations (6.10)–(6.13), first, we recall some basic facts and definitions about equilibrium thermodynamics of trivariate fluids, and next, we introduce the additional variables associated to near-equilibrium deviations, and state the conservation and constitutive laws.

6.5 Equilibrium Thermodynamics: *Brief Recap on Trivariate Fluids*

Classical equilibrium thermodynamics of a trivariate fluid asserts that all thermodynamic quantities – thermodynamic variables and thermodynamic constants – can be deduced when we know a single relationship that links three independent additive quantities: entropy, internal energy and volume. It is the so-called fundamental relation.[11] It can be written either in the original form of entropy S in function of internal energy E and volume V, or in the form of internal energy E in function of entropy S and volume V:

$$\begin{aligned} S &= S(E, V), \\ E &= E(S, V). \end{aligned} \tag{6.14}$$

Indeed, as the entropy S is postulated to be a single valued continuous and differentiable function of E and V, and moreover a monotonic increasing function of E, the fundamental relation (6.14.1) can be inverted with respect to energy, giving (6.14.2) with a function E that is also single valued, continuous and differentiable.

Knowing the fundamental relation, the pressure and temperature can be written by the formulae:

[11] See [2, 3, 5]; nice discussions of several topics of interest here, e.g., thermodynamics, equations of motion in a fluid, and averaging operations in a material, are given by Marle [5].

$$P \equiv -\left(\frac{\partial E}{\partial V}\right)_S, \quad T \equiv \left(\frac{\partial E}{\partial S}\right)_V, \tag{6.15}$$

and likewise, all other quantities can be expressed as well in terms of S, V and the partial derivatives of E.

Equation (6.15) derive immediately from the thermodynamic identity $TdS = dE + PdV$. And the latter identity is a consequence of the First Law, $đQ = dE + PdV$, stating that the internal energy can change (by dE) only by receiving work ($-PdV$) and/or heat ($đQ$), joined to the additional condition that the latter is always the form $đQ = TdS$, (the notation $đQ$ reminds that the infinitesimal heat quantity is not an exact differential). Note that, as all quantities S, E and V in (6.14) are additive, the fundamental relation is homogeneous of order 1. Hence writing (6.14.2) for a *unit mass fluid particle* which can be in translational motion, we have the so-called caloric equation:

$$\epsilon = \epsilon(s, v). \tag{6.16}$$

It gives the specific internal energy ϵ in the fluid, (i.e. internal energy per unit mass, which includes all forms of energy except the specific kinetic energy $\boldsymbol{v}^2/2$ associated with the global mass translation movement), as a function of the specific entropy s, (entropy per unit mass), and the specific volume $v = 1/\rho$. Pressure and temperature are then functions $P = P(s, v)$ and $T = T(s, v)$ given by:

$$P \equiv -\left(\frac{\partial \epsilon}{\partial v}\right)_s, \quad T \equiv \left(\frac{\partial \epsilon}{\partial s}\right)_v. \tag{6.17}$$

As all thermodynamic variables can be expressed in terms of s, v, and the partial derivatives of ϵ, there are only two independent thermodynamic variables. Any variable, e.g. temperature, can be expressed as a function of two any other independent variables, by means of equations of state characteristic of the fluid, e.g. $T = T(P, v)$, $T = T(P, \epsilon)$, etc. Contrary to the fundamental relation, the equations of state do not constitute complete knowledge of the thermodynamic properties of the fluid. Knowledge of all equations of state, however, is complete, and equivalent to the knowledge of the fundamental relation or caloric equation.

The specific heat coefficients c_P and c_V, (amount of heat needed to rise a temperature of the unit of mass by 1 Kelvin under constant pressure, or under constant volume), adiabatic exponent, and thermal expansion coefficient, are:

$$c_P \equiv T\left(\frac{\partial s}{\partial T}\right)_P, \quad c_V \equiv T\left(\frac{\partial s}{\partial T}\right)_v, \quad \gamma \equiv \frac{c_P}{c_V}, \quad \beta \equiv \frac{1}{v}\left(\frac{\partial v}{\partial T}\right)_P. \tag{6.18}$$

For a trivariate fluid β is not completely negligible, ($\beta \neq 0$), or γ is not strictly equal to one. There follows that the caloric (6.16) cannot be written in the form $\epsilon = Y(s) + Z(v)$ and none of the equations of state degenerate into relations connecting only two thermodynamic variables.

Other useful thermodynamic definitions are those of the reference adiabatic and isothermal, bulk moduli K_a and K_i (or K_0), and velocities c_a (or c_0) and c_i (or c_0', later on), in the fluid:

$$K_a \equiv \rho \left(\frac{\partial P}{\partial \rho}\right)_s, \quad c_a^2 \equiv \left(\frac{\partial P}{\partial \rho}\right)_s, \quad K_i \equiv \rho \left(\frac{\partial P}{\partial \rho}\right)_T, \quad c_i^2 \equiv \left(\frac{\partial P}{\partial \rho}\right)_T. \tag{6.19}$$

By purely thermodynamic arguments, the various constitutive constants introduced so far (6.18)–(6.19) are not independent. One can show the following general thermodynamic identities:

$$\gamma - 1 = \frac{T\beta^2 c_a^2}{c_P}, \quad c_a^2 = \gamma c_i^2. \tag{6.20}$$

These are consequences of the First Law and the fact that $đQ = TdS$. Indeed the fundamental thermodynamic requirement that $đQ/T$ is actually an exact differential, meaning that the entropy is a function of state, imposes a restriction on the analytical form of the internal energy, also implying that the caloric equation $\epsilon = \epsilon(s, v)$ is not independent of the thermal equation of state $P = P(T, v)$. In the proof of (6.20), will be involved:

(i) Maxwell's thermodynamic relations such as $-(\partial P/\partial s)_v = (\partial T/\partial v)_s$ expressing the fact that energy ϵ or other potentials such as enthalpy H (whose differential is $dH = Tds + vdP$) and the Helmholtz free energy F (whose differential is $dF = -sdT - Pdv$) are functions of state; the process of obtaining this kind of relations is as follows: let $\epsilon = \epsilon(s, v)$, then writing $d\epsilon = Tds - Pdv$ and $\partial^2\epsilon/\partial s\partial v = \partial^2\epsilon/\partial v\partial s$, we get, $-(\partial P/\partial s)_v = (\partial T/\partial v)_s$, and similarly for the other thermodynamic potentials, e.g. with the enthalpy one finds, $(\partial v/\partial s)_P = (\partial T/\partial P)_s$.

(ii) General relations of the type $(\partial v/\partial s)_P = -(\partial v/\partial P)_s(\partial P/\partial s)_v$ satisfied by any three quantities v, P, s, satisfying a single functional relationship, which again expresses the existence of equations of state; the process of obtaining these relations is as follows: let $P = P(s, v)$, then $dP = (\partial P/\partial s)_v ds + (\partial P/\partial v)_s dv$, and if $dP = 0$, then $(\partial P/\partial v)_s(dv/ds) = -(\partial P/\partial s)_v$, which also writes $(\partial v/\partial s)_P = -(\partial v/\partial P)_s(\partial P/\partial s)_v$.

The detailed proof of (6.20) is classically found in textbooks (a derivation is also available at https://www.astro.princeton.edu/~gk/A403/termo.pdf). We will always have to use the identity (6.20.1) for quantities taken in the ambient state, denoted by index 0. For simplicity, however, we do not put the index 0 on all the quantities involved. In what follows we write $c_a^2 \equiv c_0^2$ and (6.20.1) in the form of (6.5).

6.6 Derivation of the Equations of Motion

Let us now turn to the derivation of the Navier-Stokes-Fourier model equations (6.10)–(6.13) that govern the small deviations from the equilibrium state. Consider the thermodynamic total variables pressure, temperature, entropy, specific volume, density, and specific internal energy. Removing for simplicity the index "*w*", we write them in the form:

$$\begin{aligned}
\text{Pressure:} \quad & P = P_0 + p, \\
\text{Temperature:} \quad & T = T_0 + \tau, \\
\text{Specific entropy:} \quad & s = s_0 + s', \\
\text{Specific volume:} \quad & v = v_0 + v', \\
\text{Density:} \quad & \rho = \rho_0 + \rho', \\
\text{Specific internal energy:} \quad & \epsilon = \epsilon_0 + \epsilon'.
\end{aligned}$$

Because of scale separation, the fluid in the infinitesimal volumes dV used to define these quantities is very nearly in equilibrium state. Therefore, both the ambient quantities and the total variable quantities, are related by equations of state. For example, writing the caloric equation for ambient state and actual state, we have, $\epsilon_0 = \epsilon(s_0, v_0)$ and $\epsilon_0 + \epsilon' = \epsilon(s_0 + s', v_0 + v) = \epsilon(s_0, v_0) + s'\,(\partial\epsilon/\partial s)\,(s_0, v_0) + v'\,(\partial\epsilon/\partial v)\,(s_0, v_0) + \ldots$, whence in the linear approximation, $\epsilon' = s'\,(\partial\epsilon/\partial s)\,(s_0, v_0) + v'\,(\partial\epsilon/\partial v)\,(s_0, v_0)$. In that way, starting with the different state equations, and within linear approximation valid for small amplitude perturbations, many linear relations can be obtained, relating any three of the above deviatoric ("*w*") quantities, and making apparent some of the ambient quantities, and also, the fluid thermodynamic constants such as β_0, c_P and c_V. Examples of such relationships will appear below.

The general equations governing the spatiotemporal evolution of all total variables quantities – the velocity $\boldsymbol{v}(t, \boldsymbol{x})$ and the thermodynamic fields $P(t, \boldsymbol{x})$, $T(t, \boldsymbol{x})$, $\rho(t, \boldsymbol{x})$, etc. – will be derived from:

(i) General conservation laws,
(ii) Equilibrium thermodynamics,
(iii) Constitutive equations, outside the realm of equilibrium thermodynamics.

The reason why the evolution of the fields cannot be derived solely from (i) general conservation laws and (ii) equilibrium thermodynamics (despite the long-wavelength scale-separation condition $\lambda \ggg \ell$ ensuring the existence of local equilibrium states, where ℓ is a microscopic length such as mean free path), is because *the fields and, thus, the thermodynamic state of the fluid vary to some extent in space and time*, meaning that the system is, slightly, out-of-equilibrium. As later discussed with the electromagnetic analogy, taking into account the effect of these variations will already consist in taking *spatial dispersion* and *temporal dispersion* effects into account.

Due to spatial and temporal variations, irreversible transient currents are generated, aimed at bringing the system back to equilibrium. Indeed, consider two neighbouring infinitesimal particles. At each instant of time they are in slightly different

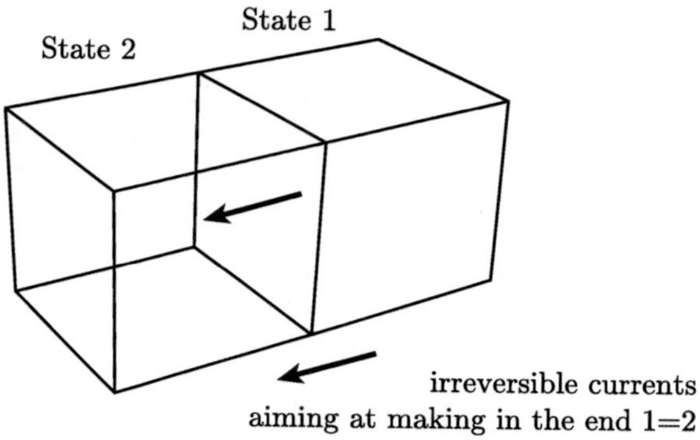

Fig. 6.4 Generation of irreversible currents

equilibrium thermodynamic states of motion (where by "thermodynamic state of motion" here we also include the information of the mean velocity), say state 1 and state 2, as sketched in Fig. 6.4.

During any physically infinitesimal period, many molecules (a thermodynamic number) pass through the boundary surface between particles 1 and 2. Particles passing in either direction carry information on the state 1 or 2, which are different; therefore, their passage is necessarily associated with an irreversible exchange of momentum and energy, aimed at suppressing the difference of state. The appearance of these transient irreversible currents is described in practice by the introduction of additional variables, related to the former by empirical constitutive laws. These laws lie outside the realm of equilibrium thermodynamics. Essentially, they must precise how much momentum and heat energy are irreversibly transferred, (resp. by so-called viscous effects and thermal conduction effects), per unit of time across the surface.

6.6.1 Conservation Laws and Equation of State

We first state the general conservation or balance laws.

Mass Conservation
It is classically found and derived in any textbook:

$$\frac{\partial \rho}{\partial t} + \partial \cdot (\rho \boldsymbol{v}) = 0.$$

After linearization, it reads

$$\frac{\partial b}{\partial t} + \partial \cdot \boldsymbol{v} = 0, \tag{6.21}$$

where $b = \rho'/\rho_0$ is the condensation.

Momentum Balance

This balance law is best written using the convective derivative, $\frac{d}{dt} = \frac{\partial}{\partial t} + \boldsymbol{v} \cdot \partial$:

$$\rho \frac{d}{dt}(v_i) = \partial_j \sigma_{ij},$$

where σ_{ij} is the excess stress tensor in the fluid, which incorporates the mentioned momentum irreversibly transferred. By integration over an infinitesimal volume dV in motion, it reads, $dm\frac{d}{dt}(v_i) = \frac{d}{dt}(v_i dm) = \oint \sigma_{ij}\hat{n}_j \, dS \, (dm = \rho dV)$. This is Newton's law, stating that the time rate of change of momentum is the result of the excess stress contact forces exerted at the particle bounding surface, (with $\hat{\boldsymbol{n}}$ the unit outward normal).

If there were present an external bulk force acting at a distance, pulling the volume dV with force $d\boldsymbol{f}$, it would be necessary to rewrite the above, as, $dm\frac{d}{dt}(v_i) = \frac{d}{dt}(v_i dm) = \oint \sigma_{ij}\hat{n}_j \, dS + df_i$. It would correspond, writing the starting equation with a bulk-source term of force, $[df_i/dV]$, added in the right-hand side. Obviously, $df_i/dV = \rho df_i/dm$, and if we count the force per unit mass, we write:

$$\rho \frac{d}{dt}(v_i) = \partial_j \sigma_{ij} + [\rho f_i] .$$

On using the expression of the convective derivative and inserting mass Conservation, it also reads

$$\frac{\partial}{\partial t}(\rho v_i) + \partial_j(\rho v_i v_j) = \partial_j \sigma_{ij} + [\rho f_i] ,$$

and after linearization

$$\rho_0 \frac{\partial}{\partial t} v_i = \partial_j \sigma_{ij} + [\rho_0 f_i] . \tag{6.22}$$

Energy Balance

This balance law is also stated in the most simple and explicit form using the convective derivative:

$$\rho \frac{d}{dt}\left(\frac{1}{2}v^2 + \epsilon\right) = -\partial_i\left(-\Sigma_{ij} v_j + q_i\right) ,$$

where $\boldsymbol{q}$ is the heat flux density due to thermal conduction, that determines the above-mentioned heat energy irreversibly transferred, and $\Sigma_{ij} = -P_0\delta_{ij} + \sigma_{ij}$ is the total variables stress tensor. By integration over an infinitesimal volume dV in motion, (and such that $\rho dV = dm$), it reads $\frac{d}{dt}\left(dm\frac{1}{2}v^2 + dm\epsilon\right) = \oint(-\Sigma_{ij}v_j + q_i)(-\hat{n}_i) \, dS$, where the surface integrals are over the bounding surface of element dV. This is the basic thermodynamic principle, stating that the time rate of change of the energy of a given fluid volume, (kinetic energy plus internal energy), is equal to the work done on it, (by the total stress contact forces exerted at the particle bounding surface), plus the heat it receives, (through this bounding surface), per unit of time.

If there were present a bulk-source of heat, directly acting in the fluid, and giving a quantity of heat $đQ$ in the volume dV during time dt, it would be necessary to rewrite the above, as, $\frac{d}{dt}\left(dm\frac{1}{2}v^2 + dm\epsilon\right) = \oint(-\Sigma_{ij}v_j)(-\hat{n}_i)\,dS + \oint q_i(-\hat{n}_i)\,dS + đQ/dt$. It would correspond writing the above starting balance law with a bulk-source term $[đQ/dVdt]$ added in the right-hand side. But, $đQ/dVdt = \rho đQ/dmdt$, and choosing to count the heat per unit of mass, i.e. noting the present $đQ/dm$ as $đQ$, we have

$$\rho\frac{d}{dt}\left(\frac{1}{2}v^2 + \epsilon\right) = -\partial_i\left(-\Sigma_{ij}v_j + q_i\right) + \left[\rho\frac{đQ}{dt}\right].$$

After linearization the energy balance equation is, (omitting the prime on ϵ)

$$\rho_0\frac{\partial\epsilon}{\partial t} = -P_0\partial_i v_i - \partial_i q_i + \left[\rho_0\frac{đQ}{dt}\right]. \tag{6.23}$$

We next introduce the equation of state, which expresses the condition of local thermodynamic equilibrium, to obtain another form of this energy balance equation.

Starting with the thermodynamic equilibrium relation, $d\epsilon = Tds - Pdv = Tds + Pd\rho/\rho^2$, substituting in (6.23), and taking into account (6.21), we find

$$\rho_0 T_0\frac{\partial s}{\partial t} = -\partial_i q_i + \left[\rho_0\frac{đQ}{dt}\right]. \tag{6.24}$$

Finally, to complete these equations and make apparent only the excess thermodynamic pressure and temperature, we write the equilibrium equations of state, $\rho = \rho(P, T)$ and $T = T(P, s)$, whose linearized versions give, resp.

$$\rho' = \left(\frac{\partial\rho}{\partial P}\right)_T p + \left(\frac{\partial\rho}{\partial T}\right)_P \tau, \quad \tau = \left(\frac{\partial T}{\partial P}\right)_s p + \left(\frac{\partial T}{\partial s}\right)_P s, \tag{6.25}$$

that is, using (6.19.3) and (6.18.4) and noting that $1/K_0 = \gamma\chi_0$,

$$b = \rho'/\rho_0 = \gamma\chi_0 p - \beta_0\tau, \quad \text{and} \quad \tau = \frac{\beta_0 T_0}{\rho_0 c_P}p + \frac{T_0}{c_P}s. \tag{6.26}$$

In the last equation, (6.18.1) has been used to evaluate $(\partial T/\partial s)_P = (T_0/c_P)$, and the remaining coefficient has been expressed using the thermodynamic identity $(\partial T/\partial P)_s = (\beta_0 T_0)/(\rho_0 c_P)$, given in Pierce [18] Sect. (6.10). Substituting (6.26.2) in (6.24), we obtain finally the following useful form of the energy balance equation:

$$\rho_0 c_P\frac{\partial\tau}{\partial t} = \beta_0 T_0\frac{\partial p}{\partial t} - \partial_i q_i + \left[\rho_0\frac{đQ}{dt}\right]. \tag{6.27}$$

To summarize, after consideration of the general conservation laws and equilibrium thermodynamic relations, we have obtained the following set, (6.28)–(6.31), of linearized equations:

$$\textbf{Mass:} \quad \frac{\partial b}{\partial t} + \partial \cdot \boldsymbol{v} = 0, \tag{6.28}$$

$$\textbf{Momentum:} \quad \rho_0 \frac{\partial v_i}{\partial t} = \partial_j \sigma_{ij} + [\rho_0 f_i], \tag{6.29}$$

$$\textbf{Equation of state:} \quad \gamma \chi_0 p = b + \beta_0 \tau, \tag{6.30}$$

$$\textbf{Energy:} \quad \rho_0 c_P \frac{\partial \tau}{\partial t} = \beta_0 T_0 \frac{\partial p}{\partial t} - \partial_i q_i + \left[\rho_0 \frac{đQ}{dt} \right], \tag{6.31}$$

with f_i, a possible external body force per unit mass, and $đQ/dt$, a possible external rate of heat supply per unit mass, acting on the fluid.

6.6.2 Constitutive Laws: Stokes and Fourier

Finally, we state irreversible-thermodynamic constitutive laws, for the evaluation of the heat flux q_i, and an additional lossy part σ'_{ij} present in the excess stresses σ_{ij}:

$$\sigma_{ij} = -p_a \delta_{ij} + \sigma'_{ij}, \qquad p_a = \chi_0^{-1} b, \tag{6.32}$$

with p_a, the fictitious excess pressure in adiabatic relation with the condensation. The additional stress σ'_{ij} owes its presence to the losses processes generated in response to the temporal and spatial variations in the fields. Using terminology inspired by the electromagnetic analogy that will later be discussed, (Sects. 6.10–6.12), the first adiabatic term would give the right result in the absence of intrinsically irreversible "polarization processes" induced by temporal and spatial variations in the fields; the additional term σ'_{ij}, is there to take into account the presence of such processes. A priori it would decompose in two parts:

$$\sigma'_{ij} = \sigma'_{\tau ij} + \sigma'_{vij}, \tag{6.33}$$

one, $\sigma'_{\tau ij}$, caused by the irreversible processes generated by the temperature variations, and the other, σ'_{vij}, caused by the irreversible processes generated by the velocity variations.

Indeed, in an inhomogeneous temperature field such as found in a wavefield, the condensation-rarefaction do not occur in exact adiabatic manner due to the small temperature exchanges which appear mainly by thermal conduction between neighbouring particles, (having slightly different instantaneous temperatures). For this reason, the natural excess pressure variable will not be the adiabatic one, $p_a = \chi_0^{-1} b$. It will be the thermodynamic excess pressure, $p(b, \tau)$, by definition related to b and

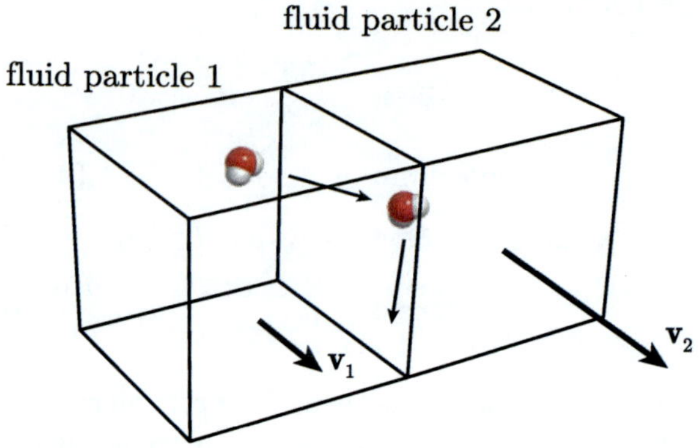

Fig. 6.5 Momentum transfer between two fluid particles 1 and 2 in contact

τ by the equation of state, (6.30).[12] This is because, as said, scale-separation here, means that the local thermodynamic equilibrium tends to be reached. Actually, as the effect of temperature spatial variations must express in $\sigma'_{\tau ij}$, we guess that this thermal stress term, once added to the adiabatic first term in (6.32), leads to the excess thermodynamic pressure, hydrostatic term:

$$-\,p_a\delta_{ij} + \sigma'_{\tau ij} = -\chi_0^{-1} b\delta_{ij} + \sigma'_{\tau ij} = -p\delta_{ij}. \tag{6.34}$$

Then the total excess stress tensor also simply writes:

$$\sigma_{ij} = -p\delta_{ij} + \sigma'_{vij}. \tag{6.35}$$

This last form (6.35) in place of (6.32), is the way the stresses are usually directly written in literature, without mentioning (6.32), (6.33), and (6.34), see e.g. Landau and Lifshitz [6]. We will soon see, in the analysis of the Fourier constitutive law which specifies the additional variable q_i, that the thermal stress tensor $\sigma'_{\tau ij}$ which allows (6.32)–(6.34) to be satisfied is:

$$\sigma'_{\tau ij} = -\kappa \frac{\gamma - 1}{\beta_0 T_0} \int_{-\infty}^{t} dt' \partial^2 \tau \delta_{ij}, \tag{6.36}$$

where κ is the coefficient of thermal conduction in Fourier's law.

The remaining σ'_{vij} in (6.33) or (6.35), is the viscous contribution, separately identified in Stokes's law. It must determine the rate of momentum transfer between two fluid particles 1 and 2 in contact, that do not have the same average velocity, as sketched in Fig. 6.5.

Indeed, over a physically infinitesimal time period, many molecules (a thermodynamic number) pass through the boundary surface between particles 1 and 2. But

[12] It will be shown that p is related to b only, by an operational nonlocal bulk modulus. The operator in the relation $p = \hat{\chi} b$ will be derived in Sect. 6.12.3

as the mean velocity in 2 is, say, greater than that in 1, this exchange is not symmetrical and particle 1 receives more momentum than does particle 2. Therefore, an irreversible exchange of momentum between the two particles is generated across the surface. This exchange creates the viscous stress tensor σ'_{vij}. It has two indices as one is defined by the direction of the normal to the considered surface, and the other is the direction defining the considered momentum component. The simplest *ansatz* is that at a given time t and position, it depends only on the first derivatives, $\partial v_i/\partial x_j$, evaluated at the same time and position, with other dependencies, (higher order spatial derivatives and time-memory effects), leading to negligible corrections.[13] With the additional observation that it should vanish if the fluid is rotating uniformly as a whole, one deduces that only the symmetric combinations $\partial v_i/\partial x_j + \partial v_j/\partial x_i$, can be involved. As a result, σ'_{vij} must be constructed with these combinations and the unit tensor δ_{ij}, the only other tensor at our disposal with isotropy assumed. There follows that the viscous irreversible current density of momentum is principally given in the form:

$$\sigma'_{vij} = \eta\left(\partial_i v_j + \partial_j v_i - \frac{2}{3}\delta_{ij}\partial_l v_l\right) + \zeta\delta_{ij}\partial_l v_l, \tag{6.37}$$

with η and ζ, two constitutive constants of the fluid, which are named the first and second viscosity. Equation (6.37) is Stokes's constitutive law.

It remains to state Fourier's constitutive law, indicating how much thermal energy is irreversibly transferred by thermal conduction. To state this law, we assume that the fluid particles 1 and 2 in contact do not have the same temperature. As before, particles 1 and 2 exchange a thermodynamic number of molecules, across their common boundary surface, over a physically infinitesimal period of time. But as the mean temperature in 2 is, say, greater than that in 1, particle 1 receives molecules having higher thermal speed than does particle 2. Therefore, a heat exchange is generated across the surface, that is described in terms of a thermal-conduction vector heat flux q_i. The simplest *ansatz* is that at a given time t and position, this vector depends only on the first derivatives $\partial\tau/\partial x_i$ evaluated at the same time and position, with other dependencies leading to negligible corrections.[14] With isotropy assumed there follows that the irreversible current density of heat transfer is principally:

$$q_i = -\kappa\partial_i\tau, \tag{6.38}$$

with κ a constitutive constant of the fluid, which is named the thermal conduction coefficient. Equation (6.38) is Fourier's constitutive law. Note that with these laws the density of entropy production can be shown to be

---

[13] In fact, these corrections should imperatively be taken into account, (which raises deep irreversible-thermodynamic questions), in such a way that the description, once reviewed in the light of the nonlocal electromagnetic analogy, remains physical and mathematically well conditioned: see end of Sect. 6.12.3.

[14] Again, the electromagnetic analogy suggests that these corrections should also be taken into account to ensure that the description remains entirely physical: see end of Sect. 6.12.3.

$$\sigma_s = -T_0^{-2} q_i \partial_i \tau + T_0^{-1} \sigma'_{vij} \partial_i v_j, \tag{6.39}$$

and by the second principle of thermodynamics which imposes $\sigma_s > 0$, we have that the introduced viscosities and thermal conduction coefficients are all positive.

Substituting Stokes's and Fourier's constitutive laws in the set (6.28)–(6.31) completed by (6.35), we finally obtain the governing linearized equations stated in (6.10)–(6.13).

We can now check our previous identifications (6.34) and (6.36). The combination of (6.12) and (6.13) gives, $\rho_0 c_P \partial_t \left(\gamma \chi_0 p - b\right) / \beta_0 = \beta_0 T_0 \partial_t p + \kappa \partial^2 \tau$, and using the first thermodynamic identity (6.20), we derive after straightforward calculation:

$$p = \chi_0^{-1} b + \kappa \frac{\gamma - 1}{\beta_0 T_0} \int_{-\infty}^{t} dt' \partial^2 \tau. \tag{6.40}$$

This expression of the thermodynamic excess pressure is coherent with (6.34), if only, we have the expression (6.36) of the thermal stresses.

We note that, in the same way as it is logical to write the momentum equation (6.29) as, $\rho_0 \partial_t v_i = \partial_j (-\chi_0^{-1} b \delta_{ij} + \sigma'_{ij})$, (using (6.32)), but it is, actually, more convenient to use the thermodynamic pressure and write it, $\rho_0 \partial_t v_i = \partial_j (-p \delta_{ij} + \sigma'_{vij})$, (using instead (6.35)), it would be logical to write, for the energy equation, $\rho_0 c_P \partial_t \tau = \beta_0 T_0 \chi_0^{-1} \partial_t b - \partial_i q'_i$, with some irreversible q'_i flux, but it is, again, more convenient to make appear the thermodynamic pressure, and use (6.31). The irreversible flux q'_i would be: $q'_i = -\gamma \kappa \partial_i \tau = \gamma q_i$, and it is preferable to deal with q_i, which is the thermal energy flux.

There is still one important point to be made here. The laws we have considered are special in that they have explicitly assumed scale separation, or sufficiently slow times variations. In reality, at very short times or very short wavelengths, the used near-equilibrium thermodynamic framework becomes unadapted and the appropriate description is expected to become much more subtle. In particular, solid-like behaviours should systematically appear in any fluid, to some extent, at very short times. This hypothesis was put forward long ago, in 1925, by the self-taught Russian theoretical physicist Yakov Il'ich Frenkel [7].

Frenkel's ideas which have long been overlooked, have recently been substantiated in liquids [8]. This suggests that, if the thermodynamic framework in which we work were broad enough, we would have to write, instead of (6.32):

$$\sigma_{ij} = -(\chi_0^{-1})_{ijkl} b_{kl} + \sigma'_{ij}, \quad b_{ij} \equiv -\frac{1}{2}\left(\partial_i a_j + \partial_j a_i\right), \quad \boldsymbol{v} \equiv \partial \boldsymbol{a}/\partial t, \tag{6.41}$$

$$(\chi_0^{-1})_{ijkl} = \chi_0^{-1} \delta_{ij} \delta_{kl} + \mu_0 \left(\delta_{ik}\delta_{jl} + \delta_{il}\delta_{jk} - \frac{2}{3}\delta_{ij}\delta_{kl} \right). \tag{6.42}$$

In the last equation, μ_0 would be a new constant, which we could denote the "Frenkel-Lamé" adiabatic shear modulus of the fluid. Its nonzero value would allow for the propagation of undamped shear waves in a very-short-times limit (times so short that the dissipative reactions expressed in the lossy σ'_{ij}, have no time to be built).

In Appendix we see how the electromagnetic-acoustic analogy is automatically, by construction, coherent with these physical views.

In a gas, the large separation between the molecules will make the modulus μ_0 a priori extraordinarily small; nevertheless, from the strict point of view of questions of principle, it should be considered nonzero, which would imply the presence of additional terms in the dissipative part σ'_{ij} of the stresses, in turn determining at least in Gibbs' sense,[15] the dynamics of propagation and attenuation of Frenkel's "elastic" waves, of extraordinarily low amplitudes. Of course, well before considering these effects in a thermodynamic extension, it would have been necessary, in the first place, to account for the molecular relaxation, and go beyond the too simplistic Stokes and Fourier's laws. A complete thermodynamic extension, we believe, would reveal the full nonlocal pattern of equations and quantities conjectured in the Appendix, inspired by the electromagnetic analogy.

6.6.3 Inherent Thermodynamic Simplifications in Navier-Stokes-Fourier's Framework

Let us now summarize and comment the above development, which has justified the Navier-Stokes-Fourier linearized-motion model equations (6.10)–(6.13). It rests on simplifications well-verified in practice: the assumption of local equilibrium allowing to use classical equilibrium thermodynamic relations and to neglect molecular relaxation phenomena, the Stokes and Fourier constitutive laws which are the simplest ones to be considered in this framework, and finally, the neglect of Frenkel's shear motions. Nevertheless, as we have just suggested, these are important simplifications from a general physical point of view. As such they can have some unwanted consequences. The motion equations obtained have an elliptic nature through the absence of Frenkel's terms and the Stokes law in the momentum equations, and a parabolic nature through Fourier's law in the energy equation. Their solutions then manifest unphysical features such as the totally diffusive nature of shear waves mode solutions,[16] and instant propagation of an initial temperature disturbance.[17]

In short, when these model equations (6.10)–(6.13) will be reexamined in the light of a general nonlocal electromagnetic analogy, (see Sects. 6.10–6.12 and Appendix), the simplifications they convey will cause defects, particularly visible in the degenerate description they will give of the shearing movements. These degeneracies could

[15] As in this context the wavelengths could be reduced to the point of becoming comparable to the mean free path, the separation of scale would not be ensured and macroscopic means would have to be taken in the Gibbs sense.

[16] See (6.66) below, this unphysical feature being related to the absence, in purely viscous shearing motions, of associated nonzero "acoustic stress field H_{ij}" – see in Appendix, after (6.205).

[17] See (6.145) below, this unphysical feature being related to the nonvanishing of the kernel $\chi_b(t, x)$, whatever the values, however large, of $|x|$, contradicting the finite velocity of propagation of physical influences.

only be totally overcome by working within an extended framework of the thermodynamics of irreversible processes, if necessary with the Gibbs mean, abandoning local equilibrium, going beyond Stokes and Fourier's laws, and introducing Frenkel's shearing motions. Nevertheless, by limiting our considerations to longitudinal movements as described by current laws, (see Sects. 6.10 and 6.12), we will deliberately avoid entering into this difficult subject, which belongs to the unclarified field of Extended Irreversible Thermodynamics far from equilibrium.

6.7 The Different Normal Modes at Fixed Real Wavevector

We detail here the different type of solutions of the model equations of motion (6.10)–(6.13). Eliminating the condensation, we have a system of five first-order-in-time differential equations for the five fields $\boldsymbol{v}(t, \boldsymbol{x})$, $p(t, \boldsymbol{x})$ and $\tau(t, \boldsymbol{x})$. Therefore, given a fixed wavevector $\boldsymbol{k}$, the solutions can be analyzed in terms of five "hydrodynamic modes" or "normal modes" solutions varying like[18]:

$$\boldsymbol{v}(t, \boldsymbol{x}) = \boldsymbol{v}_0 e^{-i\omega t + i\boldsymbol{k}\cdot\boldsymbol{x}}, \qquad p(t, \boldsymbol{x}) = p_0 e^{-i\omega t + i\boldsymbol{k}\cdot\boldsymbol{x}}, \qquad \tau(t, \boldsymbol{x}) = \tau_0 e^{-i\omega t + i\boldsymbol{k}\cdot\boldsymbol{x}}.$$

Consider first the case where the wavevector $\boldsymbol{k} = k_x\hat{\boldsymbol{x}} + k_y\hat{\boldsymbol{y}} + k_z\hat{\boldsymbol{z}}$ has real components. It is the form $\boldsymbol{k} = k\hat{\boldsymbol{n}}$, specified by a unit vector direction $\hat{\boldsymbol{n}}$ and modulus k. We must find five hydrodynamic modes solutions having this real wavevector, and complex frequencies, (to account for the damping brought by irreversible processes). To clarify these five solutions, we introduce three independent polarization directions, the longitudinal (1) direction ($\parallel \hat{\boldsymbol{n}}$), and transversal (2, 3) directions ($\perp\hat{\boldsymbol{n}}$), in which the velocity amplitude vector $\boldsymbol{v}_0$ can point, as shown in Fig. 6.6.

We write the complex frequencies as, $\omega = \omega' - i\omega''$. Hence the exponential factor will have the form $e^{-\omega'' t}e^{-i\omega' t + ik\hat{\boldsymbol{n}}\cdot\boldsymbol{x}}$. Physically, here, as the imaginary parts owe their existence to the losses terms, they are to be taken negative, i.e. ω'' positive, to ensure that the solutions fade away over time. The real parts will be either zero, corresponding to diffusive purely damped solutions, or nonzero, with the two signs possible, corresponding to right-going or left-going waves.
The five modes solutions, therefore, are as follows[19]:

Polarization 1:

$$\omega' = c_0 k - O(k^3), \qquad \omega'' = \Gamma k^2 + O(k^4),$$

[18] When using this complex notation, we understand that $\boldsymbol{v}(t, \boldsymbol{x}) = \Re\left(\boldsymbol{v}_0 e^{-i\omega t + i\boldsymbol{k}\cdot\boldsymbol{x}}\right)$, etc., systematically omitting here and elsewhere the real part symbol $\Re\,()$.

[19] The results for polarization 1 are the complex solutions, $\omega = \omega' - i\omega''$, of Kirchhoff-Langevin's dispersion equation (6.52). The results for polarization 2, 3, are the complex solutions of the vortical diffusion equation, first (6.61).

Fig. 6.6 Polarization

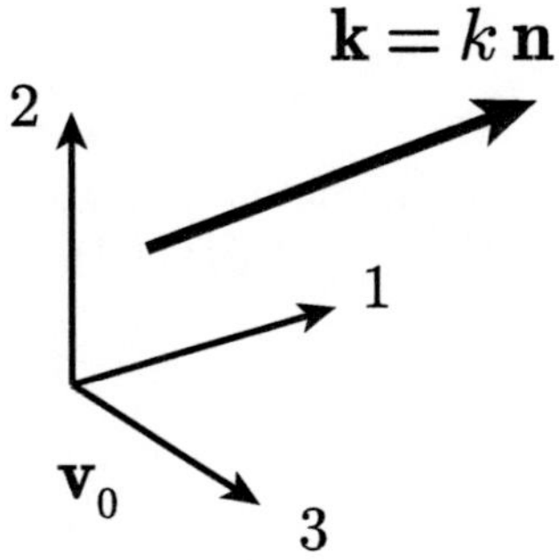

damped longitudinal wave propagating in the direction of vector $\boldsymbol{k} = k\hat{\boldsymbol{n}}$ (a so-called "acoustic wave"; it has pressure and temperature variations but almost no entropy variations),

Polarization 1:

$$\omega' = -c_0 k + O(k^3), \qquad \omega'' = \Gamma k^2 + O(k^4),$$

damped longitudinal wave of the same type (acoustic wave) propagating in the reverse direction,

Polarization 1:

$$\omega' = 0, \qquad \omega'' = \frac{\kappa}{\rho_0 c_P} k^2 - O(k^4),$$

purely damped longitudinal wave (a so-called "entropic wave"; it has entropy and temperature variations but almost no pressure variation),

Polarization 2:

$$\omega' = 0, \qquad \omega'' = \frac{\eta}{\rho_0} k^2,$$

purely damped shear wave (transversal, also called vortical; it has no pressure, temperature, or entropy variations),

Polarization 3:

$$\omega' = 0, \qquad \omega'' = \frac{\eta}{\rho_0} k^2,$$

purely damped shear wave of the same type but other polarization,

and where,

$$\Gamma \equiv \frac{1}{2\rho_0}\left[\frac{4\eta}{3}+\zeta+\kappa\left(\frac{1}{c_V}-\frac{1}{c_P}\right)\right], \tag{6.43}$$

is a small constitutive constant called the classical sound damping constant. Its order of magnitude is that of the kinematic viscosity

$$\nu \equiv \frac{\eta}{\rho_0}, \tag{6.44}$$

respectively 10^{-5} m^{2}s^{-1} and 10^{-6} m^{2}s^{-1} for air and water.

Consider next the case with real frequency and complex wavevectors. The above three kinds of waves, shear, "entropic" and "acoustic", will correspond to three different sorts of complex wavenumbers. In the following, we show how the equations of motion (6.10)–(6.13), are used to derive the characteristics of the "acoustic" and "entropic" harmonic modes; this calculation is classical and can be found in Rayleigh [9].

6.8 Some Explicit Calculations for the Longitudinal Modes

Insert (6.12) in (6.13) and use the first thermodynamic identity (6.20) and third definition (6.18) to obtain an alternative form of (6.13):

$$\partial_t \tau = \frac{\gamma-1}{\beta_0}\partial_t b + \frac{\kappa}{\rho_0 c_V}\partial^2\tau.$$

It simplifies the equations to introduce $\tau' = \beta_0\tau/(\gamma-1)$. Using this temperature variable and denoting c_0 and c_0' the adiabatic and isothermal reference sound velocities (c_a and c_i see (6.19)),[20] the (6.12) and (6.13) take the following form:

$$\frac{p}{\rho_0} = c_0'^2 b + (c_0^2 - c_0'^2)\tau', \tag{6.45}$$

and

$$\partial_t\tau' = \partial_t b + \frac{\kappa}{\rho_0 c_V}\partial^2\tau'. \tag{6.46}$$

Let us work in harmonic regime $e^{-i\omega t}$. Substituting (6.45) inside the last form of (6.11) gives an equation having the form, $\partial_t \boldsymbol{v} - \nu\partial^2\boldsymbol{v} = -\partial X$, with

$$X = \left[c_0'^2 - i\omega\frac{\zeta+\frac{\eta}{3}}{\rho_0}\right]b + (c_0^2 - c_0'^2)\tau',$$

---

[20] Which verify $\rho_0\chi_0 c_0^2 = 1$, $\rho_0\gamma\chi_0 c_0'^2 = 1$, $c_0^2 = \gamma c_0'^2$, $\gamma - 1 = \frac{c_0^2 - c_0'^2}{c_0'^2}$ and $\frac{\gamma-1}{\gamma} = \frac{c_0^2-c_0'^2}{c_0^2}$.

Equation (6.46) gives

$$b = \tau' + \frac{\kappa}{\rho_0 c_V i\omega} \partial^2 \tau'. \tag{6.47}$$

Finally, with (6.10) giving $\partial \cdot \boldsymbol{v} - i\omega b = 0$, we obtain the following velocity-temperature equations:

$$-i\omega \boldsymbol{v} - \nu \partial^2 \boldsymbol{v} = -\partial X, \qquad X = \left[c_0'^2 - i\omega \frac{\zeta + \frac{\eta}{3}}{\rho_0} \right] b + (c_0^2 - c_0'^2)\tau', \tag{6.48}$$

$$\partial \cdot \boldsymbol{v} - i\omega\tau' - \frac{\kappa}{\rho_0 c_V} \partial^2 \tau' = 0. \tag{6.49}$$

By taking the divergence of (6.48.1), and eliminating the velocity-divergence terms using (6.49), there results the following equation on temperature:

$$\begin{aligned} -\omega^2 \tau' - \left[c_0^2 - i\omega \left(\frac{\kappa}{\rho_0 c_V} + \nu + \frac{\zeta + \frac{\eta}{3}}{\rho_0} \right) \right] \partial^2 \tau' + \dots \\ - \frac{\kappa}{\rho_0 c_V i\omega} \left[c_0'^2 - i\omega \left(\nu + \frac{\zeta + \frac{\eta}{3}}{\rho_0} \right) \right] \partial^4 \tau' = 0. \end{aligned} \tag{6.50}$$

Looking for a solution having the form $e^{-i\omega t + i\boldsymbol{k}\cdot\boldsymbol{x}}$ and setting by definition

$$-\boldsymbol{k}^2 \equiv \lambda, \tag{6.51}$$

the constant λ must be the solution of the following Kirchhoff-Langevin's[21] characteristic equation:

$$\begin{aligned} -\omega^2 - \left[c_0^2 - i\omega \left(\frac{\kappa}{\rho_0 c_V} + \nu + \frac{\zeta + \frac{\eta}{3}}{\rho_0} \right) \right] \lambda + \dots \\ - \frac{\kappa}{\rho_0 c_V i\omega} \left[c_0'^2 - i\omega \left(\nu + \frac{\zeta + \frac{\eta}{3}}{\rho_0} \right) \right] \lambda^2 = 0. \end{aligned} \tag{6.52}$$

For small real k ($k(\ell \sim \nu/c_0) \lll 1$, long wavelengths), the complex frequencies ω, solutions to this equation, expand in powers of k. Writing such expansions and substituting them in (6.52), along with $\lambda = -k^2$, one obtains the three "polarization 1" solutions, (two acoustic, one entropic), given in Sect. 6.7. For real frequency, the dispersion equation (6.52) has two complex roots $\lambda_{1,2}$:

$$-\lambda_1 = k_1^2 = \frac{1}{2A}[-B + \sqrt{B^2 - 4AC}], \ -\lambda_2 = k_2^2 = \frac{1}{2A}[-B - \sqrt{B^2 - 4AC}], \tag{6.53}$$

with

---

[21] See [2] for an explanation of this denomination.

$$A = -\frac{\kappa}{\rho_0 c_V i\omega}\left[c_0'^2 - i\omega\left(\nu + \frac{\zeta + \frac{\eta}{3}}{\rho_0}\right)\right],\quad C = -\omega^2,$$
$$B = c_0^2 - i\omega\left(\frac{\kappa}{\rho_0 c_V} + \nu + \frac{\zeta + \frac{\eta}{3}}{\rho_0}\right). \tag{6.54}$$

Using by convention, square-root determinations $\Re\left(\sqrt{\ }\right) > 0$, λ_1 represents the smallest solution, mainly real, that describes acoustic waves having very small absorption, and λ_2 represents the other much larger solution,[22] mainly purely imaginary, that describes the highly damped diffusive entropic waves. The associated acoustic and entropic wavenumbers, such that the dependencies $e^{-i\omega t + ik_{1,2}x}$ represent acoustic and entropic waves propagating in the direction $+x$, are

$$k_1 = \sqrt{\frac{1}{2A}[-B + \sqrt{B^2 - 4AC}]},\quad k_2 = \sqrt{\frac{1}{2A}[-B - \sqrt{B^2 - 4AC}]}, \tag{6.55}$$

(with this choice, $\Im\left(k_{1,2}\right) > 0$, so that the waves attenuate along $+x$).

After straightforward calculations we find expansions having the form:

$$-\lambda_1 = \frac{\omega^2}{c_0^2}\left[1 + i\omega\frac{2\Gamma}{c_0^2} + \ldots\right],\quad -\lambda_2 = i\frac{\rho_0 c_P \omega}{\kappa}\left[1 + i\omega\frac{2\Gamma'}{c_0^2} + \ldots\right], \tag{6.56}$$

$$k_1 = \frac{\omega}{c_0}\left[1 + i\omega\frac{\Gamma}{c_0^2} + \ldots\right],\qquad k_2 = (1+i)\sqrt{\frac{\rho_0 c_P \omega}{2\kappa}}\left[1 + i\omega\frac{\Gamma'}{c_0^2} + \ldots\right], \tag{6.57}$$

with Γ the expression previously given in (6.43) and Γ' the expression

$$\Gamma' = \frac{\gamma - 1}{2\rho_0}\left(\frac{4\eta}{3} + \zeta - \frac{\kappa}{c_P}\right). \tag{6.58}$$

The solution for the wavenumber k_1 describes a pressure wave propagating with a soundspeed velocity always slightly less than the reference adiabatic value c_0, and which is slightly attenuated because of the small factor Γ. The solution for the wavenumber k_2 describes a mainly purely diffusive ($\Re(k) = \Im(k)$), highly attenuating entropic (or thermal) wave. The characteristic penetration length of this wave is, as seen on the exponential $\exp(ik_2 x) \cong \exp(ix\sqrt{\rho_0 c_P \omega/2\kappa})\exp(-x\sqrt{\rho_0 c_P \omega/2\kappa})$, determined by the characteristic entropic, or thermal boundary layer, length $\delta_\tau = (2\kappa/\rho_0 c_P \omega)^{1/2}$.

It can be checked that, for these two types of waves (respectively $T = 1, 2$) there are the following relations:

[22] But still small in the sense $|k_2 \ell| \lll 1$ (long wavelengths), ensuring the scale separation discussed in Sect. 6.3.

$$v_i = \frac{\omega}{k_T^2} k_{Ti} b, \quad (i = x, y, z), \quad p = \frac{1}{\gamma \chi_0} \left(1 + \frac{\gamma - 1}{1 + \gamma \dfrac{\kappa}{\rho_0 c_P} \dfrac{1}{-i\omega} k_T^2} \right) b, \quad (6.59)$$

$$\tau = \frac{1}{\beta_0} \frac{\gamma - 1}{1 + \gamma \dfrac{\kappa}{\rho_0 c_P} \dfrac{1}{-i\omega} k_T^2} b, \quad s = \frac{\gamma - 1}{\gamma} \frac{c_P}{\beta_0 T_0} \left(-1 + \frac{1}{1 + \gamma \dfrac{\kappa}{\rho_0 c_P} \dfrac{1}{-i\omega} k_T^2} \right) b$$

For the type 1, the "acoustic wave", the entropy variations are almost zero, $s \cong 0$. In effect, $(\gamma\kappa/-i\omega\rho_0 c_P)k_1^2$ is usually close to zero because of the small $\kappa/\rho_0 c_P$ value (as long as frequencies are not too high, or wavelengths, not too small). Hence the parenthesis in the expression of s almost vanishes. In the expression of p, the parenthesis almost takes the value $1 + (\gamma - 1)/1 = \gamma$, leading to pressure-density relation $p \cong \chi_0^{-1} b$, close to the adiabatic. The temperature also, is almost in adiabatic relation to the condensation, $\tau \cong b(\gamma - 1)/\beta_0$.

For the type 2, the "entropic wave", on the contrary, $(\gamma\kappa/-i\omega\rho_0 c_P)k_2^2 \cong -\gamma$. Then the entropy variations are significant:

$$\left(-1 + \frac{1}{1 + \gamma \dfrac{\kappa}{\rho_0 c_P} \dfrac{1}{-i\omega} k_2^2} \right) \cong \left(-1 + \frac{1}{1 - \gamma} \right) = \frac{\gamma}{1 - \gamma} \quad \Rightarrow \quad s \cong -\frac{c_P}{\beta_0 T_0} b.$$

The temperature variations are also non-negligible $\tau \cong b(\gamma - 1)/\beta_0(1 - \gamma) \cong -b/\beta_0$. The pressure variations are practically zero,

$$\left(1 + \frac{\gamma - 1}{1 + \gamma \dfrac{\kappa}{\rho_0 c_P} \dfrac{1}{-i\omega} k_2^2} \right) \cong 1 + \frac{\gamma - 1}{1 - \gamma} = 0 \quad \Rightarrow \quad p \cong 0.$$

Indeed, using (6.57.2) we find, for this entropic pressure, $p \cong \dfrac{1}{\chi_0} \dfrac{2i\omega\Gamma'}{(\gamma - 1)c_0^2} b$, which is very small compared to the adiabatic pressure b/χ_0. This is why the first and second type of longitudinal waves are often termed pressure waves and entropy waves, in spite of the fact that pressure and entropy variations are always simultaneously present in these two waves.

6.9 General Representation of the Fields in Harmonic Regime Without Source

Introducing acoustic and entropic potentials, ϕ_1 and ϕ_2, such as

$$\partial^2 \phi_{1,2} = \lambda_{1,2}\phi_{1,2}, \tag{6.60}$$

the general field τ', solution to the temperature equation (6.50), will be

$$\tau' = A_1\phi_1 + A_2\phi_2,$$

with $A_{1,2}$ two constants. The general velocity field $\boldsymbol{v}$ will be

$$\boldsymbol{v} = \boldsymbol{v}' + B_1\partial\phi_1 + B_2\partial\phi_2,$$

with the vortical part $\boldsymbol{v}'$ obeying the equations

$$\partial^2 \boldsymbol{v}' = \frac{-i\omega}{\nu}\boldsymbol{v}', \qquad \partial \cdot \boldsymbol{v}' = 0. \tag{6.61}$$

One expresses the coefficients B from the coefficients A by using the fact that the (6.49) is independently satisfied by terms 1 and 2. This yields

$$B_{1,2} = \left(\frac{\kappa}{\rho_0 c_V} - \frac{-i\omega}{\lambda_{1,2}}\right) A_{1,2}.$$

From the temperature field, the condensation field can be written using (6.47). The pressure field then derives using (6.45). In summary, the general form of the fields in harmonic regime and without source, is:

$$\boldsymbol{v} = \boldsymbol{v}' + \left(\frac{\kappa}{\rho_0 c_V} + \frac{i\omega}{\lambda_1}\right) A_1\partial\phi_1 + \left(\frac{\kappa}{\rho_0 c_V} + \frac{i\omega}{\lambda_2}\right) A_2\partial\phi_2, \tag{6.62}$$

$$\tau' = A_1\phi_1 + A_2\phi_2, \tag{6.63}$$

$$b = \left(1 + \frac{\kappa}{\rho_0 c_V i\omega}\lambda_1\right) A_1\phi_1 + \left(1 + \frac{\kappa}{\rho_0 c_V i\omega}\lambda_2\right) A_1\phi_2, \tag{6.64}$$

$$\frac{p}{\rho_0} = \left(c_0^2 + {c_0'}^2 \frac{\kappa}{\rho_0 c_V i\omega}\lambda_1\right) A_1\phi_1 + \left(c_0^2 + {c_0'}^2 \frac{\kappa}{\rho_0 c_V i\omega}\lambda_2\right) A_1\phi_2. \tag{6.65}$$

The wavenumber for the vortical motions (shear motions described in the part $\boldsymbol{v}'$), is purely diffusive, highly attenuated. Indeed, putting the form, $\boldsymbol{v}' = \boldsymbol{v}'_0 e^{-i\omega t + i\boldsymbol{k}_v \cdot \boldsymbol{x}}$, in (6.61.1), gives, $k_v^2 = i\omega/\nu$, or:

$$k_v = \frac{1+i}{\sqrt{2}}\left(\frac{\omega}{\nu}\right)^{1/2}. \tag{6.66}$$

The characteristic penetration length of this wave is seen on the exponential $\exp(ik_v x) = \exp(ix/\delta_v)\exp(-x/\delta_v)$. It is the characteristic shear, or viscous boundary layer, length $\delta_v = (2\nu/\omega)^{1/2}$. Because the Prandtl number, $\mathrm{Pr} \equiv \eta c_P/\kappa$, is generally of order one, (e.g. 0.71 for air), the viscous and thermal penetration lengths,

$\delta_\tau = \delta_v / \mathrm{Pr}^{1/2}$, are of a comparable order. This order is very small: for example for air at audible frequencies, typical viscous and thermal attenuation lengths are in the order of 10^{-4} m, (still very "long-wavelength" compared to the mean free path $\ell \sim \eta/\rho_0 c_0 = 4.5\ 10^{-8}$ m).

6.10 Revisiting the Longitudinal Acoustic Equations Using an Electromagnetic Analogy

We now want to make clear, that the propagation of waves in the fluid, in the presence of viscous and thermal losses, can be considered in exactly the same way as the nonlocal macroscopic propagation of electromagnetic waves in a medium. We believe that this electromagnetic analogy would assume its full beautiful form, sketched in Appendix, if it were possible to extend sufficiently the thermodynamic framework, so as to model also Frenkel's shearing movements; if we work within Navier-Stokes-Fourier model, however, the absence of Frenkel's shearing, among other shortcomings, makes the description degenerate. We mentioned previously in Sect. 6.6.3 these insufficiencies of the description. As the completion of the analogy is until now a distant perspective, in what follows, to avoid degeneration, we limit ourselves to considering the Navier-Stokes-Fourier model when shearing movements are not excited.

Recall that the momentum equation can be written, $\rho_0 \partial_t \boldsymbol{v} = -\partial p - \eta \partial \times (\partial \times \boldsymbol{v}) + (4\eta/3 + \zeta)\, \partial(\partial \cdot \boldsymbol{v})$. Focusing on longitudinal motions only, the term $-\eta\partial \times (\partial \times \boldsymbol{v})$ is not excited, and the set of equations to be revisited is

$$\partial_t b + \partial \cdot \boldsymbol{v} = 0, \tag{6.67}$$

$$\rho_0 \partial_t \boldsymbol{v} = -\partial p - \left(\frac{4\eta}{3} + \zeta\right) \partial \partial_t b + \left[\boldsymbol{f} = -\partial \mathcal{P}\right], \tag{6.68}$$

$$\gamma \chi_0 p = b + \beta_0 \tau, \tag{6.69}$$

$$\rho_0 c_P \partial_t \tau = \beta_0 T_0 \partial_t p + \kappa \partial^2 \tau. \tag{6.70}$$

In (6.68) we have included the possibility that an external density of longitudinal bulk force $\boldsymbol{f}$, that is thus given as a gradient, is acting on the fluid. We will see that these (6.67)–(6.70) can be transformed into a general pattern inspired by that of macroscopic electromagnetic equations in a homogeneous material medium, with a source $[\boldsymbol{J}]$, density of external electrical current, playing the role of source $\left[\boldsymbol{f}\right]$, density of external longitudinal body force. To prepare this way of looking at the acoustic equations of motion, we now go to macroscopic electromagnetic theory and study its formal pattern.

6.11 Macroscopic Electromagnetics

Due to the lack, to date, of a precise relationship between macroscopic electromagnetics and thermodynamics in the broad sense, (out of equilibrium), it is still not possible to perform a precise specification of some of the macroscopic quantities involved in the macroscopic equations that describe the propagation of electromagnetic waves in, say, a neutral material medium given in an ambient rest state. This point is generally not sufficiently well recognized in textbooks and literature. For example in Jackson [10, Chap. 6], there is written, about the macroscopic electromagnetic equations: "Although these equations are familiar and totally acceptable, we have yet to present a serious derivation of them from a microscopic starting point"; there follows a "microscopic derivation" of the stated equations, which is apparently intended to clarify the question, but which, in fact, remains superficial and does not address a number of difficult and embarrassing questions. This is not surprising: one could hardly expect deriving results belonging to the field of macroscopic physics proper, from scratch microscopics, (and even less so, when the microscopic level is that of semiclassical models). We try here to bring out some of the irritating pending questions, in a formal discussion. The subject will not come out of it clarified, quite the contrary, but the simple recognition of the presence of difficulties of a macroscopic thermodynamic nature in the broad sense, will be useful to us later in acoustics. Moreover, the forthcoming acoustic implementation of our discussion – in this chapter and the next – will particularly retrospectively highlight the thermodynamic nature of these difficulties, which will be circumvented in acoustics, only because of an existing (even if imperfect) thermodynamic framework, that of the Navier-Stokes-Fourier model.

6.11.1 *Nonlocal Maxwellian Pattern of the Macroscopic Equations*

Here, we are interested in the description of time variable and space variable phenomena such as occurring in macroscopic wave propagation, either corresponding to free motion (no external source), or forced motion (with external source). Although, as we have said, no complete derivations of the required macroscopic electromagnetic wave equations have ever been made from the microstructure, we will argue that, what is known at the microscopic level, suggests that, at the macroscopic level can be written a system of two field equations and two nonlocal constitutive equations:

$$\frac{\partial \boldsymbol{B}}{\partial t} = -\partial \times \boldsymbol{E}, \qquad \frac{\partial \boldsymbol{D}}{\partial t} = \partial \times \boldsymbol{H} - [\boldsymbol{J}], \qquad \boldsymbol{D} = \hat{\epsilon}\boldsymbol{E}, \qquad \boldsymbol{H} = \hat{\mu}^{-1}\boldsymbol{B}. \quad (6.71)$$

The impressed term $[\boldsymbol{J}]$ is a current density source term which we put inside brackets, as it may, or may not be applied. The quantities $\hat{\epsilon}$ and $\hat{\mu}$ are operators intrinsic to the medium, establishing action-response relations between fields, and having the same following *general nonlocal form*, whatever the source term is present, or not[23]:

$$D(t, x) = \int \epsilon(t - t', x, x') E(t', x') dt' dx', \tag{6.72}$$

$$H(t, x) = \int \mu^{-1}(t - t', x, x') B(t', x') dt' dx'. \tag{6.73}$$

Of the four electromagnetic macroscopic fields $\boldsymbol{E}$, $\boldsymbol{B}$, $\boldsymbol{D}$, $\boldsymbol{H}$, two of them, $\boldsymbol{E}$ and $\boldsymbol{B}$, will be called Lorentz's fields and directly interpreted in the next subsection as the macroscopic means of the microscopic electric and magnetic fields $\boldsymbol{e}$ and $\boldsymbol{b}$: $\boldsymbol{E} \equiv \langle \boldsymbol{e} \rangle$ and $\boldsymbol{B} \equiv \langle \boldsymbol{b} \rangle$. The other two, $\boldsymbol{D}$ and $\boldsymbol{H}$, will be called Maxwell's fields and seen to reflect the response of the medium to the preceding fields, and thus, the nature of the microstructures. They are not to be viewed as the macroscopic averages of corresponding microscopic fields. They are, in fact, *problematic to define in principle, in a material environment.* By the way, not apparent in the above Gibbs-Heaviside's standard vector notations, Lorentz's and Maxwell's fields have different tensor nature, (see Appendix[24]). Whereas $\boldsymbol{E}$ and $\boldsymbol{B}$ express in terms of true tensors, $\boldsymbol{D}$ and $\boldsymbol{H}$ express in terms of tensor densities of weight -1. The impressed densities of charges and current ρ and $\boldsymbol{J}$[25] also are tensor densities of weight -1. Like the Maxwell fields they are problematic to define in principle, in a material environment, and this is also generally obscured in literature.

The nonlocal form of the relationships (6.72) and (6.73) is essential to assure that the description is general; as it is rarely considered in the literature, it is worthwhile opening a parenthesis to comment on it in some detail.

The integration over time t', i.e. temporal nonlocality, gives rise to what is called "temporal dispersion"; the integration over space $\boldsymbol{x}'$, i.e. spatial nonlocality, gives rise to what is called "spatial dispersion". Because both integrations are accounted for, we are directly assured of the general validity of the relation (6.72) for a time-invariant medium: it would be useless to add in (6.72) an integral term built on the values $\boldsymbol{B}(t', \boldsymbol{x}')$. Indeed, as analyzed in Landau and Lifshitz [13], because the fields $\boldsymbol{E}$ and $\boldsymbol{B}$ are not totally independent but related by (6.71.1), it would be possible to view this term as an effect of spatial dispersion, already described in the first term. Concerning

[23] The dependencies are over $t - t'$, $\boldsymbol{x}$ and $\boldsymbol{x}'$, because the medium is assumed time-invariant but not translational invariant. When considering an *homogeneous unbounded medium*, the spatial dependency will be over differences $\boldsymbol{x} - \boldsymbol{x}'$. Note that the indefinite integrals become definite when considering that the kernels factors must vanish at some point – if only when the event $(t', \boldsymbol{x}')$ lies outside the relativistic cone of causality whose tip is the event $(t, \boldsymbol{x})$: $\int dt' \int d\boldsymbol{x}' \leftrightarrow \int_{-\infty}^{t} dt' \int_{|\boldsymbol{x}-\boldsymbol{x}'|<c(t-t')} d\boldsymbol{x}'$.

[24] These aspects are often overlooked in electromagnetic literature – see however Fournet [11]. See Weinberg [12] for general notions on Gauss coordinates, tensors, tensor densities, covariance, contravariance.

[25] ρ appears in another general equation, $\partial \cdot \boldsymbol{D} = \rho$, which we need not explicitly consider here, because, in the time variable, space variable regimes, it automatically follows from taking the divergence of (6.71.2), and using the conservation equation, $\partial\rho/\partial t + \partial \cdot \boldsymbol{J} = 0$.

the other nonlocal relationship (6.73), a similar justification of its generality, can be made, based on the relation established between $\boldsymbol{D}$ and $\boldsymbol{H}$ through (6.71.2), and the generality of (6.72). In brief, because they allow for spatial dispersion in addition to temporal dispersion, the relationships (6.72) and (6.73) between Lorentz's and Maxwell's fields are the most general possible ones, (including, in particular, the so-called bianisotropic constitutive relations). Often, in literature, by a convention, the temporal and spatial dispersion is introduced only in (6.72), not in (6.73), which is replaced by a direct definition-relation, $\boldsymbol{H} = \mu_0^{-1}\boldsymbol{B}$. We will comment in more detail on this convention in Sect. 6.11.3, and a similar problematics, in the acoustic analogy, will later shed useful light on it, illustrating its simple nature as an expedient.

As long as we use Cartesian coordinates and work with Gibbs-Heaviside's notation, the kernel functions $\epsilon(t-t', \boldsymbol{x}, \boldsymbol{x}')$ and $\mu^{-1}(t-t', \boldsymbol{x}, \boldsymbol{x}')$ retain the nature of second-rank tensor quantities.[26] As a rule, they are independent of the fields and only determined by the microstructure. Nevertheless, different distributions of the Lorentz fields must produce different resulting distributions of Maxwell fields $\boldsymbol{D}$ and $\boldsymbol{H}$. Precisely because the created $\boldsymbol{D}$ and $\boldsymbol{H}$ will depend on the variations of $\boldsymbol{E}$ and $\boldsymbol{B}$, all effective properties meaningful for wave propagation, such as impedances and wavenumbers, will reflect associated temporal and spatial variations of Lorentz fields.

Therefore in this sense, one may think of "temporal dispersion" as something that tells us that the effective properties of the medium depend on the time variations of the macroscopic fields; likewise, "spatial dispersion" is something that tells us that the effective properties of the medium depend on the spatial variations of the macroscopic fields.

In wave propagation problems, there are both time and space variations, therefore, temporal dispersion and spatial dispersion, should always be simultaneously present. Nevertheless, often, spatial dispersion effects are very weak: they are neglected in almost all textbooks; or else, once introduced, they are assumed to be small corrections meaningful to consider only when the wavelengths reduce sufficiently [13]. It is only very recently understood [14], that, spatial dispersion effects can be very strong effects at long wavelengths. This occurs in presence of localized resonating structures.

The same generalities will apply to acoustics. Often, spatial dispersion effects are necessary to consider only as corrections to the conventional (local) homogenization, (see next chapter), when the wavelengths reduce sufficiently to become commensurable with the microstructure characteristic lengths. But in some other instances, this time completely failing the conventional views, they can be strong effects at long wavelengths. This occurs in presence of resonances. People often speak of "local resonances" as these resonances are generally those of finite structures, localized in space. But this wording is, in-part, unfortunate: when "local resonances" are present,

[26] In general Gauss coordinates, ϵ remains second-rank but μ^{-1} becomes fourth-rank tensor; the first and last (one or two) indices, transform respectively with coefficients associated to the different points $\boldsymbol{x}$ and $\boldsymbol{x}'$.

the wave propagation physics becomes strongly nonlocal (spatially).[27] A nonlocal homogenization description will be developed in next chapter, for rigid-framed fluid-saturated materials. In this chapter, as we consider a simple fluid with no structure, the spatial dispersion effects will be weak, (but essential to allow for the propagation of other waves than the so-called acoustic one).

We close our parenthesis here. We now want to examine why it is not so evident to fix the meaning of the Maxwell fields $\boldsymbol{D}$, $\boldsymbol{H}$, and impressed source term $[\boldsymbol{J}]$. To do this, we try to sketch in the next section how these fields appear, in a material environment, and lead to the formal pattern of the above equations.

6.11.2 Passage from Microscopics to Macroscopics

To evoke the difficult nature of the questions raised by the specification of the fields $\boldsymbol{D}$, $\boldsymbol{H}$, and $[\boldsymbol{J}]$, let's start with the equations that hold true at the microscopic level. In a semi-classic description these are the well-known electromagnetic equations first written at the turn of the 20th century by Lorentz in his celebrated series of papers, and monograph, on "The theory of electrons". They comprise first, the so-called "microscopic Maxwell-Lorentz equations":

$$\partial \cdot \boldsymbol{b} = 0, \qquad \frac{\partial \boldsymbol{b}}{\partial t} + \partial \times \boldsymbol{e} = \boldsymbol{0}, \qquad \partial \cdot \boldsymbol{e} = \frac{\rho}{\epsilon_0}, \qquad \epsilon_0 \frac{\partial \boldsymbol{e}}{\partial t} = \frac{1}{\mu_0} \partial \times \boldsymbol{b} - \boldsymbol{j}. \quad (6.74)$$

We can be assured of their profound physical significance because there is a route, from Quantum Electro-Dynamics describing the photonic field, to these equations, describing electric and magnetic fields. They comprise next, the following inherently semi-classic expressions for the microscopic densities of charge ρ and current $\boldsymbol{j}$:

$$\rho = \sum_\alpha q_\alpha \delta\left[\boldsymbol{x} - \boldsymbol{x}_\alpha(t)\right], \; \boldsymbol{j} = \sum_\alpha q_\alpha \boldsymbol{v}_\alpha(t) \delta\left[\boldsymbol{x} - \boldsymbol{x}_\alpha(t)\right], \; (\boldsymbol{v}_\alpha \equiv d\boldsymbol{x}_\alpha/dt), \quad (6.75)$$

where δ is the Dirac delta, and the symbolic summation[28] $\sum_\alpha$, $\alpha = 1, 2, \ldots$, is over electrons and nuclei. They comprise finally, the following, also inherently semi-classic and symbolic equations of motion, separately obeyed by the charges moving under the action of the microscopic electric and magnetic fields:

$$\frac{d}{dt}\left(m_{0\alpha} \boldsymbol{v}_\alpha / \sqrt{1 - \frac{\boldsymbol{v}_\alpha^2}{c^2}} \right) = q_\alpha \left(\boldsymbol{e} + \boldsymbol{v}_\alpha \times \boldsymbol{b} \right), \qquad \alpha = 1, 2, \ldots \quad (6.76)$$

---

[27] In this connection, see the discussion around Fig. 7.4, next Chap. 7, Sect. 7.2.

[28] The labelling α of all charged particles is here symbolic as it makes abstraction of counting subtleties related to quantum indiscernibility of identical particles.

The justification for using delta functions in (6.75) is that electrons have no perceptible dimensions, while those of nuclei are still astonishingly small, $\sim 10^{-14}$ m. Interested in what happens at scales much larger than this, everything is supposed to happen as if charged particles were pointlike, and Maxwell-Lorentz's microscopic equations (6.74) applied everywhere in space. We take with reservation this semi-classical expression of things, which is not a complete reflection of the microscopic quantum world, and among other shortcomings, does not consider the collective aspect of electron behaviour.[29] We note that the conservation relation

$$\frac{\partial \rho}{\partial t} + \partial \cdot \boldsymbol{j} = 0, \tag{6.77}$$

can be obtained from the conservation of individual charges, $q_\alpha = Cst$, and the semi-classic picture, but more deeply, it is also a consequence of Maxwell-Lorentz's microscopic equations themselves.

Now, we observe that we are not directly concerned by microscopic equations and quantities. The fields will have rapid variations at the Ångström scale of molecular dimensions and less ($\text{Å} = 10^{-10}$ m), and we have to fix our attention not on all these irregularities but only on certain mean values that vary at a much larger ("macroscopic") scale. As Gibbs' conception of ensemble average is delicate to be made explicit here, in literature these mean values $\langle \cdot \rangle$ are generally conceived using Lorentz's conception of volume average. The idea is to smooth-out irregularities by integrating in averaging spheres or "homogenization" volumes, neither too small nor too large. The sphere radius must be sufficient to be representative of the structure of the material. Since the purpose is to get rid of the irregularities, the sphere must contain a very large number of particles. The averaging on the other hand must not obliterate the changes from point to point that can really be observed. Molecular dimensions ($\sim$ Å) are so much smaller than ordinary electromagnetic wavelengths that both conditions can be met at the same time. In homogeneous substances, a suitable macroscopic homogenization length L_h, will be on the order of 100Å, typically. In a Lorentz averaging sphere having this size, there is still on the order of 10^6 nuclei and electrons. The average can be best performed using Russakoff's signal-theory-type refinement of a convolution with an isotropic test function $w(\boldsymbol{x})$, of characteristic extent L_h, and normalized to unity, $\int d\boldsymbol{x} w(\boldsymbol{x}) = 1$, (see [15] and the discussion in Jackson [10], Chap. 6). The macroscopic volume-averaged electric field $\boldsymbol{E}$, for example, will be

$$\boldsymbol{E}(t, \boldsymbol{x}) = \langle \boldsymbol{e}(t, \boldsymbol{x}) \rangle = \langle \boldsymbol{e} \rangle (t, \boldsymbol{x}) = \int d\boldsymbol{x}' w(\boldsymbol{x}') \boldsymbol{e}(t, \boldsymbol{x} - \boldsymbol{x}'). \tag{6.78}$$

If we put a derivative symbol in front of the above integral, we can move it inside: $\partial_{x_i} \int d\boldsymbol{x}' w(\boldsymbol{x}') \boldsymbol{e}(t, \boldsymbol{x} - \boldsymbol{x}') = \int d\boldsymbol{x}' w(\boldsymbol{x}') \partial_{x_i} \boldsymbol{e}(t, \boldsymbol{x} - \boldsymbol{x}')$. Therefore the operations of space differentiation and averaging commute by construction:

[29] In this respect, see in next chapter, Appendix, the discussion about Curie-von Schweidler's law.

$$\partial_{x_i} \boldsymbol{E}(t, \boldsymbol{x}) = \partial_{x_i} \langle \boldsymbol{e}(t, \boldsymbol{x}) \rangle = \langle \partial_{x_i} \boldsymbol{e} \rangle (t, \boldsymbol{x}). \tag{6.79}$$

The commutation with time differentiation is also evident. Thus, by averaging, the microscopic Maxwell-Lorentz equations (6.74) become the following macroscopic ones

$$\partial \cdot \boldsymbol{B} = 0, \quad \frac{\partial \boldsymbol{B}}{\partial t} + \partial \times \boldsymbol{E} = \boldsymbol{0}, \quad \partial \cdot \boldsymbol{E} = \frac{\langle \rho \rangle}{\epsilon_0}, \quad \epsilon_0 \frac{\partial \boldsymbol{E}}{\partial t} = \frac{1}{\mu_0} \partial \times \boldsymbol{B} - \langle \boldsymbol{j} \rangle. \tag{6.80}$$

and the conservation relation (6.77) becomes

$$\frac{\partial \langle \rho \rangle}{\partial t} + \partial \cdot \langle \boldsymbol{j} \rangle = 0. \tag{6.81}$$

Equation (6.80.1) is, for (time-variable) wave phenomena, a consequence of (6.80.2), whereas, (6.80.3) is a consequence of (6.80.4) and the conservation of charges (6.81). Thus, in what follows, we limit our attention to (6.80.2) and (6.80.4). Comparison of these two equations with (6.71.1) and (6.71.2), suggests that Maxwell's fields $\boldsymbol{D}$ and $\boldsymbol{H}$, are introduced by the extraction from $\langle \boldsymbol{j} \rangle$ of certain contributions, that can be identified either, with "intrinsic polarization current" determined by bulk properties of the medium, or with "extrinsic" to the medium, external contribution imposed from without. That is, $\langle \boldsymbol{j} \rangle$ will have to be decomposed in a (intrinsic) polarization response current $\boldsymbol{J}_{pol}$, expressing the rearrangements of all charges (bound or free) and internal currents (circulation currents responsible for magnetism) in the medium, in relation to the presence of the macroscopic Lorentz fields,[30] and an additional (extrinsic) impressed charge current $[\boldsymbol{J}]$, imposed from the outside:

$$\langle \boldsymbol{j} \rangle = \boldsymbol{J}_{pol} + [\boldsymbol{J}]. \tag{6.82}$$

Moreover the (intrinsic) polarization current $\boldsymbol{J}_{pol}$, characteristic response of the medium to the presence of macroscopic fields $\boldsymbol{E}$ and $\boldsymbol{B}$, is supposed to decompose in two parts electric and magnetic:

$$\boldsymbol{J}_{pol} = \frac{\partial \boldsymbol{P}}{\partial t} + \partial \times \boldsymbol{M}, \tag{6.83}$$

what we call Lorentz-splitting,[31] with nonlocal operators $\hat{\chi}_E$ and $\hat{\chi}_B$, or kernels $\chi_E(t, \boldsymbol{x}, \boldsymbol{x}')$, $\chi_B(t, \boldsymbol{x}, \boldsymbol{x}')$, establishing how the "electric" and "magnetic" polarization fields $\boldsymbol{P}$ and $\boldsymbol{M}$ are determined by the electric and magnetic macroscopic fields:

[30] Say, whatever this will precisely mean, some "electric" and "magnetic" polarization rearrangements of charges and currents.

[31] The idea can be attributed to Lorentz, though we have not found a clear expression of it in original papers. Through it, the polarization current is formally decomposed in time derivative and spatial derivative terms; the part $\partial \times \boldsymbol{M}$ actually represents a contracted expression $M^{ij}_{\ ;j}$, see Appendix, with M^{ij} an antisymmetric tensor density.

$$P(t, x) = \int dt' dx' \chi_E(t - t', x, x') E(t', x'), \quad (6.84)$$

$$M(t, x) = \int dt' dx' \chi_B(t - t', x, x') B(t', x'). \quad (6.85)$$

Indeed, in (6.84) it is useless to add a similar term with the magnetic field, because, as E and B are related by (6.80.2), such a term could also be rewritten in the nonlocal integral form (6.84). Likewise, in (6.85), because of the complete nonlocal form of the relation, there is no need to add a similar term with the electric field. Therefore, when (6.82)–(6.83) are substituted in (6.80.4), the addition of the electric polarization current term $\partial P/\partial t$ to $\epsilon_0 \partial E/\partial t$ will produce the term $\partial D/\partial t$ in (6.71.2), and the subtraction of the magnetic polarization current term $\partial \times M$ to $\partial \times B/\mu_0$ will produce the term $\partial \times H$, provided D and H are interpreted as:

$$D = \epsilon_0 E + P, \qquad H = \frac{1}{\mu_0} B - M. \quad (6.86)$$

Then, (6.72) and (6.73) will apply, with

$$\epsilon(t - t', x, x') = \epsilon_0 \delta(t - t') \delta(x - x') + \chi_E(t - t', x, x'), \quad (6.87)$$

$$\mu^{-1}(t - t', x, x') = \mu_0^{-1} \delta(t - t') \delta(x - x') - \chi_B(t - t', x, x'). \quad (6.88)$$

In this way, the formal pattern of the electromagnetic macroscopic equations set out in Sect. 6.11.1 has been reconstructed and justified; it appears to be entirely general as soon as a macroscopic descrption is possible.

6.11.3 Ambiguities and a Suggested Way to Resolve Them

In this "formal derivation" of (6.71)–(6.73) we have left open two difficult questions. We have not explained, how are to be performed the successive separations (6.82) and (6.83). The first separation is clearly elusive: how do we precisely distinguish between the "intrinsic" J_{pol} and "extrinsic" $[J]$ currents? The second separation is in itself ambiguous: the rotational $\partial \times \Phi$ of an arbitrary vector potential Φ, can always be added to P, provided simultaneously, a corresponding term $\partial \Phi/\partial t$, be subtracted to M. This is not like Helmholtz's decomposition, $F = -\partial \Phi + \partial \times A$, of a vector field F, into a gradient of a scalar potential and a rotational of a vector potential, which is unambiguous when requiring the vanishing of the fields at infinity and the divergence-free nature of the vector potential.

Due to these shortcomings, we are not able to explain unambiguously, what are the electric and magnetic polarizations P and M, the Maxwell fields D and H, the polarization current J_{pol}, the impressed current $[J]$, and how to compute them in principle.

We will argue by comparison with a forthcoming acoustic implementation of a similar problematic, that we can have no answers to the above questions, in absence of thermodynamic considerations and laws for the motion of, say, the "fluid" of electronic charges in the material lattice of nuclei, considerations and laws which are however so far absent in the present state of macroscopic electromagnetics and thermodynamics.

Here we say that, if we disposed of the suitable thermodynamic knowledge, a solution to the indetermination in (6.83) would be to require that the Heaviside-Poynting vector

$$\boldsymbol{S} = \boldsymbol{E} \times \boldsymbol{H}, \tag{6.89}$$

possesses the thermodynamic meaning of the *current density of energy transported in electromagnetic form*. We will denote this, the "Heaviside-Poynting", thermodynamic identification. "In electromagnetic form" means the part of the energy current density, origin of the subsequent propagation and attenuation of the electromagnetic disturbance, and not the part, "lost" for the propagation, which is right-on converted in thermal degraded form.

If we knew what this current $\boldsymbol{S}$ is, in the sense that we would have a thermodynamic framework (so far missing) to express it, we could, by working within this framework, define $\boldsymbol{H}$ by applying (6.89), then $\boldsymbol{M}$ by applying (6.86.2). Assuming that the missing knowledge would also determine what the "impressed part" $[\boldsymbol{J}]$ is, we could then use (6.82)–(6.83) to define $\boldsymbol{P}$, and finally, by (6.86.1), define $\boldsymbol{D}$. These specifications being performed whatever the chosen source term, $[\boldsymbol{J}]$, all nonlocal operators would then be uniquely defined.

Basically, this is what we will try to do further on, in acoustics, with substituted quantities and corresponding changes. But the concrete exploitation of these ideas in electromagnetics is not feasible now, as long as a whole slice of thermodynamics, concerned with the response of electrons in matter, is missing. For the motion of a viscothermal fluid permeating a solid structure (see next chapter), or not (no solid structure, see this chapter), we dispose of the thermodynamic Navier-Stokes-Fourier equations we have stated, and this will suffice to us to progress significantly in the direction of a satisfactory nonlocal description.[32] But for the response of an "electronic fluid" (partially bounded and free) in the material nuclei lattice, we have so far nothing comparable. It is therefore not surprising that in the electromagnetic literature, the problem is not defined in the above general terms, which remain without content for the moment. Shortcuts and simplifications, which we now try to summarize, are used instead.

[32] Note that in expressing ourselves in this manner we assume that the properties of the fluid will be unaffected by its proximity to the walls of the solid; for the "electronic fluid", on the contrary, one can expect such complications to occur, and the development of an appropriate description seems to be a distant prospect.

6.11.4 Ambiguities and the Customary Way

A priori, there is in the polarization current (6.83), $\boldsymbol{J}_{pol} = \boldsymbol{J}_{bound} + \boldsymbol{J}_{cond}$, two different parts: a part $\boldsymbol{J}_{bound}$, coming from the motion of bound electrons, and a part $\boldsymbol{J}_{cond}$, coming from the motion of free electrons. The first $\boldsymbol{J}_{bound}$, includes a first contribution related to the global displacement of bound charges, to be incorporated in the term $\partial \boldsymbol{P}/\partial t$, and a second contribution not related to this but to the modifications of underlying magnetic circulation currents (e.g. in the molecules or other greater structures), to be incorporated in the $\partial \times \boldsymbol{M}$ term in (6.83). The remaining, $\boldsymbol{J}_{cond}$, will be an induced conduction current, coming from the motion of electrons not bounded to one individual molecule, and that will complete the above first contribution to produce the term $\partial \boldsymbol{P}/\partial t$ in (6.83), but also, for some structured materials, the $\partial \times \boldsymbol{M}$ term itself. Thus writing (6.82), we have

$$\langle \boldsymbol{j} \rangle = \boldsymbol{J}_{bound} + \boldsymbol{J}_{cond} + [\boldsymbol{J}]. \tag{6.90}$$

In the literature, often, by a tacite redefinition, $\boldsymbol{J}_{cond} + [\boldsymbol{J}] = \boldsymbol{J}$, of the current $\boldsymbol{J}$ which appears in the macroscopic equations, one writes (6.90) as

$$\langle \boldsymbol{j} \rangle = \boldsymbol{J}_{bound} + \boldsymbol{J}. \tag{6.91}$$

Finally, in presence of spatial dispersion and as explained in Landau and Lifshitz [13] or Agranovich and Ginzburg [16] or Melrose and McPhedran [17], one decides to put all different effects coming from the bound charges, in an abstract new $\boldsymbol{P}$ obeying the definition:

$$\boldsymbol{J}_{bound} = \frac{\partial \boldsymbol{P}}{\partial t}. \tag{6.92}$$

Then, macroscopic equations are obtained, which have the form:

$$\frac{\partial \boldsymbol{B}}{\partial t} = -\partial \times \boldsymbol{E}, \qquad \frac{\partial \boldsymbol{D}}{\partial t} = \frac{1}{\mu_0}\partial \times \boldsymbol{B} - \boldsymbol{J}, \qquad \boldsymbol{D} = \hat{\epsilon}\boldsymbol{E}, \tag{6.93}$$

with $\boldsymbol{J}$, a current, that is either taken equal to $\boldsymbol{J}_{cond}$ and obeying Ohm's law in absence of source,[33] or taken as an imposed term in presence of source,[34] and $\hat{\epsilon}$ an operator establishing a relation having the same following *general nonlocal form* whatever the $\boldsymbol{J}$ term is present or not:

$$\boldsymbol{D}(t, \boldsymbol{x}) = \int \epsilon(t - t', \boldsymbol{x}, \boldsymbol{x}')\boldsymbol{E}(t', \boldsymbol{x}')dt'd\boldsymbol{x}'. \tag{6.94}$$

[33] In the most general case this will be a nonlocal Ohm's law $\boldsymbol{J}(t, \boldsymbol{x}) = \int dt'd\boldsymbol{x}'\sigma(t - t', \boldsymbol{x}, \boldsymbol{x}')\boldsymbol{E}(t', \boldsymbol{x}')$.

[34] It also obeys the above Ohm's law, with $\boldsymbol{E}$ containing two parts, respectively associated to the forced and free motion.

In this conception, the magnetization $\boldsymbol{M}$ is, by force, defined to be zero, and the $\boldsymbol{H}$ field, is, simply, viewed as:

$$\boldsymbol{H} = \frac{1}{\mu_0}\boldsymbol{B}. \tag{6.95}$$

Contrary to the interpretation made of the term $\boldsymbol{J}$ in the set of (6.71)–(6.73), here, in the above (6.93)–(6.95), because of the presence of $\boldsymbol{J}_{cond}$ in it, the term $\boldsymbol{J}$ is not truly an external source-term. It becomes so, however, if we reincorporate the induced conduction current $\boldsymbol{J}_{cond}$ in the definition of $\boldsymbol{D}$, i.e. decide to write $\boldsymbol{J}_{pol} = \partial \boldsymbol{P}/\partial t$, instead of (6.83) or (6.92), now artificially putting all effects coming from bound and free charges, in the abstract $\boldsymbol{P}$. Anyway, whatever ones uses or not the redefinition $\boldsymbol{J}_{cond} + [\boldsymbol{J}] = \boldsymbol{J}$, certainly by setting (6.95), some artificiality is introduced in the definitions. We confirm this later in the acoustic case, where in this respect, similar problematics will arise.

6.11.5 Discussion of Our Proposal

Our idea is that there should exist a dynamical field $\boldsymbol{\mathcal{P}}$, having thermodynamic status, and such that $\boldsymbol{S} = \boldsymbol{E} \times \boldsymbol{\mathcal{P}}$ has the interpretation of current density of energy transported in electromagnetic form. Then setting $\boldsymbol{H} = \boldsymbol{\mathcal{P}}$ would allow identifying the field $\boldsymbol{H}$, removing the ambiguity in the Lorentz decomposition, and fixing the meaning of all fields and operators.

To precise this, we observe that, as discussed in textbooks, the polarization $\boldsymbol{P}$, verifies, $\partial \cdot \boldsymbol{P} = \rho_{pol}$, where $\partial \rho_{pol}/\partial t + \partial \cdot \boldsymbol{J}_{pol} = 0$. If we were to add the equation, $\partial \times \boldsymbol{P} = \boldsymbol{0}$, no ambiguity would arise for $\boldsymbol{P}$, once given the ρ_{pol}. But we are guessing here, that, in general, what should be called the electric polarization will have, $\partial \times \boldsymbol{P} \neq \boldsymbol{0}$. Let us thus denote $\boldsymbol{P}'$, the field satisfying, $\partial \cdot \boldsymbol{P}' = \rho_{pol}$, and, $\partial \times \boldsymbol{P}' = \boldsymbol{0}$. The actual $\boldsymbol{P}$, will be, $\boldsymbol{P} = \boldsymbol{P}' + \partial \times \boldsymbol{\Phi}$, (with some $\boldsymbol{\Phi}$), with the additional term, compensated by a corresponding additional one, $-\partial \boldsymbol{\Phi}/\partial t$, in $\boldsymbol{M}$. We are guessing that the requirement, $\boldsymbol{E} \times (\boldsymbol{B}/\mu_0 - \boldsymbol{M}) = \boldsymbol{E} \times \boldsymbol{\mathcal{P}}$, would then be capable to fix the appropriate term, $-\partial \boldsymbol{\Phi}/\partial t$, resolving the ambiguity of the Lorentz splitting.

For later reference, we name this process of hypothetical identification of what should be the Maxwell field $\boldsymbol{H}$, based on (6.89), Heaviside-Poynting's thermodynamic identification of the field $\boldsymbol{H}$, or in short, Heaviside-Poynting's conception of the field $\boldsymbol{H}$.

We hope that these comments will help clarify the difference between how Maxwell's fields are generally defined in the electromagnetic literature, and how they should be better defined, following (6.89), but unfortunately, in a way that is not feasible at this time, due to the absence of irreversible thermodynamic descriptions of the movement of an "electronic-fluid", in part bound and free, in material media. We note that many recent theoretical studies and experimental observations are now pointing in this direction. Electrons flowing like liquid have been reported in graphene, and notions resembling viscosity have appeared, to describe newly observed "fluidic" aspects of their collective behaviour.

In the next section we observe that a similar nonlocal pattern of relations applies to the case of the propagation of acoustic small-amplitude perturbations in our viscothermal homogeneous and unbounded fluid. In this case, however, we will not have to resort to stopgap measures of the type (6.95) expounded just above. Because in acoustics, we will have an existing irreversible thermodynamic framework – that of Navier-Stokes-Fourier equations – it will be possible to make definite use of a similarly imposed energetic condition of the type (6.89), allowing us to identify an "acoustic $\boldsymbol{H}-$field" and then to determine also the other acoustic quantities. For the reasons mentioned above (Sect. 6.6.3, and comments above (6.67) in Sect. 6.10), however, we will limit ourselves to considering the case of longitudinal movements only, (corresponding, without source, to the "acoustic" and "entropic" modes in the fluid).

Finally, we insist that the acoustic example will help us, perceive the fundamental physical insufficiencies of present-day definitions of electromagnetic Maxwell's fields $\boldsymbol{H} = \boldsymbol{B}/\mu_0$ and $\boldsymbol{D}$. It is to be hoped that in the future, a thermodynamics of the behaviour of electrons in material media will be developed, that will be capable to precise the notions of electric and magnetic polarizations $\boldsymbol{P}$ and $\boldsymbol{M}$, and polarization- and impressed-currents $\boldsymbol{J}_{pol}$ and $[\boldsymbol{J}]$, but this, again, seems to be a distant perspective.

6.12 Nonlocal Maxwellian Pattern of Longitudinal Acoustics

We now can come back to the "acoustic" longitudinal equations in a viscothermal fluid (6.67)–(6.70), and explain the formal pattern, that makes them comparable to the macroscopic electromagnetic equations discussed in the previous Sects. 6.11.1 and 6.11.2. Consider, on one hand, the two longitudinal acoustic equations (6.67)–(6.68), written with a source term $\boldsymbol{f} = -\partial\mathcal{P}$, density of external longitudinal body force:

$$\partial_t b + \partial \cdot \boldsymbol{v} = 0, \qquad \rho_0 \partial_t \boldsymbol{v} = -\partial p - \left(\frac{4\eta}{3} + \zeta\right) \partial\partial_t b + \left[\boldsymbol{f} = -\partial\mathcal{P}\right], \quad (6.96)$$

and completed by the additional thermodynamic ones (6.69)–(6.70), that tell us what the variable p is. Consider, on the other hand, the two macroscopic electromagnetic equations (6.80.2) and (6.80.4) with (6.82) inserted:

$$\frac{\partial \boldsymbol{B}}{\partial t} + \partial \times \boldsymbol{E} = \boldsymbol{0}, \qquad \epsilon_0 \frac{\partial \boldsymbol{E}}{\partial t} = \frac{1}{\mu_0} \partial \times \boldsymbol{B} - \boldsymbol{J}_{pol} - [\boldsymbol{J}], \quad (6.97)$$

and supposedly completed by additional ones, that should tell us, in the end, what the polarization current $\boldsymbol{J}_{pol}$ is. The brackets [] recall that the source terms may not be present. We say that the two sets can quite be seen in the same way. To discuss this affinity we have to avoid entering in too much details: (6.96) are directly macroscopic[35] equations involving thermodynamic results, while (6.97) are also macroscopic equations, but here we completely lack the necessary thermodynamics to understand the currents. Indeed, already because it would need giving a prohibitively large amount of information to specify the system, we cannot start at the microscopic non-thermodynamic quantum level, give the configuration, state in full the equations and find and average their solution to define $\boldsymbol{J}$ and $\boldsymbol{J}_{pol}$, (see also footnote 45 in Appendix); and we cannot, either, work directly at a thermodynamic level as it would be desirable and as is done in the acoustic case, because so far, we completely lack any irreversible-thermodynamic description of the motion of "electronic fluid" flowing in matter. Nevertheless, with this in mind, we can make appear as follows the strong affinity between the two wave-propagation problems.

6.12.1 Electromagnetic Analogy

First, to see that the philosophy of (6.96.1) can be made comparable to that of (6.97.1), we note that the macroscopic basic fields involved, $\boldsymbol{v}$, b (acoustic) and $\boldsymbol{E}$, $\boldsymbol{B}$ (electromagnetic), are "derived" fields, obtained in comparable manner from an underlying macroscopic three-vector "potential" field, which we denote $\boldsymbol{a}$ in acoustics, and $\boldsymbol{A}$ in electromagnetics.[36] In acoustics, working in the reference-frame in which the fluid is globally at rest, $\boldsymbol{a}(t, \boldsymbol{x})$ is the macroscopic displacement of fluid particles. In electromagnetics, working in the reference-frame in which the medium is globally at rest, $\boldsymbol{A}(t, \boldsymbol{x})$ is the macroscopic mean potential in Weyl's temporal gauge.[37]

The relations between the acoustic (displacement) and electromagnetic (potential) fields $\boldsymbol{a}$ and $\boldsymbol{A}$ and the derived acoustic and electromagnetic basic fields $\boldsymbol{v}$, b and $\boldsymbol{E}$, $\boldsymbol{B}$ are as follows:

$$\begin{array}{ll} \text{Acoustics:} & \text{Electromagnetics:} \\ b \equiv -\partial \cdot \boldsymbol{a}, \quad \boldsymbol{v} \equiv \dfrac{\partial \boldsymbol{a}}{\partial t}, & \boldsymbol{B} \equiv \partial \times \boldsymbol{A}, \quad \boldsymbol{E} \equiv -\dfrac{\partial \boldsymbol{A}}{\partial t}. \end{array} \tag{6.98}$$

35 The acoustic variables are from the beginning macroscopic variables, see Sect. 6.3.

36 We have $\boldsymbol{A} = \langle \boldsymbol{a} \rangle$ with $\boldsymbol{a}$ the microscopic potential associated to the fields $\boldsymbol{e}, \boldsymbol{b}$. The acoustic $\boldsymbol{a}$, in spite of his lowercase notation, is macroscopic and the correspondent of $\boldsymbol{A}$. There's no point in looking for an acoustic match for the electromagnetic $\boldsymbol{a}$.

37 The electromagnetic potential is a relativistic object $A_\mu = \langle a_\mu \rangle$ with four temporal and spatial covariant indices $\mu = 0, 1, 2, 3$; general gauge invariance of electromagnetic equations describing fields in interaction with charges, expresses in the fact that the covariant vector A_μ is determined only up to a four-gradient $\varphi_{;\mu}$; then working in the fixed reference-frame where the medium is at rest, it is possible to choose the four-gradient in such a way that the temporal component A_0 vanishes. This is Weyl's temporal gauge, in which $A_\mu = (A_0 \equiv 0, \boldsymbol{A})$. The electromagnetic fields $\boldsymbol{E}$ and $\boldsymbol{B}$, then derive in this gauge from a three-vector $\boldsymbol{A}$ with the written relations (6.98.3), (6.98.4).

The relations (6.96.1) and (6.97.1) are then seen to be, both, the tautological consequences of the way the fields b, $\boldsymbol{v}$, or $\boldsymbol{B}$, $\boldsymbol{E}$, are derived from $\boldsymbol{a}$ or $\boldsymbol{A}$ in (6.98).

The properties of the medium are not expressed in these first field equations (6.96.1) and (6.97.1); they come from the other relations, which are to be written as second field equations, constitutive relations, and other energetic (Heaviside-Poynting's) condition. If we want to see the acoustic relation (6.96.2) as a counterpart to the electromagnetic relation (6.97.2), we have an interest in putting it in the form of

$$\rho_0 \frac{\partial \boldsymbol{v}}{\partial t} = -\frac{1}{\chi_0} \partial b + \boldsymbol{f}_{pol} + [\boldsymbol{f}]. \tag{6.99}$$

The term $-\partial b/\chi_0$ compares to the term $\partial \times \boldsymbol{B}/\mu_0$. It comes from the diagonal adiabatic part $-p_a \delta_{ij}$ in the stresses (6.32), that is, from a medium response part making abstraction of losses, and the rest, denoted $\boldsymbol{f}_{pol}$ here to comply with electromagnetic notations in (6.97.2), describes an "induced" viscous and thermal "polarization" force density which is the expression of the previously mentioned, unavoidable and accompanying, irreversible exchanges of momentum and energy. Here, as the motion is assumed longitudinal, this induced "polarization" force density is simply given by, (compare (6.99) and (6.68))

$$\boldsymbol{f}_{pol} = \frac{1}{\chi_0} \partial b - \partial p - \left(\frac{4\eta}{3} + \zeta \right) \partial \frac{\partial b}{\partial t}. \tag{6.100}$$

The interpretation of the acoustic (6.99) and electromagnetic (6.97.2) equations is now made along similar physical lines of reasoning. In absence of "polarization" reactions of the media, the set of equations would be, $\partial b/\partial t = -\partial \cdot \boldsymbol{v}$, $\rho_0 \partial \boldsymbol{v}/\partial t = -\chi_0^{-1} \partial b + [\boldsymbol{f}]$, in acoustics, and $\partial \boldsymbol{B}/\partial t = -\partial \times \boldsymbol{E}$, $\epsilon_0 \partial \boldsymbol{E}/\partial t = \mu_0^{-1} \partial \times \boldsymbol{B} - [\boldsymbol{J}]$, in macroscopic electromagnetics, leading to nondispersive wave propagation with soundspeed c_0 and lightspeed c, such that, $\rho_0 \chi_0 c_0^2 = 1$, $\epsilon_0 \mu_0 c^2 = 1$. But as soon as the wavefields b, $\boldsymbol{v}$, or $\boldsymbol{B}$, $\boldsymbol{E}$, are present and variable in time and space, progressive "polarization" processes inevitably arise in response, resulting in an induced polarization force $\boldsymbol{f}_{pol}$, or an induced polarization current $\boldsymbol{J}_{pol}$, affecting in turn the variations of the fields and causing the wave dispersion.

This analogy then suggests introducing, for the polarization force $\boldsymbol{f}_{pol}$, a general representation in terms of a time-derived and space-derived term:

$$\boldsymbol{f}_{pol} = -\frac{\partial \boldsymbol{p}}{\partial t} + \partial m. \tag{6.101}$$

This will be the acoustic counterpart of the Lorentz splitting (6.83) of the macroscopic polarization current, (the "acoustic polarization" fields m and $\boldsymbol{p}$ are written in lowercase here but they have macroscopic nature). As in electromagnetics, there is an ambiguity in writing (6.101). Given some $\boldsymbol{p}$ and m satisfying (6.101), new $\boldsymbol{p}$ and m given by $\boldsymbol{p}' = \boldsymbol{p} + \partial \varphi$ and $m' = m + \partial \varphi/\partial t$ also satisfy (6.101), where φ is a scalar field. Nevertheless, if some criterium can be found to fix the scalar unknown

φ, so that unique polarizations $\boldsymbol{p}$ and m exist, we have the following consequences. Substituting (6.101) in (6.99) we arrive at

$$\frac{\partial \boldsymbol{d}}{\partial t} = \partial h + \left[\boldsymbol{f}\right], \tag{6.102}$$

with

$$\boldsymbol{d} = \rho_0 \boldsymbol{v} + \boldsymbol{p}, \qquad h = -\frac{1}{\chi_0} b + m. \tag{6.103}$$

Finally, in the same way as it has been suggested previously that macroscopic electromagnetic equations should ultimately be written in the form of (6.71)–(6.73), with the Maxwell field $\boldsymbol{H}$ to be fixed according to the Heaviside-Poynting condition that (6.89) should represent "the electromagnetic part" of the energy current density, here, it is suggested that the acoustic equations for longitudinal motions in the fluid, should ultimately be written in the following general form[38]

$$\frac{\partial b}{\partial t} + \partial \cdot \boldsymbol{v} = 0, \qquad \frac{\partial \boldsymbol{d}}{\partial t} = \partial h + \left[\boldsymbol{f} = -\partial \mathcal{P}\right], \tag{6.104}$$

$$\boldsymbol{d}(t, \boldsymbol{x}) = \hat{\rho} \boldsymbol{v}(t, \boldsymbol{x}) = \int \rho(t - t', \boldsymbol{x} - \boldsymbol{x}') \boldsymbol{v}(t', \boldsymbol{x}') dt' d\boldsymbol{x}', \tag{6.105}$$

$$\boldsymbol{h}(t, \boldsymbol{x}) = -\hat{\chi}^{-1} \boldsymbol{b}(t, \boldsymbol{x}) = -\int \chi^{-1}(t - t', \boldsymbol{x} - \boldsymbol{x}') \boldsymbol{b}(t', \boldsymbol{x}') dt' d\boldsymbol{x}', \tag{6.106}$$

with the "Maxwell field h" to be fixed according to the condition that

$$s = -\boldsymbol{v} h \tag{6.107}$$

should represent "the acoustic part" of the energy current density, or else, the *current density of energy transported in acoustic form*. Like before in the electromagnetic case, we mean by it the part of the energy current density, origin of the subsequent propagation and attenuation of a disturbance, and not the part, "lost" for the propagation, which is right-on converted in thermal degraded form. We will denote this, the "acoustic Heaviside-Poynting" thermodynamic identification of the $h-$field. With it, the $h-$field will represent an acoustic stress field, and the $\boldsymbol{d}-$field an acoustic momentum field, different from the ideal fluid expressions, $h = -\chi_0^{-1} b$, and $\boldsymbol{d} = \rho_0 \boldsymbol{v}$, because of the "acoustic polarization" processes.

Although the hypothesized Heaviside-Poynting thermodynamic identification (6.89) could not be performed, because of its current disconnection from available thermodynamics, its acoustic version (6.107) will prove to be feasible, and to remove the ambiguity inherent in Lorentz's acoustic splitting (6.101). Nevertheless, in the same way as, in electromagnetics in presence of spatial dispersion it is customary to set, $\boldsymbol{M} \equiv \boldsymbol{0}$, or, (6.95), by definition, here in acoustics it could be considered that,

[38] In (6.105), the kernels are written to depend on the difference $\boldsymbol{x} - \boldsymbol{x}'$, because we work with Cartesian coordinates and the fluid, unbounded, defines an homogeneous medium.

$m \equiv 0$, or, $h \equiv -\chi_0^{-1} b$, by definition. In what follows, after considering making the acoustic Heaviside-Poynting's thermodynamic identification (6.107), we will also have a look, for edification, at the different acoustic description resulting from this last definition, counterpart of the electromagnetic customary one.

6.12.2 Acoustic Maxwell Fields

We now consider specifying the Maxwell acoustic field h, the acoustic polarizations m and $\boldsymbol{p}$, the other Maxwell field $\boldsymbol{d}$, and thus the acoustic nonlocal operators $\hat{\rho}$ and $\hat{\chi}^{-1}$, by effecting the acoustic Heaviside-Poynting fundamental thermodynamic identification (6.107).

First, we write down an Energy Conservation-Dissipation Corollary [18], which applies to the Navier-Stokes-Fourier linearized equations:

$$\frac{\partial w}{\partial t} + \partial \cdot \boldsymbol{I} = -\mathcal{D}, \tag{6.108}$$

where

$$w = \frac{1}{2}\rho_0 v^2 + \frac{1}{2}\frac{p^2}{\rho_0 c_0^2} + \frac{1}{2}\frac{\rho_0 T_0}{c_P} s^2, \tag{6.109}$$

$$I_i = p v_i - \sigma'_{vij} v_j - \frac{\kappa}{T_0} \tau \partial_i \tau, \tag{6.110}$$

$$\mathcal{D} = \sigma'_{vij} \partial_j v_i + \frac{\kappa}{T_0} (\partial \tau)^2 . \tag{6.111}$$

Here w represents the density of disturbance energy, that is, the first term in (6.108), integrated in a control volume, represents the time rate of change of disturbance energy, whether this energy is accumulated in acoustic organised from, or other, degraded from. Vector $\boldsymbol{I}$ represents the energy-flux vector, whether the energy is transported in usable acoustic form or other degraded form. Its surface integral on the boundary of the control volume represents the rate at which disturbance energy is flowing out, in acoustic or other form. Finally, $\mathcal{D}$ represents the density of energy dissipated per unit volume and time, or in other terms, its integral in the control volume is the energy "unaccountably" lost in this volume and transformed in heat energy per unit of time. We see that $\mathcal{D} = T_0 \sigma_s$, where σ_s is the density of entropy production (6.39), in accordance with the general thermodynamic relation, $đQ = T dS$.

In covariant notations, (see Appendix), $\boldsymbol{I}$ is a contravariant vector I^i. From its Cartesian-coordinates expression (6.110) we see that I^i doesn't write in the "Heaviside-Poynting" form $-H^i_j(t,x) v^j(t,x)$. Indeed, the last thermal term in I^i will be proportional to $-\tau(t,x)\tau^{;i}(t,x)$, and this cannot be put in the form of a tensor at (t, x) contracted with the velocity at (t, x). It means that we cannot count the thermal term in (6.110) in what we call the energy-flux vector "in acoustic form" s: this term will correspond to an energy transported in a "degraded form". As regards the viscous term in $\boldsymbol{I}$, it has the right form of a tensor at (t, x) contracted with the velocity at (t, x), however we cannot count it in $\boldsymbol{s}$ because it again corresponds to energy transported in degraded form. Indeed, we observe that these two thermal and viscous contributions to the energy flux $\boldsymbol{I}$, are transformed in similar manner, $(-\frac{\kappa}{T_0}\tau\partial_i\tau \rightarrow \frac{\kappa}{T_0}\partial_i\tau\partial_i\tau, -\sigma'_{vij}v_j \rightarrow \sigma'_{vij}\partial_i v_j)$, in the two thermal conduction and viscosity contributions to the dissipated energy density $\mathcal{D}$. If the thermal term $-\frac{\kappa}{T_0}\tau\partial_i\tau$ carries only degraded energy, then it is in order also to assume that the viscous term $-\sigma'_{vij}v_j$ corresponds to the transport of degraded energy only. The remaining term $p\boldsymbol{v}$, to which corresponds no contribution in $\mathcal{D}$, must be that associated to the energy transported in usable, acoustic form. Thus, finally, we set

$$s = p\boldsymbol{v}. \tag{6.112}$$

Given the two expressions (6.107) and (6.112) for the same $\boldsymbol{s}$, it follows that we must identify the $h-$field with the (opposite) thermodynamic excess pressure:

$$h = -p. \tag{6.113}$$

This, turns out to be the translation of acoustic Heaviside-Poynting's identification of the $h-$field. It completely determines all quantities: it sets the $m-$field through using (6.103.2) and the operator $\hat{\chi}^{-1}$ through (6.106); after m is determined, it sets the $\boldsymbol{p}-$field through using (6.100), (6.101), then the $\boldsymbol{d}-$field through using (6.103.1), and then $\hat{\rho}$ through using (6.105). Let us now proceed explicitly with these determinations. Putting the acoustic Heaviside-Poynting identification (6.113) in (6.103.2) yields the following expression of the "polarization m"

$$m = -p + \frac{1}{\chi_0}b. \tag{6.114}$$

Eliminating b from using the state equation (6.69) we get

$$m = -\frac{\beta_0}{\chi_0}\left[\tau - \frac{(\gamma - 1)\chi_0}{\beta_0}p\right]. \tag{6.115}$$

Inserting the thermodynamic identity (6.5), taking the time derivative, and applying the (6.70), this is rewritten $\partial m/\partial t = -[(\gamma - 1)/\beta_0 T_0]\kappa\partial^2\tau$, or

$$m = -\frac{\gamma - 1}{\beta_0 T_0}\kappa\int_{-\infty}^{t} dt'\partial^2\tau. \tag{6.116}$$

This equation joined to (6.114) allows us to see "polarization m" as a thermal conduction part of the $h-$field or opposite pressure, which builds up due to the thermal conduction processes created in spatially variable (divergent) temperature gradients. The diagonal thermal part, $\sigma'_{\tau ij}$, in the additional lossy stress tensor (6.33), is now interpreted as being, $m\delta_{ij}$, and the h−field as being

$$h = -\frac{1}{\chi_0} b - \frac{\gamma - 1}{\beta_0 T_0} \kappa \int_{-\infty}^{t} dt' \partial^2 \tau. \tag{6.117}$$

We find back in this way (6.40), seen to express the fact that "acoustic" pressure p is not exactly in adiabatic relation with condensation, by the appearance of "polarization" processes related to thermal conduction. Putting the expression (6.114) in (6.100), (6.101) we next obtain that

$$\boldsymbol{p} = \left(\frac{4\eta}{3} + \zeta\right) \partial b = -\left(\frac{4\eta}{3} + \zeta\right) \partial(\partial \cdot \boldsymbol{a}), \tag{6.118}$$

or

$$\boldsymbol{p} = -\left(\frac{4\eta}{3} + \zeta\right) \int_{-\infty}^{t} dt' \partial(\partial \cdot \boldsymbol{v}). \tag{6.119}$$

This last equation compares with (6.116) and allows us to see "polarization $\boldsymbol{p}$" as an additional viscous momentum, which builds up in response to the spatial variations of the velocity divergence. The sign before the $\partial \boldsymbol{p}/\partial t$ term in (6.101) was chosen so that (6.119), for $\boldsymbol{p}$, has the same sign convention as (6.116), for m. Equation (6.119) also expresses as, $-\partial p_i/\partial t = \partial_j \sigma'_{vij}$, where σ'_{vij} is the viscous part in the stress tensor (6.32).[39] And $\boldsymbol{d}$, which finally is identified as

$$\boldsymbol{d} = \rho_0 \boldsymbol{v} - \left(\frac{4\eta}{3} + \zeta\right) \int_{-\infty}^{t} dt' \partial(\partial \cdot \boldsymbol{v}), \tag{6.120}$$

appears as an "acoustic momentum" that is augmented by the viscous polarization part $\boldsymbol{p}$. It represents only the part of the momentum that is transferred in reversible manner. Thus we see that, interestingly, *Heaviside-Poynting's thermodynamic identification of the h−field, leads to a natural separation of viscous effects and thermal conduction effects, the former being expressed in "viscous polarization* $\boldsymbol{p}$" (6.119), *and the latter in "thermal polarization m"* (6.116).

The laws and identifications obtained above have a somewhat unsatisfactory physical character, however, when viewed from the general nonlocal point of view. For example, looking at (6.116) or (6.119), it is obviously not reasonable in nonlocal physics, to treat on an equal footing the values of $\partial^2\tau$ or $\partial(\partial \cdot \boldsymbol{v})$ at different instants. As we see in next subsection, these identifications do not lead to an existing density

---

[39] For the case of longitudinal motions $\boldsymbol{v} = \partial\phi$ considered here, we have $\partial_i \partial_j v_j = \partial^2 v_i = \partial_i(\partial \cdot \boldsymbol{v})$, so that $\partial_j \sigma'_{vij}$ coincides with $\left(\frac{4\eta}{3} + \zeta\right) \partial_i(\partial \cdot \boldsymbol{v})$.

kernel in original space[40]; but in the Fourier space, they do correspond to definite kernels, whose expressions we will give. In Appendix we have given the more complete and general pattern of macroscopic quantities and equations, expected to hold true in presence of shear motions and when going far beyond the near-equilibrium simplifications made here. But in the best-case scenario, an attempt to construct corresponding quantities will await considerable advances in nonequilibrium thermodynamics.

6.12.3 Acoustic Nonlocal Operators

The acoustic kernels can be written in a Fourier decomposition as:

$$\begin{aligned} \rho(t-t', \boldsymbol{x}-\boldsymbol{x}') &= \int \frac{d\omega}{2\pi} \int \frac{d\boldsymbol{k}}{(2\pi)^3} \rho(\omega, \boldsymbol{k}) e^{-i\omega(t-t')+i\boldsymbol{k}\cdot(\boldsymbol{x}-\boldsymbol{x}')}, \\ \chi^{-1}(t-t', \boldsymbol{x}-\boldsymbol{x}') &= \int \frac{d\omega}{2\pi} \int \frac{d\boldsymbol{k}}{(2\pi)^3} \chi^{-1}(\omega, \boldsymbol{k}) e^{-i\omega(t-t')+i\boldsymbol{k}\cdot(\boldsymbol{x}-\boldsymbol{x}')}. \end{aligned} \tag{6.121}$$

In what follows we show how to compute the Fourier kernels coefficients $\rho(\omega, \boldsymbol{k})$ and $\chi^{-1}(\omega, \boldsymbol{k})$, even if the orginal space density kernel function $\rho(t, \boldsymbol{x})$ turns out to be not well-defined mathematically by the Stokes law, (which, interestingly, is not the case for the inverse operator $\hat{\rho}^{-1}$). In accordance with the electromagnetic analogy, the obtained expressions will be the same with and without source.

Let us first consider a calculation without source. Passing over to the Fourier transform of the equations describing longitudinal motion (6.67)–(6.70), without source ($\boldsymbol{f} = \boldsymbol{0}$), we get, on using the acoustic Heaviside-Poynting's identification (6.113)

$$-i\omega b(\omega, \boldsymbol{k}) = -i\boldsymbol{k} \cdot \boldsymbol{v}(\omega, \boldsymbol{k}), \tag{6.122}$$

$$-\rho_0 i\omega \boldsymbol{v}(\omega, \boldsymbol{k}) = i\boldsymbol{k} h(\omega, \boldsymbol{k}) - (4\eta/3 + \zeta)\, \omega \boldsymbol{k} b(\omega, \boldsymbol{k}), \tag{6.123}$$

$$-\gamma\chi_0 h(\omega, \boldsymbol{k}) = b(\omega, \boldsymbol{k}) + \beta_0 \tau(\omega, \boldsymbol{k}), \tag{6.124}$$

$$-\rho_0 c_P i\omega\tau(\omega, \boldsymbol{k}) = \beta_0 T_0 i\omega h(\omega, \boldsymbol{k}) - \kappa k^2 \tau(\omega, \boldsymbol{k}). \tag{6.125}$$

As shown by (6.123), the velocity is collinear to $\boldsymbol{k}$, and we may write $\boldsymbol{v}(\omega, \boldsymbol{k}) = v(\omega, \boldsymbol{k})\hat{\boldsymbol{n}}$, $\boldsymbol{k} = k\hat{\boldsymbol{n}}$. Equation (6.122) then is rewritten, $-i\omega b(\omega, \boldsymbol{k}) = -ikv(\omega, \boldsymbol{k})$, and, inserting this in (6.123), we get

$$\left[-\rho_0 i\omega + \left(\frac{4\eta}{3} + \zeta\right) k^2\right] \boldsymbol{v}(\omega, \boldsymbol{k}) = i\boldsymbol{k} h(\omega, \boldsymbol{k}).$$

[40] A non-convergent integral is obtained. We do not see the mathematical difficulties encountered as an indication that our nonlocal recasting of equations is not physical, but as an expression of the fact that the thermodynamic framework used is unsatisfactory, (see Sect. 6.6.3), and should be extended.

This compares directly to the equation, $-i\omega\rho(\omega, \boldsymbol{k})\boldsymbol{v}(\omega, \boldsymbol{k}) = i\boldsymbol{k}h(\omega, \boldsymbol{k})$, obtained by combining, $-i\omega\boldsymbol{d}(\omega, \boldsymbol{k}) = i\boldsymbol{k}h(\omega, \boldsymbol{k})$, with the function $\boldsymbol{d}(\omega, \boldsymbol{k}) = \rho(\omega, \boldsymbol{k})\boldsymbol{v}(\omega, \boldsymbol{k})$, which are respectively, the Fourier transform of (6.104.2) without source, and that of (6.105.1). The comparison yields the following expression of the density Fourier kernel:

$$\rho(\omega, \boldsymbol{k}) = \rho_0 + \left(\frac{4\eta}{3} + \zeta\right)\frac{k^2}{-i\omega}. \tag{6.126}$$

Next eliminating the excess temperature between (6.124) and (6.125), we find

$$-\rho_0 c_P i\omega\left(-\frac{\gamma\chi_0}{\beta_0}h(\omega, \boldsymbol{k}) - \frac{1}{\beta_0}b(\omega, \boldsymbol{k}\right) = \dots$$
$$+\beta_0 T_0 i\omega h(\omega, \boldsymbol{k}) - \kappa \boldsymbol{k}^2\left(-\frac{\gamma\chi_0}{\beta_0}h(\omega, \boldsymbol{k}) - \frac{1}{\beta_0}b(\omega, \boldsymbol{k})\right),$$

and after straightforward calculation using the thermodynamic identity (6.5) and definition (6.18.3) we get

$$h(\omega, \boldsymbol{k}) = -\chi_0^{-1}\left[1 - \frac{\gamma - 1}{\gamma}\left(1 + \frac{i\omega}{-i\omega + \dfrac{\kappa}{\rho_0 c_V}k^2}\right)\right]b(\omega, \boldsymbol{k}).$$

This compares directly to the Fourier transform of (6.106), which is given by $h(\omega, \boldsymbol{k}) = -\chi^{-1}(\omega, \boldsymbol{k})b(\omega, \boldsymbol{k})$, showing that, for the bulk-modulus Fourier kernel:

$$\chi^{-1}(\omega, \boldsymbol{k}) = \chi_0^{-1}\left[1 - \frac{\gamma - 1}{\gamma}\frac{\dfrac{\kappa}{\rho_0 c_V}k^2}{-i\omega + \dfrac{\kappa}{\rho_0 c_V}k^2}\right]. \tag{6.127}$$

From (6.119) and (6.114), the operators in the relations

$$\boldsymbol{p}(t, \boldsymbol{x}) = \int \chi_v(t - t', \boldsymbol{x} - \boldsymbol{x}')\boldsymbol{v}(t', \boldsymbol{x}')dt'd\boldsymbol{x}', \tag{6.128}$$

$$m(t, \boldsymbol{x}) = \int \chi_b(t - t', \boldsymbol{x} - \boldsymbol{x}')b(t', \boldsymbol{x}')dt'd\boldsymbol{x}', \tag{6.129}$$

inspired by (6.84) and (6.85), are found to be such that:

$$\chi_v(\omega, \boldsymbol{k}) = \left(\frac{4\eta}{3} + \zeta\right)\frac{k^2}{-i\omega}, \quad \chi_b(\omega, \boldsymbol{k}) = \chi_0^{-1}\frac{\gamma - 1}{\gamma}\frac{\dfrac{\kappa}{\rho_0 c_V}k^2}{-i\omega + \dfrac{\kappa}{\rho_0 c_V}k^2}. \tag{6.130}$$

These expressions are simple and illustrate the mentioned separation of viscous and thermal "polarization" effects, automatically performed by using the acoustic Heaviside-Poynting identification (6.113).

The consistency of the obtained expressions (6.126) and (6.127) of the kernels $\rho(\omega, \boldsymbol{k})$ and $\chi^{-1}(\omega, \boldsymbol{k})$, can be checked as follows. Passing over to the Fourier transform of (6.104)–(6.106) without source, we have

$$-i\omega b = -ikv, \qquad -i\omega d = ikh, \tag{6.131}$$

$$d = \rho(\omega, \boldsymbol{k})v, \qquad h = -\chi^{-1}(\omega, \boldsymbol{k})b, \tag{6.132}$$

from which we derive the dispersion relation

$$\rho(\omega, \boldsymbol{k})\chi(\omega, \boldsymbol{k})\omega^2 = k^2. \tag{6.133}$$

Putting in (6.133) the expressions (6.126) and (6.127), we can check that *the resulting dispersion relation is nothing but Kirchhoff-Langevin's characteristic equation* (6.52), rewritten in a nice factorized form.

In the above Fourier-transform calculations no source term was introduced. Now, if our electromagnetic analogy is fruitful, we should find that the same expressions (6.126) and (6.127) of the density and bulk-modulus Fourier kernels, are obtained in presence of an arbitrary external density of bulk longitudinal force. Basically, what we have to check is that, in presence of an external density of bulk longitudinal force $\boldsymbol{f}$ varying like $e^{-i\omega t + i\boldsymbol{k}\cdot\boldsymbol{x}}$, with ω and the components of $\boldsymbol{k}$ all independently chosen with values taken in the complex plane, the response of the fluid is still described in terms of the functions $\rho(\omega, \boldsymbol{k})$ and $\chi^{-1}(\omega, \boldsymbol{k})$ given by (6.126) and (6.127). Since ω and the components of $\boldsymbol{k}$ are now arbitrarily chosen in the complex plane, these functions actually are the analytic continuation of the preceding.

To check this, let us apply the external longitudinal body force $\boldsymbol{f} = -\partial\mathcal{P}$, $\mathcal{P} = \tilde{\mathcal{P}}e^{-i\omega t + i\boldsymbol{k}\cdot\boldsymbol{x}}$, with ω and $\boldsymbol{k}$, independent and possibly complex, (ω in the upper complex plane to ensure physical excitation). With this source inserted in the equations (6.67)–(6.70), the response fields have the form $f(t, \boldsymbol{x}) = f(\omega, \boldsymbol{k})e^{-i\omega t + i\boldsymbol{k}\cdot\boldsymbol{x}}$, and we find

$$-i\omega b(\omega, \boldsymbol{k}) = -i\boldsymbol{k}\cdot\boldsymbol{v}(\omega, \boldsymbol{k}), \tag{6.134}$$

$$-\rho_0 i\omega\boldsymbol{v}(\omega, \boldsymbol{k}) = i\boldsymbol{k}h(\omega, \boldsymbol{k}) - (4\eta/3 + \zeta)\,\omega\boldsymbol{k}b(\omega, \boldsymbol{k}) - i\boldsymbol{k}\tilde{\mathcal{P}}, \tag{6.135}$$

$$-\gamma\chi_0 h(\omega, \boldsymbol{k}) = b(\omega, \boldsymbol{k}) + \beta_0\tau(\omega, \boldsymbol{k}), \tag{6.136}$$

$$-\rho_0 c_P i\omega\tau(\omega, \boldsymbol{k}) = \beta_0 T_0 i\omega h(\omega, \boldsymbol{k}) - \kappa k^2\tau(\omega, \boldsymbol{k}), \tag{6.137}$$

which leads, after straightforward calculations, to the equations:

$$-i\omega\left[\rho_0 + \left(\frac{4\eta}{3} + \zeta\right)\frac{k^2}{-i\omega}\right]\boldsymbol{v}(\omega, \boldsymbol{k}) = i\boldsymbol{k}h(\omega, \boldsymbol{k}) - i\boldsymbol{k}\tilde{\mathcal{P}}, \tag{6.138}$$

and

$$h(\omega, \boldsymbol{k}) = -\chi_0^{-1}\left[1 - \frac{\gamma - 1}{\gamma}\frac{\dfrac{\kappa}{\rho_0 c_V}k^2}{-i\omega + \dfrac{\kappa}{\rho_0 c_V}k^2}\right] b(\omega, \boldsymbol{k}). \tag{6.139}$$

At the same time, setting in (6.104), (6.105), $\boldsymbol{f} = -\partial\{\mathcal{P} = \tilde{\mathcal{P}}e^{-i\omega t + i\boldsymbol{k}\cdot\boldsymbol{x}}\}$, and taking fields having the form, $a(t, \boldsymbol{x}) = a(\omega, \boldsymbol{k})e^{-i\omega t + i\boldsymbol{k}\cdot\boldsymbol{x}}$, we find:

$$-i\omega b(\omega, \boldsymbol{k}) = -i\boldsymbol{k}\cdot\boldsymbol{v}(\omega, \boldsymbol{k}), \qquad -i\omega\boldsymbol{d}(\omega, \boldsymbol{k}) = i\boldsymbol{k}h(\omega, \boldsymbol{k}) - i\boldsymbol{k}\tilde{\mathcal{P}}, \tag{6.140}$$

$$\boldsymbol{d}(\omega, \boldsymbol{k}) = \rho(\omega, \boldsymbol{k})\boldsymbol{v}(\omega, \boldsymbol{k}), \qquad h(\omega, \boldsymbol{k}) = -\chi^{-1}(\omega, \boldsymbol{k})b(\omega, \boldsymbol{k}), \tag{6.141}$$

where we have set, (in fact, because of convergence problem to define $\rho(t, \boldsymbol{x})$ this calculation for the density should be done with the inverse operator and kernels)

$$\begin{aligned} \rho(\omega, \boldsymbol{k}) &= \int \rho(t, \boldsymbol{x})e^{-i\omega t + i\boldsymbol{k}\cdot\boldsymbol{x}}dtd\boldsymbol{x}, \\ \chi^{-1}(\omega, \boldsymbol{k}) &= \int \chi^{-1}(t, \boldsymbol{x})e^{-i\omega t + i\boldsymbol{k}\cdot\boldsymbol{x}}dtd\boldsymbol{x}. \end{aligned} \tag{6.142}$$

By comparison of (6.140.2, 6.141.1) with (6.138), and (6.141.2) with (6.139), we see that, as intended, the expressions of the kernels (6.126) and (6.127) obtained without source still apply in the same way, in the presence of source. This justifies the analogue way of reasoning which made us directly postulate the nonlocal pattern (6.104)–(6.106) of longitudinal acoustic equations, from the nonlocal pattern of macroscopic electromagnetic equations.

To conclude, it is interesting to complete our discussion in relation to the physical-mathematical deficiencies of our thermodynamic framework that have been left in the background. We can first explicitly show that the result (6.130.2) determines a definite, but physically unsatisfactory, original-space kernel function $\chi_b(t, \boldsymbol{x})$. For the purpose of simplifying the discussion, we will make the calculation for the 1D case where the propagation is along a single direction x. In that case the dependencies over $\boldsymbol{k}$ and $\boldsymbol{x}$ are replaced by dependencies over k and x, and in (6.121), the integral $\int d\boldsymbol{k}/(2\pi)^3$ and variations $e^{i\boldsymbol{k}\cdot\boldsymbol{x}}$ are replaced by $\int dk/(2\pi)$ and e^{ikx}. In original-space the bulk modulus kernel is written as

$$\chi^{-1}(t, x) = \chi_0^{-1}\delta(t)\delta(x) - \chi_b(t, x), \tag{6.143}$$

where, according to (6.130.2)

$$\chi_b(t, x) = \chi_0^{-1}\alpha \int \frac{d\omega}{2\pi} \int \frac{dk}{2\pi} \frac{\beta k^2}{-i\omega + \beta k^2} e^{-i\omega t + ikx}, \tag{6.144}$$

with, $\alpha = (\gamma - 1)/\gamma$, and $\beta = \kappa/(\rho_0 c_V)$. Because of the exponentials, the double integral converges. Simple use of online integral calculator WolframAlpha yields the following result in closed form:

$$\chi_b(t,x) = \chi_0^{-1} \frac{\alpha}{8\sqrt{\pi}\beta^{3/2}t^{5/2}} \left(2\beta t - x^2\right) \exp\left[-\frac{x^2}{4\beta t}\right]. \tag{6.145}$$

It looks nice but is not completely physical: with finite propagation speeds, a truly physical kernel in original space would have to vanish exactly at sufficiently large values of x, (here, $|x| > c_0 t$); the above kernel doesn't respect this condition in a way characteristic of diffusive, not propagative, processes. This reveals the unsatisfactory nature of our local-thermodynamic-equilibrum framework, just extended in the most basic manner, out of equilibrium, to include Stokes' and Fourier's constitutive laws. In the same framework, if we now try to compute in similar manner the density kernel, $\rho(t,x) = \rho_0\delta(t)\delta(x) + \chi_v(t,x)$, where according to (6.130.1)

$$\chi_v(t,x) = \left(\frac{4\eta}{3} + \zeta\right) \int \frac{d\omega}{2\pi} \int \frac{dk}{2\pi} \frac{k^2}{-i\omega} e^{-i\omega t + ikx}, \tag{6.146}$$

we have an even more severe problem: the integral over k is not convergent. That over ω is defined and convergent, but only in the sense of Cauchy's Principal Value. Therefore, while Fourier's law, through (6.130.2), leads to defined but partially unsatisfactory original-space kernels, Stokes' law, through (6.130.1), is unable to produce an existing original-space density kernel. (As mentioned before this problem does not occur for the inverse density operator). Direct comparison between (6.146) and (6.144), however, shows that the integral over k would become convergent, and that over ω would become simply defined, if we extend the Stokes law in such a way to substitute, in place of the denominator $-i\omega$, the denominator $-i\omega + \beta' k^2$, with β' a new and supposedly small physical constant of same dimensions as β, i.e. $\beta' = \epsilon\left(\frac{4\eta}{3} + \zeta\right)/\rho_0$, with $\epsilon \ll 1$ a small numerical constant. By inspection it can be seen that this substitution would correspond replacing, in the motion equation, the term $\left(\frac{4\eta}{3} + \zeta\right)\partial_x^2 v$ leading to $-\left(\frac{4\eta}{3} + \zeta\right)k^2 v$, by a term $-\left(\frac{4\eta}{3} + \zeta\right)k^2 v/\left(1 + \frac{\beta' k^2}{-i\omega}\right)$, i.e. by an expansion

$$-\left(\frac{4\eta}{3} + \zeta\right)k^2 v \left(1 + \frac{-\beta' k^2}{-i\omega} + \left(\frac{-\beta' k^2}{-i\omega}\right)^2 + \left(\frac{-\beta' k^2}{-i\omega}\right)^3 + \ldots\right).$$

That is, in Stokes law it would correspond replacing the term $\left(\frac{4\eta}{3} + \zeta\right)\partial_x v$ by a more complex term with "more space and time memory"

$$\left(\frac{4\eta}{3} + \zeta\right)\left(\partial_x v + \beta' \partial_x^3 \int_{-\infty}^{t} dt' v(t',x) + \beta'^2 \partial_x^5 \int_{-\infty}^{t} dt' \int_{-\infty}^{t'} dt'' v(t'',x) + \ldots\right).$$

This modification of Stokes law would give a definite kernel in original space

$$\chi_v(t,x) = \frac{\rho_0}{\epsilon}\frac{1}{8\sqrt{\pi}\beta'^{3/2}t^{5/2}}\left(2\beta' t - x^2\right)\exp\left[-\frac{x^2}{4\beta' t}\right],$$

but of course it is not intended to be an accurate modification. It only indicates the direction of the revisions to be made: extend the description by introducing more spatial and temporal memory effects in the constitutive laws.

6.12.4 Summary: Action-Response Problem to Determine the Acoustic Kernels $\rho(\omega, k)$ and $\chi^{-1}(\omega, k)$

For later use and generalization in the next chapter, it is useful to reformulate here, what we have seen in the calculations with source in the preceding subsection. We have seen that, for arbitrary values of ω and $\boldsymbol{k}$ and when fixed by the Acoustic Heaviside- Poynting's condition (6.113), the effective nonlocal density $\rho(\omega, \boldsymbol{k})$, and the effective nonlocal bulk modulus $\chi^{-1}(\omega, \boldsymbol{k})$, can be determined by means of a simple action-response problem, in which work is performed on the system, in the form of external longitudinal bulk force $\boldsymbol{f}$ acting on it:

Action-response problem to determine $\rho(\omega, \boldsymbol{k})$ ***and*** $\chi^{-1}(\omega, \boldsymbol{k})$:

(i) Let us subject the fluid to the action of a longitudinal body force $\boldsymbol{f} = -\partial\mathcal{P}$ per unit fluid volume, deriving from a potential $\mathcal{P}$ varying as:

$$\mathcal{P} = \tilde{\mathcal{P}}e^{-i\omega t + i\boldsymbol{k}\cdot\boldsymbol{x}}, \qquad \tilde{\mathcal{P}} = Cst. \tag{6.147}$$

(ii) Let us find the response of the fluid to this action. We seek fields in the form $\boldsymbol{v} = v(\omega, \boldsymbol{k})\frac{\boldsymbol{k}}{k}e^{-i\omega t + i\boldsymbol{k}\cdot\boldsymbol{x}}$, $b = b(\omega, \boldsymbol{k})e^{-i\omega t + i\boldsymbol{k}\cdot\boldsymbol{x}}$, $p = p(\omega, \boldsymbol{k})e^{-i\omega t + i\boldsymbol{k}\cdot\boldsymbol{x}}$, $\tau = \tau(\omega, \boldsymbol{k})e^{-i\omega t + i\boldsymbol{k}\cdot\boldsymbol{x}}$, satisfying the longitudinal motion equations

$$\partial_t b + \partial\cdot\boldsymbol{v} = 0, \tag{6.148}$$

$$\rho_0\partial_t\boldsymbol{v} = -\partial p - (4\eta/3 + \zeta)\,\partial\partial_t b + \boldsymbol{f}, \tag{6.149}$$

$$\gamma\chi_0 p = b + \beta_0\tau, \tag{6.150}$$

$$\rho_0 c_P\partial_t\tau = \beta_0 T_0\partial_t p + \kappa\partial^2\tau, \tag{6.151}$$

and whose amplitudes in $(\omega, \boldsymbol{k})$ are proportional to $\tilde{\mathcal{P}}$.

(iii) There is unique solution $v(\omega, \boldsymbol{k})$, $b(\omega, \boldsymbol{k})$, $p(\omega, \boldsymbol{k})$ and $\tau(\omega, \boldsymbol{k})$, to this action-response problem, and we can easily write it by straightforward calculations. Then, the effective density $\rho(\omega, \boldsymbol{k})$ of (6.126) is obtained through the definition

$$-i\omega\rho(\omega, \boldsymbol{k})v(\omega, \boldsymbol{k}) = -ik\left(p(\omega, \boldsymbol{k}) + \tilde{\mathcal{P}}\right), \tag{6.152}$$

and the effective compressibility $\chi(\omega, \boldsymbol{k})$ of (6.127) is obtained through the definition

$$p(\omega, \boldsymbol{k}) = \chi^{-1}(\omega, \boldsymbol{k}) b(\omega, \boldsymbol{k}). \tag{6.153}$$

Note that we do not put $\boldsymbol{f}$ into bracket in (6.149), as it is here necessarily present as the forcing term creating the observed response motion. The physical content of the above procedure is just equivalent to stating that the acoustic equations can be put in the form (6.104)–(6.105), which expresses the electromagnetic analogy, with in addition, the $h-$field taken according to the acoustic Heaviside-Poynting's thermodynamic identification (6.113).

Finally, we note that the use we made here, in acoustics, of the Heaviside-Poynting thermodynamic identification, can be seen as an indirect confirmation of our analysis of macroscopic electromagnetic equations, even if we do not currently have the thermodynamics necessary to specify the corresponding notion of "energy current density carried out in electromagnetic form". Nevertheless, to finish and conclude we now examine what would have been obtained if we had transposed here, the expedients traditionally used in electromagnetic literature, (Sect. 6.11.3).

6.12.5 Acoustics Translation of the Customary Viewpoint

Using the same approach as used in conventional nonlocal electromagnetics, we would have considered, that, by definition, $m = 0$:

$$\frac{\partial b}{\partial t} + \partial \cdot \boldsymbol{v} = 0, \qquad \frac{\partial \boldsymbol{d}}{\partial t} = \partial h + \left[\boldsymbol{f}\right], \tag{6.154}$$

$$\boldsymbol{d}(t, \boldsymbol{x}) = \hat{\rho} \boldsymbol{v}(t, \boldsymbol{x}) = \int \rho(t - t', \boldsymbol{x} - \boldsymbol{x}') \boldsymbol{v}(t', \boldsymbol{x}') dt' d\boldsymbol{x}', \tag{6.155}$$

$$h(t, \boldsymbol{x}) = -\chi_0^{-1} b(t, \boldsymbol{x}). \tag{6.156}$$

The set of these formal equations (6.154)–(6.156), joined to the detailed longitudinal equations (6.67)–(6.70), allow determining the density operator in unique manner. Straightforward calculations show that the kernel coefficients

$$\rho(\omega, \boldsymbol{k}) = \int \rho(t, \boldsymbol{x}) e^{-i\omega t + i\boldsymbol{k}\cdot\boldsymbol{x}} dt d\boldsymbol{x},$$

are given by

$$\rho(\omega, \boldsymbol{k}) = \rho(\omega, \boldsymbol{k}) + \frac{k^2}{\omega^2} \left[\chi_0^{-1} - \chi^{-1}(\omega, \boldsymbol{k})\right], \tag{6.157}$$

where in the right, the quantities $\rho(\omega, \boldsymbol{k})$ and $\chi^{-1}(\omega, \boldsymbol{k})$ are those previously determined, given by (6.126) and (6.127). As before, the calculation can be done directly by passing over to the Fourier transform, without source; it can also be done with source by means of the following action-response problem.

Action-response problem to determine $\rho(\omega, \boldsymbol{k})$:

(i) Let us subject the fluid to the action of a longitudinal bulk force $\boldsymbol{f} = -\partial\mathcal{P}$ per unit fluid volume, deriving from a potential $\mathcal{P}$ varying as:

$$\mathcal{P} = \tilde{\mathcal{P}}e^{-i\omega t+i\boldsymbol{k}\cdot\boldsymbol{x}}, \qquad \tilde{\mathcal{P}} = Cst. \tag{6.158}$$

(ii) Let us find the response of the fluid to this action. We seek fields $\boldsymbol{v} = v(\omega, \boldsymbol{k})\frac{\boldsymbol{k}}{k}e^{-i\omega t+i\boldsymbol{k}\cdot\boldsymbol{x}}$, $b = b(\omega, \boldsymbol{k})e^{-i\omega t+i\boldsymbol{k}\cdot\boldsymbol{x}}$, $p = p(\omega, \boldsymbol{k})e^{-i\omega t+i\boldsymbol{k}\cdot\boldsymbol{x}}$, $\tau = \tau(\omega, \boldsymbol{k})$ $e^{-i\omega t+i\boldsymbol{k}\cdot\boldsymbol{x}}$, satisfying the longitudinal motion equations

$$\partial_t b + \partial \cdot \boldsymbol{v} = 0, \tag{6.159}$$

$$\rho_0 \partial_t \boldsymbol{v} = -\partial p - (4\eta/3 + \zeta)\, \partial\partial_t b + \boldsymbol{f}, \tag{6.160}$$

$$\gamma\chi_0 p = b + \beta_0 \tau, \tag{6.161}$$

$$\rho_0 c_P \partial_t \tau = \beta_0 T_0 \partial_t p + \kappa \partial^2 \tau, \tag{6.162}$$

and whose amplitudes in $(\omega, \boldsymbol{k})$ are proportional to $\tilde{\mathcal{P}}$.

(iii) There is unique solution $v(\omega, \boldsymbol{k})$, $b(\omega, \boldsymbol{k})$, $p(\omega, \boldsymbol{k})$ and $\tau(\omega, \boldsymbol{k})$, to this action-response problem, and we can easily write it by straightforward calculations. Then, the effective density $\rho(\omega, \boldsymbol{k})$ in (6.154)–(6.156), is obtained through the definition

$$-i\omega\rho(\omega, \boldsymbol{k})v(\omega, \boldsymbol{k}) = -ik\left(\frac{1}{\chi_0}b(\omega, \boldsymbol{k}) + \tilde{\mathcal{P}}\right). \tag{6.163}$$

The reader can easily check that this gives (6.157). The physical content of the above procedure is just equivalent to stating that the acoustic equations can be put in the form (6.104)–(6.106), which expresses the electromagnetic-acoustic analogy, with in addition, the h−field taken according to the definition

$$h = -b/\chi_0, \tag{6.164}$$

that is in fact artificial in presence of thermal conduction. Obviously, the dispersion equation that is obtained in this conception

$$\left[\rho(\omega, \boldsymbol{k}) + \frac{k^2}{\omega^2}\left(\chi_0^{-1} - \chi^{-1}(\omega, \boldsymbol{k})\right)\right]\chi_0\omega^2 = k^2, \tag{6.165}$$

is yet another way to rewrite the same Kirchhoff-Langevin's characteristic equation (6.133).

Therefore, this conception leads to exactly the same characteristic wavenumbers, but not the same characteristic impedances. Indeed this view, that mixes the different effects, (viscous, inertial, thermal, and elastic), into one abstract nonlocal density $\rho(\omega, \boldsymbol{k})$, with adiabatic compressibility, is not physical like the previous one, that places separately, inertial and viscous effects in the effective nonlocal density $\rho(\omega, \boldsymbol{k})$, and elastic and thermal effects in the effective nonlocal bulk modulus $\chi^{-1}(\omega, \boldsymbol{k})$. It provides a deep thermo-acoustic illustration of the unsatisfactory character of the current definitions in macroscopic nonlocal electromagnetics. We believe that, when generalized in acoustics to materials having finite dimensions, (as done in the next chapter as a first step, for rigid-framed and unbounded materials), the Heaviside-Poynting conception will allow us to overcome the so-called problem of "additional boundary conditions" in nonlocal media, encountered in electromagnetics [16].

Appendix: Electromagnetic-Acoustic Analogy

We present here the general form of our electromagnetic-acoustic analogy.

In electromagnetics we consider a distribution of electrons and nuclei, which can be structured at different levels, but appears homogeneous at some outer macroscopic level. Regarding the macroscopic averaging operation, we assume that we can conceive it in Lorentz's form [4], operating a spatial convolution of the considered field with a test function of suitable finite spatial extent [10, 15], or, with some advantages, in the Gibbs form of expectation values, obtained using a suitable ensemble of realizations of the environment: we imagine that the configuration of the microscopic charges, currents, as well as the structuring up to the outer macroscopic level, are appropriately varied to define a relevant average, over all the realizations.

In acoustics, we consider here (this is the subject of this chapter) a homogeneous fluid.[41] For the Lorentz conception we imagine that we have one sample, defined by one configuration of molecules, and that macroscopic averages are performed using a test function of spatial extent sufficiently large compared to mean distances between molecules and small compared to wavelengths; for the Gibbs conception we imagine that we have a suitable ensemble of configuration of molecules, and look at the expectations values over the configurations.

We consider, wave propagation of infinitesimal-amplitude electromagnetic or acoustic perturbations, either in the situation of free motion (no external source), or forced motion (an external current density or force density, is imposed, with same forcing in the different realizations when using Gibbs' conception). We work in the inertial, rest, reference-frame attached to the ambient medium. Moreover, we work here with *arbitrary 3-space Gauss coordinates* x^i, $i = 1, 2, 3$. Working with Gauss coordinates, and expressing the different equations and quantities in an invariant

[41] As just mentioned, however, the analogy extends to a medium that can be structured at different levels and becomes homogeneous at some outer macroscopic level; the next chapter (fluid permeating a rigid solid) will illustrate this.

form under a general change of these coordinates, (general covariant notations), is a valuable heuristic tool, which will help us to detect both the affinities and differences appearing between the electromagnetic and acoustic equations and quantities.

For the acoustics of a viscothermal fluid, the electromagnetic-acoustic analogy will highlight thermodynamic deficiencies in our current acoustic laws, which prevent quantities and equations from appearing entirely in a well-formed manner; for macroscopic electromagnetics, it will highlight the absence of thermodynamic foundation of the notions of electric and magnetic polarizations and Maxwell's $\boldsymbol{H}$ and $\boldsymbol{D}$ fields, also preventing the concrete expression of the complete form of quantities and equations. These insufficient aspects of the current descriptions can only be overcome when substantial progress has been made in the corresponding fields of irreversible thermodynamics. Nevertheless, in acoustics we expect that the formal pattern of quantities and equations hypothesized here for a simple fluid, will find concrate application when considering structured fluid-solid materials, that become homogeneous at an outer macroscopic scale, with nonlocalities coming principally from the structuration. For such structured materials the missing thermodynamics should not impede making, in forthcoming work, successful illustrations of the full pattern of quantities and equations postulated here, (see discussion around (6.210)).

Macroscopic Electromagnetics in Covariant Notations

Because we work in the rest reference-frame attached to the ambient medium, we are not concerned with the fact that, according to relativity theory, the electromagnetic potential is a four-vector. By using Weyl's gauge, (see footnote 37), we can set to zero its time component and thus associate, to the microscopic electromagnetic field, a covariant 3-vector potential field $a_i(t, x)$. From this 3-vector, the electric and magnetic fields, $e_i(t, x)$ and $b_{ij}(t, x)$, are derived as follows, where the semicolon ; denotes the covariant derivative in the Gauss coordinates:

$$e_i = -\frac{\partial a_i}{\partial t}, \qquad b_{ij} = \frac{1}{2}\left(a_{j;i} - a_{i;j}\right). \tag{6.166}$$

The usual pseudovector, $\boldsymbol{b} = \partial \times \boldsymbol{a}$, of the standard Gibbs-Heaviside vector notations, is obtained by working with Cartesian direct-system coordinates and setting, $b_k = 2b_{ij}$, where ijk is a circular permutation of 123; the factor of two in (6.166.2) and previous relation, is introduced for later convenience. By averaging, using a Gibbs' or Lorentz's conception, we will define corresponding macroscopic fields, $A_i = \langle a_i \rangle$, $E_i = \langle e_i \rangle$, $B_{ij} = \langle b_{ij} \rangle$. They will automatically satisfy

$$E_i = \langle e_i \rangle = -\frac{\partial A_i}{\partial t}, \qquad B_{ij} = \langle b_{ij} \rangle = \frac{1}{2}\left(A_{j;i} - A_{i;j}\right), \tag{6.167}$$

as there is direct switching (commutation) between the averaging operation $\langle\,\rangle$ and the time or spatial derivative $\partial/\partial t$ or $;i$:

$$\partial/\partial t\langle.\rangle = \langle\partial/\partial t.\rangle, \qquad \langle.\rangle;i = \langle.;i\rangle. \tag{6.168}$$

The first switching is obvious. The last is also evident with Gibbs' conception: the covariant derivative of the expectation values of the field and the expectation values of its covariant derivative are involved at the same spatial point, thus, the order of the operations can be interchanged, (with Lorentz's conception the justification is a little longer).

To obtain the macroscopic electromagnetic equations appropriate to describe time-variable space-variable phenomena, we first start with the following microscopic equations, where g_{ij} is the metric tensor of coordinates x^i, $g = |g_{ij}|$ is the determinant of the metric tensor, g^{ij} is the inverse matrix of the metric tensor, ($g_{ij}g^{jk} = \delta_i^k$, δ_i^k is the Kronecker symbol, which is a true mixed tensor), and ϵ_0 and μ_0, two scalar constants, are the electric and magnetic permittivities, (here and in the following we use Einstein's summation convention on repeated indices – one contravariant one covariant):

$$\frac{\partial b_{ij}}{\partial t} = \frac{1}{2}\left(e_{i;j} - e_{j;i}\right), \qquad \frac{\partial d^i}{\partial t} = h^{ij}_{;j} - j^i, \tag{6.169}$$

$$d^i = \epsilon_0^{ij} e_j, \qquad h^{ij} = \mu_0^{-1\,ijkl} b_{kl}, \tag{6.170}$$

$$\epsilon_0^{ij} = \epsilon_0 g^{1/2} g^{ij}, \qquad \mu_0^{-1\,ijkl} = \mu_0^{-1} g^{1/2} 2\left(g^{ik}g^{jl} - g^{il}g^{jk}\right). \tag{6.171}$$

The usual pseudovector, $\boldsymbol{h} = \boldsymbol{b}/\mu_0$, is obtained by working in Cartesian coordinates and defining, this time, $2h_k = h_{ij}$, in order that (6.169.2) assumes the known form, $\partial\boldsymbol{d}/\partial t = \partial \times \boldsymbol{h} - \boldsymbol{j}$, in Gibbs-Heaviside notations. The factor of two in the relation between h_k and h_{ij} is chosen so that (6.169.2), like the above known relation, looks simple. We note that, in order for (6.169.2) to be consistent with the additional known field equation, $d^i_{;i} = \rho$, and the conservation of charge, $\partial\rho/\partial t + j^i_{;i} = 0$, the introduced field h^{ij} must be taken antisymmetric: $h^{ij} = -h^{ji}$. We next complete these first equations by giving additional equations, specifying how the electromagnetic current j^i is related to the motion of punctual charges q_α (electrons and nuclei) located at x_α (δ is the Dirac delta function):

$$j^i = \sum_\alpha q_\alpha \frac{dx^i_\alpha}{dt}\delta\left[x - x_\alpha(t)\right], \tag{6.172}$$

and how their dynamics is related to the EM fields e and b:

$$\frac{d}{dt}\left[m_{\alpha ij}\left[\frac{dx^j}{dt}\right]_\alpha\right]=q_\alpha\left[e_i+b_{ik}\left[\frac{dx^k}{dt}\right]_\alpha\right],\ m_{\alpha ij}=\frac{g_{ij}m_{0\alpha}}{\sqrt{1-\frac{g_{kl}}{c^2}\left[\frac{dx^k}{dt}\frac{dx^l}{dt}\right]_\alpha}}. \tag{6.173}$$

where $m_{0\alpha}$ are the rest masses of the particles labelled by the index $\alpha = 1, 2, \ldots$, and c is the lightspeed constant. Finally, given the above microscopic equations, we have to average them up to the final outer macroscopic level.

Before proceeding we note that the additional equations (6.172)–(6.173) carry the non-quantum notion of "particles trajectories"; also, the summation over α is symbolic as the electrons are indistinguishable in the quantum vision. These equations are therefore written within a semi-classic, incomplete vision, which do not describe, in particular, spin and the quantum collective behaviour of electrons and nuclei. Here they only superficially play the role of additional fundamental quantum equations, which should be written to complete (6.169)–(6.171).

Aside from its flaws, the expression (6.172) allows us to see that the variance of j^i is that of a contravariant vector density of weight -1. Its contravariant variance comes from the differentials dx^i_α; its weight comes from the δ and the true scalar nature of charges q_α.[42] From (6.169.2) and (6.170) it then follows that d^i, h^{ij}, ϵ_0^{ij}, $\mu_0^{-1\,ijkl}$, are contravariant tensor densities of weight -1.

From then on, to write the (6.170)–(6.171), we only have to assume that the density fields, $d^i(t,x)$ and antisymmetric $h^{ij}(t,x)$, are, resp., directly related to the true tensor fields, $e_i(t,x)$ and $b_{ij}(t,x)$, *expressed at the same time and space position*, i.e. there is no temporal or spatial dispersion in void space. The only available tensors to construct the contravariant tensor densities of weight -1, ϵ_0^{ij} and $\mu_0^{-1\,ijkl}$, in the relations (6.170), are introduced in the problem by the arbitrary choice of coordinates. This immediately gives the expressions (6.171),[43] since we must have antisymmetry of the coefficients $\mu_0^{-1\,ijkl}$, not only over kl but also over ij. We note that the resulting ϵ_0^{ij} is automatically symmetric, $\epsilon_0^{ij} = \epsilon_0^{ji}$, and likewise, the resulting $\mu_0^{-1\,ijkl}$ is automatically symmetric under the exchange of the first and second pair of indices, $\mu_0^{-1\,ijkl} = \mu_0^{-1\,klij}$.

We now consider taking the Gibbs' or Lorentz's averages $\langle\,\rangle$ of the microscopic quantities and equations. For simplicity, it is advantageous to work with Gibbs' averages, however, the results will also be valid (but at long wavelengths only) and look the same, with Lorentz's averaging.

Because of the switching (6.168) it is immediate to average the equations (6.169)–(6.170) to write[44]:

[42] Since $\int \delta[x - x_\alpha]\, d^3x = 1$, and the differential element of volume d^3x is a scalar density of weight 1, the δ has the nature of a scalar density of weight -1.

[43] The numerical constants in (6.171) are chosen so that these relations reduce, in Cartesian coordinates and in Gibbs-Heaviside's notation, to proportionality relations, $\boldsymbol{d} = \epsilon_0 \boldsymbol{e}$, and $\boldsymbol{h} = \boldsymbol{b}/\mu_0$.

[44] We obviously have $\langle \epsilon_0^{ij} e_j \rangle = \epsilon_0^{ij} \langle e_j \rangle$, etc., as we use identical Gauss coordinates in the different realizations.

$$\frac{\partial B_{ij}}{\partial t} = \frac{1}{2}\left(E_{i;j} - E_{j;i}\right), \qquad \frac{\partial D_0^i}{\partial t} = H_{0;j}^{ij} - \langle j^i \rangle, \tag{6.174}$$

$$D_0^i = \epsilon_0^{ij} E_j, \qquad H_0^{ij} = \mu_0^{-1\,ijkl} B_{kl}, \tag{6.175}$$

where we have set

$$E_i = \langle e_i \rangle, \qquad B_{ij} = \langle b_{ij} \rangle, \qquad D_0^i = \langle d^i \rangle, \qquad H_0^{ij} = \langle h^{ij} \rangle. \tag{6.176}$$

In (6.175), the index 0 over D and H, is put to reserve notations D and H for the final macroscopic fields that will appear below, (and that are not the direct averages of underlying d and h fields). For the macroscopic current $\langle j^i \rangle$, we view it as the sum of an external impressed macroscopic current denoted $\left[J^i\right]$, (by defination the same in the different realizations), and an intrinsic macroscopic polarization current denoted J_{pol}^i:

$$\langle j^i \rangle = J_{pol}^i + \left[J^i\right]. \tag{6.177}$$

This macroscopic polarization current reflects the initiation of irreversible thermodynamic processes, trigged by the presence of the disturbance macroscopic fields E and B, zero in the equilibrium state, and nonzero in the disturbed state. It is present even if the forcing $\left[J^i\right]$ is absent. We certainly cannot truly express it from the semi-classical simplistic vision of the additional equations (6.172)–(6.173) that we have posed. We can also hardly express it by direct averaging from a complete microscopic quantum vision, because the laws that would link the macroscopic mean magnitudes together, being thermodynamic in nature, (in the sense of a macroscopic out-of-equilibrium statistics), would be difficult to obtain from the basic laws of QED.[45] Finally, as the corresponding thermodynamics of the motion of an "electronic fluid" in the lattice of nuclei, remains to be developed, a thermodynamic analysis and expression of J_{pol}^i is at present not feasible. Nevertheless, according to a division that we call Lorentz-splitting, we can always try to write this current as a spatial derivative, and a temporal derivative term:

$$J_{pol}^i = \frac{\partial P^i}{\partial t} + M_{;j}^{ij}, \tag{6.178}$$

with P^i, and antisymmetric M^{ij}, two unknown fields conventionally named the electric and magnetic polarization fields. The antisymmetry of M^{ij} is guessed on the basis of the characteristic antisymmetry observed in electromagnetic quantities and equations. Inserting (6.177)–(6.178) in (6.174.2) will provide a macroscopic equation having the form

$$\frac{\partial D^i}{\partial t} = H_{;j}^{ij} - \left[J^i\right], \tag{6.179}$$

[45] By the way, fully explaining how the transition from a quantum microscopic level to a classical macroscopic level takes place, should be a difficult task, already because of the special physical position of quantum physics which requires the classical limit for its own formulation [19].

with

$$D^i = \epsilon_0^{ij} E_j + P^i, \qquad H^{ij} = \mu_0^{-1\,ijkl} B_{kl} - M^{ij}. \tag{6.180}$$

The antisymmetry of M implies antisymmetry of the introduced H field. Because we consider infinitesimal perturbations, the macroscopic polarizations P and M appear as linear responses to the macroscopic electromagnetic field represented by E and B. There must be as a consequence, through (6.180), constitutive thermodynamic relations having, in the most general manner, the following linear form, (for small perturbations)

$$\begin{aligned} D^i(t,x) = \int_{-\infty}^{t} dt' \int d^3x' \epsilon^{ij}(t-t',x,x') E_j(t',x') + \ldots \\ \int_{-\infty}^{t} dt' \int d^3x' \kappa^{ijk}(t-t',x,x') B_{jk}(t',x'), \end{aligned} \tag{6.181}$$

$$\begin{aligned} H^{ij}(t,x) = \int_{-\infty}^{t} dt' \int d^3x' \mu^{-1\,ijkl}(t-t',x,x') B_{kl}(t',x') + \ldots \\ \int_{-\infty}^{t} dt' \int d^3x' \lambda^{ijk}(t-t',x,x') E_k(t',x'). \end{aligned} \tag{6.182}$$

The integrations over time will describe temporal dispersion effects, and those over space, spatial dispersion effects. But under the first field equation (6.174.1), which in harmonic regime implies $-i\omega B_{jk} = \left(E_{j;k} - E_{k;j}\right)/2$, it is always possible, by virtue of the presence of spatial dispersion, to incorporate in (6.181), the effect of the second integral term into the first, by appropriately redefining the nonlocal kernel function $\epsilon^{ij}(t,x,x')$. We can therefore generally simply rewrite

$$D^i(t,x) = \hat{\epsilon}^{ij} E_i(t,x) = \int_{-\infty}^{t} dt' \int d^3x' \epsilon^{ij}(t-t',x,x') E_j(t',x'). \tag{6.183}$$

Concerning (6.182), the second integral term is similarly useless, so that

$$H^{ij}(t,x) = \hat{\mu}^{-1\,ijkl} B_{kl}(t,x) = \int_{-\infty}^{t} dt' \int d^3x' \mu^{-1\,ijkl}(t-t',x,x') B_{kl}(t',x'). \tag{6.184}$$

It corresponds to expressing P_i as a nonlocal linear response to the electric field, and M_{ij} as a nonlocal linear response to the magnetic field; it justifies the denomination "electric" for P and "magnetic" for M.

There is still to be observed that the introduction of polarizations P and M, through the Lorentz-splitting (6.178), carries an essential ambiguity: an arbitrary response term $\Phi^{ij}_{;j}$, with Φ^{ij} an antisymmetric tensor of weight -1, can always be added to P^i, if, simultaneously, a term $\partial\Phi^{ij}/\partial t$ be subtracted to M^{ij}. Because of this indetermination, the fields D and H, and thus also the operators $\hat{\epsilon}$ and $\hat{\mu}^{-1}$, are ambiguous and still need to be resolved. I assert that this inherent ambiguity would be removed, i.e. all quantities above would become uniquely defined in principle, if one were able to give a precise thermodynamic content to the notion of the *current density of energy transported in usable electromagnetic form*, (by opposition to energy transported in degraded form). Given this notion, one then could require that, by definition of H^{ij}, this current density of energy be given by a Poynting-Heaviside vector S, such that

$$S^i(t,x) = -H^{ji}(t,x)E_j(t,x) = H^{ij}(t,x)E_j(t,x). \tag{6.185}$$

In the context of general nonlocal response of the media, the form of (6.185) is remarkable, in that, the current density of energy transported in electromagnetic form, *at a given time and position*, is expressed as the contraction of two fields evaluated *at the same time and position*. In fact, the corresponding remarkable choice of the field H^{ij} will determine, how is to be made the precise separation of the global effects, into the actual "time dispersion effects" per se, and "spatial dispersion effects" per se.

In electromagnetics of material media, we do not dispose of the necessary thermodynamic laws of motion of the system, comprising the electrons and solid nuclei structured lattice, to precise these notions and arrive at a definition of the electric polarization, magnetic polarization, field H^{ij}, field D^i, and operators $\hat{\epsilon}$ and $\hat{\mu}^{-1}$. By thermodynamic arguments of a general nature, however, it can be shown that $\hat{\epsilon}$ and $\hat{\mu}^{-1}$ are "generalized susceptibilities" that satisfy the general symmetry relations [13, 20]:

$$\epsilon^{ij}(t,x,x') = \epsilon^{ji}(t,x',x), \qquad \mu^{-1\,ijkl}(t,x,x') = \mu^{-1\,klij}(t,x',x). \tag{6.186}$$

Finally, in presence of spatial dispersion, it is customary in electromagnetic literature to consider that, $M^{ij} = 0$, i.e. $H^{ij} = \mu_0^{-1\,ijkl} B_{kl}$. This leads to the definition of conventional D field and ϵ kernel, certainly slightly different from the true physical ones.

We believe that, future developments crossing the boundaries of present thermodynamics, could allow the introduction of the necessary thermodynamic variables, associated to the motion of "electronic fluid" interacting with the nuclei structured lattice, from which, in particular, the electromagnetic current density of energy, S^i, could be explicitly constructed. In a sequence of successive determinations of the type illustrated in acoustics in Sects. (6.12.2-3) it would then imply, precise, and unambiguous, definition of H^{ij} field, M^{ij} field, D^i field, P^i field, operator $\hat{\epsilon}$, and operator $\hat{\mu}^{-1}$. We summarize our discussion. Macroscopic electromagnetic equations have the following covariant (invariant) pattern in all coordinate systems fixed to the ambient medium:

$$\frac{\partial B_{ij}}{\partial t} = \frac{1}{2}\left(E_{i;j} - E_{j;i}\right), \qquad \frac{\partial D^i}{\partial t} = H^{ij}_{;j} - \left[J^i\right], \tag{6.187}$$

$$D^i(t,x) = \hat{\epsilon}^{ij}E_i(t,x) = \int_{-\infty}^{t} dt' \int d^3x' \epsilon^{ij}(t-t',x,x')E_j(t',x'), \tag{6.188}$$

$$H^{ij}(t,x) = \hat{\mu}^{-1\,ijkl} B_{kl}(t,x) = \int_{-\infty}^{t} dt' \int d^3x' \mu^{-1\,ijkl}(t-t',x,x') B_{kl}(t',x'), \tag{6.189}$$

$$S^i(t,x) = H^{ij}(t,x) E_j(t,x) = -H^{ji}(t,x) E_j(t,x), \tag{6.190}$$

where a thermodynamic knowledge is assumed, (though currently missing), that could be used to construct the Heaviside-Poynting vector S^i, specifying the current density of energy transported in electromagnetic form. We name "Lorentz fields" the true tensor fields $B_{ij} = \langle b_{ij} \rangle$ and $E_i = \langle e_i \rangle$, and "Maxwell fields" the tensor density fields H^{ij} and D^i. The first are the direct averages of the corresponding microscopic fields. The second are not. They include an additional polarization part.

Let us now consider the acoustic wave propagation in our simple homogeneous fluid, (directly a macroscopic wave propagation phenomenon), and let us try to write, by analogy, the most general pattern of macroscopic equations, that would manifest both temporal dispersion and spatial dispersion.

Acoustics in Covariant Notations

In electromagnetics, working in the rest frame of the medium, we have introduced a 3-vector covariant macroscopic field $A_i(t,x) = \langle a_i \rangle(t,x)$, allowing to build the macroscopic electric and magnetic fields, E, B, according to (6.167). Here in acoustics, by analogy, we want to test the idea that, working in the rest frame of the fluid, we also have, at the macroscopic level, at least in Cartesian coordinates, an obvious 3-vector field existing at any point and time and associated with an acoustic disturbance: the macroscopic displacement $\boldsymbol{a}$ of the fluid particles.

An immediate objection to the possibility of such an analogy is that a finite displacement is not a vector when using Gauss' general coordinates. However, as we restrict here to considering infinitesimal perturbations, we can consider that the associated displacement fields a^i will behave as coordinates differentials dx^i, that is, as true contravariant vectors. We can thus try to pursue a deep electromagnetic-acoustic analogy, where the macroscopic electromagnetic *covariant* potential field A_i, will be replaced by a *contravariant* displacement field a^i. From this basic acoustic contravariant 3-vector a^i, we derive the velocity and (opposite and linearized) strain fields[46] v^i and b^{ij}:

$$v^i = \frac{\partial a^i}{\partial t}, \qquad b^{ij} = -\frac{1}{2}\left(a^{i;j} + a^{j;i}\right). \tag{6.191}$$

[46] In spite of the macroscopic nature of our quantities v and b here, we use small letters for them, because in the next chapter the medium will be structured, (with solid parts), and these quantities will have to be averaged up to some outer macroscopic level, then becoming the outer macroscopic quantities $V = \langle v \rangle$ and $B = \langle b \rangle$.

In these definitions, the characteristic electromagnetic antisymmetry, $B_{ij} = -B_{ji}$, becomes a characteristic acoustic symmetry, $b^{ij} = b^{ji}$: therefore, there is no sense to try to compare the choice of overall signs in (6.167) and (6.191). Here, we have introduced for convenience positive sign in (6.191.1) and negative sign in (6.191.2) in order that $b_i^i = b$, the condensation.[47]

We see by construction that, from (6.191), the following equation is satisfied

$$\frac{\partial b^{ij}}{\partial t} = -\frac{1}{2}\left(v^{i;j} + v^{j;i}\right). \tag{6.192}$$

It will play the role of the first electromagnetic equation (6.187).

Now, by analogy with the electromagnetic macroscopic equations (6.174), (6.175), and (6.177), we can guess that, when the fluid is forced by an arbitrary external force density $\left[F^i\right](t, x)$, its motion will be governed by equations having the form:

$$\frac{\partial b^{ij}}{\partial t} = -\frac{1}{2}\left(v^{i;j} + v^{j;i}\right), \qquad \frac{\partial D_0^i}{\partial t} = H_{0\,;j}^{ij} + F^i, \tag{6.193}$$

$$D_{0\,i} = \rho_{0\,ij} v^j, \qquad H_{0\,ij} = -\chi_{0\,ijkl}^{-1} b^{kl}, \tag{6.194}$$

$$F^i = F_{pol}^i + \left[F^i\right], \tag{6.195}$$

with, this time, we assume, a *symmetric* H_0^{ij}. Again, because of the interchange of symmetry there is no sense to compare signs in electromagnetic and acoustic equations. The signs appearing are chosen for later convenience. $\left[F^i\right]$ is the external force density (force per unit volume) impressed on the fluid, and F_{pol}^i is an induced polarization force density, which reflects the initiation of irreversible thermodynamic processes of any kind, coming in response to the presence of the disturbance macroscopic fields v and b, zero in the equilibrium state, and nonzero in the disturbed state.

With regard to force densities $\left[F^i\right]$, F_{pol}^i, we view them as contravariant vectors with a weight density of -1, (identical to that of electromagnetic current densities), and this variance comes, for example, from a comparison with a Newton's law term, $\rho \partial v^i / \partial t$, where ρ is an expression of type, $\rho = \langle \sum_\alpha m_\alpha \delta\left[x - x_\alpha(t)\right] \rangle$, the sum is on the molecules, and the masses m_α are true scalars.

Therefore, D_{0i}, H_{0ij}, $\rho_{0\,ij}$ and $\chi_{0\,ijkl}^{-1}$, are also tensor densities of weight -1. The equations (6.194) state that the "Maxwell" densities D_{0i} and H_{0ij} are, respectively, directly related to the " Loventz" true fields v^i and b^{ij}, *expressed at the same time and space position*. This is because all temporal and spatial dispersion effects, (all irreversible thermodynamic processes coming in response to the presence of the disturbance macroscopic fields), express, by definition, via the induced polarization force density F_{pol}^i. As the fluid is time invariant, homogeneous, isotropic, and invari-

[47] We recall that the lowering or raising of indices is done by contraction with the metric tensor or its inverse, (here, $b_i^i = g_{ij} b^{ij} = g^{ij} b_{ij}$).

ant in spatial reflections, the only available tensors to construct the coefficients $\rho_{0\,ij}$ and $\chi^{-1}_{0\,ijkl}$, are the metric tensor g_{ij} or its inverse g^{ij}, and the associated density of weight -1, $g^{1/2}$. Moreover, as b^{kl} is symmetric, and we assume that H_{0ij} is also symmetric, $\mu^{-1}_{0\,ijkl}$ must be symmetric, not antisymmetric, on indices ij and kl. With this exchange of symmetry, the only possible expressions are:

$$\rho_{0\,ij} = \rho_0 g^{1/2} g_{ij},\ \chi^{-1}_{0\,ijkl} = \chi_0^{-1} g^{1/2} g_{ij} g_{kl} + \mu_0 g^{1/2}\left(g_{ik}g_{jl} + g_{il}g_{jk} - \frac{2}{3} g_{ij} g_{kl}\right). \tag{6.196}$$

Note that we could have written

$$D_0^i = \rho_{0\,j}^{\,i} v^j, \qquad H_0^{ij} = -\chi_{0\,\,kl}^{-1\,ij} b^{kl},$$

in place of (6.194), and

$$\rho_{0\,j}^{\,i} = \rho_0 g^{1/2} \delta^i_j,\ \chi_{0\,\,kl}^{-1\,ij} = \chi_0^{-1} g^{1/2} g^{ij} g_{kl} + \mu_0 g^{1/2}\left(\delta^k_i \delta^l_j + \delta^l_i \delta^k_j - \frac{2}{3} g^{ij} g_{kl}\right),$$

in place of (6.196). In Cartesian coordinates the above expressions of the χ_0^{-1} tensor yield Frenkel's conjectured tensor (6.42).

The true scalar constants, ρ_0, χ_0^{-1} and μ_0, interpret as the ambient density, the adiabatic bulk modulus, and the adiabatic Frenkel-Lamé shear coefficient: remarkably, the addition of the latter coefficient is a direct result of the transformation of antisymmetry to symmetry, when we pass from electromagnetic equations to acoustic equations.

Next, in the same manner as, in electromagnetics, we assumed Lorentz's splitting of the induced polarization current, here, we assume a similar splitting of the induced polarization force:

$$F^i_{pol} = -\frac{\partial P^i}{\partial t} + M^{ij}_{;j}, \tag{6.197}$$

with P^i, and symmetric M^{ij}, some "acoustic polarization" fields, (again, because of the interchange of symmetry there is no sense to try to compare signs in (6.178) and (6.197), and the sign $-$ is introduced as inspired by (6.101)). Inserting (6.195, 6.197) in (6.193.2) will provide an equation

$$\frac{\partial D^i}{\partial t} = H^{ij}_{;j} + \left[F^i\right], \tag{6.198}$$

where

$$D_i = \rho_{0\,ij} v^j + P_i, \qquad H_{ij} = -\chi^{-1}_{0\,ijkl} b^{kl} + M_{ij}. \tag{6.199}$$

The field D^i will be an "effective acoustic momentum" field, and the field H^{ij} an "effective acoustic stress" field. As the polarizations appear in linear response to the infinitesimal acoustic field represented by v and b, there must be constitutive thermodynamic linear relations having, in the most general manner, the form

$$D_i(t,x) = \int_{-\infty}^{t} dt' \int d^3x' \rho_{ij}(t-t',x,x')v^j(t',x') + \dots \\ \int_{-\infty}^{t} dt' \int d^3x' \kappa_{ijk}(t-t',x,x')b^{jk}(t',x'), \tag{6.200}$$

$$H_{ij}(t,x) = -\int_{-\infty}^{t} dt' \int d^3x' \chi^{-1}_{ijkl}(t-t',x,x')b^{kl}(t',x') + \dots \\ \int_{-\infty}^{t} dt' \int d^3x' \lambda_{ijk}(t-t',x,x')v^k(t',x'). \tag{6.201}$$

But under the first field equation (6.193.1), which in harmonic regime implies $i\omega b^{jk} = \left(v^{j;k} + v^{k;j}\right)/2$, it is always possible, by virtue of the presence of spatial dispersion, to incorporate in (6.200) the effect of the second integral into the first, by appropriately redefining the kernel ρ_{ij}. We can therefore generally simply rewrite

$$D_i(t,x) = \hat{\rho}_{ij}v^j(t,x) = \int_{-\infty}^{t} dt' \int d^3x' \rho_{ij}(t-t',x,x')v^j(t',x'). \tag{6.202}$$

Concerning (6.201), the second integral is similarly useless, so that we can also simply rewrite

$$H_{ij}(t,x) = -\hat{\chi}^{-1}_{ijkl}b^{kl}(t,x) = -\int_{-\infty}^{t} dt' \int d^3x' \chi^{-1}_{ijkl}(t-t',x,x')b^{kl}(t',x'). \tag{6.203}$$

It amounts to say that P_i is nonlocal response to the velocity field, and M_{ij} is nonlocal response to the strain field.

Finally and again, the introduction of polarizations P^i and M^{ij} through the splitting (6.197), carries an essential ambiguity: an arbitrary response term $\Phi^{ij}_{;j}$, with Φ^{ij} a symmetric tensor of weight -1, can always be added to P^i, if, simultaneously, a term $\partial\Phi^{ij}/\partial t$ be added to M^{ij}. Because of this indetermination, the fields D and H, and thus also the operators $\hat{\rho}$ and $\hat{\chi}^{-1}$, are ambiguous and still need to be resolved. I assert that the ambiguity would be removed, i.e. all quantities above would become uniquely defined in principle, if one were able to give a precise thermodynamic content to the notion of the current density of energy transported in usable acoustic form, (by opposition to energy transported in degraded form). Given this notion, one then could require that, by definition of H^{ij}, this current density of energy be given by a "Poynting-Heaviside", or here, "Umov" vector, such that

$$S^i(t,x) = -H^i_j(t,x)v^j(t,x), \qquad (H^i_j = g_{jk}H^{ik} = g^{ik}H_{kj}). \qquad (6.204)$$

At present time, we do not dispose of the necessary thermodynamic laws of motion, to fully precise this notion and arrive at a wholly satisfactory construction of the acoustic polarizations, acoustic fields H^{ij} and D^i, and acoustic kernels $\rho_{ij}(t,x,x')$ and $\chi^{-1}_{ijkl}(t,x,x')$. By thermodynamic arguments of a general nature, however, we know that $\hat{\rho}$ and $\hat{\chi}^{-1}$ are "generalized susceptibilities" that satisfy the general symmetry relations [13, 20]:

$$\rho_{ij}(t,x,x') = \rho_{ji}(t,x',x), \qquad \chi^{-1}_{ijkl}(t,x,x') = \chi^{-1}_{klij}(t,x',x). \qquad (6.205)$$

Within the usual near-equilibrium Navier-Stokes-Fourier thermodynamic framework, what is obtained is that we must identify the effective acoustic stress field H_{ij} with $-pg^{1/2}g_{ij}$, (to respect Heaviside-Poynting's identification (6.112), and the covariant writing of the (6.113), namely, $S^i = g^{1/2}pv^i$, where p interprets here as a scalar equal to the thermodynamic pressure), but this leads to degeneracies particularly apparent in the description of shear motions: for these motions, no pressure arises and the corresponding Maxwell stress field H_{ij} is then identically equal to zero. It highlights clearly the thermodynamic insufficiencies of this framework.

We summarize our discussion. Acoustic equations have the following covariant pattern in all coordinate systems fixed to the ambient medium:

$$\frac{\partial b^{ij}}{\partial t} = -\frac{1}{2}(v^{i;j} + v^{j;i}), \qquad \frac{\partial D^i}{\partial t} = H^{ij}_{;j} + \left[F^i\right], \qquad (6.206)$$

$$D_i(t,x) = \hat{\rho}_{ij}v^j(t,x) = \int_{-\infty}^{t} dt' \int d^3x' \rho_{ij}(t-t',x,x')v^j(t',x'), \qquad (6.207)$$

$$H_{ij}(t,x) = -\hat{\chi}^{-1}_{ijkl}b^{kl}(t,x) = -\int_{-\infty}^{t} dt' \int d^3x' \chi^{-1}_{ijkl}(t-t',x,x')b^{kl}(t',x'), \qquad (6.208)$$

$$S^i(t,x) = -H^i_j(t,x)v^j(t,x), \qquad (6.209)$$

where a thermodynamic knowledge is assumed, (though currently missing), that could be used to construct the "Heaviside-Poynting" or else "Umov" vector S^i.

When the medium is structured, the analogy suggests to replace v and b by averages $V = \langle v \rangle$ and $B = \langle b \rangle$ that will be named the acoustic Lorentz fields. New acoustic Maxwell fields D and H, "Umov" vector S, and operators $\hat{\rho}$ and $\hat{\chi}^{-1}$, will appear, respecting the same equation pattern as above. The mentioned deficiencies of the thermodynamic framework will not hinder the construction of a viable nonlocal description. Indeed, the nonlocal effects resulting from the structuring of the environment will be the ones that will be important to consider first. They will be sufficiently well described in the available imperfect thermodynamic framework. An highlighting, reduced exemplification of this, will be considered in the next chapter.

The full (nonreduced) nonlocal pattern of quantities and equations will arise for structured media, e.g. media having connected fluid and solid phases. For example, consider the case where the stress-strain relations in the solid are idealized as the following lossless relations, $\sigma_{ij} = -\chi^{-1}_{sijkl} b^{kl}$, with χ^{-1}_{sijkl} given by (6.196), and χ_0^{-1} and μ_0 replaced by χ_s^{-1} and μ_s, (the constants in the solid phase). Then, the full above nonlocal pattern of equations will appear, replacing v^i and b^{ij} by $V^i = \langle v^i \rangle$ and $B^{ij} = \langle b^{ij} \rangle$, and defining the H_{ij} field through the generalized energetic "Umov" or "Heaviside-Poynting" relation, (making it a continuous field):

$$S^i(t, x) = \langle -\sigma^i_j v^j \rangle (t, x) = -H^i_j(t, x) V^j(t, x), \tag{6.210}$$

with σ_{ij}, equal to $-pg^{1/2} g_{ij}$ in the fluid, where p is the thermodynamic excess pressure. In short, it should be possible to describe any fluid-saturated poroelastic material as a suitable nonlocal elastic solid.[48] We expect that this principled position of the problems will be verified in future detailed studies.

References

1. J.W. Gibbs, *Elementary Principles in Statistical Mechanics, Developed with Especial Reference to the Rational Foundation of Thermodynamics* (Dover Publications, New York, 1902)
2. C.A. Truesdell, Precise theory of the absorption and dispersion of forced plane infinitesimal waves according to the Navier-Stokes equations. J. Ration. Mech. Anal. **2**, 643–741 (1953)
3. H.B. Callen, *Thermodynamics and an Introduction to Thermostatistics* (Wiley, New York, 1985)
4. H.A. Lorentz, *The theory of electrons and its applications to the phenomena of light and radiant heat.* A course of lectures delivered in Columbia University, New York, in March and April, 1906, Vol. 29. Leipzig, B. G. Teubner (1909)
5. C.-M. Marle, Ecoulements monophasiques en milieu poreux. Revue de l'Institut Français du Pétrole 22(10), 1471–1509 (1967). (Available online among other interesting documents at http://marle.perso.math.cnrs.fr)
6. L.D. Landau, E.M. Lifshitz, *Fluid Mechanics*, 2nd edn. (Pergamon Press, Oxford, 1987)
7. Y.I. Frenkel, *Kinetic Theory of Liquids* (Oxford University Press, Oxford, 1947)
8. D. Bolmatov, V.V. Brazhkin, K. Trachenko, The phonon theory of liquid thermodynamics. Sci. Rep. **2**, 421 (2012)
9. J.W. Strutt (Lord Rayleigh), *The Theory of Sound*, vol. 2 (Dover, New York, 1945)
10. J.D. Jackson, *Classical Electrodynamics* (Wiley, New York, 1999)
11. G. Fournet, *Électromagnétisme: à partir des équations locales*. Masson, Paris, 1985. See also G. Fournet, *Electromagnétisme. Traité génie électrique*, Techniques de l'ingénieur, D1020 (1993)
12. S. Weinberg, *Gravitation and Cosmology: Principles and Applications of the General Theory of Relativity* (Wiley, New York, 1972)
13. L.D. Landau, E.M. Lifshitz, *Electrodynamics of Continuous Media* (Pergamon Press, Oxford, 1960)

---

[48] Biot's theory [21, 22] will be a special, simplified case of this, in which at the small scale the fluid will move as an incompressible fluid and the solid will move as a whole. This limit will have the same physical position as the local-theory limit presented in Appendix of next chapter, for the case of rigid-framed materials.

14. P.A. Belov, R. Marqués, S.I. Maslovski, I.S. Nefedov, M. Silveirinha, C.R. Simovski, S.A. Tretyakov, Strong spatial dispersion in wire media in the very large wavelength limit. Phys. Rev. B **67**(11), 113103 (2003)
15. G. Russakoff, A derivation of the macroscopic Maxwell equations. Am. J. Phys. **38**(10), 1188–1195 (1970)
16. V.M. Agranovich, V.L. Ginzburg, *Spatial Dispersion in Crystal Optics and the Theory of Excitons* (Interscience Publishers, London, 1966)
17. D.B. Melrose, R.C. McPhedran, *Electromagnetic Processes in Dispersive Media - a Treatment Based on the Dielectric Tensor* (Cambridge University Press, New York, 1991)
18. A.D. Pierce, *Acoustics: An Introduction to Its Physical Principles and Applications*. (Acoustical Society of America, 1989)
19. L.D. Landau, E.M. Lifshitz, *Quantum Mechanics* (Pergamon Press, Oxford, 1965)
20. L.D. Landau, E.M. Lifshitz, *Statistical Physics* (Pergamon Press, Oxford, 1959)
21. M.A. Biot, Theory of propagation of elastic waves in a fluid-saturated porous solid. I. Low-frequency range. J. Acoust. Soc. Am. **28**(2), 168–178 (1956)
22. M.A. Biot, Theory of propagation of elastic waves in a fluid-saturated porous solid. II. Higher-frequency range. J. Acoust. Soc. Am. **28**(2), 179–191 (1956)

Chapter 7
Nonlocal Dynamic Homogenization of Fluid-Saturated Metamaterials

Denis Lafarge

Abstract The electromagnetic analogy introduced in the previous chapter is used here to construct an original macroscopic theory of sound propagation, allowing for both temporal and spatial dispersion, in fluid-saturated homogeneous porous media having arbitrary microstructure—including "metamaterials". The theory can be formulated for stationary random materials, periodic materials, and using different conceptions of the averaging operation (ensemble-average, volume-average). For simplicity, we have assumed that the structure is rigid and motionless, and the propagation occurs along a symmetry axis. The theory will have to be generalized to account for anisotropy, finite dimensions and frame deformations. In Appendix, we show that the preceding macroscopic descriptions in use in literature, leave aside spatial dispersion: this is a warning that the asymptotic two-scale homogenization method, often used to infer them, cannot be fully consistent.

7.1 Sound Propagation in Fluid-Saturated Rigid-Framed Porous Media

In the previous Chap. 6, we have studied sound propagation in a viscothermal fluid, and shown that it can be put in the form of Maxwellian nonlocal equations. Moreover, we have suggested that this Maxwellian nonlocal form of the equations should apply quite generally to media that can be described in macroscopic averaged sense. "Maxwellian nonlocal" here, means that the corresponding nonlocal pattern of the equations is similar to that which describes, (with only symmetry and variances interchanges, in the most general version detailed in the Appendix), macroscopic electromagnetic wave propagation in a medium in the presence of temporal and spatial dispersion. We now fix our attention to the propagation of sound waves when the fluid is permeating a porous structure, as illustrated in Fig. 7.1. We will show that at a suitable macroscopic level it can again be put in the form of Maxwellian nonlocal

D. Lafarge (✉)
Laboratoire d'Acoustique de l'Université du Mans, UMR CNRS 6613, Le Mans Université, Avenue Olivier Messiaen, 72085 Le Mans, France
e-mail: denis.lafarge@univ-lemans.fr

N. Jiménez et al. (eds.), *Acoustic Waves in Periodic Structures, Metamaterials, and Porous Media*, Topics in Applied Physics 143,
https://doi.org/10.1007/978-3-030-84300-7_7

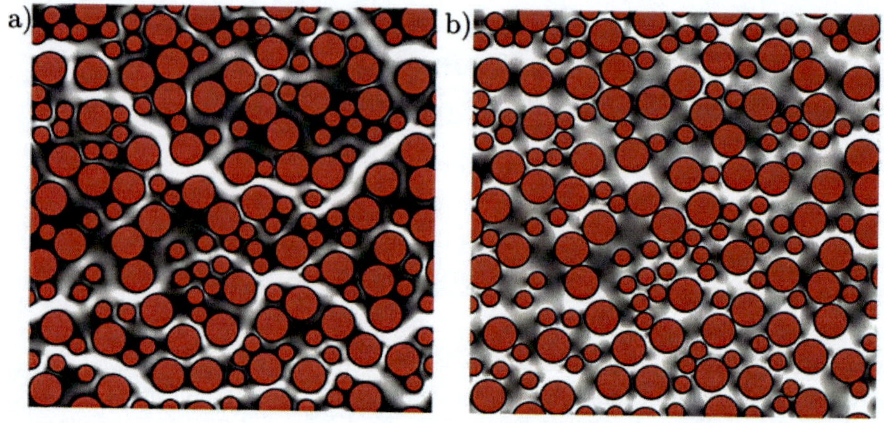

Fig. 7.1 Acoustic wave propagation in fluid-saturated rigid-framed porous media: **a** viscous-dominated low-frequency regime, **b** inertial-dominated high-frequency regime [Adapted with permission from calculations by N. Martys and E. J. Garboczi at the National Institute of Standards and Technology]

equations, and will indicate how to compute in principle the nonlocal operators. For the sake of simplicity here, we limit ourselves to materials with a, (connected or not), rigid structure, which do not move or deform during the propagation of sound in the saturating fluid. The geometry of the structure, which will determine that of the connected fluid domain, is left arbitrary except that it will appear homogeneous, (in the sense of ensemble- or volume-average), at some outer macroscopic level. In that case, and for the description of macroscopic compressional waves along a symmetry axis, we expect ending up with the same pattern of nonlocal macroscopic equations as seen in the text of Chap. 6, for the longitudinal waves in the fluid.[1]

What is well known in the literature [1], is the situation where the microstructure is simple enough to ensure that, because the macroscopic wavelengths are large relative to the dimensions of a representative volume, (scale separation), the fluid moves without divergence at the pore scale [2, 3]. This situation is generally described by applying the two-scale asymptotic method of homogenization [3, 4]. It leads to an asymptotic "local theory" description, very useful in practice. The type of geometries associated to this description is typically that represented above,[2] when the sizes of

[1] For materials with deformable structure, the expected pattern of nonlocal macroscopic equations will be the more general one seen in the Appendix of Chap. 6, but this is left for further studies.

[2] These figures are taken with permission from the electronic monograph https://concrete.nist.gov/monograph, Part III, General Random Porous Materials, Length scales relating the fluid permeability and electrical conductivity in random two-dimensional porous media, alinea: Comparison between electrical and fluid-flow problems. They represent the fluid flow, resp. in low-frequency (viscous Darcy) regime (see the velocity pattern $\boldsymbol{v}_0$ (7.189)–(7.192) in Appendix), and high-frequency (inviscid) regime (see the velocity pattern $\boldsymbol{v}_\infty$ (7.198)–(7.201) or $\boldsymbol{E}$ (7.202)–(7.205) in Appendix). The fluid flow is modelled as incompressible because long-wavelengths are assumed, *and* the microgeometry is "simple" (it does not involve widely different pore-scale lengths).

Fig. 7.2 Helmholtz resonators

grains and separations are very small compared to wavelengths. To be complete, and because of its interest in practice, we study this local theory in Appendix. We show that it is a simplification of the complete propagation problem, that makes abstraction of the spatial dispersion. For this reason, it fails, as soon as the geometry is complicated enough, (so-called metamaterials in which very different pore-scale lengths are simultaneously present), to permit that compressible movements can occur at the small-scale, in spite of scale separation. This is particularly the case when the solid includes Helmholtz resonators, as shown in Fig. 7.2, (necks and cavities have very different dimensions).

Indeed, as we will see, there is a direct link between resonances and spatial dispersion.

In what follows, by pursuing the ideas established in the Chap. 6, we develop explicitly the general principles of nonlocal theory, that will remain valid regardless of microgeometry. In comparison to the local description in Appendix, it is much more difficult to draw all the consequences from the nonlocal description. Much of this remains to be done: metamaterials will have much richer possible macroscopic behaviours than allowed by the conventional local description.

7.2 Statement of the Problem

To simplify the treatment, our medium is macroscopically homogeneous, thus boundless (see footnotes 2 and 23 in Chap. 6), and either, macroscopically isotropic, or the considered plane wave propagation occurs along a macroscopic symmetry axis x. To arrive at a precise definition of the macroscopic level, we assume that the geometric configuration is stationary random in some respects, however we will also consider the important but ambiguous case of periodic geometry. The material extends in the same stationary random way, or else periodic way, throughout all space, and is made of a solid part $\mathcal{V}_s$, shown in red colour in Fig. 7.1, (not necessarily connected), and a simply connected complementary fluid part $\mathcal{V}_f$ fully saturated with the viscothermal fluid, shown in bench of white and grey shades. The pore surface between solid and fluid is denoted $\partial\mathcal{V}$. The solid, mechanically and thermally inert, is supposed to remain perfectly still, either because it is heavy or rigid or both, or attached to external unmoving parts. Its thermal inertia also is assumed large, so that it remains at room temperature. Therefore, only the saturating fluid, which is compressible and can move, carries a disturbance. This disturbance is either due to longitudinal pressure waves coming from without, (no source in the medium), or directly created by a source-term of the type considered in the Chap. 6, (longitudinal bulk force per unit fluid volume, $\boldsymbol{f} = -\partial\mathcal{P}$, $\mathcal{P} = \tilde{\mathcal{P}}e^{-i\omega t+ikx}$, where ω and k are independently chosen[3]), and acting on the fluid. At the solid-fluid contact surfaces $\partial\mathcal{V}$, the following boundary conditions apply: the mechanical inertia of the solid and the viscous nature of the fluid, (nonzero shear viscosity), result in a no-slip cancellation condition, $\boldsymbol{v} = \mathbf{0}$, for the velocity; likewise, the thermal inertia of the solid and the thermal nature of the fluid, (nonzero thermal conduction coefficient), result in the cancellation, $\tau = 0$, of the excess temperature. In general, in response to the field coming from without, or in response to the direct source action, it will appear in the fluid, complex and phase-lagged distributions of velocities, excess temperature, excess pressure, etc.

For small-amplitude motions, their governing equations will be those of the Navier–Stokes–Fourier model discussed in Chap. 6, completed by the mentioned boundary conditions, where we put in bracket the source term, as it may, or not, be present:

$$\partial \cdot \boldsymbol{v} + \frac{\partial b}{\partial t} = 0, \qquad \text{in } \mathcal{V}_f, \tag{7.1}$$

$$\rho_0 \frac{\partial \boldsymbol{v}}{\partial t} = -\partial p + \eta \partial^2 \boldsymbol{v} + \left(\frac{\eta}{3} + \zeta\right) \partial(\partial \cdot \boldsymbol{v}) + \left[\boldsymbol{f}\right], \qquad \text{in } \mathcal{V}_f, \tag{7.2}$$

$$\gamma \chi_0 p = b + \beta_0 \tau, \qquad \text{in } \mathcal{V}_f, \tag{7.3}$$

$$\rho_0 c_P \frac{\partial \tau}{\partial t} = \beta_0 T_0 \frac{\partial p}{\partial t} + \kappa \partial^2 \tau, \qquad \text{in } \mathcal{V}_f, \tag{7.4}$$

and

$$\boldsymbol{v} = \mathbf{0}, \qquad \text{on } \partial\mathcal{V}, \tag{7.5}$$

$$\tau = 0, \qquad \text{on } \partial\mathcal{V}. \tag{7.6}$$

Note that in (7.2), while it is ultimately a matter of describing the propagation of macroscopic dilatation-compression waves in the material, we have kept the vortical viscous term $-\eta\partial \times \partial \times \boldsymbol{v}$. Indeed, whether the fluid motion is due to longitudinal pressure waves from the outside or created by a longitudinal bulk-force source term, viscous shearing movements must be taken into account, as they are automatically generated at the pore walls as a result of the application of boundary conditions (7.5).

In the Fig. 7.1, a low-frequency viscous-flow regime is represented on the left, (viscous term $\eta\partial^2\boldsymbol{v}$ much greater than inertial term $\rho_0\partial\boldsymbol{v}/\partial t$, i.e. viscous skin depth greater than pore sizes), and a high-frequency inertial-flow regime is on the right, (inertial term $\rho_0\partial\boldsymbol{v}/\partial t$ much greater than viscous term $\eta\partial^2\boldsymbol{v}$, i.e. viscous skin depth very small compared to pore sizes). In the represented domain $\mathcal{V}_f$, white colour corresponds to higher velocities, and grey colour to lower velocities. In general, the patterned, complex, and phase-lagged distributions of "high" and "low" values, will depend on both the time and the spatial variations of the macroscopic fields. The first dependence will be associated to temporal dispersion, the second to spatial

[3] Using this complex notation, we understand that $\mathcal{P} = \Re\left(\tilde{\mathcal{P}}e^{-i\omega t+i\boldsymbol{k}\cdot\boldsymbol{x}}\right)$, omitting the real part symbol $\Re\,()$.

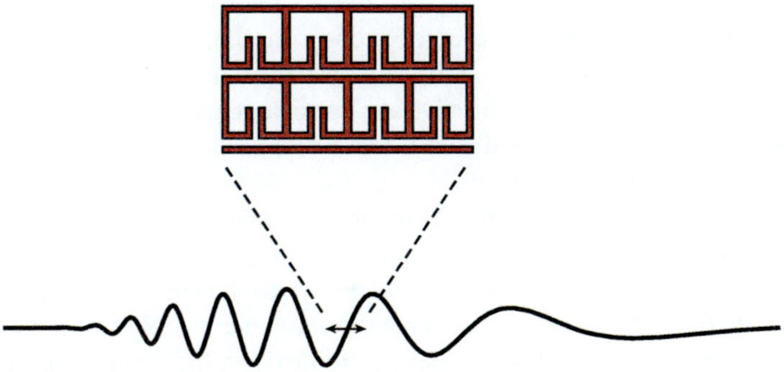

Fig. 7.3 Long-wavelength sound-field in a line of Helmholtz resonators

dispersion. Here on the two Figures 1, only the first dependence is present; the second, very feeble, is not even considered: the sound propagation is assumed to be described by the conventional approach, (local homogenization, see Appendix), which assumes *divergence-free fluid motion at the pore scale*. We mentioned that this local homogenization description falls in error in presence of Helmholtz resonators in the medium, and that this failure is linked to the failure to take spatial dispersion into account. As we saw in the Chap. 6, spatial dispersion means that the medium properties depend on the spatial variations of the external fields and it should be obvious that *this dependence cannot be described by making the simplification of locally incompressible fluid motion*.

The close relationship between the presence of resonances and the presence of spatial dispersion can be immediately understood graphically as follows. Consider a geometry with Helmholtz' resonators in succession and in which, we assume, waves having long wavelengths are established, as shown in Fig. 7.3. A Helmholtz resonance cannot occur, i.e. a significant flow cannot go to and fro an Helmholtz cavity, (red arrow in Fig. 7.4), without simultaneously being associated with significantly different flows in and out the unit cell, (black and blue arrows in Fig. 7.4), that is, without a concomitant spatial variation of the macroscopic fields. Therefore, as we cannot separate the occurrence of resonances from the occurrence of associated macroscopic spatial variations, a consistent complete treatment necessitates introducing the spatial dispersion.

The conventional description also falls in error when the wavelengths reduce and become comparable to the characteristic averaging lengths allowing to homogenize

Fig. 7.4 Link between resonances and spatial dispersion

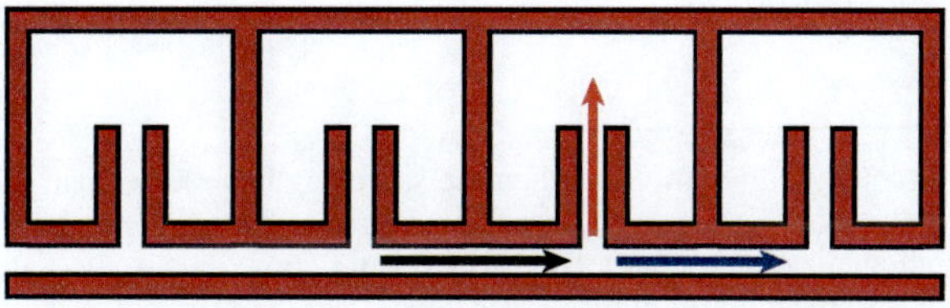

the medium. In this case also, the cornerstone simplification of the local theory, (incompressibility at the microstructure scale), is faulty, and taking spatial dispersion into account will have to be done to arrive at a consistent macroscopic theory capable to describe, in ensemble-averaged sense, the averaged fields, ("coherent fields" in multiple-scattering theory language).[4]

In short, we need a general nonlocal theory, as the local theory is insufficient at long wavelengths in resonant geometries, and in all geometries when wavelengths reduce. The generalization will have to be done outside the partly inconsistent framework of conventional two-scale homogenization. In what follows we detail how the Maxwellian nonlocal description we have introduced in the Chap. 6 in the fluid in absence of solid, and applied to longitudinal wave propagation, precisely furnishes the appropriate basis for such a generalization. This was anticipated in Appendix of Chap. 6. Here, we will explicitly show, for the case of compressional motions in fluid-saturated rigid framed materials, that the general macroscopic nonlocal theory can be put in the suggested Maxwellian nonlocal form. It will lead us to the formulation of a general Maxwellian description valid whatever microgeometries, frequencies, and wavelengths. First, in the next section and subsections, we must develop the concepts of macroscopic averages and establish their properties. While the theory is most clearly formulated for stationary random materials, in practice, however, the case of the periodical media is of interest. We will consider the two cases successively.

7.3 The Operations of Macroscopic Averaging

Remember that for the fluid we were not interested in the detailed behaviour of each molecule, but in an average macroscopic behaviour. Here, we must introduce a similar notion: we are not interested by what happens in the fluid at one particular pore-scale location, we are interested in the overall statistical macroscopic behaviour. To define it, we can use one of the two different conceptions of an average we have mentioned, Lorentz or Gibbs. The first is volume average when we are given one sample and there is scale separation between the outer scale of inhomogeneities in the medium, and the scale of considered wavelengths. The second is ensemble average when we are given an ensemble of samples, considered equivalent from a macroscopic point of view. In this case the wavelengths are not required to respect a scale separation condition, however, what is described is not what happens at a "macroscopic level" in a sample, but what happens on average in all samples, under the same conditions of excitation.

---

[4] It could be thought that when the wavelengths reduce, spatial dispersion is automatically modelled in simple microgeometries by means of the higher-order terms of the two-scale asymptotic homogenization method. But this expectation is illusory. Partially inconsequent, this method will not give meaningful higher-order terms.

Let us start with the case of materials which are in some respect stationary random and examine in more details the concept of calculating a Lorentz or Gibbs macroscopic mean.

7.3.1 *The Well-Defined Case of Stationary-Random Media*

The fields $a(t, \boldsymbol{x})$ we want to average concern what happens in the fluid phase; in the inert, immobile solid phase domain, these fields are extended to be zero. To automatically ensure it, we introduce the fluid domain indicator function $I(\boldsymbol{x})$ that takes the value 1 in the fluid and the value 0 in the solid. It will be an inherent part of all fields $a(t, \boldsymbol{x}) \equiv I(\boldsymbol{x})a(t, \boldsymbol{x})$ we consider.

7.3.1.1 The Definitions of Lorentz's and Gibbs' Averages

In Lorentz's averaging conception we are dealing with one sample; we view its indicator function I as a stationary random function of position. For I, we may think of some function devoid of a preferred origin in space, that may repeat itself but without true periodicity and with some inherent randomness. In Gibbs' averaging conception, we are given infinitely many such stationary random samples, or realizations ϖ of the medium, taken from a probability space Ω, the ensemble of which defines the homogeneous macroscopic medium in question. As an example, we can imagine that, viewing the Lorentz sample from infinitely many different origins in space, would produce a suitable Gibbs' ensemble. The fluid domain indicator function I is noted, respectively:

Lorentz Gibbs

$$I_0(\boldsymbol{x}) = \begin{cases} 1 & \boldsymbol{x} \in \mathcal{V}_f, \\ 0 & \boldsymbol{x} \in \mathcal{V}_s, \end{cases} \qquad \forall \varpi \in \Omega,\ I(\boldsymbol{x}, \varpi) = \begin{cases} 1 & \boldsymbol{x} \in \mathcal{V}_f(\varpi), \\ 0 & \boldsymbol{x} \in \mathcal{V}_s(\varpi), \end{cases} \tag{7.7}$$

(we put index 0 in Lorentz's case to insist that we have only one configuration). We first discuss Lorentz's average.

Lorentz's Average

The Lorentz volume-average $\langle a \rangle(t, \boldsymbol{x})$ of a given field $a(t, \boldsymbol{x})$ such as a fluid velocity component, the condensation, the pressure, etc., is best defined in Russakoff's manner [5] by convolution with a smooth and finite-width test function $w_{L_h}(\boldsymbol{x})$ of characteristic extent L_h, (an homogenization length, giving the size of a representative elementary volume or REV), such as e.g. $w_{L_h}(\boldsymbol{x}) = (\pi L_h^2)^{-3/2} e^{-x^2/L_h^2}$, centred at the considered location, ($\boldsymbol{x} = \boldsymbol{0}$ in the given function), and normalized to one upon

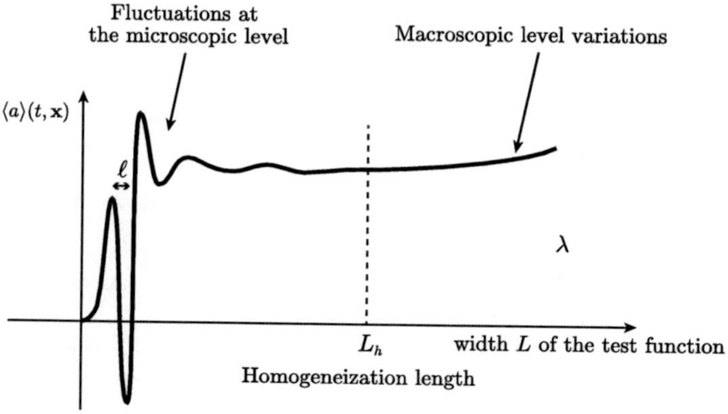

Fig. 7.5 Homogenization length for Lorentz's average

integrating over whole space. The shape of the test function is not essential, and a natural choice is that of a plateau function as discussed and plotted in Jackson [6].

The average is then defined as the following convolution product:

$$A(t, \boldsymbol{x}) = \langle a\rangle(t, \boldsymbol{x}) = \left(w_{L_h} * a\right)(t, \boldsymbol{x}) = \int d\boldsymbol{x}' w_{L_h}(\boldsymbol{x} - \boldsymbol{x}')a(t, \boldsymbol{x}'),$$
$$\int d\boldsymbol{x}' w_{L_h}(\boldsymbol{x} - \boldsymbol{x}') = 1. \tag{7.8}$$

Recall that the indicator function $I_0(\boldsymbol{x})$ is present in the field $a(t, \boldsymbol{x})$. When convenient, (see e.g. the calculation in (7.16)), it can be explicitly noted as an additional factor $I_0(\boldsymbol{x}')$, in the integral (7.8.1), with no change on the result of the integration.

If we take w_L with arbitrary L for the test function, the way the average varies in function of the size L of the averaging window, is schematized in Fig. 7.5.

The mean $\langle a\rangle(t, \boldsymbol{x})$ first presents fluctuations when L is on the order of typical small-scale distances ℓ over which the fields vary. It then smooths out and reaches a plateau when $L \sim L_h$, with $\ell \ll L_h \ll \lambda$, where λ is an estimate of macroscopic wavelengths. Finally, because of the macroscopic variations, it starts to again vary when L is further increased to become comparable to λ. The length L_h defines the suitable "homogenization" length or appropriate size of a REV. As it is assumed sufficiently large compared to ℓ, so that the average tends to a definite limit, and sufficiently small compared to typical wavelengths λ, so that the macroscopic variations of the fields are not polluting the result, the introduction of this homogenization length L_h supposes a wide scale separation $\ell \lll \lambda$.

As we use the total volume normalization (7.8.2), the mean of the characteristic function I is just the porosity ϕ, (fluid volume per unit total—fluid plus solid—volume):

$$\langle I_0\rangle = \int d\boldsymbol{x}' w_{L_h}(\boldsymbol{x} - \boldsymbol{x}')I_0(\boldsymbol{x}') = \phi. \tag{7.9}$$

Because of this normalization, the definition (7.8) of the averaging symbol $\langle\,\rangle$ interprets as a "total volume" average: porosity times the mean value in the fluid phase. The mean value in the fluid phase will be denoted by an index f putted on the mean symbol, so that for example, we will have

$$\langle \boldsymbol{v} \rangle = \phi \langle \boldsymbol{v} \rangle_f. \tag{7.10}$$

The mean operation in the fluid phase $\langle\,\rangle_f$ can be defined by a convolution as noted in (7.8.1), but with the following different normalization of the test function:

$$\int d\boldsymbol{x}' w_{L_h}(\boldsymbol{x} - \boldsymbol{x}') = 1/\phi, \tag{7.11}$$

i.e. an extra factor $1/\phi$ is put on the test function, so that as wanted

$$\langle I_0 \rangle_f = 1. \tag{7.12}$$

Gibbs' Average

Let us now consider the Gibbs-average $\langle a \rangle(t, \boldsymbol{x})$ of a given field $a(t, \boldsymbol{x}; \varpi)$ such as a fluid velocity component, the condensation, the pressure, etc., which is a function of $t, \boldsymbol{x}$ in each realization ϖ, and is extended to zero in $\mathcal{V}_s(\varpi)$. This average is simply the expectation value $\overline{a(t, \boldsymbol{x}; \varpi)}$ of the given field a, at time t, position $\boldsymbol{x}$, over all realizations ϖ in Ω:

$$A(t, \boldsymbol{x}) = \langle a \rangle(t, \boldsymbol{x}) = \overline{a(t, \boldsymbol{x}; \varpi)}. \tag{7.13}$$

Because the field $a(t, \boldsymbol{x}; \varpi)$ is extended to zero in the solid, this average over realizations (7.13) can be seen as a "total volume" average. As an example, looking at the values of the field $I(\boldsymbol{x}; \varpi)$ at a given position $\boldsymbol{x}$, and taking the realization ϖ at random, we have a probability ϕ to be in the fluid, and thus find the value 1, and a probability $1 - \phi$ to be in the solid, and thus find the value 0. Therefore

$$\langle I \rangle = \overline{I(\boldsymbol{x}; \varpi)} = \phi. \tag{7.14}$$

Also note that, to compute the Gibbs expectation values (7.13), only the fields $a(t, \boldsymbol{x}; \varpi)$ at a single (pointlike) position $\boldsymbol{x}$ in the different realizations ϖ are involved. When performing the average (7.13), we a priori deal with discontinuous functions as the functions $I(\boldsymbol{x}, \varpi)$ are discontinuous over $\boldsymbol{x}$, (for example the functions associated to the condensation or the excess pressure, extended to zero in the solid, are discontinuous at the pore-wall surfaces; velocity and excess temperature are not discontinuous because they vanish on $\partial\mathcal{V}(\varpi)$). This discontinuous nature of the fields brings unessential mathematical difficulties when willing to see how their Gibbs' averages vary in space. To avoid this, it is convenient to replace single points by infinitesimal volumes, or more precisely, replace the possibly discontinuous set of values $a(t, \boldsymbol{x}; \varpi)$ by the set of continuous mean values, $\int d\boldsymbol{x}' w_{dL}(\boldsymbol{x} - \boldsymbol{x}') a(t, \boldsymbol{x}'; \varpi)$, performed in a "physically infinitesimal" neighbourhood of $\boldsymbol{x}$, and where $I(\boldsymbol{x}'; \varpi)$ is inherent part of the field $a(t, \boldsymbol{x}'; \varpi)$. We consider dL as a much smaller length than all the lengths involved in defining geometry such as the radii of curvatures associated with pore wall surface $\partial\mathcal{V}$, and use the normalization $\int d\boldsymbol{x}' w_{dL}(\boldsymbol{x} - \boldsymbol{x}') = 1$, (at $dL = 0$, w_{dL} will be the delta-function). Following this purely technical refinement, we redefine more conveniently the Gibbs expectation value $\langle a \rangle(t, \boldsymbol{x})$ (7.13) as:

$$A(t, \boldsymbol{x}) = \langle a \rangle(t, \boldsymbol{x}) = \overline{\int d\boldsymbol{x}' w_{dL}(\boldsymbol{x} - \boldsymbol{x}') a(t, \boldsymbol{x}'; \varpi)}. \tag{7.15}$$

Finally, also willing, with Gibbs' conception, to dispose of a mean operation $\langle . \rangle_f$ having "fluid volume" instead of "total volume" normalization, we define Gibbs' fluid volume average such that $\langle a \rangle_f(t, \boldsymbol{x})$ is given by (7.15), but with normalization $\int d\boldsymbol{x}' w_{dL}(\boldsymbol{x} - \boldsymbol{x}') = 1/\phi$ of the test function w_{dL}, (at $dL = 0$, it gives for the test function the delta-function $\times$ $1/\phi$).

7.3.1.2 Commutation Relations

We now precise how the above Lorentz's and Gibbs' averaging symbols $\langle \cdot \rangle(\boldsymbol{x})$, (7.8) and (7.13) refined in the form (7.15), behave with respect to the spatial derivative symbol $\partial = \partial/\partial\boldsymbol{x}$. We show that, in general, for fields nonzero on the pore walls, the averaging do not commute with the spatial derivative.

Lorentz's Average

With Lorentz's average, the spatial derivative symbol refers to the variation of the central position $\boldsymbol{x}$ of the test function. Previously in electromagnetic theory, because the fields extended all over space, there was direct commutation between the Lorentz-Russakoff average and the spatial derivative symbol, (see the Chap. 6, Sect. 11.2, (79)). Here, because the fields are set to zero in the solid, it is convenient to substitute them explicitly, in the definition (7.8), in the form of, $a(t, \boldsymbol{x}') = I_0(\boldsymbol{x}') a(t, \boldsymbol{x}')$. As detailed below an extra surface term then generally appears in the commutation relation between average and derivative symbol:

$$
\begin{aligned}
\partial A(t,\boldsymbol{x}) = \frac{\partial}{\partial \boldsymbol{x}}\langle a\rangle(t,\boldsymbol{x}) &= \int d\boldsymbol{x}' \frac{\partial}{\partial \boldsymbol{x}} w_{L_h}(\boldsymbol{x}-\boldsymbol{x}') I_0(\boldsymbol{x}') a(t,\boldsymbol{x}'),\\
&= -\int d\boldsymbol{x}' \frac{\partial}{\partial \boldsymbol{x}'}\left[w_{L_h}(\boldsymbol{x}-\boldsymbol{x}')\right] I_0(\boldsymbol{x}') a(t,\boldsymbol{x}'),\\
&= -\int d\boldsymbol{x}' \frac{\partial}{\partial \boldsymbol{x}'}\left[w_{L_h}(\boldsymbol{x}-\boldsymbol{x}') I_0(\boldsymbol{x}') a(t,\boldsymbol{x}')\right]\\
&\quad + \int d\boldsymbol{x}' w_{L_h}(\boldsymbol{x}-\boldsymbol{x}') \frac{\partial}{\partial \boldsymbol{x}'}\left[I_0(\boldsymbol{x}')\right] a(t,\boldsymbol{x}')\\
&\quad + \int d\boldsymbol{x}' w_{L_h}(\boldsymbol{x}-\boldsymbol{x}') I_0(\boldsymbol{x}') \frac{\partial}{\partial \boldsymbol{x}'} a(t,\boldsymbol{x}').
\end{aligned}
\tag{7.16}
$$

The integral of the total derivative term vanishes because the quantity inside brackets contains the test function which quickly tends to zero at infinity. The gradient $\partial\left[I_0(\boldsymbol{x}')\right]/\partial\boldsymbol{x}'$ is a Dirac delta distribution, supported by the pore surface $\partial\mathcal{V}$, and directed along $-\hat{\boldsymbol{n}}(\boldsymbol{x}')$, where $\hat{\boldsymbol{n}}(\boldsymbol{x}')$ is the outward normal to the fluid region at position $\boldsymbol{x}'$ on this pore surface. The integral containing it, expresses as a pore-surface integral. The last integral is the definition of the Lorentz mean of the derived field, (which is extended to zero in the solid). Therefore we find:

$$
\begin{aligned}
\partial A(t,\boldsymbol{x}) =& \partial\langle a\rangle(t,\boldsymbol{x}) = \langle \partial a\rangle(t,\boldsymbol{x}) + \langle a\partial I_0\rangle(t,\boldsymbol{x}),\\
=& \int d\boldsymbol{x}' w_{L_h}(\boldsymbol{x}-\boldsymbol{x}')\partial_{\boldsymbol{x}'} a(t,\boldsymbol{x}') - \int_{\partial\mathcal{V}} d\boldsymbol{x}' a(t,\boldsymbol{x}')\hat{\boldsymbol{n}}(\boldsymbol{x}') w_{L_h}(\boldsymbol{x}-\boldsymbol{x}').
\end{aligned}
\tag{7.17}
$$

This relation is sometimes called the "averaging theorem". Note that, as the material is assumed macroscopically homogeneous, we have $\partial\phi = \boldsymbol{0}$, that is

$$
\int_{\partial\mathcal{V}} d\boldsymbol{x}' \hat{\boldsymbol{n}}(\boldsymbol{x}') w_L(\boldsymbol{x}-\boldsymbol{x}') = \boldsymbol{0}. \tag{7.18}
$$

This identity (7.18) will be useful later on. It can be obtained either by applying the averaging theorem (7.17) for $a = 1$ in the fluid, or else, by directly taking the derivative of (7.9):

$$
\begin{aligned}
\frac{\partial}{\partial \boldsymbol{x}}\phi = \int d\boldsymbol{x}' \frac{\partial}{\partial \boldsymbol{x}} w_{L_h}(\boldsymbol{x}-\boldsymbol{x}') I_0(\boldsymbol{x}') &= -\int d\boldsymbol{x}' \frac{\partial}{\partial \boldsymbol{x}'}\left[w_{L_h}(\boldsymbol{x}-\boldsymbol{x}')\right] I_0(\boldsymbol{x}'),\\
= -\int d\boldsymbol{x}' \frac{\partial}{\partial \boldsymbol{x}'}\left[w_{L_h}(\boldsymbol{x}-\boldsymbol{x}') I_0(\boldsymbol{x}')\right] &+ \int d\boldsymbol{x}' w_{L_h}(\boldsymbol{x}-\boldsymbol{x}') \frac{\partial}{\partial \boldsymbol{x}'} I_0(\boldsymbol{x}'),\\
= -\int_{\partial\mathcal{V}} d\boldsymbol{x}' \hat{\boldsymbol{n}}(\boldsymbol{x}') w_L(\boldsymbol{x}-\boldsymbol{x}') = \boldsymbol{0}.&
\end{aligned}
$$

Gibbs' Average

With Gibbs' average (7.13) refined in the form (7.15), the calculation of the spatial derivative of macroscopic averages is performed in same manner. Starting with

$$\partial A(t,\boldsymbol{x}) = \partial\langle a\rangle(t,\boldsymbol{x}) = \overline{\int d\boldsymbol{x}' \frac{\partial}{\partial \boldsymbol{x}} w_{dL}(\boldsymbol{x}-\boldsymbol{x}') I(\boldsymbol{x}';\varpi) a(t,\boldsymbol{x}';\varpi)},$$

(the I-function factor is made apparent for convenience in the calculation), using $\partial w_{dL}(\boldsymbol{x}-\boldsymbol{x}')/\partial\boldsymbol{x} = -\partial w_{dL}(\boldsymbol{x}-\boldsymbol{x}')/\partial\boldsymbol{x}'$, and integrating by parts, we obtain:

$$\begin{aligned}\partial A(t,\boldsymbol{x}) =& \partial\langle a\rangle(t,\boldsymbol{x}) = \langle\partial a\rangle(t,\boldsymbol{x}) + \langle a\partial I\rangle(t,\boldsymbol{x}),\\ =& \overline{\int d\boldsymbol{x}' w_{dL}(\boldsymbol{x}-\boldsymbol{x}')\partial a(t,\boldsymbol{x}';\varpi)}\\ &- \overline{\int_{\partial\mathcal{V}(\varpi)} d\boldsymbol{x}' a(t,\boldsymbol{x}';\varpi)\hat{\boldsymbol{n}}(\boldsymbol{x}';\varpi) w_{dL}(\boldsymbol{x}-\boldsymbol{x}')}.\end{aligned} \tag{7.19}$$

Discussion

As in Lorentz's or Gibbs' conceptions the symbols $\langle\cdot\rangle$ and ∂ obey the relations (7.17) or (7.19), they do not commute except if the surface integral terms vanish. To take different examples, the gradient of a mean field such as $B \equiv \langle b\rangle$ or $\langle p\rangle$, is, in general, not equal to the mean of the gradient, i.e.

$$\partial B \equiv \partial\langle b\rangle \neq \langle\partial b\rangle, \qquad \text{or} \qquad \partial\langle p\rangle \neq \langle\partial p\rangle. \tag{7.20}$$

Because, however, the velocity vanishes on the solid-fluid interface, (boundary condition (7.5)), the divergence of the mean velocity is always automatically the same as the mean of the velocity divergence:

$$\partial\cdot\langle\boldsymbol{v}\rangle = \langle\partial\cdot\boldsymbol{v}\rangle. \tag{7.21}$$

This remains true for an inviscid fluid: the normal component of the velocity vanishes on the fluid-solid interface as long as the solid is impenetrable, which ensures cancellation of the surface term.

In addition, there is a generic class of fields a, for which, as the surface terms happen to vanish, (while a is nonzero on the interface), the average operation commute with the derivative operation:

$$\langle\partial a\rangle = \partial\langle a\rangle. \tag{7.22}$$

This class is that of fields that would vary, (except for their systematic extension to zero in the solid), like "macroscopic fields" or "external fields". For example, the source-term field $\boldsymbol{f}$ in (7.2), or its associated potential $\mathcal{P}$, or any field a that is the macroscopic mean, $a \equiv \langle b \rangle$, of some response field b in the pore space, (and is extended to zero in the solid), are this same type of fields, verifying (7.22). Let us show it, successively using Lorentz's or Gibbs' averaging conception.

With Lorentz's average, scale separation needs to be assumed. Therefore, the fields which, (apart from their extension to zero in the solid), vary only at the macroscopic scale, present almost linear variations in the fluid within an averaging volume. In first approximation, they write, (in the fluid): $a(t, \boldsymbol{x}') = a_0 + a_1 \hat{\boldsymbol{n}}_0 \cdot \boldsymbol{x}'$, with $\hat{\boldsymbol{n}}_0$ the unit direction of the gradient, and $a_{0,1}$ two constants. As the medium is stationary random, it can then be shown that such linear variations automatically cancel the surface term:

$$\int_{\partial \mathcal{V}} d\boldsymbol{x}' a(t, \boldsymbol{x}') \hat{\boldsymbol{n}}(\boldsymbol{x}') w_{L_h}(\boldsymbol{x} - \boldsymbol{x}') =$$

$$= \int_{\partial \mathcal{V}} d\boldsymbol{x}' \left[a_0 + a_1 \hat{\boldsymbol{n}}_0 \cdot \boldsymbol{x}' \right] \hat{\boldsymbol{n}}(\boldsymbol{x}') w_{L_h}(\boldsymbol{x} - \boldsymbol{x}') = \boldsymbol{0},$$

and thus, (7.22) is satisfied. Indeed by (7.18) the first integral identically vanishes and it remains to show the additional identity

$$\int_{\partial \mathcal{V}} d\boldsymbol{x}' \left[\hat{\boldsymbol{n}}_0 \cdot \boldsymbol{x}' \right] \hat{\boldsymbol{n}}(\boldsymbol{x}') w_{L_h}(\boldsymbol{x} - \boldsymbol{x}') = \boldsymbol{0}. \tag{7.23}$$

By symmetry reasons, the mean of a purely linear variation should be a linear variation, modified by the porosity factor to account for the total volume normalization:

$$\langle a_0 + a_1 \hat{\boldsymbol{n}}_0 \cdot \boldsymbol{x}' \rangle (\boldsymbol{x}) = \phi \left(a_0 + a_1 \hat{\boldsymbol{n}}_0 \cdot \boldsymbol{x} \right).$$

Taking the spatial derivative this gives

$$\partial \langle \hat{\boldsymbol{n}}_0 \cdot \boldsymbol{x}' \rangle (\boldsymbol{x}) = \phi \hat{\boldsymbol{n}}_0.$$

But we also evidently have, by averaging the identity $\partial_{\boldsymbol{x}'} (\boldsymbol{n}_0 \cdot \boldsymbol{x}') = \boldsymbol{n}_0$,

$$\langle \partial_{\boldsymbol{x}'} (\boldsymbol{n}_0 \cdot \boldsymbol{x}') \rangle (\boldsymbol{x}) = \langle \boldsymbol{n}_0 \rangle = \phi \boldsymbol{n}_0.$$

Therefore, the averaging theorem (7.17) applied to the field $a = \boldsymbol{n}_0 \cdot \boldsymbol{x}$, yields the identity (7.23). This completes the proof and justifies (7.22) in Lorentz's averaging conception, for an arbitrary "external" field a, i.e. a field having only "long-wavelength" variations, (except for its extension to zero in solid).

With Gibbs' average, saying that the field a varies like "macroscopic" fields, means that, $a(t, \boldsymbol{x}; \varpi) \equiv I(\boldsymbol{x}; \varpi) a(t, \boldsymbol{x})$. In all realizations this field in the fluid is given by a single $a(t, \boldsymbol{x})$, and the latter writes as a superposition of exponentials $e^{i\boldsymbol{k}\cdot\boldsymbol{x}}$ with associated Fourier coefficients independent of ϖ. The Gibbs surface terms $\int_{\partial \mathcal{V}(\varpi)} d\boldsymbol{x}' a(t, \boldsymbol{x}'; \varpi) \hat{\boldsymbol{n}}(\boldsymbol{x}'; \varpi) w_{dL}(\boldsymbol{x} - \boldsymbol{x}')$ will have contributions

$\int_{\partial\mathcal{V}(\varpi)} dx' e^{ik\cdot x'}\hat{n}(x';\varpi)w_{dL}(x-x')$ in front of the Fourier coefficients and these vanish whatever the value of k. To see this, we note that, realization after realization, the exponential variations $e^{ik\cdot x'}$ automatically become slow variations in the limit $dL \to 0$. They thus factor out as the central value $e^{ik\cdot x}$ in front of an integral $\int_{\partial\mathcal{V}(\varpi)} dx'\hat{n}(x';\varpi)w_{dL}(x-x')$ which is identically zero. Indeed, at any given fixed value dL, it cancels out when making the ensemble averaging, because, as small as can be dL, the mean of the integral over realizations becomes, (when the number of realizations increases indefinitely), representative of the integral performed in the full medium, (that is, made in one sample with a value of L equal to homogenization length L_h); but this integral is the null integral (7.18), hence the cancellation of Gibbs' surface terms. It justifies (7.22) in Gibbs' conception, for an arbitrary "macroscopic" or "external" field a.

Now that the notions of Lorentz's and Gibbs' averages and some of their properties have been precised, let us return to the problem (7.1)–(7.6), alternatively stated, solved, and averaged, using Lorentz's and Gibbs' conceptions.

7.3.1.3 Ergodic Equivalences

We first argue that, at long wavelengths, the two conceptions can be used interchangeably, manifesting a property of ergodicity.

By the principle of superposition it suffices to consider and solve the problem of the response of the fluid to a longitudinal force, $f = -\partial\mathcal{P}$, given in the form of a single plane wave variation, $\mathcal{P} = \tilde{\mathcal{P}}e^{-i\omega t+ik\cdot x}$, with arbitrary ω, and k, respecting long-wavelength condition, $\Re(k_i L_h), \Im(k_i L_h) \ll 1$, $(i = 1, 2, 3)$, a prerequisite to the definition of Lorentz's average.

In the Lorentz formulation, we are given a single sample, stationary random. We call it the reference sample ϖ_0 and denote its indicator function $I(x;\varpi_0) = I_0(x)$; it determines the fluid domain $\mathcal{V}_f$, the solid-fluid interface $\partial\mathcal{V}$, and the response solution $v(t, x;\varpi_0) = v(\omega, k, x;\varpi_0)e^{-i\omega t+ik\cdot x}$, $b(t, x;\varpi_0) = b(\omega, k, x;\varpi_0)e^{-i\omega t+ik\cdot x}$, etc., of the motion equations (7.1)–(7.6), taken with the above source term. An important point is that the amplitudes in front of the exponentials, proportional to the source amplitude constant $\tilde{\mathcal{P}}$, are uniquely determined, bounded functions of x. These bounded functions are also stationary random functions.

In the Gibbs formulation, we are given an ensemble of stationary random samples $\varpi \in \Omega$, defined by their indicator functions $I(x;\varpi)$. These determine a collection of fluid domains $\mathcal{V}_f(\varpi)$, solid-fluid interfaces $\partial\mathcal{V}(\varpi)$, and response solutions $v(t, x;\varpi) = v(\omega, k, x;\varpi)e^{-i\omega t+ik\cdot x}$, $b(t, x;\varpi) = b(\omega, k, x;\varpi)e^{-i\omega t+ik\cdot x}$, etc., of the motion equations (7.1)–(7.6), with source term as above. The amplitudes in front of the exponentials, proportional to $\tilde{\mathcal{P}}$, are unique stationary random and bounded functions of x.

In the macroscopic theory we are concerned with macroscopic averages of the fields or product of fields, such as $\langle v\rangle$, $\langle b\rangle$ or $\langle pv\rangle$. As an example of the general principle we consider the mean $\langle v\rangle$.

First consider Lorentz's formulation. We have

$$\begin{aligned} \boldsymbol{V}(t,\boldsymbol{x}) = \langle \boldsymbol{v}\rangle(t,\boldsymbol{x}) &= \int d\boldsymbol{x}' w_{L_h}(\boldsymbol{x}-\boldsymbol{x}')\boldsymbol{v}(t,\boldsymbol{x}';\varpi_0),\\ &= \int d\boldsymbol{x}' w_{L_h}(\boldsymbol{x}-\boldsymbol{x}')\boldsymbol{v}(\omega,\boldsymbol{k},\boldsymbol{x}';\varpi_0)e^{-i\omega t+i\boldsymbol{k}\cdot\boldsymbol{x}'}. \end{aligned}$$

Now, as a scale separation (long-wavelength limit) is imposed, the exponential $e^{i\boldsymbol{k}\cdot\boldsymbol{x}'}$, which varies slowly in the averaging $\boldsymbol{x}'$ region defined by the test function $w_{L_h}(\boldsymbol{x}-\boldsymbol{x}')$, can be nearly replaced by its central value $e^{i\boldsymbol{k}\cdot\boldsymbol{x}}$ and extracted from the integral. Hence we will have, nearly

$$\begin{aligned} \boldsymbol{V}(t,\boldsymbol{x}) &= e^{-i\omega t+i\boldsymbol{k}\cdot\boldsymbol{x}} \int d\boldsymbol{x}' w_{L_h}(\boldsymbol{x}-\boldsymbol{x}')\boldsymbol{v}(\omega,\boldsymbol{k},\boldsymbol{x}';\varpi_0),\\ &= e^{-i\omega t+i\boldsymbol{k}\cdot\boldsymbol{x}}\langle \boldsymbol{v}\rangle_{Lor}, \end{aligned}$$

where the index Lor on the average $\langle \boldsymbol{v}\rangle$ indicates that it is Lorentz's average.

Next consider Gibbs' formulation. The mean $\langle \boldsymbol{v}\rangle$ writes

$$\begin{aligned} \boldsymbol{V}(t,\boldsymbol{x}) = \langle \boldsymbol{v}\rangle(t,\boldsymbol{x}) &= \overline{\int d\boldsymbol{x}' w_{dL}(\boldsymbol{x}-\boldsymbol{x}')\boldsymbol{v}(t,\boldsymbol{x}';\varpi)},\\ &= \overline{\int d\boldsymbol{x}' w_{dL}(\boldsymbol{x}-\boldsymbol{x}')\boldsymbol{v}(\omega,\boldsymbol{k},\boldsymbol{x}';\varpi)e^{-i\omega t+i\boldsymbol{k}\cdot\boldsymbol{x}'}}. \end{aligned}$$

In the limit $dL \to 0$, the exponential automatically factors out as the central value $e^{-i\omega t+i\boldsymbol{k}\cdot\boldsymbol{x}}$ whatever $\boldsymbol{k}$ (long or short wavelengths) and we find, this time without approximation

$$\begin{aligned} \boldsymbol{V}(t,\boldsymbol{x}) &= e^{-i\omega t+i\boldsymbol{k}\cdot\boldsymbol{x}} \overline{\int d\boldsymbol{x}' w_{dL}(\boldsymbol{x}-\boldsymbol{x}')\boldsymbol{v}(\omega,\boldsymbol{k},\boldsymbol{x}';\varpi)},\\ &= e^{-i\omega t+i\boldsymbol{k}\cdot\boldsymbol{x}}\langle \boldsymbol{v}\rangle_{Gib}, \end{aligned}$$

where the index Gib on the average $\langle \boldsymbol{v}\rangle$ reminds that it is Gibbs' average.

Now, invoking the stationary random character of the geometries, it should be the same thing to perform the ensemble-average $\langle \boldsymbol{v}\rangle_{Gib}$ or to perform the volume-average $\langle \boldsymbol{v}\rangle_{Lor}$, because, as small as can be dL, the mean of the integral over realizations becomes representative of the full-medium Lorentz integral, when the number of realizations increases indefinitely. It means that, at long-wavelengths, the Gibbs ensemble average will be equivalent to the Lorentz average in one realization; the two will be used interchangeably.

We can go a step further. While Lorentz's method of averaging loses its physical meaning when the wavelengths reduce sufficiently, the Gibbs method remains feasible. As mentioned, in this case the macroscopic theory will not describe what happens in one sample, but what happens on average in the ensemble of realizations.

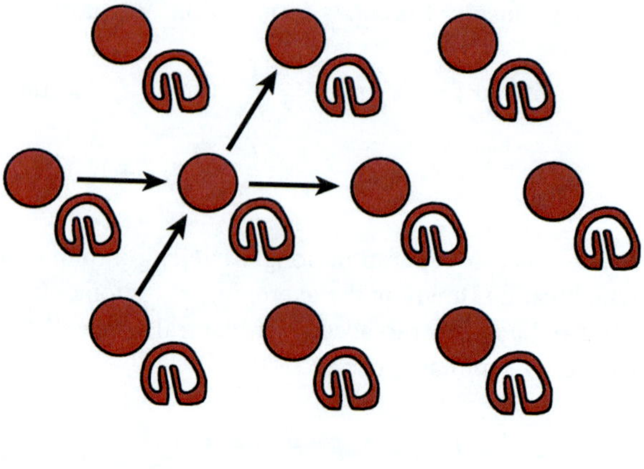

Fig. 7.6 Periodic sample ϖ_0

Now, whatever $\boldsymbol{k}$ is long- or short-wavelengths, for the fields created by the potential $\mathcal{P} = \tilde{\mathcal{P}} e^{-i\omega t + i\boldsymbol{k}\cdot\boldsymbol{x}}$, we always have the right to write Gibbs' average as

$$\boldsymbol{V}(t, \boldsymbol{x}) = \langle \boldsymbol{v} \rangle (t, \boldsymbol{x}) = e^{-i\omega t + i\boldsymbol{k}\cdot\boldsymbol{x}} \langle \boldsymbol{v} \rangle_{Gib} = e^{-i\omega t + i\boldsymbol{k}\cdot\boldsymbol{x}} \langle \boldsymbol{v} \rangle_{Lor}. \tag{7.24}$$

because the ergodicity property $\langle \boldsymbol{v} \rangle_{Gib} = \langle \boldsymbol{v} \rangle_{Lor}$ seen above is not linked to the size of wavelengths but to the stationary random nature of the geometry. Therefore, in general, when making Gibbs' average, we can always choose to use Lorentz's average to perform the average of the tilde part of the fields.[5]

7.3.2 *The Ambiguous Case of Periodic Media*

We now consider the case of periodic materials as well. Obviously, the idea of a periodicity clashes with the preceding idea of a macroscopic homogeneity obtained through underlying uniform randomness. In this context, similar averaging operations and properties as mentioned in the previous section, can now only be obtained *cum grano salis*. Because of the periodicity, ambiguities will appear in the definitions, which will complicate the presentation (Fig. 7.6).[6]

[5] By "tilde part", we mean, to take the example of the velocity field which writes $\boldsymbol{v}(t, \boldsymbol{x}; \varpi) = \boldsymbol{v}(\omega, \boldsymbol{k}, \boldsymbol{x}; \varpi) e^{-i\omega t + i\boldsymbol{k}\cdot\boldsymbol{x}}$, the amplitude in the right-hand side, in front of the exponential; usually it is denoted with a tilde which we suppress here for the simplicity of the notation. In the equality $\langle \boldsymbol{v} \rangle_{Gib} = \langle \boldsymbol{v} \rangle_{Lor}$, the $\boldsymbol{v}$ are the tilde parts. At short wavelengths $\langle \boldsymbol{v} \rangle_{Gib}(t, \boldsymbol{x})$ and $\langle \boldsymbol{v} \rangle_{Lor}(t, \boldsymbol{x})$, with $\boldsymbol{v}$ the original space velocities, are not the same; the last average $\langle \boldsymbol{v} \rangle_{Lor}(t, \boldsymbol{x})$ in general will have no precise utility or significance.

[6] Ultimately, however, the theory will have to deal with media having finite dimensions. In this case, true periodicity will be lost, which will hopefully help mitigate the mentioned ambiguities.

7.3.2.1 The Definitions of Lorentz's and Gibbs' Averages

In Lorentz's averaging conception, we are, (as we make abstraction of finite dimensions), dealing with one indefinite sample ϖ_0 of the medium, which is now assumed periodic. Thus its indicator function $I(\boldsymbol{x}, \varpi_0) = I_0(\boldsymbol{x})$ verifies:

$$I_0\left(\boldsymbol{x} + \sum_{i=1}^{3} m_i \boldsymbol{a}_i\right) = I_0(\boldsymbol{x}), \quad \forall \boldsymbol{x}, \quad \forall m_{1,2,3} \in \mathbb{Z},$$

with $\boldsymbol{a}_{1,2,3}$, one of the possible equivalent choices of three primitive translation vectors, leaving unchanged the sample. Also willing to introduce a Gibbs' averaging conception, we would like to have infinitely many periodic samples, or realizations ϖ of the "same macroscopic medium", taken from a probability space Ω, the ensemble of which defines the homogeneous macroscopic medium in question. A natural idea that comes to mind is that at the macroscopic level we will not pay attention to the underlying positioning in space of the crystal, so that the "different realizations ϖ of the same macroscopic medium" can be defined here as produced by the infinitely many random translations of the original realization ϖ_0. To characterize them we introduce a random translation vector $\boldsymbol{X}$

$$\boldsymbol{X} = \sum_{i=1}^{3} \xi_i \boldsymbol{a}_i,$$

with each of the ξ_i, a random variable uniformly distributed in $\left[-\frac{1}{2}, \frac{1}{2}\right]$. Accordingly, the fluid domain periodic indicator function will be:

Lorentz	Gibbs	
		(7.25)
$I_0(\boldsymbol{x}) = \begin{cases} 1 & \boldsymbol{x} \in \mathcal{V}_f \\ 0 & \boldsymbol{x} \in \mathcal{V}_s \end{cases}$	$\forall \varpi \in \Omega,\ I(\boldsymbol{x}, \varpi) = \begin{cases} 1 & \boldsymbol{x} \in \mathcal{V}_f(\varpi) \\ 0 & \boldsymbol{x} \in \mathcal{V}_s(\varpi) \end{cases}$	(7.26)
$I_0(\boldsymbol{x}) = I_0(\boldsymbol{x} + \sum_{i=1}^{3} m_i \boldsymbol{a}_i)$	$I(\boldsymbol{x}, \varpi) = I_0(\boldsymbol{x} - \boldsymbol{X}), \quad \boldsymbol{X} = \sum_{i=1}^{3} \xi_i \boldsymbol{a}_i$	(7.27)
$m_{1,2,3} \in \mathbb{Z}$	$\xi_{1,2,3} \in \left[-\frac{1}{2}, \frac{1}{2}\right]$	(7.28)

Lorentz's Average

Because the underlying geometry is periodic, we can define it by giving $I_0(\boldsymbol{x})$ in restricted regions, and next, complete the rest by duplication: we can give $I_0(\boldsymbol{x})$ in the irreducible region

$$\Delta_{111} = \left\{ \boldsymbol{x} \mid \boldsymbol{x} = \sum_{i=1}^{3} \xi_i \boldsymbol{a}_i,\ \xi_{1,2,3} \in \left[-\frac{1}{2}, \frac{1}{2} \right] \right\}, \tag{7.29}$$

and complete the rest by requiring the periodicity, $I_0(\boldsymbol{x}) = I_0(\boldsymbol{x} + \sum_{i=1}^{3} m_i \boldsymbol{a}_i)$, $m_{1,2,3} \in \mathbb{Z}$. We can also define it by giving $I_0(\boldsymbol{x})$ in the more extended non-irreducible region

$$\Delta_{M_1 M_2 M_3} = \left\{ \boldsymbol{x} \mid \boldsymbol{x} = \sum_{i=1}^{3} \xi_i M_i \boldsymbol{a}_i,\ \xi_i \in \left[-\frac{1}{2}, \frac{1}{2} \right] \right\}, \quad M_{1,2,3} \in \mathbb{N}_{\neq 0}, \tag{7.30}$$

and complete the rest by requiring the lower periodicity, $I_0(\boldsymbol{x}) = I_0(\boldsymbol{x} + \sum_{i=1}^{3} m_i M_i \boldsymbol{a}_i)$, $m_{1,2,3} \in \mathbb{Z}$. This is a first instance of the ambiguities that will appear below.

Proceeding as before, it is convenient to perform the average of a given field, $a(t, \boldsymbol{x}) \equiv I_0(\boldsymbol{x}) a(t, \boldsymbol{x})$, by convolution with a finite-width test function. To play the role of the test function, the periodic nature of the geometry suggests taking a slot function $w_{M_1 M_2 M_3}(\boldsymbol{x})$ centred at $\boldsymbol{x} = \boldsymbol{0}$, equal to the inverse cell volume $1/V_{M_1 M_2 M_3} = 1/M_1 \boldsymbol{a}_1 \cdot (M_2 \boldsymbol{a}_2 \times M_3 \boldsymbol{a}_3)$, if $\boldsymbol{x} \in \Delta_{M_1 M_2 M_3}$, and equal to zero, if $\boldsymbol{x} \notin \Delta_{M_1 M_2 M_3}$, as shown in Fig. 7.7.

With this test function, Lorentz's average is thus defined as:

$$A(t, \boldsymbol{x}) = \langle a \rangle (t, \boldsymbol{x}) = \int d\boldsymbol{x}' w_{M_1 M_2 M_3}(\boldsymbol{x} - \boldsymbol{x}') a(t, \boldsymbol{x}'),$$
$$\int d\boldsymbol{x}' w_{M_1 M_2 M_3}(\boldsymbol{x} - \boldsymbol{x}') = 1. \tag{7.31}$$

where as before we can, if we wish, make apparent the function $I_0(\boldsymbol{x}')$ in the integrand (7.31.1). We dispense from noting the choice of $M_1 M_2 M_3$ on A and $\langle . \rangle$ to lighten the notation.

We later clarify the reason for multivocity, (i.e. why we might be interested in taking $M_1 M_2 M_3 \neq 111$). We note that, as we use the total volume normalization (7.31.2), the mean of the characteristic function I is the porosity:

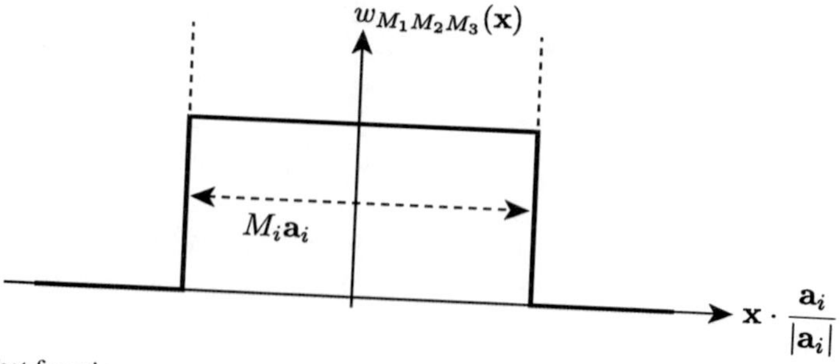

Fig. 7.7 Test function

$$\langle I_0 \rangle = \int d\boldsymbol{x}' w_{M_1 M_2 M_3}(\boldsymbol{x} - \boldsymbol{x}') I_0(\boldsymbol{x}') = \phi. \tag{7.32}$$

Because of this normalization, the averaging symbol $\langle \cdot \rangle$ in (7.31) interprets as one ("$M_1 M_2 M_3$") "total volume" average, i.e., porosity times the ("$M_1 M_2 M_3$") mean value in the fluid phase: $\langle \cdot \rangle = \phi \langle \cdot \rangle_f$.

Gibbs' Average

Just as the definition of the Lorentz mean given above is multiple, so is the Gibbs mean that we are defining now. At first, not considering multivocity, we would define Gibbs' average as:

$$A(t, \boldsymbol{x}) = \langle a \rangle (t, \boldsymbol{x}) = \overline{a(t, \boldsymbol{x}; \varpi)},$$

where we extend the field to zero in the solid and the overline is average over the random realizations ϖ. These are defined by applying the random translations $\boldsymbol{X} = \sum_{i=1}^{3} \xi_i \boldsymbol{a}_i$ to the reference configuration ϖ_0, so that $I(\boldsymbol{x}; \varpi) = I(\boldsymbol{x} - \boldsymbol{X}; \varpi_0) = I_0(\boldsymbol{x} - \boldsymbol{X})$. Introducing multivocity, however, we consider instead writing the random translations in the form of, $\boldsymbol{X}_{M_1 M_2 M_3} = \sum_{i=1}^{3} \xi_i M_i \boldsymbol{a}_i$, with some choice for integers M_i. It means that the realization ϖ will have for indicator function, $I(\boldsymbol{x}, \varpi) = I_0(\boldsymbol{x} - \boldsymbol{X}_{M_1 M_2 M_3})$. To remember this and make appear the choice of integers M_i in the definition, (we again dispense from noting it on A and $\langle . \rangle$), we denote finally the corresponding Gibbs average:

$$A(t, \boldsymbol{x}) = \langle a \rangle (t, \boldsymbol{x}) = \overline{a(t, \boldsymbol{x}; \varpi)}_{M_1 M_2 M_3}. \tag{7.33}$$

As before, it is convenient to rewrite the definition in a form that allows easy expression of the spatial derivative of the mean. To effect this refinement we introduce Δ, a region centred at $\boldsymbol{x} = \boldsymbol{0}$:

$$\Delta = \left\{ \boldsymbol{x} \mid \boldsymbol{x} = \sum_{i=1}^{3} \xi_i \epsilon \boldsymbol{a}_i, \quad \xi_i \in \left[-\frac{1}{2}, \frac{1}{2} \right] \right\}. \tag{7.34}$$

and that is considered in the limit $\epsilon \to 0$ where it is vanishingly small.

We introduce also the corresponding test function w_ϵ, which equals the inverse volume $1/V_\Delta = 1/\epsilon \boldsymbol{a}_1 \cdot (\epsilon \boldsymbol{a}_2 \times \epsilon \boldsymbol{a}_3)$ of this region if $\boldsymbol{x} \in \Delta$, and is zero if $\boldsymbol{x} \notin \Delta$. Then using the form of (7.15) we have the following definition of Gibbs' average:

$$A(t, \boldsymbol{x}) = \langle a \rangle (t, \boldsymbol{x}) = \overline{\int d\boldsymbol{x}' w_\epsilon (\boldsymbol{x} - \boldsymbol{x}') a(t, \boldsymbol{x}'; \varpi)}_{M_1 M_2 M_3}. \tag{7.35}$$

It has total volume normalization because of $\int d\boldsymbol{x}' w_\epsilon (\boldsymbol{x} - \boldsymbol{x}') = 1$.

In case we need an average $\langle a\rangle_f$ with fluid volume normalization, we just multiply by $1/\phi$ the test functions, ($w_{M_1M_2M_3}$ for Lorentz, w_ϵ for Gibbs), so that, again, $\langle . \rangle = \phi\langle . \rangle_f$. We now clarify the reason of the multivocity.

Reason of the Multivocity

The averaging operations are destined to be used to perform averages of fields, (or product of fields), solutions to one problem of type (7.1)–(7.6) stated for one periodic sample and at long wavelengths (Lorentz), or solutions to an ensemble of problems of type (7.1)–(7.6) for the collection of translated samples and at arbitrary wavelengths (Gibbs). For the present discussion we will assume that an impressed source term, $\boldsymbol{f} = -\partial(\tilde{\mathcal{P}}e^{-i\omega t+i\boldsymbol{k}\cdot\boldsymbol{x}})$, is present.[7] In response to the source term, fields varying like $\boldsymbol{v}(t,\boldsymbol{x}) = \boldsymbol{v}(\omega,\boldsymbol{k},\boldsymbol{x})e^{-i\omega t+i\boldsymbol{k}\cdot\boldsymbol{x}}$, $b(t,\boldsymbol{x}) = b(\omega,\boldsymbol{k},\boldsymbol{x})e^{-i\omega t+i\boldsymbol{k}\cdot\boldsymbol{x}}$, etc., will appear in the fluid, (omitting the realization argument ϖ_0 or ϖ). In the stationary random case, the response-amplitudes in front of the exponentials were uniquely fixed by the condition to be bounded fields. Now in the periodic case, this condition becomes insufficient to uniquely fix them: *we can have different choices for the periodicities.* Along the direction of vector $\boldsymbol{a}_1$ for example, we can require that the solutions amplitudes verify $\boldsymbol{v}(\omega,\boldsymbol{k},\boldsymbol{x}+\boldsymbol{a}_1) = \boldsymbol{v}(\omega,\boldsymbol{k},\boldsymbol{x})$, etc. We can instead require that they do not verify this, but verify $\boldsymbol{v}(\omega,\boldsymbol{k},\boldsymbol{x}+2\boldsymbol{a}_1) = \boldsymbol{v}(\omega,\boldsymbol{k},\boldsymbol{x})$, etc., and so on, with increasing value of periodicity. Thus, to unambiguously fix the response solutions, we have to precise what are the minimal periodicities of the (tilde-)amplitudes we are selecting among the different possible ones, i.e. specify the minimal integers, $M_1M_2M_3 \in (\mathbb{N}_{\neq 0})^3$, that will be such that, $\forall \boldsymbol{x}$, $\boldsymbol{v}(\omega,\boldsymbol{k},\boldsymbol{x}+M_i\boldsymbol{a}_i) = \boldsymbol{v}(\omega,\boldsymbol{k},\boldsymbol{x})$, for $i = 1, 2, 3$, and as soon as $N_i < M_i$, $\exists \boldsymbol{x}$, $\boldsymbol{v}(\omega,\boldsymbol{k},\boldsymbol{x}+N_i\boldsymbol{a}_i) \neq \boldsymbol{v}(\omega,\boldsymbol{k},\boldsymbol{x})$. Evidently, to perform macroscopic averages on a solution determined by such a choice, the corresponding "$M_1M_2M_3$" Lorentz or Gibbs average will have to be employed. This is the reason of the previous multivoked definitions. It is intrinsically connected with the unbounded nature of the periodic geometries considered. Having made this point, the discussion of commutation relations and ergodic equivalences, follows without difficulty.

7.3.2.2 Commutation Relations

Using the definitions of Lorentz's and Gibbs' averages, (7.31) and (7.33)–(7.35), it is easy to rewrite what has been stated in the stationary random case.

[7] In its absence, the same problematics also arises for the specification of the possible normal waves, with however additional technicalities as $\boldsymbol{k}$ then interprets as a Bloch wave-vector, and is thus determined only up to the addition of a reciprocal lattice vector, $\boldsymbol{K} = \sum_{i=1}^{3} m_i\boldsymbol{b}_i$, (with $\boldsymbol{b}_i \cdot \boldsymbol{a}_j = 2\pi\delta_{ij}$), which induces a corresponding ambiguity in the definition of (tilde-)amplitudes.

Lorentz's Average

Rewriting (7.17) we have the following commutation relation or averaging theorem:

$$\begin{aligned}\partial A(t,\boldsymbol{x}) &= \partial\langle a\rangle(t,\boldsymbol{x}) = \langle\partial a\rangle(t,\boldsymbol{x}) + \langle a\partial I_0\rangle(t,\boldsymbol{x}),\\ &= \int d\boldsymbol{x}' w_{M_1M_2M_3}(\boldsymbol{x}-\boldsymbol{x}')\partial a(t,\boldsymbol{x}') - \int_{\partial\mathcal{V}} d\boldsymbol{x}' a(t,\boldsymbol{x}')\hat{\boldsymbol{n}}(\boldsymbol{x}')w_{M_1M_2M_3}(\boldsymbol{x}-\boldsymbol{x}').\end{aligned} \tag{7.36}$$

The relation (7.18), that expressed macroscopic homogeneity, can be written here

$$\int_{\partial\mathcal{V}} d\boldsymbol{x}'\hat{\boldsymbol{n}}(\boldsymbol{x}')w_{M_1M_2M_3}(\boldsymbol{x}-\boldsymbol{x}') = \boldsymbol{0}. \tag{7.37}$$

Gibbs' Average

For Gibbs' average the previous commutation relation (7.19) now writes

$$\begin{aligned}\partial A(t,\boldsymbol{x}) =& \partial\langle a\rangle(t,\boldsymbol{x}) = \langle\partial a\rangle(t,\boldsymbol{x}) + \langle a\partial I\rangle(t,\boldsymbol{x}),\\ =& \overline{\int d\boldsymbol{x}' w_\epsilon(\boldsymbol{x}-\boldsymbol{x}')\partial a(t,\boldsymbol{x}';\varpi)}_{M_1M_2M_3}\\ &- \overline{\int_{\partial\mathcal{V}(\varpi)} d\boldsymbol{x}' a(t,\boldsymbol{x}';\varpi)\hat{\boldsymbol{n}}(\boldsymbol{x}';\varpi)w_\epsilon(\boldsymbol{x}-\boldsymbol{x}')}_{M_1M_2M_3}.\end{aligned} \tag{7.38}$$

Discussion

The discussion to be done here is the same as before, *mutatis mutandis*. Therefore we will not repeat it.

7.3.2.3 Ergodic Equivalences

Here we can explicitly check the ergodic equivalences, ensuring that the Gibbs ensemble average can always be done in terms of a Lorentz volume average.

The periodic Lorentz medium is defined by the indicator periodic function $I_0(\boldsymbol{x})$ of reference configuration ϖ_0, *and*, a choice of the integers M_1, M_2, M_3. When subjected to the action of the external force $\boldsymbol{f} = -\partial\mathcal{P}$ specified by a potential, $\mathcal{P} = \tilde{\mathcal{P}}e^{-i\omega t+i\boldsymbol{k}\cdot\boldsymbol{x}}$, the response fields, $\boldsymbol{v} = \boldsymbol{v}(\omega,\boldsymbol{k},\boldsymbol{x};\varpi_0)e^{-i\omega t+i\boldsymbol{k}\cdot\boldsymbol{x}}$, $b = b(\omega,\boldsymbol{k},\boldsymbol{x};\varpi_0)e^{-i\omega t+i\boldsymbol{k}\cdot\boldsymbol{x}}$, etc., are set, uniquely, by the condition that the tilde amplitudes[8] are proportional to the source tilde amplitude $\tilde{\mathcal{P}}$, and are periodic functions,

[8] Meaning the amplitudes in front of the exponentials; as always we omit the tilde symbol to lighten the writing.

$f(\boldsymbol{x} + M_i \boldsymbol{a}_i) = f(\boldsymbol{x})$, $\forall \boldsymbol{x}$, $i = 1, 2, 3$, with the minimal periodicities specified by the integers M_1, M_2, M_3.

An associated Gibbs medium will be the ensemble of randomly translated samples $\varpi \in \Omega$, each defined by the indicator periodic function

$$I_0(\boldsymbol{x} - \boldsymbol{X}_{M_1 M_2 M_3}), \ \text{with } \boldsymbol{X}_{M_1 M_2 M_3} = \sum_{i=1}^{3} \xi_i M_i \boldsymbol{a}_i, \ \text{and } \xi_i \in [-1/2, 1/2].$$

As fixed by $M_1 M_2 M_3$, the random translations are over several periods because they will serve to average fields having tilde-amplitudes varying with corresponding periodicities. Indeed, when subjected to the action of the same external force as above, the response fields we consider, $\boldsymbol{v} = \boldsymbol{v}(\omega, \boldsymbol{k}, \boldsymbol{x}; \varpi)e^{-i\omega t + i\boldsymbol{k}\cdot\boldsymbol{x}}$, $b = b(\omega, \boldsymbol{k}, \boldsymbol{x}; \varpi)$ $e^{-i\omega t + i\boldsymbol{k}\cdot\boldsymbol{x}}$, etc., will be set as before, uniquely, by the condition that the tilde amplitudes are proportional to source tilde amplitude $\tilde{\mathcal{P}}$, and are periodic functions, $f(\boldsymbol{x} + M_i \boldsymbol{a}_i) = f(\boldsymbol{x})$, having the minimal periodicities specified by M_1, M_2, M_3.

Consider then Gibbs' average

$$\boldsymbol{V}(t, \boldsymbol{x}) = \langle \boldsymbol{v} \rangle (t, \boldsymbol{x}) = e^{-i\omega t + i\boldsymbol{k}\cdot\boldsymbol{x}} \overline{\boldsymbol{v}(\omega, \boldsymbol{k}, \boldsymbol{x}; \varpi)}_{M_1 M_2 M_3}. \tag{7.39}$$

In evaluating it, there is the following relation that can be used between $\boldsymbol{v}(\omega, \boldsymbol{k}, \boldsymbol{x}; \varpi)$ and $\boldsymbol{v}(\omega, \boldsymbol{k}, \boldsymbol{x}; \varpi_0)$:

$$\boldsymbol{v}(\omega, \boldsymbol{k}, \boldsymbol{x}; \varpi) = \boldsymbol{v}(\omega, \boldsymbol{k}, \boldsymbol{x} - \boldsymbol{X}_{M_1 M_2 M_3}; \varpi_0). \tag{7.40}$$

To see it, let us apply the translation $\boldsymbol{X}_{M_1 M_2 M_3}$ to the reference configuration ϖ_0, then obtaining the translated configuration ϖ whose indicator function is $I(\boldsymbol{x}, \varpi) = I_0(\boldsymbol{x} - \boldsymbol{X}_{M_1 M_2 M_3})$. In translated coordinate axes $\boldsymbol{y}$ related to the $\boldsymbol{x}$ by $\boldsymbol{x} = \boldsymbol{y} + \boldsymbol{X}_{M_1 M_2 M_3}$, this translated configuration ϖ is the same as ϖ_0 in the axes $\boldsymbol{x}$, that is, $I(\boldsymbol{y} + \boldsymbol{X}_{M_1 M_2 M_3}; \varpi) = I_0(\boldsymbol{y}; \varpi_0)$. The response fields would then be the same, that is, $\boldsymbol{v}(t, \boldsymbol{x}; \varpi) = \boldsymbol{v}(\omega, \boldsymbol{k}, \boldsymbol{y}; \varpi_0)e^{-i\omega t + i\boldsymbol{k}\cdot\boldsymbol{y}}$, if the source potential $\tilde{\mathcal{P}}e^{-i\omega t + i\boldsymbol{k}\cdot\boldsymbol{x}}$ had also been "displaced", so as to write $\tilde{\mathcal{P}}e^{-i\omega t + i\boldsymbol{k}\cdot\boldsymbol{y}}$ and have amplitude $\tilde{\mathcal{P}}$ at $\boldsymbol{y} = \boldsymbol{0}$, (new position, in the present Gibbs realization ϖ, of the material that was in $\boldsymbol{x} = \boldsymbol{0}$ in the reference configuration ϖ_0). But the source term, independent of realization, is not displaced. It possesses now a multiplicative factor $e^{i\boldsymbol{k}\cdot\boldsymbol{X}_{M_1 M_2 M_3}}$, giving it the value $\tilde{\mathcal{P}}e^{i\boldsymbol{k}\cdot\boldsymbol{X}_{M_1 M_2 M_3}}$ at the new origin $\boldsymbol{y} = \boldsymbol{0}$. Hence correcting for this factor, we conclude that in the translated configuration, with source unchanged, we have $\boldsymbol{v}(t, \boldsymbol{x}; \varpi) = \boldsymbol{v}(\omega, \boldsymbol{k}, \boldsymbol{y}; \varpi_0)e^{-i\omega t + i\boldsymbol{k}\cdot\boldsymbol{y}} e^{i\boldsymbol{k}\cdot\boldsymbol{X}_{M_1 M_2 M_3}}$, that is, $\boldsymbol{v}(t, \boldsymbol{x}; \varpi) = \boldsymbol{v}(\omega, \boldsymbol{k}, \boldsymbol{x} - \boldsymbol{X}_{M_1 M_2 M_3}; \varpi_0)e^{-i\omega t + i\boldsymbol{k}\cdot\boldsymbol{x}}$, which is the indicated relation (7.40).

Inserting (7.40) in (7.39) we then have

$$\boldsymbol{V}(t, \boldsymbol{x}) = \langle \boldsymbol{v} \rangle (t, \boldsymbol{x}) = e^{-i\omega t + i\boldsymbol{k}\cdot\boldsymbol{x}} \overline{\boldsymbol{v}(\omega, \boldsymbol{k}, \boldsymbol{x} - \boldsymbol{X}; \varpi_0)}_{M_1 M_2 M_3}. \tag{7.41}$$

The overline, average over realizations, amounts here in distributing the random displacement $\boldsymbol{X}_{M_1 M_2 M_3}$ uniformly in a "unit cell" region $\Delta_{M_1 M_2 M_3}$ given by

$$\Delta_{M_1M_2M_3} = \left\{ \boldsymbol{x} \mid \boldsymbol{x} = \sum_{i=1}^{3} \xi_i M_i \boldsymbol{a}_i, \quad \xi_i \in \left[-\frac{1}{2}, \frac{1}{2} \right] \right\}. \tag{7.42}$$

As the field is extended to zero in the solid, we have

$$\overline{\boldsymbol{v}(\omega, \boldsymbol{k}, \boldsymbol{x} - \boldsymbol{X}_{M_1M_2M_3}; \varpi_0)} = \frac{1}{V_{\Delta_{M_1M_2M_3}}} \int_{\Delta_{M_1M_2M_3}} d\boldsymbol{X} \times \cdots$$
$$\cdots \boldsymbol{v}(\omega, \boldsymbol{k}, \boldsymbol{x} - \boldsymbol{X}; \varpi_0). \tag{7.43}$$

By inspection, we can see that this is the same as the Lorentz mean

$$\langle \boldsymbol{v}(\omega, \boldsymbol{k}, \boldsymbol{x}; \varpi_0) \rangle_{Lor} = \int d\boldsymbol{x}' w_{M_1M_2M_3}(\boldsymbol{x} - \boldsymbol{x}') \boldsymbol{v}(\omega, \boldsymbol{k}, \boldsymbol{x}'; \varpi_0), \tag{7.44}$$

with $w_{M_1M_2M_3}(\boldsymbol{x})$ the slot function represented in Fig. 7.7. Finally, it shows us that the Gibbs mean, is, whatever the chosen $\boldsymbol{k}$:

$$\begin{aligned} \boldsymbol{V}(t, \boldsymbol{x}) =& \langle \boldsymbol{v} \rangle (t, \boldsymbol{x}) = e^{-i\omega t + i\boldsymbol{k}\cdot\boldsymbol{x}} \overline{\boldsymbol{v}(\omega, \boldsymbol{k}, \boldsymbol{x}; \varpi)}, \\ =& e^{-i\omega t + i\boldsymbol{k}\cdot\boldsymbol{x}} \overline{\boldsymbol{v}(\omega, \boldsymbol{k}, \boldsymbol{x} - \boldsymbol{X}; \varpi_0)} = e^{-i\omega t + i\boldsymbol{k}\cdot\boldsymbol{x}} \langle \boldsymbol{v}(\omega, \boldsymbol{k}, \boldsymbol{x}; \varpi_0) \rangle_{Lor}, \end{aligned}$$

where we suppressed mention of the retained choice $M_1M_2M_3$. Therefore, apart from an undesirable multiplicity, we have demonstrated in the periodic case, for our macroscopic averaging operations, the same properties as we asserted before in the stationary random case. Whatever the wavelengths, the Gibbs mean can always be performed in terms of Lorentz mean directly made on the tilde amplitudes. At long wavelengths, both averages are interchangeable.

7.4 Macroscopic Equations and Definition of the Acoustic H-Field from Electromagnetic Analogy

We now address the question of formulating the general macroscopic equations describing the compressional-dilatational wave propagation in a macroscopically homogeneous rigid-framed porous medium saturated with a viscothermal fluid.

In Lorentz's conception we are given one sample and the theory we are to formulate is intended to describe long-wavelength sound propagation in this sample. In the Gibbs' conception we are given an ensemble of samples and the theory is intended to describe ensemble-averaged fields, whatever the frequencies and wavelengths.

In what follows, to simplify the discussion, we limit ourselves to considering wave propagation or excitation along a single axis x, which is also assumed to be a macroscopic symmetry axis. Precisely, because the external actions $\boldsymbol{f}$ in (7.2) are directed along x, a symmetry axis, we can assume that :

$$f = -\partial \mathcal{P}, \quad \mathcal{P} = \int \frac{d\omega}{2\pi}\frac{dk}{2\pi}\tilde{\mathcal{P}}(\omega, k)e^{-i\omega t+ikx}, \quad \langle \boldsymbol{v}\rangle \parallel \hat{\boldsymbol{x}}, \quad \langle p\boldsymbol{v}\rangle \parallel \hat{\boldsymbol{x}}. \tag{7.45}$$

The most general fields we consider here will have a "forced" part coming in response to the impressed excitation f, and a "free" part that superposes, (solution to the homogeneous equations (7.1)–(7.6) with f suppressed), and that can be viewed as the result of the presence of an "external" incident field, coming from without, in the directions $\pm x$.

We recall first, the results we have obtained in Chap. 6. We recall in Sect. (7.4.1), the pattern of nonlocal equations in the homogeneous viscothermal fluid itself, without solid, and written for compressional motions along axis x to comply with the above-mentioned restrictions; with no ambiguity, we do not indicate the indice x on v, d, and f. We recall next, in Sect. (7.4.2), the pattern of the equations in macroscopic nonlocal electromagnetics, written for propagation/excitation along a principal axis x of a general homogeneous structured material. Finally, in Sect. (7.4.3), we pass to the pattern of the macroscopic equations we would like to write, by analogy, for the propagation/excitation of compressional waves along macroscopic axis x in our homogeneous, structured fluid/solid medium.

7.4.1 Unbounded Fluid (Longitudinal Motions)

In the unbounded fluid, the corresponding pattern of nonlocal acoustic equations found in Chap. 6, was as follows:

Field equations

$$\frac{\partial b}{\partial t} + \partial_x v = 0, \qquad \frac{\partial d}{\partial t} = \partial_x h + [f], \tag{7.46}$$

where

$$f = -\partial_x \mathcal{P}, \qquad \mathcal{P} = \int \frac{d\omega}{2\pi}\frac{dk}{2\pi}\tilde{\mathcal{P}}(\omega, k)e^{-i\omega t+ikx}. \tag{7.47}$$

Constitutive relations

$$d(t, x) = \hat{\rho} v(t, x) = \int \rho(t - t', x - x')v(t', x')dt'dx', \tag{7.48}$$

$$h(t, x) = -\hat{\chi}^{-1} b(t, x) = -\int \chi^{-1}(t - t', x - x')b(t', x')dt'dx'. \tag{7.49}$$

Definition of the h-field

$$h = -p. \tag{7.50}$$

The last "acoustic Heaviside-Poynting" identification is obtained as a result of setting, at the same time

$$s = -vh, \tag{7.51}$$

$$s = vp. \tag{7.52}$$

The first equation, where s is the "energy current density transported in acoustic form", is to be viewed as satisfied by principle, by definition of what will be the h-field. The second, which specifies s, is obtained by using the irreversible thermodynamics of the Navier–Stokes–Fourier model.

7.4.2 Macroscopic Electromagnetics

In the structured homogeneous electromagnetic medium, the corresponding pattern of nonlocal electromagnetic equations found in Chap. 6, was as follows:

Field equations

$$\frac{\partial \boldsymbol{B}}{\partial t} + \partial \times \boldsymbol{E} = \boldsymbol{0}, \qquad \frac{\partial \boldsymbol{D}}{\partial t} = \partial \times \boldsymbol{H} - [\boldsymbol{J}], \tag{7.53}$$

$$\text{where} \qquad \boldsymbol{E} = \langle \boldsymbol{e} \rangle, \qquad \boldsymbol{B} = \langle \boldsymbol{b} \rangle, \tag{7.54}$$

$$\boldsymbol{J} = J\hat{\boldsymbol{x}}, \qquad J = \int \frac{d\omega}{2\pi}\frac{dk}{2\pi}\tilde{\mathcal{J}}(\omega, k)e^{-i\omega t + ikx}. \tag{7.55}$$

Constitutive relations

$$\boldsymbol{D}(t, x) = \hat{\epsilon}\boldsymbol{E}(t, x) = \int \epsilon(t - t', x - x')\boldsymbol{E}(t', x')dt'dx', \tag{7.56}$$

$$\boldsymbol{H}(t, x) = \hat{\mu}^{-1}\boldsymbol{B}(t, x) = \int \mu^{-1}(t - t', x - x')\boldsymbol{B}(t', x')dt'dx'. \tag{7.57}$$

Definition of the $\boldsymbol{H}$-field

$$\boldsymbol{H} = \text{thermodynamic field } \boldsymbol{\mathcal{P}}. \tag{7.58}$$

The last would result from setting, at the same time:

$$\boldsymbol{S} = \boldsymbol{E} \times \boldsymbol{H}, \tag{7.59}$$

$$\boldsymbol{S} = \boldsymbol{E} \times \boldsymbol{\mathcal{P}}. \tag{7.60}$$

The first equation, where $\boldsymbol{S}$ is the "energy current density transported in electromagnetic form", is to be viewed as satisfied by principle, by definition of what will be the $\boldsymbol{H}$-field. To be written, the second, which specifies $\boldsymbol{S}$, would require an irreversible thermodynamic description of the behaviour of charges and currents in material media. Currently, we cannot properly define the field $\boldsymbol{H}$, because, we do not have this irreversible thermodynamic description.

7.4.3 Macroscopic Acoustics (Fluid-Saturated Rigid-Framed Porous Medium)

By analogy and extension, we now write the pattern of required equations for the propagation of macroscopic compressional waves in our structured fluid/solid medium, along a macroscopic symmetry axis. Since it is only in the fluid part that the movement is not null, new technicalities intervene in the presentation where one can for example choose to make ϕ factors appear or not. We will therefore start with some preliminary points regarding our definitions.

There is assumed to be an external longitudinal bulk force, $\boldsymbol{f}(\boldsymbol{x}) = f(\boldsymbol{x})\hat{\boldsymbol{x}}$, acting in the fluid, such that, $\boldsymbol{f}(\boldsymbol{x}) = -\partial_x \mathcal{P}(\boldsymbol{x})$, $f(\boldsymbol{x}) = -\partial_x \mathcal{P}(\boldsymbol{x})$, $\mathcal{P}(\boldsymbol{x}) = \int \frac{d\omega}{2\pi}\frac{dk}{2\pi} \tilde{\mathcal{P}}(\omega, k) e^{-i\omega t + ikx}$, if $\boldsymbol{x}$ is in the fluid, and $\boldsymbol{f}(\boldsymbol{x}) = \boldsymbol{0}$, $\mathcal{P}(\boldsymbol{x}) = 0$, if $\boldsymbol{x}$ is in the solid. We note that the fields $\boldsymbol{f}(\boldsymbol{x})$, $f(\boldsymbol{x})$ and $\mathcal{P}(\boldsymbol{x})$, have the nature of the "macroscopic impressed fields" or "external fields" discussed in Sect. 7.3.1: they vary independently of the geometry except for their conventional extension to zero in the solid. Therefore in particular we will have, using the commutation relation (7.22), $\langle f \rangle = \langle -\partial_x \mathcal{P} \rangle = -\partial_x \langle \mathcal{P} \rangle$. Now, for later convenience, we introduce different notations for the macroscopic means $\langle f \rangle$ and $\langle \mathcal{P} \rangle$. For the first, we chose to denote it $\langle f \rangle \equiv \phi F$, then adopting for $F = \langle f \rangle_f$ a fluid-volume-average normalization. For $\langle \mathcal{P} \rangle$ we chose to write, by language abuse

$$\langle \mathcal{P} \rangle = \phi \mathcal{P}, \tag{7.61}$$

denoting with *same* letter $\mathcal{P}$ the field, macroscopic proper, given by $\mathcal{P}(\boldsymbol{x}) = \mathcal{P}(x) = \int \frac{d\omega}{2\pi}\frac{dk}{2\pi} \tilde{\mathcal{P}}(\omega, k) e^{-i\omega t + ikx}$, (with same coefficients $\tilde{\mathcal{P}}$ as above), *whatever the values* $\boldsymbol{x}$, x, *(including* $\boldsymbol{x}$ *in the solid)*. In fact, this language abuse is made possible by the "external field" nature of the $\mathcal{P}$ field. Owing to this nature of the $\mathcal{P}$ field, this interpretation of the letter $\mathcal{P}$ in the right side of (7.61) is evident to see with Gibbs' averaging conception. The same writing (7.61) can also easily be obtained with Lorentz's averaging, with reasonings of the type previously seen in Sect. 7.3.1.2 to justify the cancellation (7.23). We do not use it for $\langle f \rangle = \phi F$, (that could in same manner be denoted $\langle f \rangle = \phi f$), just to have a capital letter for the external force in our macroscpic equations. In this way we have the following notation, whose form is simple:

$$F \equiv \frac{\langle f \rangle}{\phi} = \frac{\langle -\partial_x \mathcal{P} \rangle}{\phi} = \frac{-\partial_x \langle \mathcal{P} \rangle}{\phi} = -\partial_x \mathcal{P}. \tag{7.62}$$

With these preliminaries made, the juxtaposition of the two sets of equations, acoustic (7.46)–(7.51) in the unbounded fluid, and electromagnetic (7.53)–(7.60) in the material, suggests that for compressional macroscopic motions along symmetry axis x in the fluid-saturated porous medium, we should have macroscopic nonlocal equations taking the following form:

Field equations

$$\frac{\partial B}{\partial t} + \partial_x V = 0, \qquad \frac{\partial D}{\partial t} = \partial_x H + [F], \tag{7.63}$$

where

$$V = \langle v \rangle, \qquad B = \langle b \rangle, \tag{7.64}$$

$$F = -\partial_x \mathcal{P}, \qquad \mathcal{P} = \int \frac{d\omega}{2\pi}\frac{dk}{2\pi}\tilde{\mathcal{P}}(\omega, k)e^{-i\omega t + ikx}. \tag{7.65}$$

Constitutive relations

$$D(t,x) = \hat{\rho}V(t,x) = \int \rho(t-t', x-x')V(t',x')dt'dx', \tag{7.66}$$

$$H(t,x) = -\hat{\chi}^{-1}B(t,x) = -\int \chi^{-1}(t-t', x-x')B(t',x')dt'dx'. \tag{7.67}$$

Definition of the H-field

$$-H\langle v \rangle = \langle pv \rangle. \tag{7.68}$$

The last is obtained as a result of setting, at the same time, for the "macroscopic acoustic part of the energy current density":

$$S = -\langle v \rangle H, \tag{7.69}$$

$$S = \langle s \rangle = \langle vp \rangle. \tag{7.70}$$

Equation (7.70) is an interpretation, made possible by the existence of thermodynamic concepts, of what is the macroscopic acoustic part of the energy current density $\boldsymbol{S} = S\hat{\boldsymbol{x}}$: it is the mean of pore-scale acoustic part of the energy current density, $\boldsymbol{S} = \langle \boldsymbol{s} \rangle$, and this gives a usable definition, because we have a thermodynamic background to identify $\boldsymbol{s}$: $\boldsymbol{s} = p\boldsymbol{v}$, where p is the thermodynamic pressure. Equation (7.69), written as an "acoustic Heaviside-Poynting" relation that must be satisfied by definition of what will be the acoustic H-field, then gives the identification (7.68) which is a usable definition of this acoustic H-field. It is in fact nothing but the customary definition of lumped acoustic pressure discussed in Pierce [7], Sect. 7.2 Lumped-Parameters Models. In Sect. 7.7, after entering in the detail of the micro-macro passage, we will see how the above definition (7.68) of macroscopic H-field, is a powerful statement capable to fix the macroscopic properties of the medium, i.e. the operators $\hat{\rho}$ and $\hat{\chi}^{-1}$. Before this, however, we want to examine what other definitions we could make, if we were to proceed like in conventional electromagnetics, or if we were to use, as is done in local theory, (see Appendix), a macroscopic pressure defined by direct volume averaging.

7.5 Macroscopic Equations: Other Points of View

First recall that the customary point of view in electromagnetics of spatially dispersive media, found e.g. in Landau and Lifshitz [8], Agranovich and Ginzburg [9], Melrose and McPhedran [10], and which we view as a stopgap measure, consists in setting $\boldsymbol{M} = \boldsymbol{0}$, in the presence of spatial dispersion, i.e. write by definition:

$$\boldsymbol{H} = \frac{\boldsymbol{B}}{\mu_0}, \tag{7.71}$$

and put all effects in an effective electric susceptibility $\hat{\epsilon}$. It leads to write the nonlocal equations in a form seen in Chap. 6 and recalled in next subsection.

7.5.1 *Macroscopic Electromagnetics—Customary Point of View*

Following the customary point of view, the pattern of nonlocal electromagnetic equations is set as follows:

Field equations

$$\frac{\partial \boldsymbol{B}}{\partial t} + \partial \times \boldsymbol{E} = \boldsymbol{0}, \qquad \frac{\partial \boldsymbol{D}}{\partial t} = \frac{1}{\mu_0}\partial \times \boldsymbol{B} - [\boldsymbol{J}], \tag{7.72}$$

where

$$\boldsymbol{E} = \langle \boldsymbol{e} \rangle, \qquad \boldsymbol{B} = \langle \boldsymbol{b} \rangle, \tag{7.73}$$

$$\boldsymbol{J} = J\hat{\boldsymbol{x}}, \qquad J = \int \frac{d\omega}{2\pi}\frac{dk}{2\pi}\tilde{\mathcal{J}}(\omega, k)e^{-i\omega t + ikx}. \tag{7.74}$$

Constitutive relation

$$\boldsymbol{D}(t, x) = \hat{\epsilon}\boldsymbol{E}(t, x) = \int \epsilon(t - t', x - x')\boldsymbol{E}(t', x')dt'dx'. \tag{7.75}$$

In a similar manner, it would perfectly be possible here, to formulate the acoustic nonlocal equations by introducing a conventional H-field, different from that of (7.68). Let us take two examples.

7.5.2 Acoustics Translation of the Customary Point of View in Electromagnetics

The formulation that would be the exact counterpart of the above customary electromagnetic point of view, would consist in setting $M = 0$, i.e. write by definition, (the occurrence of porosity factor here, is as in (7.62), see also Sect. 7.6):

$$H = -\frac{B}{\chi_0 \phi}. \tag{7.76}$$

This would lead to write the nonlocal acoustic equations in the form:

Field equations

$$\frac{\partial B}{\partial t} + \partial_x V = 0, \qquad \frac{\partial D}{\partial t} = -\frac{1}{\chi_0 \phi}\partial_x B + [F], \tag{7.77}$$

$$\text{where} \qquad V = \langle v \rangle, \qquad B = \langle b \rangle, \tag{7.78}$$

$$F = -\partial_x \mathcal{P}, \qquad \mathcal{P} = \int \frac{d\omega}{2\pi}\frac{dk}{2\pi}\tilde{\mathcal{P}}(\omega, k)e^{-i\omega t + ikx}. \tag{7.79}$$

Constitutive relation

$$D(t, x) = \hat{\rho} V(t, x) = \int \rho(t - t', x - x')V(t', x')dt'dx'. \tag{7.80}$$

This point of view sets the bulk modulus operator $\hat{\chi}^{-1}$ to a constant $1/\phi\chi_0$, determined by the adiabatic value. It is clearly an artificial replacement.

7.5.3 Acoustics Formulation in Terms of Volume-Averaged Pressure

Still another definition, inspired this time by what is done within the local homogenization, would be to use in the macroscopic equations the ordinary macroscopic pressure, $P = \langle p \rangle_f$, defined by a fluid-volume-average. Using this variable as the opposite H-field, we would have the following formulation:

Field equations

$$\frac{\partial B}{\partial t} + \partial_x V = 0, \qquad \frac{\partial D}{\partial t} = -\partial_x P + [F], \tag{7.81}$$

$$\text{where} \qquad V = \langle v \rangle, \qquad B = \langle b \rangle, \tag{7.82}$$

$$F = \partial_x \mathcal{P}, \qquad \mathcal{P} = \int \frac{d\omega}{2\pi}\frac{dk}{2\pi}\tilde{\mathcal{P}}(\omega, k)e^{-i\omega t+ikx}. \tag{7.83}$$

Constitutive relations

$$D(t, x) = \hat{\rho}V(t, x) = \int \rho(t - t', x - x')V(t', x')dt'dx', \tag{7.84}$$

$$P(t, x) = \hat{\chi}^{-1}B(t, x) = \int \chi^{-1}(t - t', x - x')B(t', x')dt'dx'. \tag{7.85}$$

Definition of the macroscopic stress field H and pressure field P

$$H = -P = -\langle p\rangle/\phi = -\langle p\rangle_f. \tag{7.86}$$

This point of view, however, sets undue importance to the direct volume-average of the pressure $\langle p\rangle_f$. As soon as the pressure is significantly distributed at the pore-scale, this average is not a very meaningful macroscopic variable. A distributed pore-scale pressure occurs principally in two instances: long wavelengths with complex geometry producing local resonances, and short wavelengths, (described within a Gibbs conception). In further work, the interest of using, preferentially the definition (7.68) in these cases should be studied, in particular, when the finite dimensions of the media are taken into account. The conception (7.68) would have to generalize smoothly and we should see that it is advantageous to use, compared to others.

Indeed, consider a slab of material and compare how the H-variables of the different conceptions will behave at a boundary, $x = Cst$, between the material and the fluid, and in its vicinity.

Consider first the artificial definition $H = -B/\chi_0\phi$. There is no reason that $\chi_0\phi$ be a meaningful compressibility in the material, especially in presence of resonances. Therefore, we expect that this variable H can quickly vary in the vicinity of the boundary, and be discontinuous at it. Likewise, consider the definition $H = -\langle p\rangle_f$. In presence of local Helmholtz resonances in the material, the pressure will be rapidly variable on the small scale, and there is no reason that its direct mean shouldn't exhibit rapid variation at the boundary, and be discontinuous at it. This is at variance with the indirect mean or "lumped" variable H, in the "acoustic Heaviside-Poynting" definition (7.68), $S = \langle pv\rangle = -H\langle v\rangle$. Indeed, as we have continuity of normal flow $\langle v\rangle$, and we can also expect continuity of normal acoustic power flow S, this definition (7.68) generates a H-field continuous at the material boundary, and that will not vary rapidly in its vicinity. That is why it should lead to natural determination of the density and compressibility operators, when the finite dimensions are taken into account. In particular, not only appropriate to describe characteristic wavenumbers, it would give also appropriate characteristic impedances (see Sect. 7.9).

With these remarks in mind, the deficiencies of the acoustic conceptions, $H = -B/\chi_0\phi$, and $H = -\langle p\rangle_f$, cast a harsh light on the present day definitions of electromagnetic Maxwell fields $\boldsymbol{H}$ and $\boldsymbol{D}$, generally effected in presence of spatial dispersion through setting $\boldsymbol{H} = \boldsymbol{B}/\mu_0$.

7.6 Derivation of the Macroscopic Equations by a Micro-Macro Passage

In preceding Sects. 7.4 and 7.5, by following an electromagnetic analogy, we proposed different specific patterns for the macroscopic acoustic equations. Referring to conventional ideas in electromagnetics we were led to the pattern (7.76)–(7.80); referring to other conventional ideas in acoustics we were led to the pattern (7.81)–(7.86). Finally, by deepening the analogy itself, with new perspectives that we defended and that concerned as much electromagnetics, irreversible-thermodynamics, and acoustics, we were brought to the pattern (7.63)–(7.68), that we consider is the preferable one, physically.

We now scrutinize how microscopic equations (7.1)–(7.6) applied for longitudinal motions along x, supposed to be a macroscopic symmetry axis, indeed lead, after averaging, to macroscopic equations that can be put in the given various patterns (7.63)–(7.68) or (7.76)–(7.80) or (7.81)–(7.86). It will also lead us clarify, next, how the respective operators can be in principle derived, from microstructure.

The "micro-macro" passage, from microscopic to macroscopic equations is as follows.[9]

Using Lorentz's conception, we write the equations (7.1)–(7.6) in one single realization of our stationary-random or periodic medium, say ϖ_0, we solve them, and then we make a "micro-macro" transition by taking a Lorentz volume average. Using Gibbs' conception, we write the equations in an arbitrary realization ϖ, we solve them, and make a "micro-macro" transition by taking the Gibbs ensemble average over realizations.

In this process, when we take the average of the (7.1) and apply the general commutation relation (7.21), we directly obtain the first macroscopic field equation:

$$0 = \frac{\partial\langle b\rangle}{\partial t} + \langle\partial\cdot\boldsymbol{v}\rangle = \frac{\partial\langle b\rangle}{\partial t} + \partial\cdot\langle\boldsymbol{v}\rangle \;\to\; \frac{\partial B}{\partial t} + \partial\cdot\boldsymbol{V} = 0.$$

[9] "Micro" do not refer here to any level comparable to that of molecules and electrons in electromagnetics. It refers to the inner macroscopic level where equations of fluid mechanics apply and the material appears inhomogeneous, by opposition to the outer macroscopic level, defined in Lorentz's or Gibbs' sense. Note also that, in the general reasonings made here, as well as in the two preceding sections, we work in the stationary-random case to make the discussion definite and avoid the ambiguities encountered in the periodic case.

Here, as we restrict to considering free wave propagation along symmetry axis x, or else, source-driven problem with external density of force directed along this axis, the velocity has only nonzero component x noted V:

$$\frac{\partial B}{\partial t} + \partial_x V = 0. \tag{7.87}$$

The second macroscopic field equation, which will be either (7.63.2) with (7.68), or, (7.77.2), or, (7.81.2) with (7.86), is not at all obtained in the same direct manner, by averaging. It involves an additional crucial step, playing the role of the Lorentz splitting seen in electromagnetics, and which is related to the choice of definition of the H-field. Let us detail it now.

Based on the rewrite in the unbounded fluid, (6.99), Sect. 6.12.1 of the Chap. 6, we begin here by rewriting (7.2) as follows

$$\rho_0 \frac{\partial \boldsymbol{v}}{\partial t} = -\frac{1}{\phi \chi_0} \partial \langle b \rangle + \boldsymbol{f}_{pol} - [\partial \mathcal{P}], \tag{7.88}$$

with

$$\boldsymbol{f}_{pol} = \frac{1}{\phi \chi_0} \partial \langle b \rangle - \partial p + \eta \partial^2 \boldsymbol{v} + \left(\frac{\eta}{3} + \zeta\right) \partial \left(\partial \cdot \boldsymbol{v}\right). \tag{7.89}$$

This induced force is the expression of processes provoked by the presence of the macroscopic perturbation, the existence of viscous and thermal losses, and the solid interface on which the boundary conditions (7.5)–(7.6) apply. Its precise value $\boldsymbol{f}_{pol}$ is fixed by the expression (7.89), and the considered solution of the complete system of equations (7.1)–(7.6).[10] In the right-hand side of (7.88), the first term is chosen so that, after averaging, $(\langle \frac{1}{\phi\chi_0} \partial \langle b \rangle \rangle = \frac{1}{\chi_0} \partial \langle b \rangle)$, this Equation compares well with the electromagnetic equation obtained by combining the (6.80.4) and (6.82) of Sect. 6.11.2 of the Chap. 6:

$$\epsilon_0 \frac{\partial \boldsymbol{E}}{\partial t} = \frac{1}{\mu_0} \partial \times \boldsymbol{B} - \boldsymbol{J}_{pol} - [\boldsymbol{J}]. \tag{7.90}$$

Indeed, by taking the macroscopic mean (Lorentz or Gibbs) of (7.88), and noting that the direct switching relationship (7.22) applies to the fields $a = \langle b \rangle$ and $a = \partial \mathcal{P}$, which only vary at the macroscopic scale, we find

$$\rho_0 \frac{\partial \langle \boldsymbol{v} \rangle}{\partial t} = -\frac{1}{\phi \chi_0} \partial \langle \langle b \rangle \rangle + \langle \boldsymbol{f}_{pol} \rangle - [\partial \langle \mathcal{P} \rangle].$$

As we extend the fields to zero in the solid, we have $\langle \langle b \rangle \rangle = \phi \langle b \rangle$, and $\langle \mathcal{P} \rangle = \phi \mathcal{P}$, and thus

---

[10] This solution is not unique because it is made of a unique forced response proportional to the excitation, and a nonunique source-free solution which depends on a possible incident field, coming from without: see comment after (7.45).

$$\rho_0 \frac{\partial \langle \boldsymbol{v} \rangle}{\partial t} = -\frac{1}{\chi_0} \partial \langle b \rangle + \langle \boldsymbol{f}_{pol} \rangle - [\phi \partial \mathcal{P}] .$$

Finally, by setting

$$\phi \boldsymbol{F}_{pol} \equiv \langle \boldsymbol{f}_{pol} \rangle = \frac{1}{\chi_0} \partial \langle b \rangle - \langle \partial p \rangle + \langle \eta \partial^2 \boldsymbol{v} + \left(\frac{\eta}{3} + \zeta \right) \partial \left(\partial \cdot \boldsymbol{v} \right) \rangle, \tag{7.91}$$

the averaged equation reads

$$\rho_0 \frac{\partial \langle \boldsymbol{v} \rangle}{\partial t} = -\frac{1}{\chi_0} \partial \langle b \rangle + \phi \boldsymbol{F}_{pol} - [\phi \partial \mathcal{P}] . \tag{7.92}$$

Now we proceed by considering that this (7.92) is a meaningful acoustic counterpart of the electromagnetic equation (7.90). In electromagnetics we have formally decomposed, in the Lorentz splitting—see the Chap. 6, Sect. 6.11.2, (6.83), the macroscopic polarization current density $\boldsymbol{J}_{pol}$ in (7.90) in two macroscopic terms, one given by temporal derivatives, the other by spatial derivatives. Here, as we have previously done with success in the unbounded fluid—see the Chap. 6, Sect. 6.12.1, (6.101), we consider that the polarization force density $\phi \boldsymbol{F}_{pol}$ in (7.92), given by (7.91), is formally decomposed in two macroscopic terms, one given by temporal derivatives, the other by spatial derivatives[11]:

$$\phi \boldsymbol{F}_{pol} = -\phi \frac{\partial \boldsymbol{P}}{\partial t} + \phi \partial M, \tag{7.93}$$

what we call an acoustic Lorentz splitting. As we restrict to a macroscopic motion along x, we have $\boldsymbol{F}_{pol} = F_{pol}\hat{\boldsymbol{x}}$, $\boldsymbol{P} = P\hat{\boldsymbol{x}}$, (with "polarization" P not to be confused with the mean pressure), and the above decomposition writes

$$F_{pol} = -\frac{\partial P}{\partial t} + \partial_x M. \tag{7.94}$$

There are nonlocal operators $\hat{\chi}_V$ and $\hat{\chi}_B$, or kernels $\chi_V(t, x)$ and $\chi_B(t, x)$, establishing how the "polarization fields P and M" are determined by the velocity and condensation macroscopic fields[12]:

[11] Note that if we were not to assume isotropy or propagation along a symmetry axis, we would define symmetric tensors H-field and M-field, through writing $\left(F_{pol}\right)_i = \frac{-\partial P_i}{\partial t} + \partial_j M_{ji}$, and $S_i = -H_{ij} V_j = \langle p v_i \rangle$; in the electromagnetic case we have $S_i = -H_{ji} E_j$ for the Poynting vector (not paying attention to the variances), and, $\left(J_{pol}\right)_i = \frac{\partial P_i}{\partial t} - \partial_j M_{ji}$, which looks the same, with however the important difference that H_{ij} and M_{ij} are antisymmetric instead of symmetric tensors.

[12] Recall that we limit ourselves here for simplicity to macroscopically homogeneous media (unbounded media), so that we have difference-kernels, i.e. kernels which depend on the (Cartesian-coordinates) difference $x - x'$.

$$P(t,x) = \int dt'dx'\chi_V(t-t',x-x')V(t',x'), \tag{7.95}$$

$$M(t,x) = \int dt'dx'\chi_B(t-t',x-x')B(t',x'). \tag{7.96}$$

In (7.95) there is no need to add a similar term with the condensation field, because, as V and B are related by (7.87), such a term could also be rewritten in the integral form (7.95). Likewise, in (7.96), because of the complete form of the dispersion, there is no need to add a similar term with the velocity field. Substituting (7.94) in (7.92) we get

$$\rho_0 \frac{\partial \langle v \rangle}{\partial t} + \phi \frac{\partial P}{\partial t} = -\frac{1}{\chi_0}\partial_x \langle b \rangle + \phi \partial_x M - [\phi \partial_x \mathcal{P}], \tag{7.97}$$

which, by setting

$$D = \frac{\rho_0}{\phi}\langle v \rangle + P = \frac{\rho_0}{\phi}V + P, \quad H = -\frac{1}{\chi_0 \phi}\langle b \rangle + M = -\frac{1}{\chi_0 \phi}B + M, \tag{7.98}$$

takes the form

$$\begin{aligned} \frac{\partial D}{\partial t} &= \partial_x H + [F], \\ F = -\partial_x \mathcal{P} \quad \mathcal{P} &= \int \frac{d\omega}{2\pi}\frac{dk}{2\pi}\tilde{\mathcal{P}}(\omega,k)e^{-i\omega t + ikx}. \end{aligned} \tag{7.99}$$

And finally, saying that we necessarily have nonlocal relations of type (7.95) and (7.96), is the same as saying that we have nonlocal relations of type:

$$D(t,x) = \hat{\rho}V(t,x) = \int \rho(t-t',x-x')V(t',x')dt'dx', \tag{7.100}$$

$$H(t,x) = -\hat{\chi}^{-1}B(t,x) = -\int \chi^{-1}(t-t',x-x')B(t',x')dt'dx', \tag{7.101}$$

with the connection

$$\rho(t-t',x-x') = \frac{\rho_0}{\phi}\delta(t-t')\delta(x-x') + \chi_V(t-t',x-x'), \tag{7.102}$$

$$\chi^{-1}(t-t',x-x') = \frac{1}{\chi_0 \phi}\delta(t-t')\delta(x-x') - \chi_B(t-t',x-x'). \tag{7.103}$$

In this way, the formal pattern (7.63)–(7.67) of the macroscopic acoustics equations is now evidenced.

Nevertheless, as the decomposition (7.93) is not unique, (we can add an arbitrary term $\partial \Phi$ to the polarization $\boldsymbol{P}$, if we simultaneously add a term $\frac{\partial \Phi}{\partial t}$ to polarization M), the description is not unique. It remains to be shown that by using an additional condition of definition of the H-field, which will be either (7.68), or (7.76), or else (7.86), to take up the various possibilities successively considered previously, we arrive at unique definitions of the relationships between the "Maxwell" fields M or H and the "Lorentz" field B, on one hand, and the "Maxwell" fields P or D and the "Lorentz" field V, on the other hand, so that the associated nonlocal operators, respectively, $\hat{\chi}_B$ or $\hat{\chi}^{-1}$, and $\hat{\chi}_V$ or $\hat{\rho}$, are determined in a unique way from the microstructure.

Of the various determinations, the one obtained with the "acoustic Heaviside-Poynting" identification (7.68), we believe, will be the most advantageous. We start with it.

7.7 Action-Response Problem to Determine the Nonlocal Dynamic Operators from Microstructure

We show here how the definition (7.68) uniquely determines in principle the operators, $(\hat{\chi}_B, \hat{\chi}^{-1})$, and $(\hat{\chi}_V, \hat{\rho})$, in a direct generalization of the previous "action-response problem" 6.12.4 in the unbounded fluid.

We consider the saturating fluid response to an harmonic excitation density of force along x, $f = -\partial_x \mathcal{P}, \mathcal{P} = \tilde{\mathcal{P}} e^{-i\omega t + ikx}$, in the fluid, paying our attention, either directly at the macroscopic level, or initially at the microscopic level, and then at macroscopic level after taking an average. For consistency, both considerations will have to lead to the same macroscopic response. This, in conjunction with the definition (7.68), will fix the Fourier coefficients of the kernels, $\rho(\omega, k)$ and $\chi^{-1}(\omega, k)$, in a unique way.

In the first, directly macroscopic consideration, we write by principle the following macroscopic equations:

$$\frac{\partial B}{\partial t} + \partial_x V = 0, \qquad \frac{\partial D}{\partial t} = \partial_x H - \partial_x \mathcal{P}, \tag{7.104}$$

$$D(t, x) = \hat{\rho} V(t, x) = \int \rho(t - t', x - x') V(t', x') dt' dx', \tag{7.105}$$

$$H(t, x) = -\hat{\chi}^{-1} B(t, x) = -\int \chi^{-1}(t - t', x - x') B(t', x') dt' dx', \tag{7.106}$$

where the H-field is set so that the "Heaviside-Poynting" definition (7.68)

$$\langle pv \rangle(t, x) = -\langle v \rangle(t, x) H(t, x), \tag{7.107}$$

will be satisfied.

The kernels can be represented in Fourier-transforms as

$$\rho(t-t',x-x') = \int \frac{d\omega}{2\pi}\frac{dk}{2\pi}\rho(\omega,k)e^{-i\omega(t-t')+ik(x-x')},$$
$$\chi^{-1}(t-t',x-x') = \int \frac{d\omega}{2\pi}\frac{dk}{2\pi}\chi^{-1}(\omega,k)e^{-i\omega(t-t')+ik(x-x')}. \tag{7.108}$$

As the excitation source term is here taken as a single Fourier-component

$$F = -\partial_x \mathcal{P} = -ik\tilde{\mathcal{P}}e^{-i\omega t+ikx}, \tag{7.109}$$

the macroscopic response-fields are sought in the form, (as usual, by language abuse, we dispense noting the tilde on the fields)

$$V = Ve^{-i\omega t+ikx}, \quad \text{etc.} \tag{7.110}$$

The above equations then directly express as

$$-i\omega B = -ikV, \qquad -i\omega D = ikH - ik\tilde{\mathcal{P}}, \tag{7.111}$$
$$D = \rho(\omega,k)V, \qquad H = -\chi^{-1}(\omega,k)B, \tag{7.112}$$

and we find that

$$\rho(\omega,k) = \frac{ik\left(H-\tilde{\mathcal{P}}\right)}{-i\omega V}, \quad \chi^{-1}(\omega,k) = -\frac{H}{B}. \tag{7.113}$$

In the second consideration, starting at the microscopic level, we write the microscopic action-response problem to be solved, (7.1)–(7.6), with the excitation $\boldsymbol{f} = -\partial_x \mathcal{P}\hat{\boldsymbol{x}}$, $-\partial_x \mathcal{P}$ given by (7.109) in the fluid, and we look for its solution with fields varying like

$$\boldsymbol{v} = \boldsymbol{v}(\omega,k,\boldsymbol{x})e^{-i\omega t+ikx}, \quad \text{etc.} \tag{7.114}$$

If the geometry is stationary random, the solution amplitudes $\boldsymbol{v}(\omega,k,\boldsymbol{x})$, etc., are uniquely fixed by the condition to be stationary random, (and proportional to the excitation amplitude); if it is periodic, they are fixed by the condition to be periodic, (and proportional to the excitation amplitude), with an arbitrary choice to be made, on the minimal periodicities of the tilde-fields.[13]

If we are to use Lorentz's averaging conception we solve the above in one single realization; if we are to use Gibbs' average, we solve it in the different realizations, and for short, above and in what follows, we do not mention the presence of the realization argument ϖ in the amplitudes: $\boldsymbol{v}(\omega,k,\boldsymbol{x}) \to \boldsymbol{v}(\omega,k,\boldsymbol{x};\varpi)$, etc., and in the fluid and pore-surface domains: $\mathcal{V}_f, \partial\mathcal{V} \to \mathcal{V}_f(\varpi), \partial\mathcal{V}(\varpi)$.

The macroscopic velocity $V = Ve^{-i\omega t+ikx}$, and macroscopic condensation, $B = Be^{-i\omega t+ikx}$, in the first consideration, are the macroscopic averages of the (x com-

[13] In that case, here, we only have to precise the integer M_1, fixing the retained period along x.

ponent of) velocity and condensation in the second consideration, i.e.

$$\begin{aligned} V = Ve^{-i\omega t+ikx} =& \langle \boldsymbol{v}(\omega,k,\boldsymbol{x})e^{-i\omega t+ikx}\rangle \cdot \hat{\boldsymbol{x}} \\ =& \langle \boldsymbol{v}(\omega,k,\boldsymbol{x})\rangle \cdot \hat{\boldsymbol{x}}e^{-i\omega t+ikx} = \langle v(\omega,k,\boldsymbol{x})\rangle e^{-i\omega t+ikx}, \end{aligned} \tag{7.115}$$

$$B = Be^{-i\omega t+ikx} = \langle b(\omega,k,\boldsymbol{x})e^{-i\omega t+ikx}\rangle = \langle b(\omega,k,\boldsymbol{x})\rangle e^{-i\omega t+ikx}, \tag{7.116}$$

(these equalities are exactly obtained with Gibbs' averaging, with no restriction on k; with Lorentz's averaging, it is assumed that the macroscopic wavelength $\lambda = 2\pi/k$ is large compared to the homogenization length). Applying the definition (7.107) in which we substitute, (the notation $+c.c.$ adds the complex conjugate)

$$\begin{aligned} p &= \frac{1}{2}\left(p(\omega,k,\boldsymbol{x})e^{-i\omega t+ikx} + c.c.\right), \\ v &= \frac{1}{2}\left(v(\omega,k,\boldsymbol{x})e^{-i\omega t+ikx} + c.c.\right), \\ H &= \frac{1}{2}\left(H(\omega,k)e^{-i\omega t+ikx} + c.c.\right), \end{aligned} \tag{7.117}$$

and noting that the exponentials can be extracted from the averaging symbols in (7.107), (for Lorentz's average, long-wavelength regime is assumed in extracting the exponentials; for Gibbs' average, the extraction is general), we derive, after identification of the terms having same exponentials:

$$\langle p(\omega,k,\boldsymbol{x})v(\omega,k,\boldsymbol{x})\rangle = -\langle v(\omega,k,\boldsymbol{x})\rangle H(\omega,k). \tag{7.118}$$

Requiring then the compatibility of the above two considerations—directly macroscopic, and microscopically averaged—there follows that the kernels operators, by definition, can be computed after solving the microscopic action-response problem (7.1)–(7.6), averaging, and plugging in the definitions (7.113) the following values:

$$V = \langle v(\omega,k,\boldsymbol{x})\rangle, \qquad B = \langle b(\omega,k,\boldsymbol{x})\rangle, \qquad H = -\frac{\langle p(\omega,k,\boldsymbol{x})v(\omega,k,\boldsymbol{x})\rangle}{\langle v(\omega,k,\boldsymbol{x})\rangle}. \tag{7.119}$$

In brief, an "action-response problem" to determine the nonlocal $\rho(\omega,k)$ and $\chi^{-1}(\omega,k)$ is obtained, that can be summarized as follows, (we do not mention the presence of the realization argument, fixed argument ϖ_0 for Lorentz's conception, variable argument ϖ for Gibbs' conception, in the amplitudes and in the fluid and pore-surface domains).

Action-response problem to determine $\rho(\omega, k)$ ***and*** $\chi^{-1}(\omega, k)$:

(i) Let us subject the permeating fluid to the action of a longitudinal bulk force $\boldsymbol{f}$ per unit fluid volume, deriving from a potential $\mathcal{P}$, and varying as follows in the fluid:

$$\boldsymbol{f} = f\hat{\boldsymbol{x}} = -\partial_x \mathcal{P}\hat{\boldsymbol{x}}, \qquad \mathcal{P} = \tilde{\mathcal{P}} e^{-i\omega t + ikx}, \qquad \tilde{\mathcal{P}} = Cst. \tag{7.120}$$

(ii) Let us find the response of the permeating fluid to this action. We seek fields $\boldsymbol{v} = \boldsymbol{v}(\omega, k, \boldsymbol{x})e^{-i\omega t+ikx}$, $b = b(\omega, k, \boldsymbol{x})e^{-i\omega t+ikx}$, $p = p(\omega, k, \boldsymbol{x})e^{-i\omega t+ikx}$, $\tau = \tau(\omega, k, \boldsymbol{x})e^{-i\omega t+ikx}$, satisfying the motion equations

$$\partial \cdot \boldsymbol{v} + \frac{\partial b}{\partial t} = 0, \qquad \text{in } \mathcal{V}_f, \tag{7.121}$$

$$\rho_0 \frac{\partial \boldsymbol{v}}{\partial t} = -\partial p + \eta \partial^2 \boldsymbol{v} + \left(\frac{\eta}{3} + \zeta\right) \partial(\partial \cdot \boldsymbol{v}) + \boldsymbol{f}, \qquad \text{in } \mathcal{V}_f, \tag{7.122}$$

$$\gamma \chi_0 p = b + \beta_0 \tau, \qquad \text{in } \mathcal{V}_f, \tag{7.123}$$

$$\rho_0 c_P \frac{\partial \tau}{\partial t} = \beta_0 T_0 \frac{\partial p}{\partial t} + \kappa \partial^2 \tau, \qquad \text{in } \mathcal{V}_f, \tag{7.124}$$

and

$$\boldsymbol{v} = \boldsymbol{0}, \qquad \text{on } \partial\mathcal{V}, \tag{7.125}$$

$$\tau = 0, \qquad \text{on } \partial\mathcal{V}, \tag{7.126}$$

and whose tilde-amplitudes in $(\omega, k, \boldsymbol{x})$ are proportional to $\tilde{\mathcal{P}}$.

(iii) There is unique solution $\boldsymbol{v}(\omega, k, \boldsymbol{x})$, $b(\omega, k, \boldsymbol{x})$, $p(\omega, k, \boldsymbol{x})$ and $\tau(\omega, k, \boldsymbol{x})$, to this action-response problem. We denote $v(\omega, k, \boldsymbol{x}) = \boldsymbol{v}(\omega, k, \boldsymbol{x}) \cdot \hat{\boldsymbol{x}}$. Then according to (7.113), the effective density $\rho(\omega, k)$ and effective bulk modulus $\chi^{-1}(\omega, k)$ are obtained through the definitions

$$\rho(\omega, k) = \frac{ik\left(H - \tilde{\mathcal{P}}\right)}{-i\omega V}, \quad \chi^{-1}(\omega, k) = -\frac{H}{B}, \tag{7.127}$$

where we plug the values (7.119) of V, B, and H:

$$V = \langle v(\omega, k, \boldsymbol{x})\rangle, \ B = \langle b(\omega, k, \boldsymbol{x})\rangle, \ H = -\frac{\langle p(\omega, k, \boldsymbol{x}) v(\omega, k, \boldsymbol{x})\rangle}{\langle v(\omega, k, \boldsymbol{x})\rangle}. \tag{7.128}$$

According to (7.102) and (7.103), the operators $\hat{\chi}_V$, $\hat{\rho}$, and $\hat{\chi}_B$, $\widehat{\chi^{-1}}$, have Fourier kernels which verify the relations

$$\rho(\omega, k) = \frac{\rho_0}{\phi} + \chi_V(\omega, k), \tag{7.129}$$

and

$$\chi^{-1}(\omega, k) = \frac{1}{\chi_0 \phi} - \chi_B(\omega, k). \tag{7.130}$$

Therefore, the above determination of $\hat{\rho}$ and $\hat{\chi}^{-1}$ is also a determination of $\hat{\chi}_V$ and $\hat{\chi}_B$.

Finally, we note that the physical content of the above procedure is just equivalent to stating that the acoustic equations can be put in the form (7.63)–(7.67) which expresses the electromagnetic-acoustic analogy, with in addition, the H-field, taken according to the "acoustic Heaviside-Poynting" identification, (7.68). We view this identification as a final deepening of the electromagnetic analogy, even if, at present, in macroscopic electromagnetics, we have no clue on the corresponding notion of "energy current density carried out in electromagnetic form".

7.8 Other Points of View

If we were to use the other points of view, sketched in Sect. 7.5, we would write the following.

7.8.1 Acoustics Translation of the Customary Point of View in Electromagnetics

Considering that the (7.76)–(7.80) obtained by setting $M \equiv 0$ would apply, we would write, instead of (7.111)–(7.112):

$$-i\omega B = -ikV, \qquad -i\omega D = ikH - ik\tilde{\mathcal{P}}, \tag{7.131}$$

$$D = \rho(\omega, k)V, \qquad H = -\chi^{-1}(\omega, k)B = -\frac{B}{\chi_0 \phi}, \tag{7.132}$$

and it would give

$$\rho(\omega, k) = \frac{-ik\left(\dfrac{B}{\chi_0\phi} + \tilde{\mathcal{P}}\right)}{-i\omega V}, \quad \chi^{-1}(\omega, k) = \frac{1}{\chi_0 \phi}. \tag{7.133}$$

The action-response problem to determine $\rho(\omega, k)$ would be as follows.

Action-response problem to determine $\rho(\omega, k)$:

(i) Let us subject the permeating fluid to the action of a longitudinal bulk force $\boldsymbol{f}$ per unit fluid volume, deriving from a potential $\mathcal{P}$, and varying as follows in the fluid:

$$\boldsymbol{f} = f\hat{\boldsymbol{x}} = -\partial_x \mathcal{P}\hat{\boldsymbol{x}}, \qquad \mathcal{P} = \tilde{\mathcal{P}} e^{-i\omega t + ikx}, \qquad \tilde{\mathcal{P}} = Cst. \tag{7.134}$$

(ii) Let us find the response of the permeating fluid to this action.
We seek fields $\boldsymbol{v} = \boldsymbol{v}(\omega, k, \boldsymbol{x})e^{-i\omega t+ikx}$, $b = b(\omega, k, \boldsymbol{x})e^{-i\omega t+ikx}$, $p = p(\omega, k, \boldsymbol{x})e^{-i\omega t+ikx}$, $\tau = \tau(\omega, k, \boldsymbol{x})e^{-i\omega t+ikx}$, satisfying the motion equations

$$\partial \cdot \boldsymbol{v} + \frac{\partial b}{\partial t} = 0, \qquad \text{in } \mathcal{V}_f, \tag{7.135}$$

$$\rho_0 \frac{\partial \boldsymbol{v}}{\partial t} = -\partial p + \eta \partial^2 \boldsymbol{v} + \left(\frac{\eta}{3} + \zeta\right) \partial(\partial \cdot \boldsymbol{v}) + \boldsymbol{f}, \qquad \text{in } \mathcal{V}_f, \tag{7.136}$$

$$\gamma \chi_0 p = b + \beta_0 \tau, \qquad \text{in } \mathcal{V}_f, \tag{7.137}$$

$$\rho_0 c_P \frac{\partial \tau}{\partial t} = \beta_0 T_0 \frac{\partial p}{\partial t} + \kappa \partial^2 \tau, \qquad \text{in } \mathcal{V}_f, \tag{7.138}$$

and

$$\boldsymbol{v} = \boldsymbol{0}, \qquad \text{on } \partial\mathcal{V}, \tag{7.139}$$

$$\tau = 0, \qquad \text{on } \partial\mathcal{V}, \tag{7.140}$$

and whose amplitudes in $(\omega, k, \boldsymbol{x})$ are proportional to $\tilde{\mathcal{P}}$.

(iii) There is unique solution $\boldsymbol{v}(\omega, k, \boldsymbol{x})$, $b(\omega, k, \boldsymbol{x})$, $p(\omega, k, \boldsymbol{x})$ and $\tau(\omega, k, \boldsymbol{x})$, to this action-response problem. We denote $v(\omega, k, \boldsymbol{x}) = \boldsymbol{v}(\omega, k, \boldsymbol{x}) \cdot \hat{\boldsymbol{x}}$. Then according to (7.133.1), the effective density $\rho(\omega, k)$ is obtained through the definition

$$\rho(\omega, k) = \frac{-ik\left(\dfrac{B}{\chi_0 \phi} + \tilde{\mathcal{P}}\right)}{-i\omega V}, \tag{7.141}$$

where we plug the following values of V and B

$$V = \langle v(\omega, k, \boldsymbol{x}) \rangle, \quad B = \langle b(\omega, k, \boldsymbol{x}) \rangle. \tag{7.142}$$

This formulation—whose electromagnetic counterpart is just that ordinarily used in literature on spatial dispersion [8–10]—fixes the bulk modulus of the material to an artificial constant value (7.133.2) determined by the adiabatic bulk modulus of the fluid, independently of ω and k. But this constant does not appear quite meaningful, in the present acoustic context, where the pressure can be distributed at the pore scale, (with its direct mean, not necessarily being the meaningful macroscopic variable, see below), and thermal exchanges can occur. It suggests that in electromagnetics, the counterpart setting $\boldsymbol{H} = \boldsymbol{B}/\mu_0$, common in the presence of spatial dispersion, will also be found to be an unappropriate definition, in some respect and some cases.

Finally, let us conclude with the case where, as is done in local theory, we identify the H-field with the volume-averaged opposite pressure (7.86).

7.8.2 *Acoustics Formulation in Terms of Volume-Averaged Pressure*

Considering that the (7.81)–(7.86) would apply, we would have no change in (7.111)–(7.112), and obtain as before the relations (7.113). The difference would be that, for H, we would have to use the artificial definition, $H = -\langle p \rangle_f$. Therefore, the action-response problem to determine $\rho(\omega, k)$ and $\chi^{-1}(\omega, k)$ would be as follows.

Action-response problem to determine $\rho(\omega, k)$ ***and*** $\chi^{-1}(\omega, k)$:

(i) Let us subject the permeating fluid to the action of a longitudinal bulk force $\boldsymbol{f}$ per unit fluid volume, deriving from a potential $\mathcal{P}$, and varying as follows in the fluid:

$$\boldsymbol{f} = f\hat{\boldsymbol{x}} = -\partial\mathcal{P}, \qquad \mathcal{P} = \tilde{\mathcal{P}}e^{-i\omega t+ikx}, \qquad \tilde{\mathcal{P}} = Cst. \tag{7.143}$$

(ii) Let us find the response of the permeating fluid to this action. We seek fields $\boldsymbol{v} = \boldsymbol{v}(\omega, k, \boldsymbol{x})e^{-i\omega t+ikx}$, $b = b(\omega, k, \boldsymbol{x})e^{-i\omega t+ikx}$, $p = p(\omega, k, \boldsymbol{x})e^{-i\omega t+ikx}$, $\tau = \tau(\omega, k, \boldsymbol{x})e^{-i\omega t+ikx}$, satisfying the motion equations

$$\partial \cdot \boldsymbol{v} + \frac{\partial b}{\partial t} = 0, \qquad \text{in } \mathcal{V}_f, \tag{7.144}$$

$$\rho_0\frac{\partial \boldsymbol{v}}{\partial t} = -\partial p + \eta\partial^2\boldsymbol{v} + \left(\frac{\eta}{3} + \zeta\right)\partial(\partial \cdot \boldsymbol{v}) + \boldsymbol{f}, \qquad \text{in } \mathcal{V}_f, \tag{7.145}$$

$$\gamma\chi_0 p = b + \beta_0\tau, \qquad \text{in } \mathcal{V}_f, \tag{7.146}$$

$$\rho_0 c_P\frac{\partial \tau}{\partial t} = \beta_0 T_0\frac{\partial p}{\partial t} + \kappa\partial^2\tau, \qquad \text{in } \mathcal{V}_f, \tag{7.147}$$

and

$$\boldsymbol{v} = \boldsymbol{0}, \qquad \text{on } \partial\mathcal{V}, \tag{7.148}$$

$$\tau = 0, \qquad \text{on } \partial\mathcal{V}, \tag{7.149}$$

and whose amplitudes in $(\omega, k, \boldsymbol{x})$ are proportional to $\tilde{\mathcal{P}}$.

(iii) There is unique solution $\boldsymbol{v}(\omega, k, \boldsymbol{x})$, $b(\omega, k, \boldsymbol{x})$, $p(\omega, k, \boldsymbol{x})$ and $\tau(\omega, k, \boldsymbol{x})$, to this action-response problem. We denote $v(\omega, k, \boldsymbol{x}) = \boldsymbol{v}(\omega, k, \boldsymbol{x}) \cdot \hat{\boldsymbol{x}}$. Then, the effective density $\rho(\omega, k)$ and effective bulk modulus $\chi^{-1}(\omega, k)$ are obtained through the definitions

$$\rho(\omega, k) = \frac{ik\left(H - \tilde{\mathcal{P}}\right)}{-i\omega V}, \quad \chi^{-1}(\omega, k) = -\frac{H}{B}, \tag{7.150}$$

where we plug the following values of V, B, and H

$$V = \langle v(\omega, k, \boldsymbol{x})\rangle, \quad B = \langle b(\omega, k, \boldsymbol{x})\rangle, \quad H = -\frac{\langle p(\omega, k, \boldsymbol{x})\rangle}{\phi}. \tag{7.151}$$

This determination has no counterpart in electromagnetics in the absence of the hypothesized thermodynamic laws associated to the behaviour of polarization charges and currents in matter. When the pressure is distributed at the pore scale because of long-wavelengths local resonances, or short wavelengths, there is no reason that its direct volume average would be the meaningful variable to define an effective bulk modulus, (see again end of Sect. 7.5.3). To repeat ourselves, we expect that the two preceding formulations, contrary to the first (7.104)–(7.107), will not lead to natural generalization when the finite dimensions of materials will be taken into account.

7.9 Characteristic Wavenumbers and Impedances

A characteristic feature of a nonlocal effective medium theory is that it allows for the propagation of several normal waves at a given angular frequency ω. Here, as we focus on the macroscopic propagation along a symmetry axis x, we can have a series of normal waves varying like $e^{-i\omega t+ikx}$, with ω and k related by the dispersion equation:

$$\rho(\omega, k)\chi(\omega, k)\omega^2 = k^2. \tag{7.152}$$

At given real ω, as our medium is lossy, the imaginary parts of the wavenumbers $k(\omega)$ solutions to (7.152), will be positive, $\Im[k(\omega)] > 0$, for the waves propagating in the direction $+x$, (that can be created by a source in the direction $-x$). This ensures that these waves are damped. Depending on the case, the waves propagating in the direction $+x$ can have positive or negative sign of $\Re[k(\omega)]$, corresponding to positive or negative phase velocity, $c_\varphi(\omega) = \omega/\Re[k(\omega)]$.

In the well-defined stationary-random case, assuming that these solutions $k(\omega)$ are nondegenerate, they can be labelled

$$k = k_n(\omega), \tag{7.153}$$

with a discrete label $n = 1, 2, \ldots$, which orders them from the least-attenuated mode, to the highly attenuated ones, giving, $0 < \Im[k_1(\omega)] < \Im[k_2(\omega)] < \cdots$.

The normal-mode fields will have the form, (we do not indicate the realization argument, fixed argument ϖ_0 for Lorentz's conception, variable argument ϖ for

Gibbs' conception), $\boldsymbol{v}(t, \boldsymbol{x}) = \tilde{\boldsymbol{v}}_n(\omega, \boldsymbol{x})e^{-i\omega t + ik_n(\omega)x}$, $b(t, \boldsymbol{x}) = \tilde{b}_n(\omega, \boldsymbol{x})e^{-i\omega t + ik_n(\omega)x}$, $p(t, \boldsymbol{x}) = \tilde{p}_n(\omega, \boldsymbol{x})e^{-i\omega t + ik_n(\omega)x}$, etc., with unique stationary-random amplitude functions, $\tilde{\boldsymbol{v}}_n(\omega, \boldsymbol{x})$, $\tilde{b}_n(\omega, \boldsymbol{x})$, $\tilde{p}_n(\omega, \boldsymbol{x})$, etc., presenting when n increases, more and more rapid variations at the small scale.

To the nth normal-mode solution is associated an H-field varying like $H(t, \boldsymbol{x}) = \tilde{H}_n(\omega)e^{-i\omega t + ik_n(\omega)x}$, and having, resp., characteristic amplitudes, $\tilde{H}_n(\omega) = -\langle \tilde{p}_n(\omega, \boldsymbol{x})\tilde{\boldsymbol{v}}_n(\omega, \boldsymbol{x})\rangle \cdot \hat{\boldsymbol{x}} / \langle \tilde{\boldsymbol{v}}_n(\omega, \boldsymbol{x})\rangle \cdot \hat{\boldsymbol{x}}$, in the acoustic Heaviside-Poynting conception (7.63)–(7.68), or $\tilde{H}_n(\omega) = -\chi_0^{-1}\langle \tilde{b}_n(\omega, \boldsymbol{x})\rangle$, in the acoustic conception (7.76)–(7.80) counterpart of conventional electromagnetics, or $\tilde{H}_n(\omega) = -\langle \tilde{p}_n(\omega, \boldsymbol{x})\rangle$, in the acoustic conception (7.81)–(7.86) inspired by the local-theory usual definition of macroscopic pressure as a volumic mean. These conceptions, in turn, lead to define frequency-dependent, characteristic macroscopic modal impedances $(Z_c)_n(\omega)$, all given by, in the different cases

$$(Z_c)_n(\omega) = \frac{-\tilde{H}_n(\omega)}{\langle \tilde{\boldsymbol{v}}_n(\omega, \boldsymbol{x})\rangle \cdot \hat{\boldsymbol{x}}}, \tag{7.154}$$

as well as associated frequency-dependent characteristic modal density and bulk-modulus functions, all given by

$$\rho_n(\omega) = \frac{k_n(\omega)}{\omega}(Z_c)_n(\omega), \qquad \chi_n^{-1}(\omega) = \frac{\omega}{k_n(\omega)}(Z_c)_n(\omega). \tag{7.155}$$

In the presence of resonances, these modal functions may have much more general and complicated behaviours than the simple relaxational-ones, described in the Appendix for the $\rho(\omega)$ and $\chi^{-1}(\omega)$ functions of local theory.

The different formulations we have given of the nonlocal equations and operators, namely based on (7.63)–(7.68) or (7.76)–(7.80) or (7.81)–(7.86), lead to the same wavenumbers but different impedances, densities and compressibilities. Subsequent work, we believe, will show, (in particular when considering inhomogeneous materials), that the latter quantities are best defined in the formulation (7.63)–(7.68) using Heaviside-Poynting's identification (7.68).

In the periodic case we will obtain comparable results, with however ambiguities. There will be direct relation between the normal modes in the considered macroscopic homogeneous medium, and the so-called Bloch modes in one periodic realization. General properties will have to be closely examined in further work. We anticipate that the normal mode wavenumbers defined by the macroscopic Gibbs nonlocal homogeneous medium will allow attributing unambiguously and successively, definite Brillouin zones to a given Bloch-wave, when frequency increases.

7.10 Conclusions

By following an electromagnetic analogy introduced in the Chap. 6, we have shown that the general equations governing sound propagation in a direction x of symmetry, in macroscopically homogeneous rigid-framed fluid-saturated porous metamaterials, assume the nonlocal Maxwellian pattern of (7.63)–(7.67).

In these equations, V and B are the mean (volume- or ensemble-averaged) x-velocity and condensation, and D and H, related to the former fields by nonlocal constitutive operator relations, can be defined in different manners, leading to different definitions of the operators.

We suggested that there is a physically preferred choice, which will be to choose the H-field according to the "acoustic Heaviside-Poynting" identification (7.68). In future work, this statement will have to be supported by explicitly showing the advantages of this identification (7.68).

In this connection we should mention an imprecision made in preliminary works on the present nonlocal theory [11–14]. The fact that the electromagnetic analogy led directly to the simultaneous definition of the two operators density $\hat{\rho}$ and bulk-modulus $\hat{\chi}^{-1}$, by solving a single action-response problem, (7.120)–(7.128), where the medium is subjected to an external longitudinal force, was missed in these works. Instead of being deduced from the analogy, the action-response procedures for calculating density and bulk-modulus operators were independently postulated, based on what was observed to be true in the unbounded fluid. For density, it led to the correct action-response procedure. But for bulk-modulus, it turns out that the proposed procedure, [11], (71), is a slightly faulty one, which is working correctly only when k is a characteristic wavenumber of the medium. This error could not be detected in the verifications that were made in subsequent work [12–14], because the latter were restricted to considering only the Bloch modes, for which k was, by force, one of the characteristic wavenumbers. When the forcing is made at ω and k equal to a characteristic wavenumber solution $k_n(\omega)$ to the dispersion equation (7.152), a resonance occurs, meaning that a finite response is produced by a vanishingly small forcing. In that case, the faulty procedure [11], (71), gives back the present (7.127.2), because the forcing amplitudes disapppear.

In future work, we will have to clarify if and how the same operators can also be obtained by giving heat instead of doing work.

Finally, we recall that, much remains to be done to generalize the description in the case where the medium has finite dimensions, is anisotropic and poroelastic, and to express all the consequences of nonlocalities.

Appendix: Local Dynamic Homogenization of Rigid-Framed Fluid-Saturated Porous Materials

In some geometries and at long wavelengths, rather than trying to solve the very complex system of coupled (7.1)–(7.6), we can break it down and solve it into independent and simplified pieces, encapsulating the main effects. In the density operator will be put inertial and viscous effects, in the compressibility operator, elastic and thermal effects, evaluated in a corresponding special local limit. At the macroscopic level where one writes, for macroscopic motion along principal axis x[14]:

$$\frac{\partial \langle b \rangle}{\partial t} + \partial_x \langle v \rangle = 0, \quad \frac{\partial D}{\partial t} = \partial_x H + [F], \tag{7.156}$$

this will mean using a simplified conception of the fields D and H, that will correspond to introducing the approximations:

$$-H\langle v \rangle = \langle pv \rangle \cong \langle p \rangle_f \langle v \rangle, \tag{7.157}$$

(note that, as the wavelengths are large, the mean operation $\langle\,\rangle$ can be conceived in Lorentz's manner), and simultaneously, discarding spatial dispersion:

$$D(t,x) = \hat{\rho}\langle v \rangle(t,x) \cong \int_{-\infty}^{t} \rho(t-t')\langle v \rangle(t',x)dt', \tag{7.158}$$

$$H(t,x) = -\hat{\chi}^{-1}\langle b \rangle(t,x) \cong -\int_{-\infty}^{t} \chi^{-1}(t-t')\langle b \rangle(t',x)dt'. \tag{7.159}$$

This simplified conception will be justified when the geometries are simple, characterized by one typical pore size, meaning that the fluid motion is practically divergence-free at the pore scale. Usually, it is obtained at first order by applying the two-scale asymptotic homogenization method [3, 4], however, the corresponding process does not appear to be truly consistent. Here we will get the same results "with our hands", from the simplification that the spatial dispersion phenomena are absent, and its corollary here, the divergence-free nature of fluid motion at the pore scale.

Dynamic Viscous and Thermal Tortuosities and Permeabilities, and Definition of the Local Density $\rho(\omega)$ and Compressibility $\chi(\omega)$

Here, we reason for convenience with external, long-wavelength force present. Analogous considerations will hold without it, still assuming long-wavelengths.

For the external force F, it suffices to consider a single exponential form, $\boldsymbol{F} = F\hat{\boldsymbol{x}} = -\partial_x \mathcal{P}\hat{\boldsymbol{x}}$, with $\mathcal{P} = \frac{1}{2}(\tilde{\mathcal{P}}e^{-i\omega t+ikx} + c.c.)$, and $\tilde{\mathcal{P}}$ a complex constant. Whenever convenient and without notice, we work in complex representation, e.g. $\mathcal{P} = \tilde{\mathcal{P}}e^{-i\omega t+ikx}$, omitting the real part symbol $\Re\,()$. A long-wavelength limit will be considered, i.e. $kL_h \to 0$, where L_h is a homogenization length.[15]

[14] Generalization to anisotropic materials presents no difficulty.

[15] k is set as $2\pi/\lambda$, with λ taken to be on the same order as the macroscopic wavelength in the medium, (which is unique within local theory).

We will first observe that, as there are no local resonances in the simple geometries considered, the pressure in the fluid must be everywhere, very close to its average value, $\langle p \rangle_f$, where $\langle \rangle_f$ is the Lorentz average in the fluid at the given point. Let us indeed write the pressure in terms of its mean and deviatoric part, at the given position:

$$p = \langle p \rangle_f + \delta p, \qquad \langle \delta p \rangle_f = 0. \tag{7.160}$$

Considering that the gradient of the two terms are comparable, and since $\langle p \rangle_f \sim e^{ikx}$ varies over macroscopic distances $\lambda = 2\pi/k$, whereas δp varies over small-scale distances ℓ, (an estimate of the pore size[16]), we write $\langle p \rangle_f/\lambda \sim \delta p/\ell$. Consequently, the order of magnitude of the deviatoric part δp is that of the mean part $\langle p \rangle_f$, times the very small ratio ℓ/λ. In porous materials used for noise control at audible frequencies, this ratio is typically very small, on order of 10^{-4}, coherent with considering a limit $\ell/\lambda \to 0$. Hence, because of the huge scale separation, we have that

$$p \cong \langle p \rangle_f, \quad \langle pv \rangle \cong \langle p \rangle_f \langle v \rangle, \quad H \cong -\langle p \rangle_f. \tag{7.161}$$

In particular, for the distribution of excess temperature in a representative elementary volume (REV) around a given position x_0, we can consider, instead of the profile of the exact solution of the (7.1)–(7.6), the approximate profile, generated in the following simplified problem, where the pressure field is, in the REV, assimilated to its average part or $(-H)$-part $\cong \langle p \rangle_f$, calculated at the central position:

$$\rho_0 c_P \frac{\partial \tau}{\partial t} = \beta_0 T_0 \frac{\partial \langle p \rangle_f}{\partial t} + \kappa \partial^2 \tau, \qquad \text{in } \mathcal{V}_f, \tag{7.162}$$

$$\langle p \rangle_f = \text{spatial constant}, \qquad \text{in } \mathcal{V}_f, \tag{7.163}$$

$$\tau = 0, \qquad \text{on } \partial\mathcal{V}. \tag{7.164}$$

With $\langle p \rangle_f = -[\tilde{H} e^{ikx_0}]e^{-i\omega t}$, the field τ in the REV is uniquely fixed by the constant $-[\tilde{H} e^{ikx_0}]$ and the frequency. The field τ solution to the above problem, can be used to define a response function $\alpha'(\omega)$, known as "dynamic thermal tortuosity" [15], determined by the microgeometry, and such that, by definition

$$\rho_0 c_P \alpha'(\omega) \frac{\partial \langle \tau \rangle_f}{\partial t} = \beta_0 T_0 \frac{\partial \langle p \rangle_f}{\partial t}. \tag{7.165}$$

Equivalently, one defines a "dynamic thermal permeability" $k'(\omega)$ by setting [3]

$$\phi \langle \tau \rangle_f = \frac{k'(\omega)}{\kappa} \beta_0 T_0 \frac{\partial \langle p \rangle_f}{\partial t}. \tag{7.166}$$

---

[16] Note that the idealization that there is mainly "one" pore-size, excludes the presence of Helmholtz resonators: resonators involve widely different sizes in their necks and cavities.

The denominations of "tortuosity" and "permeability" are not quite appropriate here, but they come from an analogy with corresponding viscous functions, seen below. Obviously, both functions are related by

$$k'(\omega)\alpha'(\omega) = \frac{\nu'\phi}{-i\omega}, \tag{7.167}$$

where

$$\nu' \equiv \frac{\kappa}{\rho_0 c_P} = \nu/\,\mathrm{Pr}, \quad \nu \equiv \frac{\eta}{\rho_0}. \tag{7.168}$$

Taking the fluid-average of the equation of state, and its time derivative, we write

$$\gamma\chi_0\frac{\partial\langle p\rangle_f}{\partial t} = \frac{\partial\langle b\rangle_f}{\partial t} + \beta_0\frac{\partial\langle\tau\rangle_f}{\partial t}. \tag{7.169}$$

Inserting (7.165) in (7.169) and using the thermodynamic identity seen in the Chap. 6, Sect. 6.2, (6.5), this gives the relation, (after removal of the time derivatives)

$$\chi_0\left[\gamma - \frac{\gamma-1}{\alpha'(\omega)}\right]\langle p\rangle_f = \langle b\rangle_f. \tag{7.170}$$

Therefore in accordance with (7.161) and (7.159), we find a relation having the form, in harmonic regime

$$H = -\chi^{-1}(\omega)\langle b\rangle, \tag{7.171}$$

with

$$\chi(\omega) = \phi\chi_0\left[\gamma - \frac{\gamma-1}{\alpha'(\omega)}\right] = \phi\chi_0\left[\gamma - (\gamma-1)\frac{-i\omega}{\nu'\phi}k'(\omega)\right]. \tag{7.172}$$

This gives the Fourier coefficients of the kernel function $\chi(t)$ in (7.159). It represents a dynamic compressibility, function of frequency because of the thermal exchanges between fluid and solid. This is often written as a relation

$$\chi_0\beta(\omega)\frac{\partial\langle p\rangle_f}{\partial t} = -\partial\cdot\langle \boldsymbol{v}\rangle_f, \tag{7.173}$$

with the function $\beta(\omega)$

$$\beta(\omega) = \frac{\chi(\omega)}{\phi\chi_0} = \gamma - \frac{\gamma-1}{\alpha'(\omega)} = \gamma - (\gamma-1)\frac{-i\omega}{\boldsymbol{v}'(\phi)}\boldsymbol{k}'(\omega) \tag{7.174}$$

a normalized dynamic compressibility, modifying the adiabatic value [3, 15].

In defining the above excess temperature pattern τ, whose average served us to compute the wanted response functions, we made abstraction of the spatial variations of $\langle p\rangle_f$ in the representative volume. Exactly the same average $\langle\tau\rangle_f$ and hence response functions, would have been obtained, however, if we had also taken into account, to first order, the variation of $\langle p\rangle_f$ in the representative volume: to first order the quantity possesses a linear variation, which doesn't contribute, by symmetry, to the mean excess temperature.

To obtain the Fourier coefficients of the kernel function $\rho(t)$, we proceed in similar manner. To compute the velocity pattern in a REV around a given position x_0, we consider that the fluid is subjected to a spatially-constant macroscopic force

$$\partial H + \boldsymbol{F} \cong -\partial\langle p\rangle_f + \boldsymbol{F} = \left[-ik\left(\tilde{\mathcal{P}} + \langle\tilde{p}\rangle_f\right)e^{ikx_0}\right]e^{-i\omega t}\hat{\boldsymbol{x}}, \tag{7.175}$$

equal to the sum of external force $\boldsymbol{F}$ and macroscopic pressure-gradient force. That is, instead of the exact solution of (7.1)–(7.6), we consider the approximated velocity profile generated in the following simplified problem where $\partial H + \boldsymbol{F}$ is introduced as the constant (7.175):

$$\rho_0\frac{\partial \boldsymbol{v}}{\partial t} = -\partial(\delta p) + \eta\partial^2\boldsymbol{v} + \partial H + \boldsymbol{F}, \qquad \text{in } \mathcal{V}_f, \tag{7.176}$$

$$\delta p = \text{bounded, stationary random field}, \qquad \text{in } \mathcal{V}_f, \tag{7.177}$$

$$\partial\cdot\boldsymbol{v} = 0, \qquad \text{in } \mathcal{V}_f, \tag{7.178}$$

$$\boldsymbol{v} = \boldsymbol{0}, \qquad \text{on } \partial\mathcal{V}. \tag{7.179}$$

The condition that δp is a stationary random field, applies to stationary random geometries; in periodic geometries it is to be replaced by the condition that δp is a periodic field. Unambiguously here, because of the incompressibility condition and related spatial constancy of the excitation, the periodicity can be taken as the smallest possible. The solution field $\boldsymbol{v}$ is uniquely fixed by the amplitude in brackets $[\cdot]$ in (7.175), the condition set on δp (bounded field), the frequency, and the macroscopic direction $\hat{\boldsymbol{x}}$. There is however an arbitrary constant in the field δp, and it can be noted that it is fixed to the same value by requiring either $\langle\delta p\rangle = 0$ or $\langle\boldsymbol{v}\delta p\rangle = \boldsymbol{0}$.[17] The solution field $\boldsymbol{v}$ to the above problem, can be used to define a response function $\alpha(\omega)$, known as "dynamic viscous tortuosity", determined by the microgeometry, and such that, by definition [2]

$$\rho_0\alpha(\omega)\frac{\partial\langle\boldsymbol{v}\rangle_f}{\partial t} = -\partial\langle p\rangle_f + F\hat{\boldsymbol{x}}. \tag{7.180}$$

Equivalently, one defines a "dynamic viscous permeability" $k(\omega)$ by setting

$$\phi\langle\boldsymbol{v}\rangle_f = \frac{k(\omega)}{\eta}\left[-\partial\langle p\rangle_f + F\hat{\boldsymbol{x}}\right]. \tag{7.181}$$

Both functions are related by

[17] Actually it means that within local theory, and as also expressed in (7.161), we do not see the difference between the two definitions $H = -\langle p\rangle_f$ and $H = -\langle pv\rangle/\langle v\rangle$ of the H-field.

$$k(\omega)\alpha(\omega) = \frac{\nu\phi}{-i\omega}. \tag{7.182}$$

This can now be compared with the equation obtained by combining (7.156.2) and (7.158):

$$\rho(\omega)\frac{\partial\langle v\rangle}{\partial t} = \partial_x H + F. \tag{7.183}$$

Recalling (7.10) and (7.161) we see that

$$\rho(\omega) = \frac{\rho_0\alpha(\omega)}{\phi} = \frac{\eta}{-i\omega k(\omega)}. \tag{7.184}$$

It represents a dynamic density, function of frequency because of the viscous effects. For later use, we observe that the dynamic tortuosities $\alpha(\omega)$ and $\alpha'(\omega)$ are related as follows, to the velocity and excess temperature patterns $\boldsymbol{v}$ and τ, (where the star denotes complex conjugate):

$$\frac{\alpha(\omega)}{\phi} = \frac{\langle \boldsymbol{v}\cdot\boldsymbol{v}^*\rangle}{\langle\boldsymbol{v}\rangle\cdot\langle\boldsymbol{v}^*\rangle} + \frac{\nu}{-i\omega}\frac{\langle -\boldsymbol{v}\cdot\partial^2\boldsymbol{v}^*\rangle}{\langle\boldsymbol{v}\rangle\cdot\langle\boldsymbol{v}^*\rangle}, \tag{7.185}$$

and

$$\frac{\alpha'(\omega)}{\phi} = \frac{\langle\tau\tau^*\rangle}{\langle\tau\rangle\langle\tau^*\rangle} + \frac{\nu'}{-i\omega}\frac{\langle -\tau\partial^2\tau^*\rangle}{\langle\tau\rangle\langle\tau^*\rangle}. \tag{7.186}$$

To see this, we take the dot product of (7.176.1) with $\boldsymbol{v}$, and perform volume average:

$$-i\omega\rho_0\langle\boldsymbol{v}\cdot\boldsymbol{v}\rangle = -\langle\boldsymbol{v}\cdot\partial(\delta p)\rangle + \eta\langle\boldsymbol{v}\cdot\partial^2\boldsymbol{v}\rangle + \langle\boldsymbol{v}\cdot(\partial_x H + F)\hat{\boldsymbol{x}}\rangle.$$

The term $\langle\boldsymbol{v}\cdot\partial(\delta p)\rangle$ vanishes after integration by part, owing to incompressibility, no-slip condition, and the stationary random nature of the fields (periodic nature, in periodic geometries). The term $\langle\boldsymbol{v}\cdot(\partial_x H + F)\hat{\boldsymbol{x}}\rangle$ factorizes as $(\partial_x H + F)\langle\boldsymbol{v}\rangle\cdot\hat{\boldsymbol{x}}$ since $(\partial H + F)$ is treated as a spatial constant, (in so doing, spatial dispersion effects are discarded). This gives, dividing by $\langle\boldsymbol{v}\rangle\cdot\hat{\boldsymbol{x}}$,

$$-i\omega\rho_0\frac{\langle\boldsymbol{v}\cdot\boldsymbol{v}\rangle}{\langle\boldsymbol{v}\rangle\cdot\hat{\boldsymbol{x}}} = \frac{\eta\langle\boldsymbol{v}\cdot\partial^2\boldsymbol{v}\rangle}{\langle\boldsymbol{v}\rangle\cdot\hat{\boldsymbol{x}}} + (\partial_x H + F),$$

from which we obtain the representation:

$$\frac{\alpha(\omega)}{\phi} = \frac{\langle\boldsymbol{v}\cdot\boldsymbol{v}\rangle}{\langle\boldsymbol{v}\rangle^2} + \frac{\nu}{-i\omega}\frac{\langle -\boldsymbol{v}\cdot\partial^2\boldsymbol{v}\rangle}{\langle\boldsymbol{v}\rangle^2}. \tag{7.187}$$

The given form (7.185) is then obtained with the same reasoning, if we work with the complex conjugate of the starting equations, and account for the fact that the fields are stationary. The calculation leading to (7.186) is made in analogous manner, by multiplying the complex conjugate of (7.162) by τ, averaging, and taking into account the stationary random nature of the field.

Finally, knowing $\alpha(\omega)$ or $k(\omega)$, and $\alpha'(\omega)$ or $k'(\omega)$, the effective density and compressibility, are given by (7.184) and (7.172).

Low-Frequency Parameters: k_0, α_0, k'_0, α'_0

In the low-frequency limit, $\omega \to 0$, where viscous effects are "relaxed", viscous shearing motions have time to fully develop in the fluid. It corresponds to a limit of large viscous skin depths $\delta_v = (2\eta/\rho_0\omega)^{1/2}$ with respect to characteristic pore sizes ℓ, i.e. $\omega \ll \ell^2\rho_0/2\eta$. The dynamic permeability tends to Darcy's permeability k_0, and the dynamic tortuosity is mainly purely positive imaginary:

$$\alpha(\omega) \to \frac{\nu\phi}{-i\omega k_0} + \alpha_0, \tag{7.188}$$

with a correction α_0 which is Norris' inertial factor T_0 [16]. We can denote $\boldsymbol{v}_0$ the fluid velocity pattern which appears in d.c. permanent flow ($\omega = 0$):

$$\mathbf{0} = -\partial\delta p_0 + \eta\partial^2\boldsymbol{v}_0 + \left[-\partial_x\langle p\rangle_f + F\right]\hat{\boldsymbol{x}}, \quad \text{in } \mathcal{V}_f \tag{7.189}$$

$$\partial\cdot\boldsymbol{v}_0 = 0, \quad \text{in } \mathcal{V}_f \tag{7.190}$$

$$p_0 = \text{stationary random}, \quad \text{in } \mathcal{V}_f \tag{7.191}$$

$$\boldsymbol{v}_0 = \mathbf{0}, \quad \text{on } \partial\mathcal{V} \tag{7.192}$$

A comparison of (7.188) with (7.185) taken in the limit $\omega \to 0$ shows that

$$\frac{\phi}{k_0} = \frac{\langle -\boldsymbol{v}_0\cdot\partial^2\boldsymbol{v}_0\rangle_f}{\langle\boldsymbol{v}_0\rangle_f^2}, \quad \alpha_0 = \frac{\langle\boldsymbol{v}_0^2\rangle_f}{\langle\boldsymbol{v}_0\rangle_f^2}, \tag{7.193}$$

where we have suppressed complex conjugates as in this "Poiseuille" limit the velocities at different positions in the pores are all in phase. Parameters k_0 and α_0, resp. dimension of surface and dimensionless, are named static or d.c. permeability and tortuosity.

Likewise, in the low-frequency limit, $\omega \to 0$, where thermal effects are "relaxed", thermal exchanges between solid and fluid have time to occur completely. It corresponds to a limit of large thermal skin depths $\delta_\tau = (2\eta/\rho_0\,\mathrm{Pr}\,\omega)^{1/2}$ with respect to characteristic pore sizes ℓ, i.e. $\omega \ll \ell^2\rho_0\,\mathrm{Pr}\,/2\eta$. The solid, which generally has large thermal inertia, imposes its steady ambient temperature to the fluid. The condensation-rarefactions occur in isothermal regime and the fluid "effective bulk modulus" is equal to the isothermal modulus K_0, equal to P_0 the ambient pressure, meaning that $\beta(\omega) \to \gamma$, the adiabatic constant of the saturating fluid. In this limit, corrections to the leading isothermal behaviour are easily obtained by using (7.174) and by noting that the dynamic thermal permeability $k'(\omega)$ tends to Lafarge's d.c. thermal permeability k'_0, (equal to the inverse trapping constant, see [3, 17]), and the dynamic thermal tortuosity is mainly purely positive imaginary:

$$\alpha'(\omega) \to \frac{\nu'\phi}{-i\omega k'_0} + \alpha'_0. \tag{7.194}$$

We can denote τ_0 the fluid excess temperature pattern which appears in d.c. regime, (where d.c. is here only in the sense $\omega \to 0$, and not $\omega = 0$):

$$0 = \kappa \partial^2 \tau_0 + \beta_0 T_0 \partial \langle p \rangle_f / \partial t, \qquad \text{in } \mathcal{V}_f, \tag{7.195}$$

$$\tau_0 = 0, \qquad \text{on } \partial\mathcal{V}. \tag{7.196}$$

A comparison of (7.194) with (7.186) shows that

$$\frac{\phi}{k'_0} = \frac{\langle -\tau_0 \partial^2 \tau_0 \rangle_f}{\langle \tau_0 \rangle_f^2}, \quad \alpha'_0 = \frac{\langle \tau_0^2 \rangle_f}{\langle \tau_0 \rangle_f^2}, \tag{7.197}$$

where we have as before suppressed complex conjugates as the temperature variations in the pores are all in phase in this d.c. limit.

High-Frequency Parameters: α_∞, Λ, Λ'

In the opposite high-frequency limit, $\omega \to \infty$, where the viscous effects are "frozen", viscous shearing motions only have time to develop in an immediate vicinity of the pore walls. It corresponds to a limit of small viscous skin depths, $\omega \gg \ell^2 \rho_0 / 2\eta$. The fluid velocity tends to the ideal-fluid flow pattern $\boldsymbol{v}_\infty$, such that:

$$\rho_0 \frac{\partial \boldsymbol{v}_\infty}{\partial t} = -\partial \delta p_\infty + \left[-\partial_x \langle p \rangle_f + F \right] \hat{\boldsymbol{x}}, \qquad \text{in } \mathcal{V}_f, \tag{7.198}$$

$$\partial \cdot \boldsymbol{v}_\infty = 0, \qquad \text{in } \mathcal{V}_f, \tag{7.199}$$

$$\delta p_\infty = \text{stationary random}, \qquad \text{in } \mathcal{V}_f, \tag{7.200}$$

$$\boldsymbol{v}_\infty \cdot \hat{\boldsymbol{n}} = 0, \qquad \text{in } \partial\mathcal{V}. \tag{7.201}$$

excepted for a vanishingly small viscous boundary layer region at the pore walls, and for an additional potential-flow perturbation in the bulk, of comparatively vanishingly small amplitude.

The above leading order ideal-fluid flow pattern can be described in terms of the scaled electric field $\boldsymbol{E}$ that appears in the pores, when the fluid is conducting, the solid is insulating, and a unit macroscopic electric field $1\hat{\boldsymbol{x}}$ is imposed, (see [17]):

$$\begin{aligned} \boldsymbol{E} &= -\partial \Phi + \hat{\boldsymbol{x}}, && \text{in } \mathcal{V}_f, && (7.202)\\ \partial \cdot \boldsymbol{E} &= 0, && \text{in } \mathcal{V}_f, && (7.203)\\ \Phi &= \text{stationary random}, && \text{in } \mathcal{V}_f, && (7.204)\\ \boldsymbol{E} \cdot \hat{\boldsymbol{n}} &= 0, && \text{on } \partial\mathcal{V}. && (7.205) \end{aligned}$$

Obviously,

$$\boldsymbol{v}_\infty = \frac{\left[-\partial_x \langle p \rangle_f + F\right]}{-i\omega\rho_0} \boldsymbol{E}, \quad \delta p_\infty = \left[-\partial_x \langle p \rangle_f + F\right] \Phi. \tag{7.206}$$

The actual fluid velocity in the high-frequency limit, will be

$$\boldsymbol{v} = \boldsymbol{v}_\infty + \delta \boldsymbol{v}, \tag{7.207}$$

with $\boldsymbol{v}_\infty \propto \boldsymbol{E}$ the above ideal-fluid profile, and $\delta \boldsymbol{v} = \delta \boldsymbol{v}_p + \delta \boldsymbol{v}_v$ made of two parts: the vortical boundary-layer part, $\delta \boldsymbol{v}_v$, such that $\eta \partial^2 \boldsymbol{v} = \eta \partial^2 \delta \boldsymbol{v}_v \neq \boldsymbol{0}$, rapidly decreasing away from the pore-walls, and the additional small-amplitude bulk potential-flow perturbation, $\delta \boldsymbol{v}_p$, $\eta \partial^2 \boldsymbol{v}_p = \boldsymbol{0}$, with no contribution to the Laplacian viscous term $\eta \partial^2 \boldsymbol{v}$.

As regards the fluid excess-temperature pattern in the high-frequency limit, $\omega \to \infty$, where thermal exchanges are "frozen", (small thermal skin depths, $\omega \gg \ell^2 \rho_0 c_P / 2\kappa$), and again because we make the important simplification that the fluid motion is that of an incompressible fluid at the pore scale, it tends to the adiabatic constant profile τ_∞, such that:

$$\rho_0 c_P \frac{\partial \tau_\infty}{\partial t} = \beta_0 T_0 \frac{\partial \langle p \rangle_f}{\partial t}, \qquad \text{in } \mathcal{V}_f, \tag{7.208}$$

or

$$\tau_\infty = \frac{\beta_0 T_0}{\rho_0 c_P} \langle p \rangle_f, \qquad \text{in } \mathcal{V}_f, \tag{7.209}$$

excepted for a vanishingly small thermal boundary layer region at the pore walls. The actual excess-temperature in the high-frequency limit, will be:

$$\tau = \tau_\infty + \delta \tau, \tag{7.210}$$

with $\delta\tau$, a diffusive part near the pore-walls, associated with $\kappa \partial^2 \tau = \kappa \partial^2 \delta \tau \neq 0$, and rapidly decreasing away from the pore-walls.

As shown below, in the present simplification a consistent boundary-layer calculation gives

$$\alpha(\omega) = \alpha_\infty \left[1 + \left(\frac{\nu}{-i\omega}\right)^{1/2} \frac{2}{\Lambda} + \cdots \right], \ \alpha'(\omega) = \alpha'_\infty \left[1 + \left(\frac{\nu'}{-i\omega}\right)^{1/2} \frac{2}{\Lambda'} + \cdots \right], \tag{7.211}$$

when the pore-walls $\partial\mathcal{V}$ are assumed locally plane, where:

$$\alpha_\infty = \frac{\langle \boldsymbol{v}_\infty^2 \rangle_f}{\langle \boldsymbol{v}_\infty \rangle_f^2}, \qquad \alpha'_\infty = 1, \tag{7.212}$$

$$\frac{2}{\Lambda} = \frac{\int_{\partial\mathcal{V}} \boldsymbol{v}_\infty^2 dS}{\int_{\mathcal{V}_f} \boldsymbol{v}_\infty^2 dV}, \quad \frac{2}{\Lambda'} = \frac{\int_{\partial\mathcal{V}} dS}{\int_{\mathcal{V}_f} dV}, \tag{7.213}$$

where $\int_{\partial\mathcal{V}} dS$ denotes integral on the pore-walls, and $\int_{\mathcal{V}_f} dV$ integral in the fluid volume. Here, α_∞ is the ideal-fluid or electrical tortuosity ($\langle \boldsymbol{E} \rangle_f \cdot \boldsymbol{e} = 1/\alpha_\infty$), Λ is Johnson's viscous characteristic length [2] and Λ' is Allard's thermal characteristic length [18].

To show (7.211)–(7.213), we rectify, in the presence of thermal effects, a reasoning given in [2], (see [19] for a different, more involved but direct calculation). It consists in requiring the compatibility of two sound attenuation calculations, one directly macroscopic and the other starting at the microscopic level. Looking for plane-wave solutions varying as $e^{-i\omega t + iq(\omega)x}$, of the macroscopic motion equations (7.156)–(7.159), (without the source term), we find a dispersion equation

$$\rho_0 \alpha(\omega) \chi_0 \beta(\omega) \omega^2 = q(\omega)^2. \tag{7.214}$$

On a macroscopic scale, the intensity will decay like $e^{-2q''x}$, with $q'' = \Im q(\omega)$. With asymptotics (7.211), the attenuation constant q'' is found to be:

$$q'' = \frac{\omega}{c_0}\sqrt{\alpha_\infty}\left[\frac{1}{\Lambda}\sqrt{\frac{\nu}{2\omega}} + (\gamma - 1)\frac{1}{\Lambda'}\sqrt{\frac{\nu'}{2\omega}}\right] + \cdots \tag{7.215}$$

On the other hand, starting at the microscopic scale, with a classical reasoning given in Landau and Lifshitz [20], p. 299, q'' can be related to the velocity and excess-temperature fields $\boldsymbol{v}(\boldsymbol{x})$ and $\tau(\boldsymbol{x})$:

$$q'' = \frac{\overline{|\dot{E}_{mech}|}}{2\overline{S_0}}, \tag{7.216}$$

where $\overline{S_0}$ is the time-averaged acoustic energy flux calculated by making abstraction of the effect of losses,[18] and $\overline{\dot{E}_{mech}}$ is the time-averaged rate of energy dissipation per unit total volume $V = V_f/\phi$. The mean acoustic energy flux $\overline{S_0}$ is estimated as $c\overline{E}$ where c is the speed of sound without losses, $c = c_0/\sqrt{\alpha_\infty}$, and $\overline{E}$ is the mean acoustic energy per unit total volume, equal to twice the mean kinetic or potential energy per unit total volume:

[18] Equation (7.216) is a first order calculation, which will give the leading term in (7.215). The losses determine the numerator so they need not be taken into account in the denominator.

$$\overline{S_0} = \frac{c_0}{\sqrt{\alpha_\infty}} 2 \frac{1}{V} \int_{V_f} \frac{1}{2} \rho_0 \overline{\boldsymbol{v}_\infty^2(t, \boldsymbol{x})} dV = \frac{c_0}{\sqrt{\alpha_\infty}} \frac{1}{V} \int_{V_f} \frac{1}{2} \rho_0 |\boldsymbol{v}_\infty(\boldsymbol{x})|^2 dV$$
$$= \frac{c_0}{\sqrt{\alpha_\infty}} 2 \frac{1}{V} \int_{V_f} \frac{1}{2} \chi_0 \overline{\langle p_\infty \rangle_f^2(t, \boldsymbol{x})} dV = \frac{c_0}{\sqrt{\alpha_\infty}} \frac{1}{V} \int_{V_f} \frac{1}{2} \chi_0 |\langle p_\infty(\boldsymbol{x}) \rangle_f|^2 dV, \tag{7.217}$$

where in the right are complex amplitudes, and we have added index ∞ to recall high-frequency asymptotic limit.

The rate of energy dissipation per unit volume $\dot{E}_{mech}$ is [20]:

$$\dot{E}_{mech} = -\frac{\kappa}{T_0} \frac{1}{V} \int (\partial \tau)^2 dV + \cdots$$
$$- \frac{1}{2} \eta \frac{1}{V} \int \left(\frac{\partial v_i}{\partial x_k} + \frac{\partial v_k}{\partial x_i} - \frac{2}{3} \delta_{ik} \frac{\partial v_l}{\partial v_l} \right)^2 dV - \zeta \frac{1}{V} \int (\partial \cdot \boldsymbol{v})^2 dV,$$

(this is the volume integral of the dissipation rate $-\mathcal{D}$, see the Chap. 6, (6.111)), and after average over a cycle and replacement of the real quantities by their complex amplitudes, it reads

$$\overline{\dot{E}_{mech}} = -\frac{\kappa}{T_0} \frac{1}{2} \frac{1}{V} \int |\partial \tau|^2 dV - \frac{1}{4} \eta \frac{1}{V} \int |\frac{\partial v_i}{\partial x_k} + \frac{\partial v_k}{\partial x_i}|^2 dV, \tag{7.218}$$

where the incompressibility of the fluid on the size scale of the pores has been accounted for. By integrations by parts, we have:

$$\int |\partial \tau|^2 dV = \int \frac{\partial}{\partial x_k} \left(\tau \frac{\partial \tau^*}{\partial x_k} \right) dV - \int \tau \partial^2 \tau^* dV,$$
$$\int \left| \frac{\partial v_i}{\partial x_k} + \frac{\partial v_k}{\partial x_i} \right|^2 dV = 2 \int \frac{\partial}{\partial x_k} \left(v_i \frac{\partial v_i^*}{\partial x_k} \right) dV + 2 \int \frac{\partial}{\partial x_k} \left(v_i \frac{\partial v_k^*}{\partial x_i} \right) dV + \cdots$$
$$- 2 \int \left(v_i \frac{\partial}{\partial x_i} \frac{\partial v_k^*}{\partial x_k} \right) dV - 2 \int v_i \partial^2 v_i^* dV.$$

In the two expressions, only the last integrals subsist: for the total derivative integrals, their vanishing comes from the stationary nature of the fields and the vanishing of velocity and excess temperature on $\partial \mathcal{V}$; for the above third integral, its vanishing comes from incompressibility. The remaining two integrals come from the viscous and thermal boundary layers where $\partial^2 v_i^*$ and $\partial^2 \tau^*$, nonzero, are rapidly decreasing. To evaluate these integrals, a high-frequency limit is considered, in which the pore-surface interface $\partial \mathcal{V}$ is assumed to be locally plane at the scale of the relevant boundary-layer thicknesses, resp. $\delta = (2\nu/\omega)^{1/2}$ for the velocity field, and $\delta' = (2\nu'/\omega)^{1/2}$ for the excess temperature field. In this case, we find by a classic calculation, (boundary-layer profile near a flat surface [20], p. 91), that the excess-temperature and velocity fields have the following form, where $\boldsymbol{x}_w$ denotes a position on $\partial\mathcal{V}$, and ξ is a coordinate along the normal, $(-\hat{\boldsymbol{n}}\xi = \boldsymbol{x} - \boldsymbol{x}_w$, with $\hat{\boldsymbol{n}}$, the outward normal to the fluid region at $\boldsymbol{x}_w)$:

$$\boldsymbol{v}(\boldsymbol{x}) = \boldsymbol{v}_\infty(\boldsymbol{x}_w)\left[1 - e^{ik_v\xi}\right], \quad \tau(\boldsymbol{x}) = \tau_\infty(\boldsymbol{x}_w)\left[1 - e^{ik_\tau\xi}\right], \tag{7.219}$$

and $k_v = (1+i)/\delta$ and $k_\tau = (1+i)/\delta'$ are the shear and entropic wavenumbers, (6.66) and (6.57.2), of the Chap. 6. Substituting (7.219) in the integrals and making the integrations $\int dV$ in the form of $\int_{\partial\mathcal{V}} dS \int_0^\infty d\xi$, we find

$$\begin{aligned}\overline{\dot{E}_{mech}} = &- \frac{1}{T_0 2V}\left(\frac{1}{2}\omega\kappa\rho_0 c_P\right)^{1/2}\int_{\partial\mathcal{V}} |\tau_\infty|^2 dS + \cdots \\ &- \frac{1}{2V}\left(\frac{1}{2}\omega\eta\rho_0\right)^{1/2}\int_{\partial\mathcal{V}} |\boldsymbol{v}_\infty(\boldsymbol{x}_w)|^2 dS.\end{aligned} \tag{7.220}$$

Then dividing the two terms by the appropriate (kinetic-energy or potential-energy) form of (7.217), we get, by (7.216), the following expression of q'':

$$\begin{aligned}q'' = &\frac{1}{T_0 2V}\left(\frac{1}{2}\omega\kappa\rho_0 c_P\right)^{1/2}\frac{\int_{\partial\mathcal{V}} |\tau_\infty(\boldsymbol{x}_w)|^2 dS}{2\frac{c_0}{\sqrt{\alpha_\infty}}\frac{1}{V}\int \frac{1}{2}\chi_0 |\langle p_\infty(\boldsymbol{x})\rangle_f|^2 dV} + \cdots \\ &+ \frac{1}{2V}\left(\frac{1}{2}\omega\eta\rho_0\right)^{1/2}\frac{\int_{\partial\mathcal{V}} |\boldsymbol{v}_\infty(\boldsymbol{x}_w)|^2 dS}{2\frac{c_0}{\sqrt{\alpha_\infty}}\frac{1}{V}\int \frac{1}{2}\rho_0 |\boldsymbol{v}_\infty(\boldsymbol{x})|^2 dV} + \cdots\end{aligned} \tag{7.221}$$

According to (7.209), $|\langle p_\infty\rangle_f|^2$ can be replaced by $|\tau_\infty|^2\,(\beta_0 T_0/\rho_0 c_P)^2$, so that, after using the general thermodynamic relation Equation (5) of the Chap. 6, we obtain

$$q'' = \frac{\omega}{c_0}\sqrt{\alpha_\infty}\left[\frac{\int_{\partial\mathcal{V}} |\boldsymbol{v}_\infty|^2 dS}{2\int_{\mathcal{V}_f} |\boldsymbol{v}_\infty|^2 dV}\sqrt{\frac{\nu}{2\omega}} + (\gamma - 1)\frac{\int_{\partial\mathcal{V}} |\tau_\infty|^2 dS}{2\int_{\mathcal{V}_f} |\tau_\infty|^2 dV}\sqrt{\frac{\nu'}{2\omega}}\right] + \cdots \tag{7.222}$$

Comparison with (7.215) shows that this expression of q'' justifies the asymptotics (7.211), with expressions (7.212) and (7.213) of the parameters. The $|.|$ are useless and can be removed, as the different velocities and temperatures are in phase in a REV. The factor of τ_∞^2 do not appear in the characteristic thermal length, because, as it is a pore-scale constant, it can be removed from numerator and denominator.[19]

[19] In nonlocal theory, there will be a generalization to be made, and when the pressure is distributed at the pore scale, the factors $|\tau_\infty|^2$ should be maintained since they are not constant.

Intermediate Frequencies: Constraints, and Simple Model-Functions

At intermediate frequencies, the functions $\alpha(\omega)$ and $\alpha'(\omega)$ will interpolate smoothly between the low-frequencies and high-frequencies behaviours (7.188), (7.194) and (7.211). Indeed, if we write these functions in the form:

$$\alpha(\omega) = \frac{\eta\phi}{-i\omega\rho_0 k_{app}(\omega)} + \alpha_{app}(\omega),\ \alpha'(\omega) = \frac{\kappa\phi}{-i\omega\rho_0 c_P k'_{app}(\omega)} + \alpha'_{app}(\omega), \tag{7.223}$$

the introduced new functions $k_{app}(\omega)$, $\alpha_{app}(\omega)$, $k'_{app}(\omega)$, $\alpha'_{app}(\omega)$, are apparent tortuosities and permeabilities in intermediate regime, given by (see (7.185) and (7.186)):

$$\frac{\phi}{k_{app}} = \frac{\langle -\boldsymbol{v}\cdot\partial^2\boldsymbol{v}^\star\rangle_f}{\langle\boldsymbol{v}\rangle_f\cdot\langle\boldsymbol{v}^\star\rangle_f}, \qquad \alpha_{app} = \frac{\langle\boldsymbol{v}\cdot\boldsymbol{v}^\star\rangle_f}{\langle\boldsymbol{v}\rangle_f\cdot\langle\boldsymbol{v}^\star\rangle_f}, \tag{7.224}$$

$$\frac{\phi}{k'_{app}} = \frac{\langle -\tau\partial^2\tau^\star\rangle_f}{\langle\tau\rangle_f\langle\tau^\star\rangle_f}, \qquad \alpha'_{app} = \frac{\langle\tau\tau^\star\rangle_f}{\langle\tau\rangle_f\langle\tau^\star\rangle_f}, \tag{7.225}$$

where $\boldsymbol{v}$ and τ are the complex patterns of the velocity and excess temperature fields appearing at frequency ω, *and we can show that, because of the divergence-free nature of the pore-scale motion, these new functions are monotonic strictly decreasing functions of frequency* [21]:

$$\frac{dk_{app}}{d\omega} < 0, \quad \frac{d\alpha_{app}}{d\omega} < 0, \quad \frac{dk'_{app}}{d\omega} < 0, \quad \frac{d\alpha'_{app}}{d\omega} < 0. \tag{7.226}$$

These behaviours are related to the condition, (expressing the divergence-free motion), that the singularities, poles, and zeros, of the functions $\alpha(\omega)$, $k(\omega)$, and $\alpha'(\omega)$, $k'(\omega)$, of complex ω, are located on the negative imaginary ω axis, (see [2, 3, 17]). This is a much more severe condition than that imposed solely by causality. It excludes resonant behaviours because the effect of spatial nonlocalities is not accounted for, in the modeling. There follows that there are relatively simple analytical models of the frequency dependence of functions $\alpha(\omega)$ and $\alpha'(\omega)$, which allow them to be represented at any real frequency, with reasonable precision, in terms of the physical constants of the fluid and the preceding low- and high-frequency geometric parameters.

To express the characteristic relaxation these functions present, it is appropriate to consider their inverses $1/\alpha(\omega)$ and $1/\alpha'(\omega)$. These inverses pass from the values $[\alpha^{-1}]_0 = [\alpha'^{-1}]_0 = 0$ at $\omega = 0$, to the values $[\alpha^{-1}]_\infty = 1/\alpha_\infty$ and $[\alpha'^{-1}]_\infty = 1$ at $\omega = \infty$. Thus we write them in terms of relaxation functions $\chi(\omega)$ and $\chi'(\omega)$, relaxing from 1, at $\omega = 0$, in relaxed state, to 0, at $\omega = \infty$, in frozen state:

$$\frac{1}{\alpha(\omega)} = [\alpha^{-1}]_\infty + \left([\alpha^{-1}]_0 - [\alpha^{-1}]_\infty\right)\chi(\omega) = \frac{1}{\alpha_\infty}(1-\chi(\omega)),$$
$$\frac{1}{\alpha'(\omega)} = [\alpha'^{-1}]_\infty + \left([\alpha'^{-1}]_0 - [\alpha'^{-1}]_\infty\right)\chi'(\omega) = 1-\chi'(\omega). \tag{7.227}$$

Based on the property that the poles and zeros are on the negative imaginary axis, and the limiting behaviours (7.188), (7.194), and (7.211), must be satisfied, it can be anticipated that the following simple expressions of the relaxation functions $\chi(\omega)$ and $\chi'(\omega)$, will provide a very satisfactory full-frequency description, (with discrepancies in the transition region, less than a few per cent):

$$\frac{1}{\chi(\omega)} = 1 - \frac{ix}{S(x)}, \qquad \frac{1}{\chi'(\omega)} = 1 - \frac{ix'}{S'(x')}, \tag{7.228}$$

$$S(x) = 1 - \mathrm{P} + \mathrm{P}\sqrt{1 - \frac{ixM}{2\mathrm{P}^2}}, \qquad S'(x') = 1 - \mathrm{P}' + \mathrm{P}'\sqrt{1 - \frac{ix'M'}{2\mathrm{P}'^2}}, \tag{7.229}$$

$$x = \frac{\omega}{\omega_v}, \qquad x' = \frac{\omega}{\omega_\tau}, \tag{7.230}$$

$$\omega_v = \frac{\nu}{Fk_0}, \quad \left(F = \frac{\alpha_\infty}{\phi}\right), \qquad \omega_\tau = \frac{\nu'}{F'k'_0}, \quad \left(F' = \frac{1}{\phi}\right), \tag{7.231}$$

$$M = \frac{8Fk_0}{\Lambda^2}, \qquad M' = \frac{8F'k'_0}{\Lambda'^2}, \tag{7.232}$$

$$\mathrm{P} = \frac{M}{4\left(\dfrac{\alpha_0}{\alpha_\infty} - 1\right)}, \qquad \mathrm{P}' = \frac{M'}{4(\alpha_0 - 1)}. \tag{7.233}$$

The quantities ω_v and ω_τ, are characteristic viscous and thermal, pulsation rollover frequencies, between low and high frequency regimes. The purely geometrical quantities M, P, M' and P', are dimensionless, viscous and thermal, form factors of order one, that will determine the precise shape of the viscous and thermal relaxation. The notation M is borrowed from [2]. The notation P (P for Pride), refers to the paper [22] rectified in [15]. When P is set to the value 1, one obtains the model of $\alpha(\omega)$ proposed by Johnson et al. [2]. When M' and P' are set to the value 1, one obtains the model of $\beta(\omega)$ proposed by Allard and Champoux [18]; when only $\mathrm{P}' = 1$, the model proposed by Lafarge et al. [3].

The behaviours described by these functions closely remind behaviours observed in electromagnetics of dielectric and ionic conductors. With forms factors taken around 1, the above relaxation forms of the functions $\chi(\omega)$ and $\chi'(\omega)$, best represented using Cole-Cole (also named Argand) plots ($\Im\chi(\omega)$ versus $\Re\chi(\omega)$), are roughly comparable to a Davidson-Cole relaxation with exponent 1/2: $\chi, \chi' \approx 1/(1-i\omega\Theta)^{1/2}, 1/(1-i\omega\Theta')^{1/2}$. When the M M' are small, the relaxation tends to Debye's relaxation $\chi, \chi' \approx 1/(1-i\omega\Theta), 1/(1-i\omega\Theta')$, excepted at sufficiently high frequencies. Notably, also, it can be observed that the characteristic high-frequency limits (7.211) imply, for the relaxation functions, the limiting behaviours,

$\omega \to \infty, \chi, \chi' \approx (M\omega_v/-2i\omega\text{P}^2)^{1/2}, (M'\omega_\tau/-2i\omega\text{P}'^2)^{1/2}$. For the associated time-domain functions $\chi(t)$ and $\chi'(t)$, this give laws of the empirical "Curie-von Schweidler" power-law type [23, 24]: $\chi(t)$ and $\chi'(t) \propto t^{-1/2}$, for the typical decrease of any field, at small times, just after an excitation. This fractional power-law type of behaviour has caused interrogations in electromagnetics; it has been interpreted as indicative of a many-body problem, revealing collective behaviour of electrons. Here, it is interpreted in terms of the existence of lossy boundary layers; following [17] we could show that it expresses in terms of fractal accumulation of viscous or thermal relaxation times, near zero, with dimension 1/2 (see [17], Appendix C). With pore-walls taken cusped with some fractality, instead of being locally plane, we would expect obtaining a similar relaxational behaviour, but with different values for the exponent (because of the different fractal dimension of the accumulation), leading in particular to the more general form of the Curie-von Schweidler law: $\chi(t)$, $\chi'(t) \propto t^{-n}, t^{-n'}, 0 < n \; and \; n' < 1$.

Finally, let us recall the situation in electromagnetics of non-ferromagnetic materials, when one considers that the dispersion effects on the magnetic susceptibility $\mu(\omega)$ are small relativistic effects, of second order on the small parameter $\beta = v/c$, (with v an estimate of electronic velocities in molecules, and c the speed of light). In this case, because of the smallness of β and quadratic (β^2) nature of relativistic effects, the low-frequency and high-frequency values of $\mu(\omega)$ coincide, $\mu(\omega) \cong \mu_0$, and no dispersion intervenes. There is a similar situation here when the permeating fluid is a liquid. In a simple fluid there is the general thermodynamic identity (Chap. 6, (6.5)):

$$\frac{\beta_0^2 T_0 c_0^2}{c_P} = \gamma - 1, \tag{7.234}$$

where β_0 is the thermal expansion coefficient of the fluid, T_0 is ambient temperature, c_0 is the adiabatic velocity of sound and c_P is the heat coefficient at constant pressure. It turns out that in a liquid, $\gamma \cong 1$, because the deviation $\gamma - 1$ is a quadratic effect on the thermal expansion coefficient β_0, which is very small. Therefore, for a liquid-saturated material *described by the present local theory*, (i.e. when the material has a sufficiently *simple microstructure*), the low-frequency and high-frequency bulk-moduli values coincide, $\beta(\omega) \cong 1$, and no dispersion intervenes on the compressibility.

References

1. J.-F. Allard, N. Atalla, *Propagation of Sound in Porous Media: Modelling Sound Absorbing Materials*, 2nd edn (Wiley, New York, 2009)
2. D.L. Johnson, J. Koplik, R. Dashen, Theory of dynamic permeability and tortuosity in fluid-saturated porous media. J. Fluid Mech. **176**, 379–402 (1987)
3. D. Lafarge, P. Lemarinier, J.-F. Allard, V. Tarnow, Dynamic compressibility of air in porous structures at audible frequencies. J. Acous. Soc. Amer. **102**(4), 1995–2006 (1997)

4. R. Burridge, J.B. Keller, Poroelasticity equations derived from microstructure. J. Acous. Soc. Amer. **70**(4), 1140–1146 (1981)
5. G. Russakoff, A derivation of the macroscopic Maxwell equations. Amer. J. Phys. **38**(10), 1188–1195 (1970)
6. J.D. Jackson, *Classical Electrodynamics* (Wiley, New York, 1999)
7. A.D. Pierce, *Acoustics: An Introduction to its Physical Principles and Applications* (Acoustical Society of America, 1989)
8. L.D. Landau, E.M. Lifshitz, *Electrodynamics of Continuous Media* (Pergamon Press, Oxford, 1960)
9. V.M. Agranovich, V.L. Ginzburg, *Spatial Dispersion in Crystal Optics and the Theory of Excitons* (Interscience Publishers, London, 1966)
10. D.B. Melrose, R.C. McPhedran, *Electromagnetic Processes in Dispersive Media* (Cambridge University Press, New York, 1991)
11. D. Lafarge, N. Nemati, Nonlocal Maxwellian theory of sound propagation in fluid-saturated rigid-framed porous media. Wave Motion **50**(6), 1016–1035 (2013)
12. N. Nemati, D. Lafarge, Check on a nonlocal Maxwellian theory of sound propagation in fluid-saturated rigid-framed porous media. Wave Motion **51**(5), 716–728 (2014)
13. N. Nemati, A. Kumar, D. Lafarge, N.X. Fang, Nonlocal description of sound propagation through an array of Helmholtz resonators. Comptes Rendus Mecanique **343**(12), 656–669 (2015)
14. N. Nemati, Y.E. Lee, D. Lafarge, A. Duclos, N.X. Fang, Nonlocal dynamics of dissipative phononic fluids. Phys. Rev. B **95**(22), 224304 (2017)
15. J. Kergomard, D. Lafarge, J. Gilbert, Transients in porous media: exact and modelled time-domain Green's functions. Acta Acustica United Acustica **99**(4), 557–571 (2013)
16. A.N. Norris, On the viscodynamic operator in Biot's equations of poroelasticity. J Wave Mat Interact **1**, 365–380 (1986)
17. M. Avellaneda, S. Torquato, Rigorous link between fluid permeability, electrical conductivity, and relaxation times for transport in porous media. Phys. Fluids A: Fluid Dyn. **3**(11), 2529–2540 (1991). see also, D. Lafarge, Comments on "Rigorous link between fluid permeability, electrical conductivity, and relaxation times for transport in porous media". Phys. Fluids A **5**, 500 (1993)
18. Y. Champoux, J.-F. Allard, Dynamic tortuosity and bulk modulus in air-saturated porous media. J. Appl. Phys. **70**(4), 1975–1979 (1991)
19. A. Cortis, D.M.J. Smeulders, J.L. Guermond, D. Lafarge, Influence of pore roughness on high-frequency permeability. Phys. Fluids **15**(6), 1766–1775 (2003)
20. L.D. Landau, E.M. Lifshitz, *Fluid Mechanics*, 2nd edn. (Pergamon Press, Oxford, 1987)
21. C. Boutin, C. Geindreau, Estimates and bounds of dynamic permeability of granular media. J. Acoust. Soc. Amer. **124**(6), 3576–3593 (2008)
22. S.R. Pride, F.D. Morgan, A.F. Gangi, Drag forces of porous-medium acoustics. Phys. Rev. B **47**(9), 4964 (1993)
23. J. Curie, Recherches sur le pouvoir inducteur spécifique et sur la conductibilité des corps cristallisés. Annales de Chimie et de Physique, 6ième sèrie, t XVII:385 (1889)
24. E.R.v. Schweidler, Studien über die Anomalien im Verhalten der Dielektrika. Annalen der Physik **329**(14), 711–770 (1907)

Chapter 8
Numerical Methods for Modelling and Simulation of Porous Materials

An Introduction Based on Finite Elements

Peter Goransson and Olivier Dazel

Abstract We introduce methods for modelling and simulation of anisotropic poroelastic materials in the frequency domain. Starting from the equations formulated by Biot, in their anisotropic form, we derive two different symmetric weak forms together with the boundary conditions that has to be satisfied. We employ a mixed displacement-pressure formulation and solutions are obtained by applying the finite element method to the proposed weak forms. In order to illustrate the use of the finite element method, we highlight some particular aspects related to simulations where poroelastic materials are involved. These include convergence of the discretised solution and boundary conditions at interfaces between poroelastic materials and solids/fluids. Results are given for some selected application examples of foam and plate combinations as well as a poroelastic foam with embedded inclusions.

8.1 Introduction

In flexible porous materials with open cells, the vibroacoustic energy is carried both through an airborne path, i.e. the sound pressure waves propagating through the fluid in the pores, and through an structure borne path, i.e. the elastic stress waves carried through the solid frame of the material. These waves are strongly coupled, i.e. they simultaneously propagate in both the fluid and the solid frame but with different strengths and relative phase. A characteristic of this coupled wave propagation, is that the vibroacoustic energy is dissipated and converted into heat as the wave travels through the material. To model such a dynamic behaviour, we use the elastic properties (i.e. stiffness controlled by material, topology, geometry, interfaces),

P. Goransson (✉)
KTH Royal Institute of Technology, Stockholm, Sweden
e-mail: pege@kth.se

O. Dazel
Laboratoire d'Acoustique de l'Université du Mans, UMR CNRS 6613, Le Mans Université, Avenue Olivier Messiaen, 72085 Le Mans, France
e-mail: olivier.dazel@univ-lemans.fr

N. Jiménez et al. (eds.), *Acoustic Waves in Periodic Structures, Metamaterials, and Porous Media*, Topics in Applied Physics 143,
https://doi.org/10.1007/978-3-030-84300-7_8

the viscoelastic properties (i.e. solid damping controlled by material, geometry), the acoustic properties (which are governed by the fluid medium) and the viscoacoustic properties (i.e. fluid damping controlled by geometry, topology, interfaces).

All these dissipation mechanisms are in general functions of frequency and furthermore, vary with frequency in strength and character. For a given situation, the balance between energy dissipated through vibration of the solid frame and changes in the acoustic pressure varies with the topological arrangement, choice of material properties, geometrical dimensions, interfacial conditions, etc. Traditionally, and also in the present work, the dynamic behaviour of porous materials is described in terms of macroscopic, space averaged quantities, such as acoustic pressure, elastic stress, solid and fluid displacements. In the current work parts of the modelling paradigm of porous foams, known in the literature as Biot's theory, will be briefly reviewed from an acoustics and vibrations perspective.

Traditional modelling and simulation of poroelastic materials (PEM) in vibroacoustic applications have for a long time been based on the assumption of isotropic elastic and acoustic properties, i.e. the materials possess a high degree of symmetry in their constitutive properties. However, it is well known that, due to the manufacturing processes involved, real PEMs are anisotropic to a certain degree. Examples where this holds are foamed polymers, fibrous wools etc. We will not review the large number of different methods that are available to solve problems involving isotropic PEM, for this we point to the recent review in [1]. Here we will instead start from Biot's equations in their anisotropic form, from which we derive and solve two different symmetric weak forms together with the boundary conditions that have to be satisfied in different configurations. Although there are a number of different choices of variables that can be made, the weak forms are here set up in terms of a mixed displacement-pressure formulation. The solutions are obtained by applying the finite element (FE) method.

To illustrate different aspects of modelling in applications where PEMs are involved, we solve some problems and results are given for selected application examples of a foam and plate combination. We highlight particular aspects related to convergence of the discretised solutions, boundary conditions at interfaces between poroelastic materials and solids/fluids, as well as solutions involving periodic boundary conditions and Bloch waves.

Our objective is to give a general introduction to the modelling of anisotropic PEM, however, without going into too much detail of the FE method as such, as this is beyond the scope of the present work. We will illustrate the effects of anisotropy with examples from the literature and we will show some intricate aspects of solving problems using simpler material models but still complex configurations. Our hope is that this will serve as a point of entry to the world of numerical modelling of these materials and an inspiration to go deeper into the subject.

8.2 Biot's Equations

The modelling considered here, is based on continuum mechanics at a mesoscopic scale, i.e. on homogenised fields of the porous materials. This is only possible if there exists a Representative Volume Element (RVE) whose size is sufficiently small in comparison with the wavelengths but also sufficiently large compared to the characteristic size of the heterogeneities to be representative. The motion of the solid (resp. fluid) part is averaged over the RVE and the homogenised media are called the solid and the fluid phase respectively.

The partial differential equations (PDEs) governing an anisotropic PEM involve (analogously to an isotropic material) coupling between the solid phase and the fluid phase in several different ways and the equation terms used to represent these interactions will be introduced below. The derivations take as a starting point the works by Biot [2–6]. Note that here but a brief introduction will be given, for more details see the literature cited throughout the text and the references provided therein. We define the necessary symbols as they are introduced in the derivation.

We start with some definitions and useful relations, in the form of the constitutive and the momentum equations. Recalling that the constitutive relations describe how the stresses relate to the deformation gradients, in both the solid and the fluid, there are coupling terms between the two phases related to the dilatation (compression) and deviation (shear). Similarly, the momentum equations which express the balance between internal forces, relate the gradients in the stress fields to the corresponding accelerations. These will be referred to as first order equations. We will identify frequency dependent quantities with a tilde symbol.

The expressions presented in this section originate from the theoretical framework proposed by Biot, [2, 5]. We will not recall the original equations stated by Biot as there are a large number of publications where they may be studied in close detail, see e.g. [7]. Instead most of the derivations that follow are inspired by [8], where the focus was on weak forms and FE, here presented in a slightly modified form.

In the original Biot theory, the modelling is based on the fluid homogenised displacements of the solid phase u_i^s and of the fluid phase u_i^f. The two corresponding stress tensors are σ^s and σ^f, which respectively represent the stress tensors of the solid and fluid phases, together forming the total stress $\sigma^t = \sigma^s + \sigma^f$. A few years after the original formulation, Biot proposed a modification to the original theory which was required to handle the case of inhomogeneous materials, in particular when the porosity is a function depending on space. The key was to replace the fluid displacement u_i^f by the relative flow $w_i = \phi(u_i^f - u_i^s)$, where ϕ is the open porosity.

From the basic relations proposed by Biot, a number of different representations may be proposed. For numerical modelling and solutions based on the FE method, where the computational effort required is strongly dependent on the number of *dof*, most of them keep the solid displacement and introduce the pressure as dependent field variables. This is directly related to one of the key properties of σ^f, i.e. that it is scalar and proportional to the acoustic pressure, through the relation $\sigma_{ij}^f = -\phi p \delta_{ij}$.

As will be shown below, using this, the fluid displacement and the fluid stress tensor may then be eliminated. This then leads to a new set of slightly modified solid displacements and stresses.

8.2.1 Constitutive Laws

In the following we will discuss the constitutive laws, here written in Cartesian tensor component notation, with Einstein's summation convention implied for repeated indices. Thus, with i, j the component ordinal numbers in Cartesian co-ordinate system, x_i with $i = 1, 2, 3$, $(.)_{,i} = \frac{\partial(.)}{\partial x_i}$ is the partial derivative with respect to co-ordinate x_i. δ_{ij} is the Kronecker's delta.

We begin with the constitutive laws for the solid,

$$\sigma^s_{ij} = \left(\hat{C}_{ijkl} + \frac{\tilde{Q}_{ij}\tilde{Q}_{kl}}{\phi^2 \tilde{K}_{eq}}\right)\varepsilon^s_{kl} + \tilde{Q}_{ij}u^f_{k,k}, \tag{8.1}$$

and for the fluid,

$$\sigma^f_{ij} = \left(\tilde{Q}_{kl}\varepsilon^s_{kl} + \phi^2 \tilde{K}_{eq}u^f_{k,k}\right)\delta_{ij}, \tag{8.2}$$

where ε^s_{kl} is the solid frame Cauchy strain tensor and $\hat{C}_{ijkl}$ corresponds to the *in-vacuo* Hooke tensor of the solid phase. As shown in [8], K_{eq} is the scalar fluid compressibility modulus:

$$\tilde{K}_{eq} = \frac{K_s}{1 - \phi - K_s\tilde{C}_{ijkl}d_{ij}d_{kl} + \phi K_s/K_f}, \quad d_{ij} = -\frac{\varepsilon^s_{ij}}{p} \tag{8.3}$$

where d_{ij} is the unjacketed frame compressibility compliance tensor. Furthermore, $\tilde{Q}_{ij}$ is the dilatational coupling tensor:

$$\tilde{Q}_{ij} = \frac{\left[(1-\phi)\,\delta_{ij} - \tilde{C}_{ijkl}d_{kl}\right]\phi K_s}{1 - \phi - K_s\tilde{C}_{ijkl}d_{ij}d_{kl} + \phi K_s/K_f}. \tag{8.4}$$

Two other stress tensors may also be considered. The total stress tensor σ^t and the *in-vacuo* stress tensor of the solid phase, which corresponds to the stress in the absence of fluid: $\hat{\sigma}^s_{ij} = \hat{C}_{ijkl}\varepsilon^s_{kl}$. These are linked by the following relations:

$$\sigma^t_{ij} = \sigma^s_{ij} + \sigma^f_{ij} = \hat{\sigma}_{ij} - \tilde{\gamma}'_{ij}p, \quad , \tag{8.5}$$

where $\tilde{\gamma}'_{ij}$ is an elastic coupling coefficient,

$$\tilde{\gamma}'_{ij} = \phi \left(\delta_{ij} + \frac{\tilde{Q}_{ij}}{\phi^2 \tilde{K}_{eq}} \right), \tag{8.6}$$

which may be used to express the link between the solid and *in-vacuo* stress tensors:

$$\sigma^s_{ij} = \hat{\sigma}^s_{ij} - \left(\tilde{\gamma}'_{ij} - \phi\delta_{ij}\right) p. \tag{8.7}$$

The pressure p is related to the divergence of the fluid displacement u^W,

$$p = -\tilde{K}_{eq} u^W_{j,j}, \tag{8.8}$$

where u^W is a combination of the solid and fluid displacement defined as,

$$u^W_j = \phi \left(u^f_j + \frac{\tilde{Q}_{jl}}{\phi^2 \tilde{K}_{eq}} u^s_l, \right) \tag{8.9}$$

which is equivalent to,

$$u^f_j = \frac{u^W_j}{\phi} - \frac{\tilde{Q}_{jl}}{\phi^2 \tilde{K}_{eq}} u^s_l. \tag{8.10}$$

Thus, using (8.10), we may replace the fluid displacement by a combination of u^W and the gradient of the pressure.

8.2.2 Momentum Equations

The momentum equations were given in [8] as:

$$\sigma^s_{ij,j} = -\omega^2 \tilde{\rho}^{11}_{ij} u^s_j - \omega^2 \tilde{\rho}^{12}_{ij} u^f_j, \tag{8.11}$$

and

$$\sigma^f_{ij,j} = -\omega^2 \tilde{\rho}^{12}_{ij} u^s_j - \omega^2 \tilde{\rho}^{22}_{ij} u^f_j. \tag{8.12}$$

As in the original Biot's formulation, these equations involve three complex densities. In order to simplify the expressions, especially as now we have tensors (as opposed to scalars for an isotropic material) that will be inverted and multiplied, we choose to express all of them through the dynamic tortuosity, $\tilde{\alpha}_{ij}$,

$$\tilde{\rho}^{22}_{ij} = \phi\rho_0\tilde{\alpha}_{ij}, \quad \tilde{\rho}^{12}_{ij} = \phi\rho_0 \left(\delta_{ij} - \tilde{\alpha}_{ij}\right), \quad \tilde{\rho}^{11}_{ij} = \rho_1 + \phi\rho_0 \left(\tilde{\alpha}^{\infty}_{ij} - \delta_{ij}\right), \tag{8.13}$$

with

$$\tilde{\alpha}_{ij} = \alpha^{\infty}_{ij} - j\frac{\tilde{b}_{ij}}{\omega}, \tag{8.14}$$

and α_{ij}^{∞} being the geometric tortuosity tensor and $\tilde{b}_{ij}$ the viscous drag tensor. One may now take advantage of the scalar property of the fluid stress tensor, to express the fluid displacement u^f as a function of the solid displacement and the pressure. This leads to a second relation which allows us to replace the fluid displacement:

$$u_j^f = \frac{\hat{\alpha}_{jk} p_{,k}}{\phi \rho_0 \omega^2} + (\delta_{jk} - \hat{\alpha}_{jk}) u_k^s, \tag{8.15}$$

where $\hat{\alpha}_{jk}$ is the inverse of the dynamic tortuosity tensor ($\hat{\alpha}_{ik}\tilde{\alpha}_{kj} = \delta_{ij}$).

From this, we may rewrite the solid momentum equation as:

$$\hat{\sigma}_{ij,j}^s + \tilde{\gamma}_{ij} p_{,j} = -\omega^2 \tilde{\rho}_{ij} u_j^s, \tag{8.16}$$

with

$$\tilde{\gamma}_{ij} = \phi \hat{\alpha}_{ij} - \tilde{\gamma}'_{ij}, \tag{8.17}$$

$$\tilde{\rho}_{ij} = \rho_1 + \phi \rho_0 (\delta_{ij} - \hat{\alpha}_{ij}), \tag{8.18}$$

where $\tilde{\gamma}_{ij}$ is a coupling factor which involves dynamic effects through the dynamic tortuosity and elastic coupling through $\tilde{\gamma}'_{ij}$. Note that $\tilde{\rho}_{ij}$ is the solid apparent density.

In a similar way the fluid displacement may be eliminated from the fluid momentum equation using the displacement u_i^W:

$$-\phi^2 p_{,i} = -\omega^2 \phi \rho_0 \left(\delta_{ij} - \tilde{\alpha}_{ij} + \frac{\tilde{\alpha}_{ik} \tilde{Q}_{kj}}{\phi^2 \tilde{K}_{eq}} \right) u_j^s - \omega^2 \rho_0 \tilde{\alpha}_{ij} u_j^W, \tag{8.19}$$

We may condense this relation through rewriting it as,

$$-\frac{\phi^2}{\rho_0 \omega^2} \hat{\alpha}_{ij} p_{,j} = -\tilde{\gamma}_{ij} u_j^s - u_i^W, \tag{8.20}$$

The advantage of considering u_i^W is that it may now be eliminated by taking the divergence and introducing (8.8),

$$-\frac{\phi^2}{\rho_0 \omega^2} \hat{\alpha}_{ij} p_{,ji} = -\tilde{\gamma}_{ij} u_{j,i}^s + \frac{p}{\tilde{K}_{eq}}. \tag{8.21}$$

We have with these relations completed the transformation from fluid stress tensor and displacements, to pressure and solid displacements for the anisotropic PEM modelling.

8.3 Weak Forms

There are a number of different ways of setting up the weak forms for an anisotropic PEM. Although in principle equal, slightly different choices may be made in the derivations. These choices introduce boundary terms that may lead to natural, homogeneous coupling conditions that may be of interest in certain applications. Frequently we face modelling situations where a PEM is placed next to a solid panel, or when it is in direct contact to an air domain, but there are of course other configurations that could be considered as well. Here, two mixed displacement formulations will be presented, one which naturally couples with an acoustic air domain, see [9] for the isotropic PEM modelling, (here referred to as the air formulation or PEM1) and one which naturally couples with a solid in the case where the solid displacements are continuous and the normal relative flow is zero, see [10] for the isotropic PEM modelling, (here referred to as the bonded solid formulation or PEM2).

8.3.1 PEM1 Weak Formulation

To state the weak forms, we introduce the test functions v_i^s and q. We start with the solid momentum equations, (8.16), and multiply by v_i^s, integrate over the porous domain Ω of boundary Γ_p and perform a partial integration of the solid stress gradient term. This gives,

$$\int_{\Omega} v_{i,j}^s \hat{C}_{ijkl} u_{k,l}^s - \omega^2 v_i^s \tilde{\rho}_{ij} u_j^s + v_i^s \tilde{\gamma}_{ij} p_{,j} \, \mathrm{d}\Omega = \int_{\Gamma} v_i^s \hat{\sigma}_{ij}^s n_j \, \mathrm{d}\Gamma. \tag{8.22}$$

Similarly we multiply equation (8.21) by q, integrate over the porous domain Ω_p and perform a partial integration of the term involving the second order gradient of the pressure,

$$\int_{\Omega} \frac{\phi^2}{\rho_0 \omega^2} \hat{\alpha}_{ij} p_{,j} q_{,i} - \frac{pq}{\tilde{K}_{eq}} + \tilde{\gamma}_{ij} u_{j,i}^s q \, \mathrm{d}\Omega = \int_{\Gamma} \frac{\phi^2}{\rho_0 \omega^2} \hat{\alpha}_{ij} p_{,j} n_i q \, \mathrm{d}\Gamma. \tag{8.23}$$

As will be shown later, these two equations naturally couples with a weak form representing air through the boundary integral on their respective RHS. It will in the following be referred to as PEM1.

8.3.2 PEM2 Weak Formulation

The objective of the second formulation is to modify the weak form in order to introduce slightly different boundary integral terms that naturally couples with a solid attached panel. We start by rewriting the third term on the LHS of (8.22), using (8.18) and performing a partial integration, as

$$\int_{\Omega} v_i^s \tilde{\gamma}_{ij} p_{,j} \, d\Omega = \int_{\Omega} \phi v_i^s \hat{\alpha}_{ij} p_{,j} \, \mathrm{d}\Omega + \tilde{\gamma}'_{ij} \left(\int_{\Omega} v^s_{i,j} p \, d\Omega - \int_{\Gamma} v_i^s \delta_{ij} n_j p \, \mathrm{d}\Gamma \right).$$

This allows us to rewrite the weak form of the solid part as,

$$\int_{\Omega} v^s_{i,j} \hat{C}_{ijkl} u^s_{k,l} - \omega^2 v_i^s \tilde{\rho}_{ij} u_j^s + \phi v_i^s \hat{\alpha}_{ij} p_{,j} + \tilde{\gamma}'_{ij} v^s_{i,j} p \, d\Omega = \int_{\Gamma} v_i^s \hat{\sigma}^t_{ij} n_j \, \mathrm{d}\Gamma, \quad (8.24)$$

Similarly, for the third term of the LHS of (8.23), we may write,

$$\int_{\Omega} \tilde{\gamma}_{ij} u^s_{j,i} q \, d\Omega = \int_{\Omega} \phi \hat{\alpha}_{ij} u^s_{j,i} q \, d\Omega + \left(\int_{\Omega} \tilde{\gamma}'_{ij} u_i^s q_{,j} \, d\Omega - \int_{\Gamma} p \delta_{ij} v_i^s n_j \, \mathrm{d}\Gamma \right) d\Gamma,$$

and the weak form of the fluid is then,

$$\int_{\Omega} \frac{\phi^2}{\rho_0 \omega^2} \hat{\alpha}_{ij} p_{,i} q_{,j} - \frac{pq}{\tilde{K}_{eq}} + \tilde{\gamma}'_{ij} u^s_{j,i} q + \phi \hat{\alpha}_{ij} u^s_{j,i} q \, d\Omega = \int_{\Gamma} q w_j n_j q \, \mathrm{d}\Gamma. \quad (8.25)$$

In these forms, the above two equations naturally couples with a weak form representing a solid through the boundary integral on their respective RHS. This formulation will in the following be referred to as PEM2.

8.3.3 *Elastic Solid*

In order to prepare for the discussion of the coupling between different types of domains in more detail, we introduce the weak form for the elastic structure as

$$\int_{\Omega_p} v^e_{i,j} C^e_{ijkl} u^e_{k,l} - \rho_e \omega^2 v_i^e u_j^e \, \mathrm{d}\Omega = \int_{\Gamma} v_i^s \hat{\sigma}^e_{ij} n^e_j \, \mathrm{d}\Gamma, \quad (8.26)$$

where u^e and v^e are respectively the unknown elastic solid displacements and the corresponding test fields. Furthermore, $\hat{\sigma}^e_{ij} = C^e_{ijkl} u^e_{k,l}$ is the stress tensor of the elastic structure, ρ_e is its density and n^e is the outgoing normal of the elastic domain.

8.3.4 Air

Finally, to conclude this part discussing weak forms, the air cavity weak form is,

$$\int_{\Omega_i} \frac{p^a_{,j} q^a_{,j}}{\rho_a \omega^2} - \frac{p^a q^a}{K_a} \, \mathrm{d}\Omega = \int_{\Gamma_i} \frac{1}{\rho_a \omega^2} q^a p^a_{,j} n^a_j \, \mathrm{d}\Gamma, \tag{8.27}$$

where p_a is the acoustic pressure in the air domain, K_a is the bulk modulus of air, ρ_a the ambient density of air, n^a is the outgoing normal of the acoustic domain and q_a the test field.

8.4 Discrete Linear System

To obtain a FE based solution of a problem involving, among others PEMs, we start from the weak forms presented in (8.3). The steps required to obtain the discretisation are the same as for elastic structures and acoustic domains etc., and we will only recall the PEM case here.

First we divide the domain into a finite number of, non-overlapping domains, commonly referred to as elements, denoted by Ω_e with $e = 1 \ldots N_e$. These elements form a mesh and in each of them the PDEs are transformed into algebraic equations, which are approximations to the exact solution. The elements are preferably simple in shape, thus requiring a finite set of polynomial shape functions to sufficiently accurate describe the variation of the field variables in the sub-domain. Typical functions used are nodal or bubble shape functions along edges, faces and in the interior. Some examples (Lobatto shape functions) are given in the following.

With simple polynomial shape functions, the element equations may be numerically integrated, resulting in elementary matrices that approximate the spatial variation, and then assembled into a larger system of algebraic equations, which then may be solved numerically and the solution obtained may be used to obtain relevant results such as pressures, displacements, etc.

8.4.1 Elementary Matrices

Assuming that we have N_d shape functions in an element e, a physical field $f^{(e)}$ may be approximated by:

$$f^{(e)}(\mathbf{x}) \approx \sum_{k=1}^{N_d} \varphi_k(\mathbf{x}) f_k^{(e)} = [\boldsymbol{\varphi}(\mathbf{x})]\mathbf{f}^{(e)}. \tag{8.28}$$

The shape functions $\varphi_k(\mathbf{x})$ approximate the spatial dependence of the fields, and usually the test fields, v_i^s and q, are approximated in a similar manner. Each shape function is associated to a degree of freedom f_k^e which will be an unknown of the global FE problem. The shape functions and degrees of freedom of one element can be respectively gathered in a matrix $[\boldsymbol{\varphi}(\mathbf{x})]$ and a vector $\mathbf{f}^e$.

The key characteristic of a poroelastic material is that four fields should be discretised (three displacement fields and one pressure fields). We then have, for the pressure:

$$p^{(e)}(\mathbf{x}) \approx [\boldsymbol{\varphi}(\mathbf{x})]\mathbf{p}^{(e)}, \tag{8.29}$$

and for the displacements, where each direction is discretised independently,

$$\begin{Bmatrix} u_1^{(e)}(\mathbf{x}) \\ u_2^{(e)}(\mathbf{x}) \\ u_3^{(e)}(\mathbf{x}) \end{Bmatrix} = \underbrace{\begin{bmatrix} [\boldsymbol{\varphi}(\mathbf{x})] & 0 & 0 \\ 0 & [\boldsymbol{\varphi}(\mathbf{x})] & 0 \\ 0 & 0 & [\boldsymbol{\varphi}(\mathbf{x})] \end{bmatrix}}_{[\boldsymbol{\varphi}_{\mathbf{u}}(\mathbf{x})]} \mathbf{u}^{(e)}, \quad \mathbf{u}^{(e)} = \begin{Bmatrix} \mathbf{u}_1^{(e)} \\ \mathbf{u}_2^{(e)} \\ \mathbf{u}_3^{(e)} \end{Bmatrix}. \tag{8.30}$$

The dynamics of the response of the PEM in one element is then approximated in terms of elementary vectors $\mathbf{u}^{(e)}$ and $\mathbf{p}^{(e)}$. Note that here we have chosen to use the same shape functions for the pressure and the three displacements fields. However, this is not a requirement as different shape functions (or different orders) can be considered for each unknown.

8.4.2 Discretisation of the Weak Forms in One Element

The discretisation of the 8 volume integral terms (4 for the solid and four for the pressure), that appear for both the PEM1 and the PEM2 formulations, is assumed to be the same. Most of the integrals involved are standard for any FE problem and will not be detailed here. Instead, we will present the discretisation for the terms which are typical for PEM modelling, i.e. the corresponding volumetric coupling between the solid and fluid phases. Note that in this section, we mix the summation convention and matrix forms and it should be clear from the context when one or the other is used. We start with the discretisation of the third term of the LHS of (8.24):

$$\int_{\Omega_{(e)}} v_i^s \hat{\alpha}_{ij} p_{,j} \, \mathrm{d}\Omega \approx \int_{\Omega_{(e)}} \mathbf{v}_i^{(e)T} [\boldsymbol{\varphi}_{\mathbf{u}}(\mathbf{x})]^T \hat{\alpha}_{ij} [\boldsymbol{\varphi}_{,j}(\mathbf{x})] \mathbf{p}^{(e)} \, \mathrm{d}\Omega. \tag{8.31}$$

Note that the term corresponding to the test field has been transposed, this allows us to separate the vectors $\mathbf{v}^{(e)}$ and $\mathbf{p}^{(e)}$ from the integral. This leads to the definition of a volume coupling matrix $[\mathbf{C}_{(e)}]$ which is a $(3N_d \times N_d)$ matrix as:

$$[\mathbf{C}^1_{(e)}] = \int_{\Omega_{(e)}} [\boldsymbol{\varphi}_{\mathbf{u}}(\mathbf{x})]^T \hat{\alpha}_{ij} [\boldsymbol{\varphi}_{,j}(\mathbf{x})] \, \mathrm{d}\Omega. \tag{8.32}$$

The superscript 1 is introduced to distinguish this particular form of the coupling matrix from other coupling terms, that will be discussed below. The discretisation of the coupling term over an element will then lead to:

$$\int_{\Omega_{(e)}} v_i^s \hat{\alpha}_{ij} p_{,j} \, \mathrm{d}\Omega \approx \mathbf{v}^{(e)^T} [\mathbf{C}^1_{(e)}] \mathbf{p}^{(e)}. \tag{8.33}$$

Similarly for the fourth term of the LHS of (8.24), another $(3N_d \times N_d)$ matrix $[\mathbf{C}^2_{(e)}]$ is introduced as:

$$\int_{\Omega_{(e)}} \tilde{\gamma}'_{ij} v^s_{i,j} p \, \mathrm{d}\Omega \approx \mathbf{v}^{(e)^T} [\mathbf{C}^2_{(e)}] \mathbf{p}^{(e)}, \tag{8.34}$$

Finally there is yet one more coupling term, i.e. the third one in the LHS of (8.22), which is discretised as:

$$\int_{\Omega_{(e)}} v_i^s \hat{\gamma}_{ij} p_{,j} \, \mathrm{d}\Omega \approx \mathbf{v}^{(e)^T} [\mathbf{C}^3_{(e)}] \mathbf{p}^{(e)}. \tag{8.35}$$

Without going into further details of the derivations, we apply the same methodology to the other terms which are common between the two different formulations:

$$\int_{\Omega_{(e)}} v^s_{i,j} \hat{C}_{ijkl} u^s_{k,l} \, \mathrm{d}\Omega \approx \mathbf{v}^{(e)^T} [\mathbf{K}_{(e)}] \mathbf{u}^{(e)}, \quad \int_{\Omega_{(e)}} v_i^s \tilde{\rho}_{ij} u_j^s \, \mathrm{d}\Omega \approx \mathbf{v}^{(e)^T} [\mathbf{M}_{(e)}] \mathbf{u}^{(e)},$$

$[\mathbf{K}_{(e)}]$ and $[\mathbf{M}_{(e)}]$ are the elementary stiffness and mass matrices. For terms relative to the fluid, one has:

$$\int_{\Omega_{(e)}} \frac{\phi^2}{\rho_0} \hat{\alpha}_{ij} p_{,j} q_{,j} \, \mathrm{d}\Omega \approx \mathbf{q}^{(e)^T} [\mathbf{H}_{(e)}] \mathbf{p}^{(e)}, \quad \int_{\Omega_{(e)}} \frac{pq}{\tilde{K}_{eq}} \, \mathrm{d}\Omega \approx \mathbf{q}^{(e)^T} [\mathbf{q}_{(e)}] \mathbf{p}^{(e)},$$

$[\mathbf{H}_{(e)}]$ and $[\mathbf{Q}_{(e)}]$ are the elementary kinetic and compression energy matrices which are also common to the two weak forms. Note that all integrals above are in practice computed for a standard reference element, and scaled to the actual element geometry.

The full elementary dynamic matrix $[\mathbf{A}_{(e)}]$ for the PEM1 formulation, is then

$$[\mathbf{A}_{(e)}] = \left[\begin{array}{c|c} [\mathbf{K}_{(e)}] - \omega^2 [\mathbf{M}_e] & [\mathbf{C}^3_{(e)}] \\ \hline [\mathbf{C}^3_{(e)}]^T & \dfrac{[\mathbf{H}_{(e)}]}{\omega^2} - [\mathbf{Q}_{(e)}] \end{array}\right]. \tag{8.36}$$

For the PEM2 formulation where $\{\mathbf{u}, \mathbf{p}\}$ naturally couples to a solid, we have

$$[\mathbf{A}'_{(e)}] = \left[\begin{array}{c|c} [\mathbf{K}_{(e)}] - \omega^2[\mathbf{M}_e] & [\mathbf{C}^1_{(e)}] + [\mathbf{C}^2_{(e)}] \\ \hline [\mathbf{C}^1_{(e)}]^T + [\mathbf{C}^2_{(e)}]^T & \dfrac{[\mathbf{H}_{(e)}]}{\omega^2} - [\mathbf{Q}_{(e)}] \end{array}\right]. \tag{8.37}$$

We see that the two formulations differ by the coupling terms between the solid and the fluid, but otherwise have the same structure. Thus, in the following parts of the procedure, as most of the steps are common between the two formulations, we have decided not to overload the notation and instead to use a generic form as,

$$[\mathbf{A}_{(e)}] = \begin{bmatrix} [\mathbf{A}^{uu}_{(e)}] & [\mathbf{A}^{up}_{(e)}] \\ [\mathbf{A}^{pu}_{(e)}] & [\mathbf{A}^{pp}_{(e)}] \end{bmatrix}. \tag{8.38}$$

8.4.3 Assembly

The global system matrices, required to complete the modelling for a complex problem consisting of many elements, are obtained by summation of elementary matrices and the method is the same for the two formulations. This is in principle not different from any other FE solution, but some extra attention needs to be paid to the matching between different degrees of freedom to their proper neighbours. As the vector of degrees of freedom for a particular element $\{\mathbf{u}^{(e)}, \mathbf{p}^{(e)}\}$ is a sub-vector of the global vector of degrees of freedom $\{\mathbf{u}, \mathbf{p}\}$, we need to establish a correspondence. There are several ways of doing this, one is to introduce a boolean matrix $[\mathbf{L}^{(e)}]$ of dimension $(4N_d \times 4N_g)$ where $4N_g$ corresponds to the total number of degrees of freedom. We then have:

$$\begin{Bmatrix} \mathbf{u}^{(e)} \\ \mathbf{p}^{(e)} \end{Bmatrix} = [\mathbf{L}_e] \begin{Bmatrix} \mathbf{u} \\ \mathbf{p} \end{Bmatrix}. \tag{8.39}$$

The global dynamic matrix then reads:

$$[\mathbf{A}] = \sum_{e=1}^{N} [\mathbf{L}^{(e)}]^T [\mathbf{A}_{(e)}][\mathbf{L}^{(e)}] = \begin{bmatrix} [\mathbf{A}^{uu}] & [\mathbf{A}^{up}] \\ [\mathbf{A}^{pu}] & [\mathbf{A}^{pp}] \end{bmatrix}. \tag{8.40}$$

Note that this way of approaching the assembly is purely formal and just used to explain the principles involved. Indeed, it can be shown that the multiplication $[\mathbf{L}^{(e)}]^T[\mathbf{A}_{(e)}][\mathbf{L}^{(e)}]$ is a matrix of size $(4N_g \times 4N_g)$ with a submatrix corresponding to $[\mathbf{A}_{(e)}]$ at the position corresponding to the *dof* of the element. The global matrix is then composed of four blocks which are the assembly of the elementary ones.

8.5 Coupling Between Domains

We will in the following discuss how a FE-PEM domain is assembled into a more complex arrangement of different media, such as a solid and a fluid. As part of this we will also illustrate the difference between the two weak formulations derived.

8.5.1 Coupling with an Air Domain

First we consider the problem of coupling between an acoustic domain and a poroelastic one, see Fig. 8.1 where the two domains are artificially separated to clearly illustrate the coupling between the two. The degrees of the acoustic domain correspond to the pressure. At the boundary Γ between the two domains, a compatible mesh is considered and the common pressure degrees of freedom are denoted $\mathbf{p}_\Gamma$, hence the *dof* of the two media can be partitioned as $\{\mathbf{u}, \mathbf{p}_i, \mathbf{p}_\Gamma\}$ for the PEM where $\mathbf{p}_i$ correspond to the vector of pressure in the interior of the domain. For the fluid domain, the *dof* are similarly $\{\mathbf{p}_\Gamma, \mathbf{p}_i^a\}$.

We now partition the generic global dynamic PEM matrix (8.40) into three blocks in order to introduce the partitioning between the boundary and interior *dof*. Applying the same partitioning to the test fields, the weak form is then approximated by:

$$\begin{bmatrix}\mathbf{v}^T & \mathbf{q}_i{}^T & \mathbf{q}_\Gamma{}^T\end{bmatrix}\begin{bmatrix}[\mathbf{A}^{uu}] & [\mathbf{A}_i^{up}] & [\mathbf{A}_\Gamma^{up}] \\ [\mathbf{A}_i^{pu}] & [\mathbf{A}_{ii}^{pp}] & [\mathbf{A}_{i\Gamma}^{pp}] \\ [\mathbf{A}_\Gamma^{pu}] & [\mathbf{A}_{\Gamma i}^{pp}] & [\mathbf{A}_{\Gamma\Gamma}^{pp}]\end{bmatrix}\begin{Bmatrix}\mathbf{u} \\ \mathbf{p}_i \\ \mathbf{p}_\Gamma\end{Bmatrix}. \tag{8.41}$$

The weak form for the air cavity is given in (8.27). Note that the outgoing normal n^a is in the opposite direction of n^p. The volume integrals, for the air domain may then be discretised as discussed above and a dynamic acoustic matrix, which may also be partitioned between boundary and interior *dof*, is then:

$$\begin{bmatrix}\mathbf{q}_\Gamma{}^T \mathbf{q}_i^{aT}\end{bmatrix}\begin{bmatrix}[\mathbf{A}_{\Gamma\Gamma}^a] & [\mathbf{A}_{\Gamma i}^a] \\ [\mathbf{A}_{i\Gamma}^a] & [\mathbf{A}_{ii}^a]\end{bmatrix}\begin{Bmatrix}\mathbf{p}_\Gamma \\ \mathbf{p}_i^a\end{Bmatrix}. \tag{8.42}$$

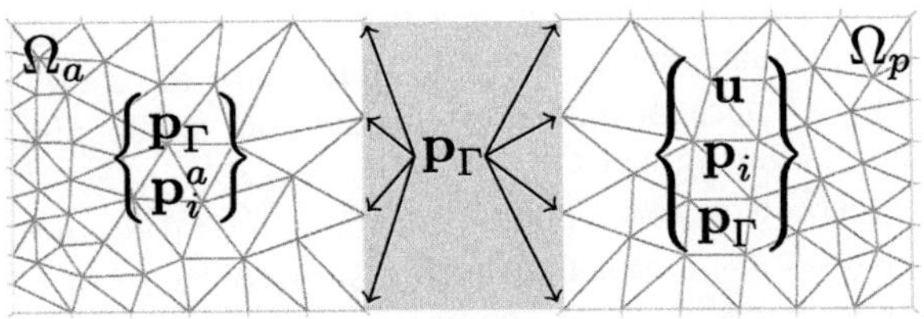

Fig. 8.1 Coupling of acoustic and poroelastic FE models

Adding the two discretised weak forms, air and PEM, leads to:

$$\left[\mathbf{v}^T\ \mathbf{q}_i{}^T\ \mathbf{q}_\Gamma{}^T\ \mathbf{q}_i^{aT}\right]\begin{bmatrix}[\mathbf{A}^{uu}] & [\mathbf{A}_i^{up}] & [\mathbf{A}_\Gamma^{up}] & [\mathbf{0}]\\ [\mathbf{A}_i^{pu}] & [\mathbf{A}_{ii}^{pp}] & [\mathbf{A}_{i\Gamma}^{pp}] & [\mathbf{0}]\\ [\mathbf{A}_\Gamma^{pu}] & [\mathbf{A}_{\Gamma i}^{pp}] & [\mathbf{A}_{\Gamma\Gamma}^{pp}]+[\mathbf{A}_{\Gamma\Gamma}^{a}] & [\mathbf{A}_{\Gamma i}^{a}]\\ [\mathbf{0}] & [\mathbf{0}] & [\mathbf{A}_{i\Gamma}^{a}] & [\mathbf{A}_{ii}^{a}]\end{bmatrix}\begin{Bmatrix}\mathbf{u}\\ \mathbf{p}_i\\ \mathbf{p}_\Gamma\\ \mathbf{p}_i^a\end{Bmatrix}. \tag{8.43}$$

We can deduce from this expression that, as for the assembly of elements, the global matrix which is the assembly of the two domains may be done by summing the overlapping parts of the dynamic matrices.

So far we have treated the volumetric terms in the weak forms and, as pointed out before, there is no difference between the two up to now. What remains is to derive the interface terms, and how they should be discretised. Here we have to distinguish between the two formulations, PEM1 and PEM2, as we will show below. However, let us first consider the interface relations between the different physical fields at the common boundary. Along Γ the pressure has to be continuous, the normal displacement of air has to be equal to the normal total displacement of the porous medium and, finally, as the solid frame of the PEM is free, the *in-vacuo* solid frame stress of the PEM has to vanish:

$$p^a = p, \quad \hat{\sigma}_{ij}^s n_j = 0, \quad \frac{p^a_{,j} n_j}{\rho_a \omega^2} = u_n^a = u_n^t = \frac{\phi^2 \hat{\alpha}_{ij} p_{,j} n_i}{\rho_0 \omega^2}. \tag{8.44}$$

If we now consider the case of the PEM1 formulation, that naturally should couple with air, the sum of the boundary terms associated to the PEM are,

$$\int_{\Gamma_i} v_i^s \hat{\sigma}_{ij}^s n_j + \frac{\phi^2}{\rho_0 \omega^2} \hat{\alpha}_{ij} p_{,j} n_i \, \mathrm{d}\Gamma. \tag{8.45}$$

The first term is zero as the *in-vacuo* stress is null at the interface. Concerning the second one, let us combine the terms relative to the cavity and the PEM in one integral expression. We then have:

$$\int_\Gamma \left(\frac{\phi^2}{\rho_0 \omega^2} \hat{\alpha}_{ij} p_{,j} n_i q - \frac{1}{\rho_a \omega^2} q^a p^a_{,j} n_j \right) \mathrm{d}\Gamma, \tag{8.46}$$

where the minus sign for the air term is due to the orientation of the normals. As we are only dealing with the case of compatible meshes, q^a and q are identical and we can deduce the nullity of this integral by the continuity of normal displacements, (8.44). Hence, for this $\{\mathbf{u}, \mathbf{p}\}$ formulation, the coupling with an acoustic domain is natural (which means that there is no surface term). The degrees of freedom of the global problem correspond to the merging of the acoustic pressures in the air domain $\mathbf{p}^a$ and the displacement $\mathbf{u}$ and pressure $\mathbf{p}$ of the PEM. We see that the porous material and the acoustic air have common degrees of freedom.

For the PEM2 formulation, the boundary terms are

$$\int_{\Gamma}\left(v_i^s\hat{\sigma}_{ij}^t n_j + q w_j n_j - \frac{1}{\rho_a\omega^2} q^a p_{,j}^a n_j\right) \mathrm{d}\Gamma. \tag{8.47}$$

However, as the *in-vacuo* stress tensor is zero, one has that $\hat{\sigma}_{ij}^t = -p\delta_{ij}$ and due to the continuity of pressure along Γ, we can replace the total stress by the pressure of air. In addition

$$q w_j n_j - \frac{1}{\rho_a\omega^2} q^a p_{,j}^a n_j = q^a\left(u_j^t - u_j^s - u_j^a\right) n_j = -q^a u_j^s n_j. \tag{8.48}$$

The combined boundary terms are then:

$$-\int_{\Gamma} p^a v_j^s n_j + q^a u_j^s n_j \,\mathrm{d}\Gamma. \tag{8.49}$$

This term is a standard fluid structure interaction couple and its discretisation leads to a coupling matrix $[\boldsymbol{\Gamma}_{fs}]$ between the solid displacement of the porous material and the pressure in air which should be added to the global matrix. The system matrix will then be:

$$\begin{bmatrix} [\mathbf{A}^{uu}] & [\mathbf{A}_i^{up}] & [\mathbf{A}_\Gamma^{up}] - [\boldsymbol{\Gamma}_{fs}] & [\mathbf{0}] \\ [\mathbf{A}_i^{pu}] & [\mathbf{A}_{ii}^{pp}] & [\mathbf{A}_{i\Gamma}^{pp}] & [\mathbf{0}] \\ [\mathbf{A}_\Gamma^{pu}] - [\boldsymbol{\Gamma}_{fs}]^T & [\mathbf{A}_{\Gamma i}^{pp}] & [\mathbf{A}_{\Gamma\Gamma}^{pp}] + [\mathbf{A}_{\Gamma\Gamma}^a] & [\mathbf{A}_{\Gamma i}^a] \\ [\mathbf{0}] & [\mathbf{0}] & [\mathbf{A}_{i\Gamma}^a] & [\mathbf{A}_{ii}^a] \end{bmatrix}. \tag{8.50}$$

The coupling between an air domain and this form of the $\{\mathbf{u}, \mathbf{p}\}$ formulation is thus not natural as we need the additional term $[\boldsymbol{\Gamma}_{fs}]$.

8.5.2 Coupling with an Elastic Solid

The coupling of a PEM domain and an elastic structural domain can be done in a similar way. The weak form for the elastic solid structure is given by (8.26). Note that, n^e in each point along the interface is in a direction opposite to the one of the poroelastic material. The interface relations at such an interface are:

$$\sigma_{ij}^e = \hat{\sigma}_{ij} - p\delta_{ij}, \quad u_j^e = u_j^s. \tag{8.51}$$

Let us first consider the PEM2 formulation, which we have previously claimed to naturally couple with a solid. The boundary terms are

$$\int_{\Gamma}\left(v_i^s\sigma_{ij}^t n_j + q w_j n_j - v_i^e\hat{\sigma}_{ij}^e n_j\right) \mathrm{d}\Gamma. \tag{8.52}$$

Similarly to the previous case compatible meshes are considered, and we then have $v_i^e = v_i^s$. In addition $w_j n_j = (u_j^t - u_j^s) n_j = 0$. Hence the boundary integral is zero meaning that the coupling between an elastic structure and this PEM2 formulation is natural. Similarly to the coupling between PEM1 and an air domain, the dynamic equation system is obtained by summing the matrix terms corresponding to common *dof* along Γ.

For the PEM1 formulation in Sect. 8.3.1, the surface term is given as:

$$\int_\Gamma \left(v_i^s \left(\hat{\sigma}_{ij} - \hat{\sigma}_{ij}^e \right) n_j + q \frac{\phi^2}{\rho_a \omega^2} \hat{\alpha}_{ij} p_{,j} n_i \right) \mathrm{d}\Gamma. \tag{8.53}$$

The interface conditions of interest are

$$\hat{\sigma}_{ij} - \sigma_{ij}^e = p \delta_{ij}, \quad u_i^e n_i = \frac{\phi^2}{\rho_a \omega^2} \hat{\alpha}_{ij} p_{,j} n_i. \tag{8.54}$$

The boundary integral is then finally:

$$- \int_\Gamma \left(p v_j^s n_j + q u_j^s n_j \right) \mathrm{d}\Gamma, \tag{8.55}$$

which is similar to (8.49), and the assembled equation system is obtained by adding the *dof* corresponding to the Γ and adding the coupling terms in a way similar to what is shown in (8.50).

8.5.3 Coupling Through a Thin Airgap Interface

Quite often, the interface between a PEM and a solid is such that the two are not in full (elastic) contact but are still close to each other, i.e. they may be thought of as separated by a thin airgap. This is a case which either may be modelled as described in Sect. 8.5.1, in case the airgap must be meshed as a separate domain, or in cases where this could be complicated (due to e.g. geometry) it may be modelled through an approximate boundary condition as presented below (Fig. 8.2). We use the superscript o to distinguish between the acoustic air domain previously discussed, Sect. 8.5.1, and the present domain which will only be used in an intermediate step.

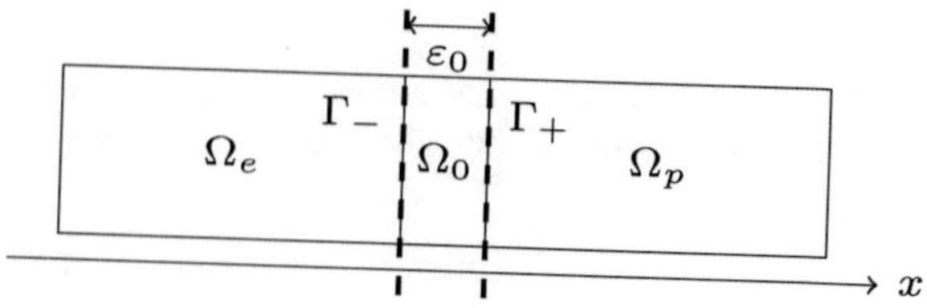

Fig. 8.2 Zoom on the region with a thin airgap separating the PEM domain from the elastic domain

At the boundary where the PEM and the airgap domain meet, here identified as Γ^-, the total stress in the PEM must equal the acoustic pressure in the airgap because of a zero *in-vacuo* stress,

$$\sigma_{ij}^{tot} n_j^p = -p^o; \quad x_i \in \Gamma^-, \tag{8.56}$$

and the pressure and the displacements have to be continuous,

$$p = p^o; \quad x_i \in \Gamma^-, \tag{8.57}$$

$$u_i^o n_i^o = (1-\phi)\, u_i^s n_i^p + \phi u_i^f n_i^p; \quad x_i \in \Gamma^-. \tag{8.58}$$

At the opposite side of the airgap domain, Γ^+, the elastic solid displacements has be equal to the acoustic displacements,

$$u_i^e n_i^e = u_i^o n_i^e; \quad x_i \in \Gamma^+, \tag{8.59}$$

and the solid stress has to be equal to the acoustic pressure in the airgap,

$$\sigma_{ij}^e n_j^e = -p^o; \quad x_i \in \Gamma^+. \tag{8.60}$$

If we now assume that the width ε_o of the airgap is very small, we can also assume that the fluid in the airgap is incompressible and the pressure thus constant in the direction normal to the boundary, $p^o = p$. Noting that $n_j^p = -n_j^o$ and assuming that ε_o goes to zero, the surface integrals pertaining to the PEM2 formulation that naturally couples to a solid, together results in,

$$\int_{\Gamma_+} \left(p n_i^p v_i^s + u_i^s n_i^p q\right) d\Gamma - \int_{\Gamma_-} \left(p n_i^p v_i^e + u_i^e n_i^p q\right) d\Gamma. \tag{8.61}$$

The boundary integrals appearing in (8.61), are also in this case standard fluid-structure interaction area integrals, here denoted by Γ_{fs}, and the assembled equation system may be written as,

$$\begin{bmatrix} [\mathbf{A}^e] & [\mathbf{0}] & -[\boldsymbol{\Gamma}_{fs}^-] \\ [\mathbf{0}] & [\mathbf{A}^{uu}] & [\mathbf{A}^{up}] + [\boldsymbol{\Gamma}_{fs}^+] \\ -[\boldsymbol{\Gamma}_{fs}^-]^T & [\mathbf{A}^{pu}] + [\boldsymbol{\Gamma}_{fs}^+]^T & [\mathbf{A}^{pp}] \end{bmatrix}, \tag{8.62}$$

where we have omitted the partitioning into interior and boundary *dof* for clarity (Fig. 8.3).

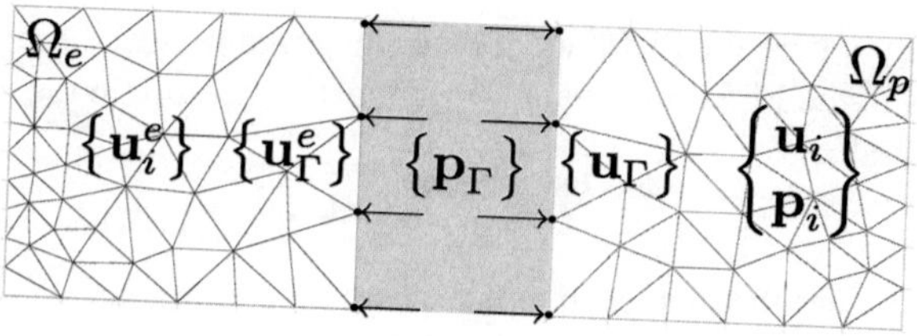

Fig. 8.3 Coupling of elastic and poroelastic FE models using an approximate representation of a thin airgap. Arrows symbolically indicate the pressure loading on the elastic solid nodes and the porous solid *dof* at the boundary

8.5.4 Coupling with Bloch Waves

We now proceed to the case where a PEM domain is coupled with a Bloch wave. Here a 2D problem is considered for a unit cell of period D and with periodic boundary conditions. In the present case, this problem is composed of two domains, an acoustic air and the second one is PEM. The PEM boundary is excited by an incident plane wave in the air domain. The components of the wave vector are $k_x = k_0 \sin(\theta)$ and $k_z = k_0 \sin(\theta)$, with k_0 the wave number in air and θ the angle of the incident wave. In the air domain, the physical fields can be represented as a superposition of the incident wave and reflected Bloch waves. The pressure p^a may then be expressed as:

$$p^a(x,z) = e^{j(k_x x + k_z z)} + \sum_{l=-\infty}^{+\infty} e^{j(k_x(l)x - k_z(l)z)} R_l, \tag{8.63}$$

where R_l are the amplitudes of the reflected Bloch waves, $k_x(l)$ and $k_z(l)$ are the x and z components, respectively in the wave vector. They are given by

$$k_z(j) = \sqrt{k_0^2 - k_x(l)^2}, \quad k_x(l) = k_x + \frac{2\pi l}{D}. \tag{8.64}$$

From Euler's equation, we can deduce the displacement of air in the z direction:

$$u_a^- = \frac{j}{\rho_a \omega^2} \left(k_z e^{j(k_x x + k_z z)} + \sum_{l=-\infty}^{+\infty} k_z(l) e^{j(k_x(l)x - k_z(l)z)} R_l \right). \tag{8.65}$$

The porous material in the FE domain is modelled by the PEM1 formulation. We recall, the boundary integral on the interface:

$$I_{FE} = \int_{\Gamma_b} v_i^s(x) \hat{\sigma}_{ij}^s(x) n_j + \frac{\phi^2}{\rho_0 \omega^2} \hat{\alpha}_{ij} p_{,j}(x) n_i q(x) \, \mathrm{d}\Gamma = 0. \tag{8.66}$$

In order to couple the two domains, one should consider the interface relations on Γ_b, the continuity of the pressure, of the fluid displacement and the vanishing *in-vacuo* stress:

$$p^a(x) = p(x), \quad \hat{\sigma}^s_{iz}(x) = 0, \quad u^a_z(x) = u^t_z(x) = \frac{\phi^2}{\rho_0\omega^2}\hat{\alpha}_{zj}p_{,j}(x). \tag{8.67}$$

Using these relations, the integral I_{FE} in (8.66) may be rewritten. First, the solid term of the integral is zero due to the nullity of the *in-vacuo* stress. Concerning the pressure, it may be rewritten using the continuity of the normal displacement and the expression (8.65). We then have:

$$I_{FE} = \frac{j}{\rho_a\omega^2}\int_{\Gamma_b}\left(-k_z e^{-jk_x x} - \sum_{l=-\infty}^{+\infty} k_z(l)e^{-jk_x(l)x}R_l\right)q(x)\,\mathrm{d}\Gamma. \tag{8.68}$$

We may then deduce from this expression that the boundary term of the weak form can be expressed as a function of the amplitudes of the Bloch waves. The method then consists in considering these amplitudes as unknowns and then to add them to the degrees of freedom of the FE problem. It is first necessary to truncate the infinite sum in (8.68). The infinite sum is then approximated by considering that $l = -N_b .. + N_b$, where N_b is determined by considering classical criteria to truncate Bloch waves. The integral can be rewritten in a more condensed form:

$$I_{FE} = -\frac{jk_z}{\rho_a\omega^2}I(0) + \sum_{l=-N}^{N}\frac{jk_z(l)}{\rho_a\omega^2}I(l)R_l, \tag{8.69}$$

with

$$I(l) = \int_{\Gamma_b} e^{-jk_x(l)x}q(x)\,\mathrm{d}\Gamma, \tag{8.70}$$

and $I(l)$ is the integral of the shape function weighted by an exponential term associated with the Bloch wave. It may be discretised in a way similar to the one presented above. As it is the product of an exponential and a polynomial, its value may be computed analytically by successive integration by parts. The first term in the right hand side of (8.69) will then lead to the excitation vector. The second one will lead to a block [**B**] that will be associated to variational *dof* **q** and to the vector of unknown reflection coefficients **B**.

As $2N_b + 1$ new unknowns have been introduced, it is necessary to combine them with new constraints. This can be easily done by considering the continuity of pressure (which is the only relation in (8.66) which was not considered up to now in the derivation). First, we consider an index $m \in \mathbb{R}$, the projection of the pressure on $e^{jk_x(m)x}$ then leads to:

$$\int_{\Gamma_b} p^a(x)e^{jk_x(m)x}\,\mathrm{d}\Gamma = D\delta_{m0} + DR_m. \tag{8.71}$$

We can then discretise this relation by replacing the pressure in air by the degrees of freedom on the boundary. The discretisation of the first term in the right hand side will then lead to an excitation term and the other ones leads to a block $[\mathbf{B}']$ to be inserted in the global matrix. The final linear system is then

$$\begin{bmatrix} [\mathbf{A}^{uu}] & [\mathbf{A}^{up}] & [\mathbf{0}] \\ [\mathbf{A}^{pu}] & [\mathbf{A}^{pp}] & [\mathbf{B}] \\ [\mathbf{0}] & [\mathbf{B}'] & D[\mathbf{I}_{2N_b+1}] \end{bmatrix} \begin{Bmatrix} \mathbf{u} \\ \mathbf{p} \\ \mathbf{R} \end{Bmatrix} = \begin{Bmatrix} \mathbf{0} \\ \mathbf{F} \\ \mathbf{1}_0 \end{Bmatrix}. \tag{8.72}$$

$[\mathbf{I}_{2N_b+1}]$ is the identity matrix and $\mathbf{1}_0$ is the vector of $\mathbb{R}^{2N_b+1}$ whose first component is one and all the others are zero.

8.6 Application Examples

One of the characteristic properties of modelling of PEMs, is the intricate interactions both throughout the domain as well as at the boundary to other porous and non-porous domains. These interactions quite often tend to control the discretisation and the meshing, thus driving the size of the resulting algebraic systems to be solved. There are several reasons for this, e.g.:

- Wavelengths of the Biot waves are generally shorter than those of the acoustic or the solid waves, see [11–13]. To meet adequate mesh criteria, usually a larger mesh density or number of *dof* are required in the porous domain, as compared to the fluid or solid domains.
- The poroelastic domain is meshed by volume elements, requiring a discretisation in the 3 spatial directions (for 3D problems), unlike the elements of thin structures, such as plates or shells, which in many cases of practical interest may be meshed in 2 directions.
- The number of degrees of freedom per node of the poroelastic elements is at least 4, which together with the volume modelling required, tend to increase the model size as well as the bandwidth of the problem, [14].

From the above points we realise that the mixed-pressure formulation, having only 4 degrees of freedom per node, is interesting for many applications and we will in the following show some examples of varying complexity of its use in different modelling situations. These examples are chosen in order to illustrate the effects of boundary conditions between an PEM and a solid, as well as the effects of anisotropy, in the behaviour of different multi-layered configurations.

8.6.1 Convergence Aspects

In this section, some aspects related to the convergence of the FE model are presented. These are illustrated through a simple case, see Fig. 8.4, which consists of a 1D problem of a foam layer. The thickness of the layer is equal to 10 cm. On the left boundary, the foam is excited by plane wave of amplitude 1. On the right boundary, the porous layer is fixed and the solid displacement as well as the normal gradient of the pressure are both zero. A reflected wave with an amplitude R is then created. As R is not known, it is added to the degrees of freedom of the FE model by the way of the technique presented in Sect. 8.5.4. For this simple problem an analytical expression of the surface impedance can be derived [7, 15] from which the reflection coefficient R_t can be deduced. This analytical solution will be considered as the reference when the convergence is discussed in the following. The layer is modelled by the PEM1 formulation and the mesh consists of n element Ω_n of equal dimension. On each reference element (corresponding to $\xi \in]-1, 1[$), the solid displacement and the pressure are discretised with Lobatto shape functions. The first two orders correspond to linear shape functions:

$$\varphi_0(\xi) = \frac{1-\xi}{2}, \quad \varphi_1(\xi) = \frac{1+\xi}{2}. \tag{8.73}$$

For the higher orders $k > 1$ the shape functions are the Lobatto shape functions

$$\varphi_k(\xi) = \sqrt{k - \frac{1}{2}} \int_1^{xi} L_{k-1}(s)\, \mathrm{d}s, \quad k > 2p \tag{8.74}$$

where $L_k(\xi)$ are the Legendre polynomials and p is the order of the approximation. In each element, we then have p degrees of freedom for both the solid displacement and pressure.

We present the convergence at a single frequency, here 946 Hz, which corresponds to the resonance frequency of the solid frame. The results shown are representative also for other frequencies and the conclusions are thus frequency independent. The analysis is done by fixing the order p and refining the mesh by increasing the number

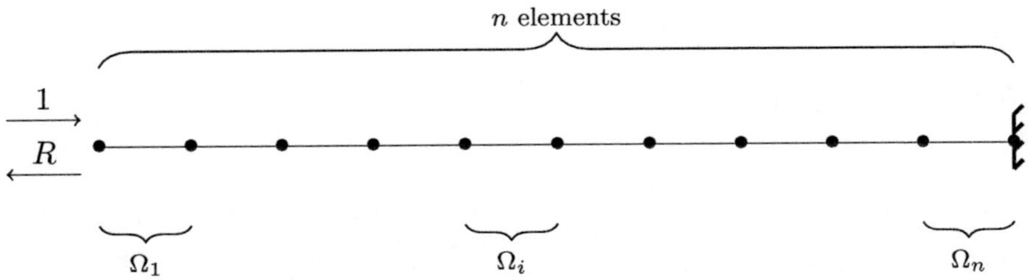

Fig. 8.4 Configuration used for the convergence analysis

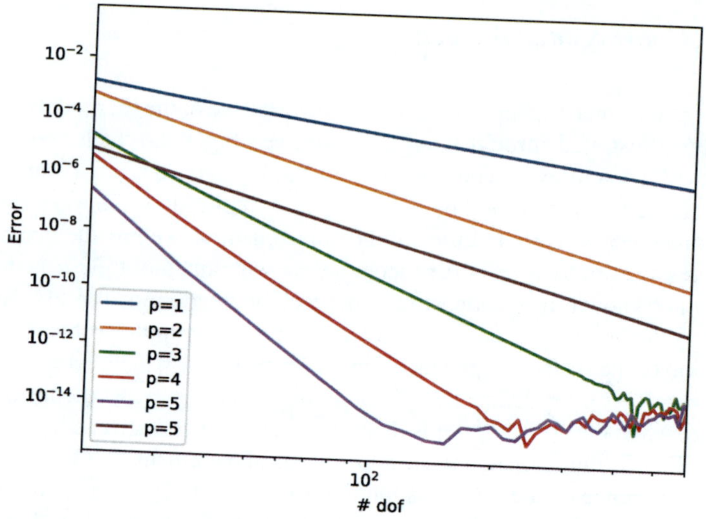

Fig. 8.5 Convergence curves for several orders

of element n of the mesh. The error ε is simply defined as the difference between the analytically and numerically computed reflection coefficients:

$$\varepsilon = |R - R_t|. \tag{8.75}$$

The result are presented in Fig. 8.5 for each order p between 1 and 5. The error is plotted as a function of the number of degrees of freedom and the results are displayed in logarithmic scale. The results obtained for a PEM follows the same trends as can be observed for other kinds of media. Starting with the case $p = 1$ i.e. linear elements, we can see that the error decreases with the number of elements but the convergence is rather slow. For $p = 2$, i.e. quadratic elements, the convergence rate is higher in the sense that the slope of the error is larger. As a more general observation, the slope is increasing with increasing order p, indicating a considerable potential for using higher order elements. However, it is not straightforward to deduce which are the optimal orders of the elements as the sparseness of the matrix system is reduced for high p. For more details, the reader can consult [8].

8.6.2 *FE Cases in 1D*

For the 1D example, the objectives are to illustrate and study some characteristic behaviours of the solutions to the discretised weak forms. Here we will use the PEM2 formulation to investigate the influence of contact between a PEM and an interfacing solid, which is a well-known and critical integration effect in industrial applications.

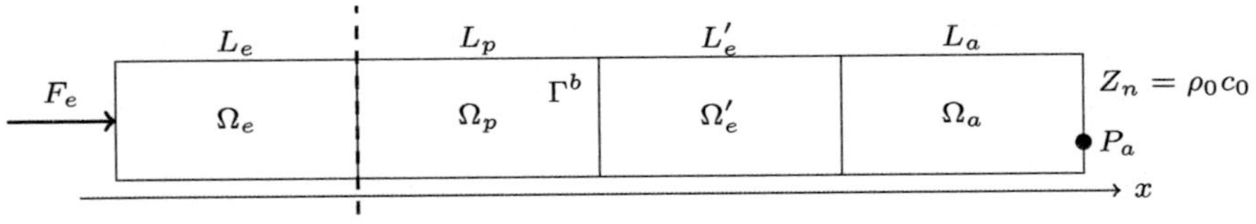

Fig. 8.6 Multilayer setup, studied as 1D example. Excitation applied to solid at the leftmost boundary and reflection free termination of acoustic domain at rightmost boundary. Along $\Gamma^{b/ub}$, either full contact or complete separation is assumed, along Γ^b full contact

The background is the difficulty in ensuring a well-defined contact between the PEM and e.g. a panel surface, as a multi-layered sound package component is manufactured. It also poses a challenging aspect in building a simulation model of such components. In addition, we emphasise that the purpose here is not to show realistic configurations and their performance, but to give the reader a deeper understanding into certain modelling aspects that should be kept in mind when building more complex and elaborate models. Thus, the physical dimensions (here thicknesses) are slightly exaggerated to allow for the visualisation to be more clear.

We take as the starting point the multilayer setup shown in Fig. 8.6, where an arrangement with two solid domains with a PEM in between, and an acoustic domain with an reflection-free termination is presented. For this example, the properties of the solid and the PEM, are given in Table 8.1. Note that we are using second order polynomials for the shape functions, see also Fig. 8.4 for a discussion of this aspect. We assume that in one case the solids are in full contact to the PEM at both boundaries, $\Gamma^{b/ub}$ and Γ^b, and in the other case that one of the solid boundaries, $\Gamma^{b/ub}$, is separated from the PEM through an airgap.
In the following we will focus on:

- The modelling as such, we will show how the airgap can be modelled either through meshing the thin acoustic domain separating the solid and the PEM, or through the approximate modelling principle taken from [16] and presented in Sect. 8.5.3.
- The effects of the airgap, illustrated through the transmission performance as well as the actual resulting fields computed at a particular frequency,

When modelling of a combination of a PEM and a solid is concerned, two extreme cases for handling of the boundary conditions may be considered. Either we choose to have full contact, or full separation with a more or less well-defined distance between the boundaries. In the case of full contact, we need to introduce the boundary conditions specified in Sect. 8.5.2. For the PEM2 formulation, this amounts to vanishing boundary integrals on both sides of the PEM, and we thus solve a problem that is equivalent in form to Sect. 8.5.2.

If we now instead assume that there is an airgap, separating the PEM from the solid, with a finite thickness, of e.g. 0.1 mm, and that we as a first choice mesh it with acoustic FE. In this case we need to introduce the conditions stated in Sect. 8.5.1 at the $\Gamma^{b/ub}$ boundary to the PEM, thus solving a problem that is equivalent in form to (8.55).

Table 8.1 Parameters of the materials used for the application examples. Note isotropic equivalent values are also given for the anisotropic properties of the foam in Sect. 8.6.4

Elastic materials				
Parameter	Units	Solid (Sect. 8.6.2)		Panel (Sect. 8.6.4) Inclusion (Sect. 8.6.3)
Density	$kg.m^{-3}$	2700		7800
Young's modulus	GPa	72		200
Loss factor	–	0.001		0.001
Poisson ratio	–	0.3		0.3
Isotropic foams				
Parameter	Units	Sect. 8.6.2	Sect. 8.6.3	Sect. 8.6.4 Iso equi.
Thickness	mm	100	20	42
Viscous charac. length	μm	21	214	110
Thermal charac. length	μm	50	214	740
Porosity	–	0.992	0.989	0.98
Density of frame	$kg.m^{-3}$	50	6.1	22.1
Loss factor	–	0.1	0.21	0.
Young's modulus	kPa	120	56.4	**100**
Poisson ratio	–	0.4	0.21	**0.4**
Flow resistivity	$kPa.s.m^{-2}$	26.5	8.6	**60**
Tortuosity	–	1.2	1.0	**1.3**
Anisotropic foam used in Sect. 8.6.4				
Parameter	Units	Value		
Elastic moduli, xx	kPa	40		
", yy	kPa	89 KPa		
", zz	kPa	300 KPa		
", xy	kPa	33 KPa		
", xz	kPa	37 KPa		
", yz	kPa	131 KPa		
Shear moduli, xz	kPa	26 KPa		
Shear moduli, xz	kPa	21 KPa		
Shear moduli, xz	kPa	26 KPa		
Tortuosity in x, y, z	–	1.2, 1.2, 1.5		
Flow resistivity in x, y, z	$kPa.s.m^{-2}$	37.5, 37.5, 67.5		

We could as well decide to use the approximate boundary condition, Sect. 8.5.3, in case we need to introduce area integral terms. These introduces the forcing of the acoustic pressure in the PEM on the solid displacements of the PEM as well as to the solid displacements of the solid. In this case we solve a problem equivalent in principle to (8.62).

8.6.2.1 Effects of Bonding 1D

It is well-known that the presence of an airgap could have a strong influence on the transmission, depending on the properties of the actual PEM. With a 0.1 mm airgap we get the results in the left part of Fig. 8.7 which are compared to the fully bonded results in the same figure. The differences between the two are clear and increases over the frequency range shown. There are several reasons for the two solutions to be different from each other, see e.g. [7, 16], one quite obvious is the change in total stiffness as the solid and the PEM are joined together. This can be observed through the slight shift of the first resonance in (8.7) (a). As the PEM is quite soft in comparison to the solid, this effect is not very large. However, for the resonance at 1850 Hz, the effect of the airgap is more significant as it tends to reduce the transmission through the panel in comparison to the bonded case. Clearly, presence of an airgap is affecting the vibroacoustic behaviour in a quite complex way which is difficult to predict a priori.

To illustrate this further, we show the effects of varying the airgap thickness in the right part of Fig. 8.7. We measure the effect in terms of the relative difference to a very thin gap of 0.01 mm thickness. Clearly, for thinner airgaps the dependence upon the airgap thickness of the transmission characteristics is reduced. For airgaps less than 1 mm, the difference in the transmitted pressure at P_a is small. For larger airgaps the transmission characteristics have a strong dependence on the actual thickness as

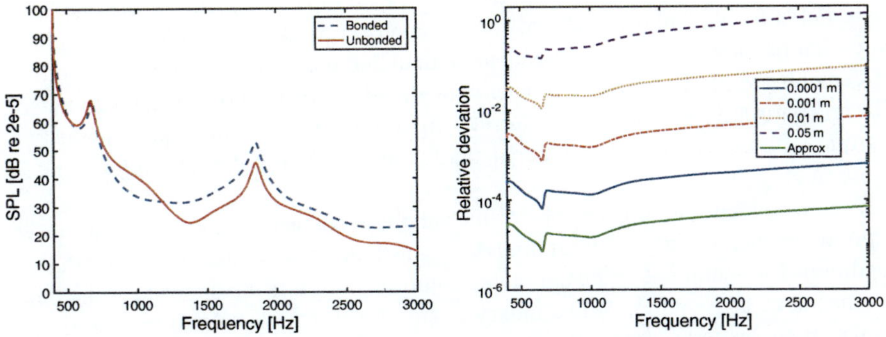

Fig. 8.7 Left: Effect of coupling conditions, bonded or unbonded, on transmitted pressure at point P_a, see Fig. 8.6, through the multilayered arrangement. Right: Relative difference in transmitted pressure at point P_a, for different airgap sizes, with 10^{-5} m used as reference gap thickness

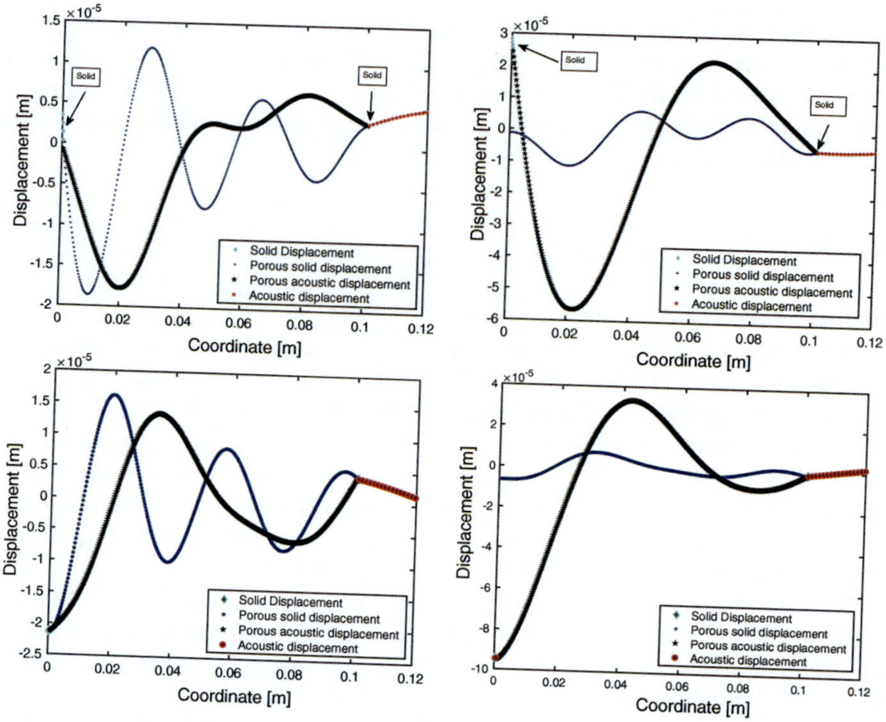

Fig. 8.8 Illustration of effects of bonding/unbonding, close-up of displacement fields inside PEM in a multilayer arrangement, real parts (top) and imaginary parts (bottom) evaluated at resonance 1850 Hz. Left: Bonded. Right: Unbonded and using approximate boundary condition

could have been expected. From Fig. 8.7, we also see that the approximate modelling represents the very thin airgap coupling well.

Let us now investigate what the effects of the airgap are in terms of the displacement fields for the solid frame and for the acoustic fluid in the PEM, see Fig. 8.8. Note that the acoustic displacements are calculated using (8.15). The different displacements are shown for the two cases considered above, i.e. bonded and unbonded. The real parts are shown in the two top sub-figures and the imaginary parts in the two bottom, with the computed displacements for the bonded to the left and unbonded to the right.

For the bonded solution, the solid frame and the acoustic fluid displacements move in phase and are identical at both boundaries, as indeed prescribed by the boundary conditions for bonding. For the unbonded solution, it is quite the opposite as the fluid displacement, at the unbonded boundary, is significantly higher than the solid frame displacement, in particular for the imaginary part. Furthermore, they are not in phase and the solid displacement is lower than the acoustic displacement at $\Gamma^{b/ub}$. Indeed, from Fig. 8.9, we see that the relative displacement $w_i = \phi(u_i^f - u_i^s)$, is higher for the unbonded case which implies a higher level of dissipation as compared to the bonded case, see (8.11)–(8.14).

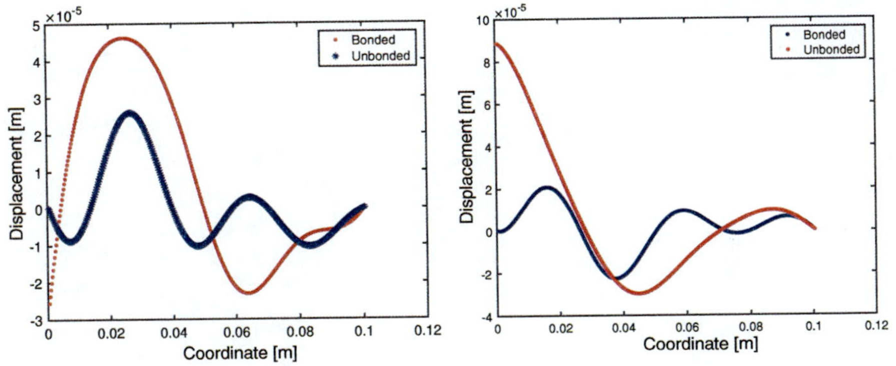

Fig. 8.9 Illustration of effects of bonding/unbonding, close-up of relative displacement fields inside PEM in a multilayer arrangement, Left: real part, right: imaginary part evaluated at resonance 1850 Hz

To conclude on this example, the approximate unbonded boundary condition represents a thin airgap well, and is an interesting alternative for complex 3D models, where the thin airgap modelling may lead to numerical problems related to elements having unacceptable aspect ratios, and where the approximation eliminates the need for meshing of the airgap through the use of boundary coupling terms instead.

8.6.3 Simulation of a Metaporous Material

A more complicated problem to solve could be a porous matrix with an inclusion. This kind of arrangement, which may involve all the different kinds of boundary conditions introduced here, is sometimes called metaporous and a 2D example is presented in Fig. 8.10. The current example is composed of a porous material Ω_p in a rectangular domain with a circular inclusion Ω_i in the center of the domain. Along Γ_t the porous medium is bonded to a rigid surface and it is excited by a plane wave on the opposite boundary, Γ_b. As this inclusion is supposed to be a part of an array of such inclusions, periodic boundary conditions are assigned along Γ_r and Γ_l boundaries, i.e. the arrangement is in fact of infinite extent in the horizontal direction. These kinds of composite structures have been the subject of extensive research in the last decade. For more details, the reader can refer to [17–22] and from these appreciate the huge variety of configurations that may be considered.

Due to the plane-wave excitation and the periodic boundary conditions, the fields in the semi-infinite lower domain are the superposition of the incident field and reflected Bloch waves. Periodic boundary conditions can be taken into account by applying the method presented in Sect. 8.6.4 and the amplitudes of the Bloch waves can be added to the FE *dof* with the method presented in Sect. 8.5.4. As the condition on Γ_t corresponds to a rigid boundary, it leads to natural boundary conditions for both PEM1 and PEM2.

Fig. 8.10 Case of a metaporous material

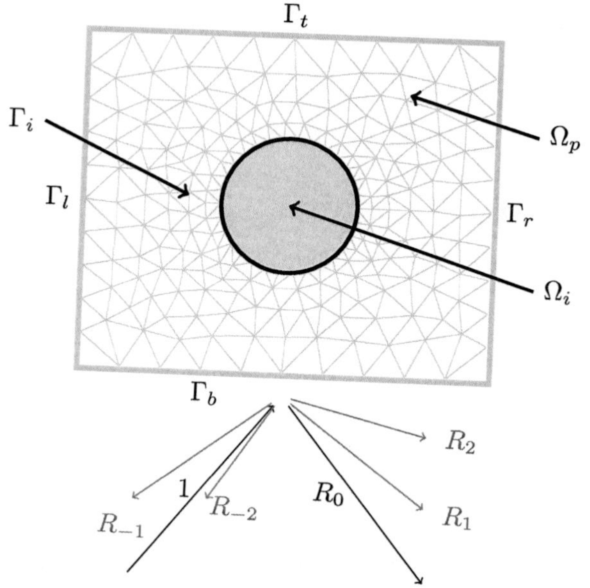

Let us consider an example presented in [22]. The porous matrix is composed of foam II and has dimensions 2 cm by 2 cm. The inclusion is made of a steel material with a radius r, that will be varied in the results shown below. In this example we then have three different media: poroelastic, elastic and fluid (incident medium which is modelled by the Bloch wave superposition). Regardless of the choice of the porous formulation (PEM1 or PEM2) one of the couplings is not natural. In the results that are presented, we have chosen to use the PEM1 formulation but the results are similar for PEM2. As we have discussed previously, we then need to include a fluid-structure coupling term at the boundary Γ_i between the porous structure and the elastic inclusion. The global problem is then composed of 5×5 blocks as:

$$\begin{bmatrix} [\mathbf{A}^{ee}] & [\mathbf{A}^{e\Gamma}] & [\mathbf{A}^{eu}] & -[\boldsymbol{\Gamma}^{up}] & [\mathbf{0}] \\ [\mathbf{A}^{\Gamma e}] & [\mathbf{A}^{\Gamma\Gamma}] & [\mathbf{A}^{\Gamma u}] & [\mathbf{A}^{\Gamma p}] & [\mathbf{0}] \\ [\mathbf{A}^{ue}] & [\mathbf{A}^{u\Gamma}] & [\mathbf{A}^{uu}] & [\mathbf{A}^{up}] & [\mathbf{0}] \\ -[\boldsymbol{\Gamma}^{pu}] & [\mathbf{A}^{p\Gamma}] & [\mathbf{A}^{pu}] & [\mathbf{A}^{pp}] & [\mathbf{B}] \\ [\mathbf{0}] & [\mathbf{0}] & [\mathbf{0}] & [\mathbf{B}'] & D[\mathbf{I}_{2N_b+1}] \end{bmatrix} \begin{Bmatrix} \mathbf{u}_e \\ \mathbf{u}_\Gamma \\ \mathbf{u} \\ \mathbf{p} \\ \mathbf{R} \end{Bmatrix} = \begin{Bmatrix} \mathbf{0} \\ \mathbf{0} \\ \mathbf{0} \\ \mathbf{0} \\ \mathbf{1}_0 \end{Bmatrix}. \tag{8.76}$$

The porous matrix and the inclusion share displacement degrees of freedom $\mathbf{u}_\Gamma$ and $\mathbf{u}_e$ and $\mathbf{u}$ respectively denotes the interior degrees of freedom of the inclusion and the porous material. The upper left 3×3 corresponds to the dynamic matrix associated to the displacements. $[\mathbf{A}^{\Gamma\Gamma}]$ is assembled as the sum of a contribution from the inclusion and one of the porous material. As the PEM1 formulation does not lead to a natural coupling between fluid structure terms $[\boldsymbol{\Gamma}^{up}]$ and $[\boldsymbol{\Gamma}^{pu}]$ are introduced which are applied on the displacement of the elastic inclusion and the pressure of

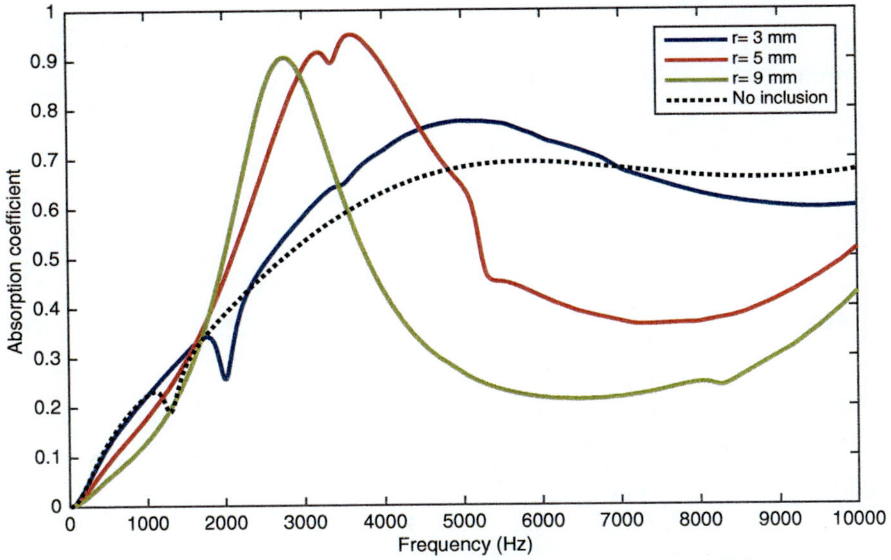

Fig. 8.11 Absorption for several different sizes of the inclusion

the porous material. Finally, the excitation introduces coupling terms [**B**] and [**B**′] between the pressure of the porous material and the reflexion coefficients. In the case of PEM1, there is no coupling term with the solid displacement.

In Fig. 8.11 the absorption coefficient for a selection of three values of the radius of the inclusion, is presented. They can be compared to the case without inclusion which is also presented. The effect of elasticity of the frame can by a localised decrease of the absorption which can be observed for example 1400 Hz in the case of an homogeneous layer 2050 Hz for $r = 3$ mm. The effect of the inclusion is to create lower frequency resonances which can increase the absorption for low frequencies. This subject is the topic of intensive researches in the last 5 years and the reader can have more details in [17, 21, 22].

The key point is that all these configurations can be modelled by the FE method. For this configuration it is also possible to derive a model based on the multiple scattering theory [22]. It is then possible to deduce an error between the two methods which is presented in Fig. 8.12. This error is mainly governed by the the mesh. In the present simulation, the mesh corresponds to 6 quadratic elements by wavelength for the shear wave. We can see that the error is mostly lower than 1%, the maximum of this error being at the resonance frequencies of the structure. The quality of the FE simulation is thus very good.

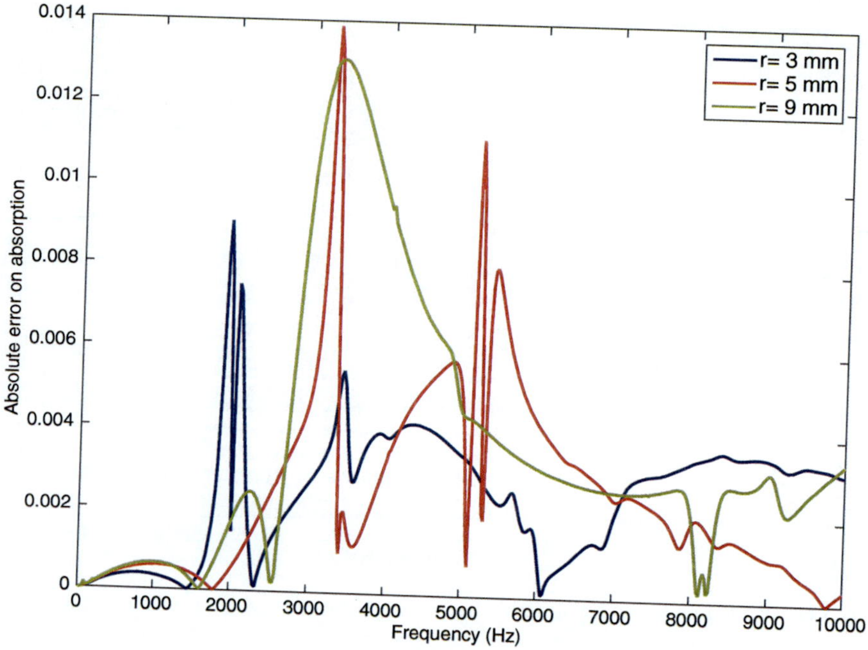

Fig. 8.12 Error between the FE method and the multiple scattering technique

8.6.4 3D Anisotropic Modelling

The effects of anisotropy of a PEM will be illustrated for a 3D example, employing the same different types of boundary conditions as discussed in the 1D case, Sect. 8.6.2. Here we use a model representing a multilayered panel, two face sheets and a PEM in between, see Fig. 8.13. Along one interface the PEM is fully bonded and along the opposite an airgap is introduced. In order to highlight the effects of anisotropy in the resulting deformations, results for an isotropic equivalent foam are also shown.

Let us first focus on the difference between the isotropic and the anisotropic PEM deformations. A significant shear deformation of the PEM is taking place in the latter, see Fig. 8.14a–b, and c–d, respectively. This is not seen in the isotropic equivalent PEM model, as there are only small Poisson's ratio effects inducing lateral deformations, and the deformation of the PEM core is almost uniaxial. This is in contrast to the anisotropic PEM model, where we see complicated shear deformations with short spatial wave lengths, in particular for the bonded case in Fig. 8.14a.

The lateral deformations are also noticeable in the presence of an airgap, see Fig. 8.14c, d. However, the difference between the isotropic and the anisotropic PEM models is slightly less. This is partially related to the release of the solid displacements at the upper panel interface, as now the coupling between the panel and the PEM

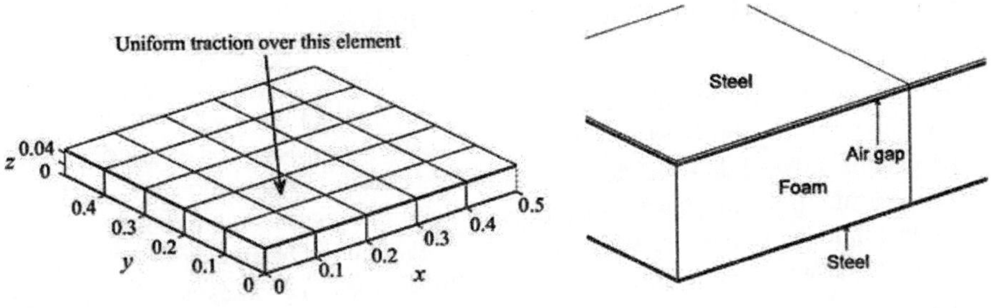

Fig. 8.13 Model problem used in 3D example of simulation of anisotropic foam formulation, parameters used in the simulations may be found in Table 8.1. For comparison, an airgap of thickness 0.001m is introduced at the upper interface between the PEM core and the face plate. Reprinted from [23] with permission

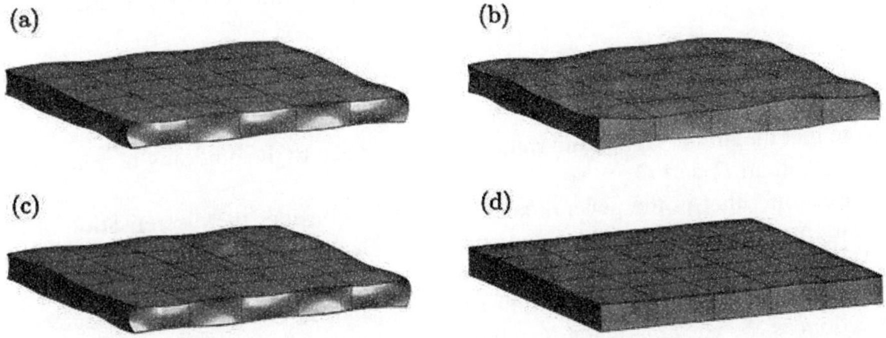

Fig. 8.14 Deformations 400 Hz. Light grey is foam core, dark is solid panel. Left: anisotropic, Right isotropic. **a** and **b** Bonded configuration, **c** and **d** Unbonded configuration. Reprinted from [23] with permission

is via the pore pressure, see Sect. 8.5.3. However, also for the airgap case, there are short wave length shear deformations in the PEM core, which are not present in the isotropic model.

We conclude from this, without giving any detailed explanations, that different waves in the anisotropic PEM are excited when the different boundary conditions are applied. This is in agreement with the simpler model studied in Sect. 8.6.2 and has been further discussed in [24] using a transfer matrix solution approach.

Acknowledgements This chapter is based upon work from COST Action DENORMS CA15125, supported by COST (European Cooperation in Science and Technology). The authors would like to acknowledge the funding and support by the following institutions: KTH Royal Institute of Technology, Centre for ECO2 Vehicle Design; University of Le Mans, Le Mans Acoustique. The authors would also like to thank Thomas Weisser for providing results relative to [22].

Appendix - Periodic Boundary Conditions

Here we review a technique for introduction of periodic boundary conditions, which is not specific to poroelasticity but included for reference as it is used in one of the examples. Note that the steps taken in this method is thus the same for the two $\{\mathbf{u}, \mathbf{p}\}$ formulations discussed here. If periodicity may be assumed, then the involved physical fields have a phase shift $\delta = e^{-jk_xL}$ between the boundaries of the unit cell, Γ_l and Γ_r. If we consider the fields of the PEM2 formulation as an illustration, we have:

$$p(\Gamma_r) = \delta p(\Gamma_l)\, u_i(\Gamma_r) = \delta u_i(\Gamma_l)\, \hat{\sigma}^t_{ij}(\Gamma_r) = \phi\hat{\sigma}^t_{ij}(\Gamma_l)\, p_{,j}(\Gamma_r) = \delta p_{,j}(\Gamma_l). \quad (8.77)$$

We can then deduce a relation between the surface integrals on the two boundaries:

$$\int_{\Gamma_r} v^s_i\hat{\sigma}^t_{ij}n_j + \frac{\phi^2}{\rho_0\omega^2}\hat{\alpha}_{ij}p_{,j}n_iq\,\mathrm{d}\Gamma = -\delta\int_{\Gamma_l} v^s_i\hat{\sigma}^t_{ij}n_j + \frac{\phi^2}{\rho_0\omega^2}\hat{\alpha}_{ij}p_{,j}n_iq\,\mathrm{d}\Gamma. \quad (8.78)$$

Note that the minus sign in the right hand side is due to the outgoing normal which is opposite in Γ_l and Γ_r.

Recalling that in the general case, we need to consider the discretisation of the weak form including the volume integrals in the matrix and the integrals relative to Γ_l and Γ_r in the left hand side. Furthermore, we will distinguish between *dof* on the left and right boundaries of the unit cell as $\mathbf{X}_l$ and $\mathbf{X}_r$ and the interior degrees of freedom as $\mathbf{X}_i$. The *dof* vector, $\mathbf{X}$, will be partitioned accordingly and the discretised system may be written as:

$$\begin{bmatrix} [\mathbf{A}_{ll}] & [\mathbf{A}_{li}] & [\mathbf{A}_{lr}] \\ [\mathbf{A}_{il}] & [\mathbf{A}_{ii}] & [\mathbf{A}_{ir}] \\ [\mathbf{A}_{rl}] & [\mathbf{A}_{ri}] & [\mathbf{A}_{rr}] \end{bmatrix} \begin{bmatrix} \mathbf{X}_l \\ \mathbf{X}_i \\ \mathbf{X}_r \end{bmatrix} = \begin{bmatrix} \mathbf{F}_l \\ \mathbf{0}_i \\ \mathbf{F}_r \end{bmatrix}. \quad (8.79)$$

We now assume that the mesh interpolations on Γ_l and Γ_r are compatible, and thus the following relations hold:

$$\mathbf{X}_r = \delta\mathbf{X}_l, \quad \mathbf{F}_r = -\delta\mathbf{F}_l. \quad (8.80)$$

To proceed we need to eliminate $\mathbf{F}_l$ and this is achieved by suitable linear combinations of the parts of the matrix as follows. In the last row, we substitute tbF_r by $\delta\mathbf{F}_l$ and then multiply the whole row by δ^* (which is equal to $1/\delta$). This gives us an expression for $\mathbf{F}_l$ which may be used to rewrite the first row. The third row is replaced by the condition of periodicity on the *dof* , i.e. ($\mathbf{X}_r = \delta\mathbf{X}_l$). The system may then be reformulated as:

$$\begin{bmatrix} [\mathbf{A}_{ll}] + \delta^*[\mathbf{A}_{rl}] & [\mathbf{A}_{li}] + \delta^*[\mathbf{A}_{ri}] & [\mathbf{A}_{lr}] + \delta^*[\mathbf{A}_{rl}] \\ [\mathbf{A}_{il}] & [\mathbf{A}_{ii}] & [\mathbf{A}_{ir}] \\ \delta[\mathbf{I}_{rr}] & [\mathbf{0}] & -[\mathbf{I}_{rr}] \end{bmatrix} \begin{bmatrix} \mathbf{X}_l \\ \mathbf{X}_i \\ \mathbf{X}_r \end{bmatrix} = \begin{bmatrix} \mathbf{0}_l \\ \mathbf{0}_i \\ \mathbf{0}_r \end{bmatrix}, \quad (8.81)$$

which only has known quantities and furthermore the set of degrees of freedom is the same before and after the imposition of the periodicity. The latter is a key advantage with the above approach.

References

1. E. Deckers, S. Jonckheere, D. Vandepitte, W. Desmet, Modelling techniques for vibro-acoustic dynamics of poroelastic materials. Arch Comput Methods Eng **22**, 183–236 (2015)
2. M.A. Biot, Theory of elasticity and consolidation for a porous anisotropic solid. J Appl Phys **26**(2), 182–185 (1955)
3. M.A. Biot, Theory of propagation of elastic waves in a fluid-saturated porous solid. i. low-frequency range. J Acoust Soc Am **28**(2), 168–178 (1956)
4. M.A. Biot, Theory of stress-strain relations in anisotropic viscoelasticity and relaxation phenomena. J Appl Phys **25**(11), 1385–1391 (1954)
5. M.A. Biot, Theory of deformation of a porous viscoelastic anisotropic solid. J Appl Phys **27**(5), 459–467 (1956)
6. M.A. Biot, Mechanics of deformation and acoustic propagation in porous media. J Appl Phys **33**(4), 1482–1498 (1962)
7. J.F. Allard, N. Atalla, *Propagation of Sound in Porous Media. Modelling Sound Absorbing Materials*, 2nd edn. (Wiley, 2009)
8. N.-E. Hörlin, P. Göransson, Weak, anisotropic symmetric formulations of biot's equations for vibro-acoustic modelling of porous elastic materials. Int J Numer Methods Eng **84**(12), 1519–1540 (2010)
9. N. Atalla, R. Panneton, P. Debergue, A mixed displacement-pressure formulation for poroelastic materials. J Acoust Soc Am **104**, 1444–1452 (1998)
10. N. Atalla, M. Hamdi, R. Panneton, Enhanced weak integral formulation for the mixed (u, p) poroelastic equations. J Acoust Soc Am **109**, 3065–3068 (2001)
11. J.F. Allard, O. Dazel, J. Descheemaeker, N. Geebelen, L. Boeckx, W. Lauriks, Rayleigh waves in air saturated axisymmetrical soft porous media. J Appl Phys **106**(1) (2009)
12. P. Göransson, Tailored acoustic and vibrational damping in porous solids - engineering performance in aerospace applications. Aerosp Sci Technol **12**, 26–41 (2008)
13. P. Goransson, Acoustic and vibrational damping in porous solids. Philos Trans R Soc A - Math Phys Eng Sci **364**, 89–108 (2006)
14. O.C. Zienkiewicz, R.L. Taylor, O.C. Zienkiewicz, R.L. Taylor, *The Finite Element Method*, vol. 3 (McGraw-hill London, 1977)
15. O. Dazel, B. Brouard, C. Depollier, S. Griffiths, An alternative Biot's displacement formulation for porous materials. J Acoust Soc Am **121**(6), 3509–16 (2007)
16. P. Goransson, A 3-D, symmetric, finite element formulation of the Biot equations with application to acoustic wave propagation through an elastic porous medium. Int J Numer Methods Eng **41**(1), 167–192 (1998)
17. J.-P. Groby, B. Brouard, O. Dazel, B. Nennig, L. Kelders, Enhancing rigid frame porous layer absorption with three-dimensional periodic irregularities. J Acoust Soc Am **133**(2), 821–831 (2013)
18. C. Lagarrigue, J.P. Groby, V. Tournat, O. Dazel, O. Umnova, Absorption of sound by porous layers with embedded periodic arrays of resonant inclusions. J Acoust Soc Am **134**(6), 4670–4680 (2013)
19. J.-P. Groby, C. Lagarrigue, B. Brouard, O. Dazel, V. Tournat, B. Nennig, Using simple shape three-dimensional rigid inclusions to enhance porous layer absorption. J Acoust Soc Am **136**(3), 1139–1148 (2014)

20. J.-P. Groby, C. Lagarrigue, B. Brouard, O. Dazel, V. Tournat, B. Nennig, Enhancing the absorption properties of acoustic porous plates by periodically embedding Helmholtz resonators. J Acoust Soc Am **137**(1), 273–280 (2015)
21. C. Lagarrigue, J.-P. Groby, O. Dazel, V. Tournat, Design of metaporous supercells by genetic algorithm for absorption optimization on a wide frequency band. Appl Acoust **102**, 49–54 (2016)
22. T. Weisser, J.-P. Groby, O. Dazel, F. Gaultier, E. Deckers, S. Futatsugi, L. Monteiro, Acoustic behavior of a rigidly backed poroelastic layer with periodic resonant inclusions by a multiple scattering approach. J Acoust Soc Am **139**(2), 617–629 (2016)
23. P. Göransson, N.-E. Hörlin, Vibro-acoustic modelling of anisotropic porous elastic materials: a preliminary study of the influence of anisotropy on the predicted performance in a multi-layer arrangement. Acta Acustica united with Acustica **96**(2), 258–265 (2010)
24. J.P. Parra Martinez, O. Dazel, P. Göransson, J. Cuenca, Acoustic analysis of anisotropic poroelastic multilayered systems. J Appl Phys **119**(8), 084907 (2016)

Part III
Industrial Applications of Porous Media and Acoustic Metamaterials

Chapter 9
Industrial Applications I

A General Perspective

Arnaud Duval and Fabien Chevillotte

Abstract The main goal of these industrial chapters is to help non-specialist engineers throughout the industry to take robust noise treatments optimization decisions without being obliged to become experts themselves. Another goal is also to help "low hanging fruit" technology transfers between various fields of acoustic industrial application domains, from aeronautic to automotive industry for example or from automotive to building industry or the other way around.

9.1 Introduction

The ambition of these industrial chapters on acoustic package optimization methods is to give insights into noise treatments key optimization levers in various industries and foster technology transfers between industries. In the following, well-known physical phenomena will be reminded and their applications illustrated in either automotive, aeronautic or building industrial cases. Without going into sensitive information or specific patents, each industry has gathered canonic optimized solutions that are often not known or fully understood by other industries, or not used at all out of a specific domain. These "low-hanging" fruits are the actual targets of the approach aiming at bringing innovations and competitiveness to our industries.

The main goal of these industrial chapters is to help non-specialists engineers throughout the industry to take robust noise treatments optimization decisions without being obliged to become experts themselves as much as possible. Optimizing porous treatments requires understanding the noise source characteristics and the noise propagation path as well as identifying which dissipation mechanisms one should foster: absorption, insulation or vibration damping? To achieve a signifi-

A. Duval (✉)
TREVES Product, Services & Innovation, 2–4 rue Emile Arquès, Reims, France
e-mail: arnaud.duval@treves-group.com

F. Chevillotte
MATELYS - Research Lab., Bât. B, 7 Rue des Maraîchers, Vaulx-En-Velin, France
e-mail: fabien.chevillotte@matelys.com

N. Jiménez et al. (eds.), *Acoustic Waves in Periodic Structures, Metamaterials, and Porous Media*, Topics in Applied Physics 143,
https://doi.org/10.1007/978-3-030-84300-7_9

cant noise reduction, a link between the ideal porous media properties and a feasible porous standard or meta-material has to be established, e.g. using micro-macro approaches. In order to determine these feasible microstructure morphologies, it is necessary to understand what each manufacturing process permits to control. Indeed, manipulating the porous media macro-parameters of the Biot-Johnson-Champoux-Allard (Biot-JCA) theory for material optimization, such as the porosity ϕ, airflow resistivity σ, tortuosity α_∞, viscous and thermal characteristic lengths Λ and Λ', is not an easy task for the noise, vibration and harshness (NVH) engineer, especially when communicating with chemistry or process engineers involved in material manufacturing [1–3].

Before discussing each specific industrial noise treatments problems, a general review of standard NVH problems will be presented focusing on absorption, insulation and damping optimization key levers of typical felts or foams poroelastic materials, and/or viscoelastic materials.

9.2 Absorption Optimization

Optimizing the absorption of a porous material requires accounting from the very beginning for the final use and installation of this material as well as the multiple constraints it will have to cope with: thickness, temperature, fluids, flammability, stiffness etc.

Indeed, the coupling parameters of the Biot-JCA model describing the viscothermal dissipation are important to optimize (rigid porous material model), but the density (for limp material model) as well as the thickness are also essential parameters, especially for insulation problem. If mechanical resonances of the frame occur in the frequency range of interest, the structural parameters will have to be taken into account as well (full Biot-JCA model for poroelastic material).

The full Biot-JCA model is composed of three groups of parameters [3]:

- The fluid phase acoustical parameters.
- The solid phase mechanical parameters.
- The coupling parameters between the fluid and solid phase.

The parameters of the fluid phase, usually the air, are normally well-known. The solid frame and coupling parameters need to be obtained using characterization methods for the full Biot-JCA model of the porous media. The required parameters are summarized in Fig. 9.1.

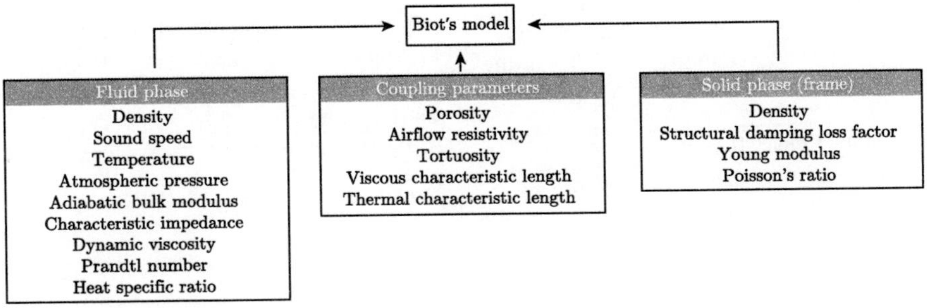

Fig. 9.1 Groups of parameters of the Biot-JCA model (4 elastic parameters, 5 parameters for the JCA model)

9.2.1 *The Coupling Parameters: Johnson-Champoux-Allard (JCA) Model*

For industrial applications, the following 5 coupling parameters of the Johnson-Champoux-Allard model are usually sufficient:

The porosity, ϕ, is the ratio of the air (fluid) volume (V_f) and the total volume (V_t) of porous material:

$$\phi = \frac{V_f}{V_t}. \tag{9.1}$$

The porosity should be kept high in order to get large viscothermal dissipation and, thus, a high value of the absorption coefficient, e.g., in standard cases, $\phi > 0.8$.

The static airflow resistivity, σ, describes the losses due to viscosity at low frequencies. Below a given frequency, the viscous boundary layers are equal or progressively thicker than the characteristic sizes of the pores. Viscous frictions dominates the acoustic wave propagation in the fluid. The airflow resistivity is the indicator of the strength of friction force. The airflow resistance $R_s = \sigma \cdot h$, in Ns/m^3, where h is the thickness in (m), should be as high as possible while maintaining a good porosity, but should be kept below $R_s < 2000$ Ns/m^3 anyway, in order to avoid saturation in standard cases (absorption coefficient decreases).

The tortuosity, α_∞, describes the complex trajectory of the fluid particles imposed by the frame. This inertial parameter brings a correction to the fluid effective density at high frequencies. For all materials $\alpha_\infty \geq 1$, and describes either an oblique path or a path with varying cross-section:

- For fibrous material with thermoplastic bi-component fibers binding or in the absence of thermoset resin binding $\alpha_\infty = 1$.
- For foam material with open membranes (high reticulation) after typical free expansion $1.1 < \alpha_\infty < 1.4$.

Table 9.1 Possible variation of the coupling parameters for classical open cell porous materials

Parameter	Porosity ϕ	Tortuosity α_∞	Airflow resistivity σ [kNsm^{-4}]	Viscous length Λ [μm]	Thermal length Λ' [μm]
Minimal value	0.7	1	1	5	10
Maximal value	0.99	4.5	5000	1000	2000

- For foam material with semi-closed membranes (low reticulation) after typical injection molding expansion or relatively closed cell free expansion $1.5 < \alpha_\infty < 2.5$.

A famous exception from these rules is the melamine foam, whose microstructure is similar to a 3D microfiber material with 4 to 6 μm diameter ligaments and with a $\alpha_\infty = 1$.

The viscous characteristic length, Λ, describes the strength of viscoinertial forces at high frequencies. The viscous characteristic length is about the dimension of the radius of the smallest pores and necks (semi-closed membranes), where viscous losses dominate thermal losses. i.e., the dimension of the pore constrictions.

The thermal characteristic length, Λ', characterizes the thermal exchanges between the two phases at high frequencies which take place near the surfaces of the pores. The thermal characteristic length is about the dimension of the radius of the biggest pores, where the thermal exchanges are the strongest. The thermal characteristic length Λ' depends only on the geometry of the pores, for example:

- For random oriented open cell materials $\Lambda' > \Lambda$ with a Λ'/Λ ratio of 3, usually.
- For horizontally aligned fibers $\Lambda' > \Lambda$ with a Λ'/Λ ratio of 2.
- For almost closed cell foams the $\Lambda' > \Lambda$ and the ratio Λ'/Λ can reach very high values up to 4 or 5 and beyond [4].

The following table shows the possible variation of the coupling parameters for classical open cell porous materials (Table 9.1).

Figure 9.2 illustrates the dominating frequency zones of each JCA coupling parameters of the Biot-JCA model and thus instructs which parameter should be optimized in the frequency range of interest. As the thickness alone is shifting the absorption coefficient towards the low frequencies, the JCA parameters, should be optimized with a targeted thickness in mind.

The difficulty of application of the aforementioned recommendations lies in the inter-dependency of the JCA parameters. The first task of the optimization is to determine a feasible microstructure morphological path that can be mastered by the manufacturing process, as well as the absorption performance target. Then, a micro-macro model (analytical, empirical or numerical) should be applied in order

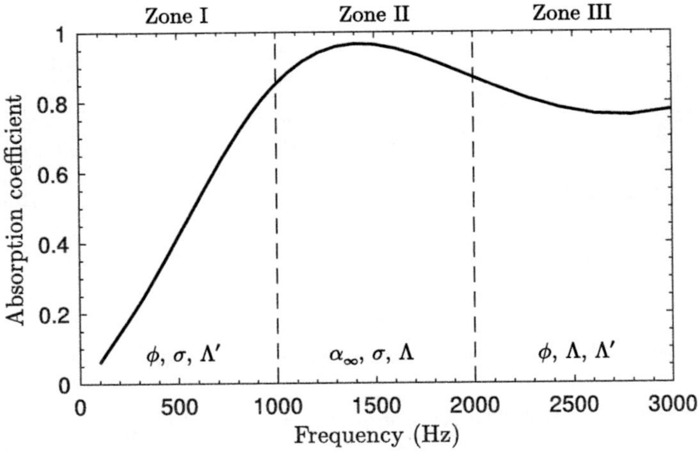

Fig. 9.2 Dominating frequency zones of the Biot-JCA parameters [3]

to get the set of Biot-JCA parameters corresponding to each microstructure, with which macro-performance descriptors, such as the absorption coefficient or surface impedance can be computed. Plotting surface responses enables finding an optimal or better a range of optimum microstructures achieving the target performance [5].

9.2.2 Optimizing Felts or Fibrous Materials for Absorption

The fibrous porous material case is the simplest one, in the sense that the cell morphology is easy to control. This is especially true when using noble "sized controlled" fibers and when adjusting the compression rate (throat size) for fibers that are laid horizontally in one direction. The latter is easy to achieve with carding/napping/thermofixing felt production processes, like Polyethylene terephthalate (PET) felts. Typical fibers where one can control the diameter size are synthetic or mineral fibers such as PET, Polypropylene (PP), acrylic, glass fibers, etc. All recycled felts manufactured using, for example, cotton waste (also called "shoddy"), contain randomly oriented fibres and so the statistical characteristics of the microstructure are required (of about 2.5 dtex meaning 1g/10km = 25 μm on average, which is too large) [6, 7].

		Compressed felt reduction
	Thickness	20 mm
Geometrical	Porosity (ϕ)	↘
	Thermal Characteristic Length (Λ)	↘
Transport	Airflow Resistivity (σ)	↗↗
	Tortuosity (α_∞)	↗
	Viscous Characteristic Length (Λ')	↘

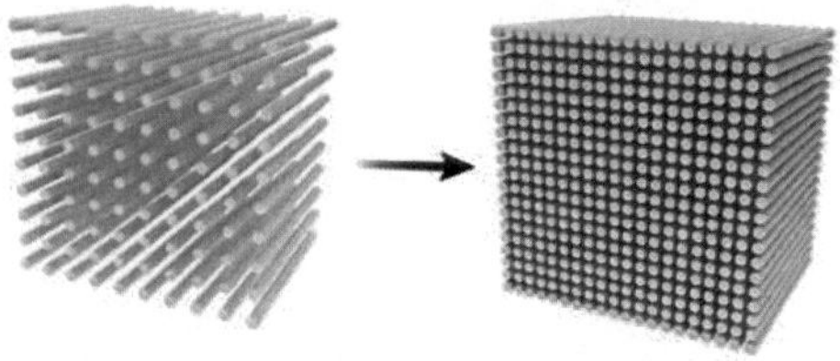

Fig. 9.3 Standard felt compression morphological path

9.2.2.1 Standard Felt Compression Morphological Path

Figure 9.3 shows the standard felt compression morphological path, which allows an interesting airflow resistivity increase. The size of the fibers is not changing here and thus the density is increasing as well, which has negative implications for weight reduction but, even worse, the porosity is decreasing! Whenever specific stiffness or mass barrier effects (hybrid stiff concept, see after) are not required, this optimization morphological path should be avoided.

9.2.2.2 Microfiber Felt Homothetic Reduction Morphological Path

Figure 9.4 illustrates the optimization scheme for absorption properties that has been applied for 20 years using microfibers having diameters between 10 μm down to an ideal 1 μm (approximately 1 dtex down to 0.1 dtex for PET fibers). This scheme allows to achieve either performance improvement at the same mass per unit area (higher compression rate even at iso-thickness, also used for resistive screens) or weight reduction of 40–50% typically compared to "shoddy", i.e., less compression rate: homothetic reduction.

		PET felt homothetic reduction
	Thickness	20 mm
Geometrical	Porosity (ϕ)	→
	Thermal Characteristic Length (Λ)	↘
Transport	Airflow Resistivity (σ)	↗
	Tortuosity (α_∞)	→
	Viscous Characteristic Length (Λ')	↘

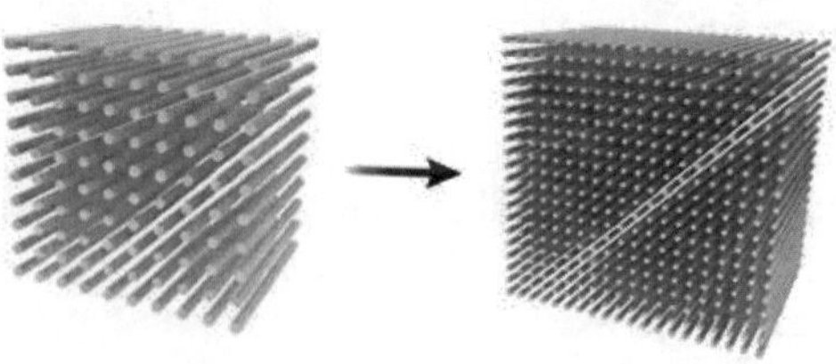

Fig. 9.4 Microfiber felt homothetic reduction morphological path

9.2.3 *Optimizing Foam Materials for Absorption*

9.2.3.1 Open Foam Homothetic Reduction Morphological Path

It is remarkable that in slab foams morphologies with almost no membranes the cell sizes are changing homothetically [8]. The macro-parameters evolve roughly as summarized Fig. 9.5, with almost constant porosity and tortuosity, in a very similar way as those of fiber felts. Most of classical open porous polyurethane (PUR) foams have rather large cells compared to melamine foams (or microfiber felts) with triangular concave cross-section shapes, while presenting low density (15 kg/m^3) and high stiffness.

This stiffness is also limiting such low density open porous PUR foams applications to absorption purposes alone, while being significantly less absorptive than melamine foams or microfiber felts (from PET or glass).

9.2.3.2 Foam Membrane Closure & Homothetic Reduction Morphological Path

These open porous and coarse microstructure morphologies were the standard 15 to 10 years ago. The introduction of membranes interconnecting the pores as well as reducing the size of the PUR foam cells homothetically, as illustrated in Fig. 9.6, has eliminated the expensive melamine foams from the European automotive engine compartments as well as endangered the position of glass microfiber felts (which have a good 6 μm fiber diameter though) thanks to a significant weight reduction.

The noise treatment density has been reduced from 30 kg/m^3 down to 15 kg/m^3 or even 12 kg/m^3 [9]. As reported in Ref. [4] and summarized in Fig. 9.6, the introduction of membranes closing the pores has a very strong influence on the airflow resistivity leading to an increase in tortuosity, which is excellent for absorption properties in the middle frequency range especially. These very light slab foams remain very stiff and thus are good self-supportive absorbers but, at the same time, are very bad poroelastic decouplers behind heavy layers for insulation purposes.

		Foam homothetic reduction
Geometrical	Thickness	20 mm
	Porosity (ϕ)	→
	Thermal Characteristic Length (Λ)	↘
Transport	Airflow Resistivity (σ)	↗
	Tortuosity (α_∞)	→
	Viscous Characteristic Length (Λ')	↘

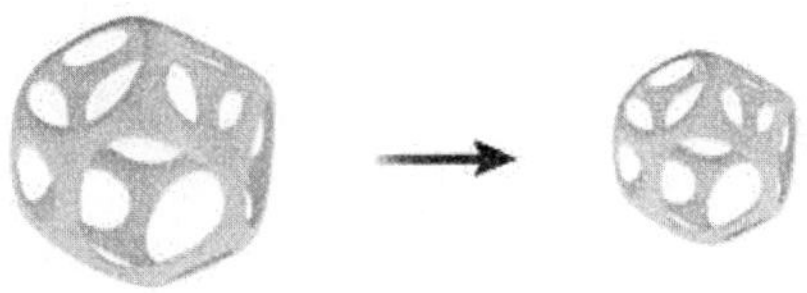

Fig. 9.5 Open foam homothetic reduction morphological path

		Foam membrane closure
Geometrical	Thickness	20 mm
	Porosity (ϕ)	→
	Thermal Characteristic Length (Λ)	↘
Transport	Airflow Resistivity (σ)	↗↗
	Tortuosity (α_∞)	↗↗
	Viscous Characteristic Length (Λ')	↘↘

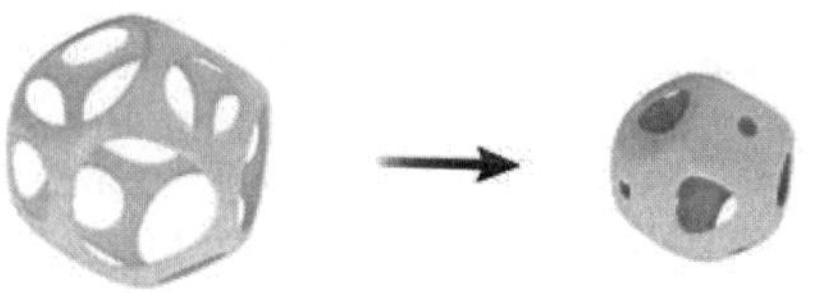

Fig. 9.6 Foam membrane closure & homothetic reduction morphological path

9.2.3.3 Control of the Foam Microstructure Depending of Sample Thickness and Excitation

A recent study has investigated the effect of the 3D morphology on the sound absorption of foams [5]. A body-centered cubic (BCC) arrangement is employed in order to mimic the behaviour of a wide range of foams, from low to high porosity, as shown in Fig. 9.7. The selected single number rating in this study is the sound absorption average (SAA) over one-third octave bands between 125 and 4000 Hz ($SAA_{125-4000}$).

It has been shown that a specific throat size R_t, see Fig. 9.7, maximizing the sound absorption can be found for each thickness and each excitation. This parameter is known as the main influencing factor parameters since it controls the airflow resistivity. Figure 9.8 shows the influence of the throat size R_t on the sound absorption coefficient for a 25 mm-thick sample, a pore size $R_p = 230\ \mu m$, with a rigid backing for normal incidence (NI) or diffuse field (DF) excitation. These specific radii are shown in Fig. 9.9 for several sample thicknesses in the range 10–50 mm. The ranges of radii enabling to reach 95% of the maximum $SAA_{125-4000}$ rating are illustrated

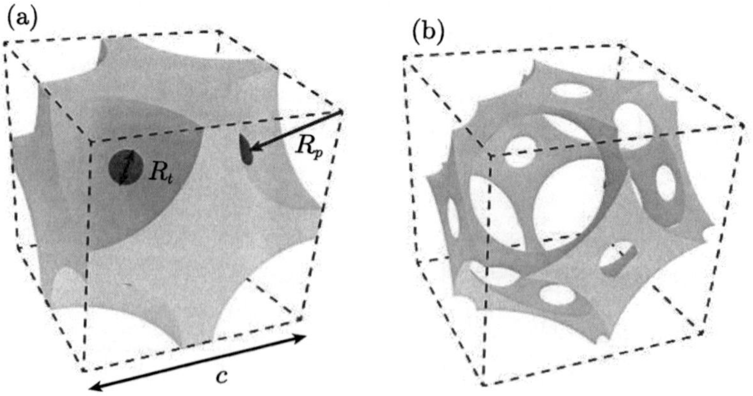

Fig. 9.7 Illustration of cells: **a** Porosity $\phi = 0.7$ **b** Porosity $\phi = 0.96$. the parameter c is the width of the cell, R_p is the pore size, and R_t is the throat size

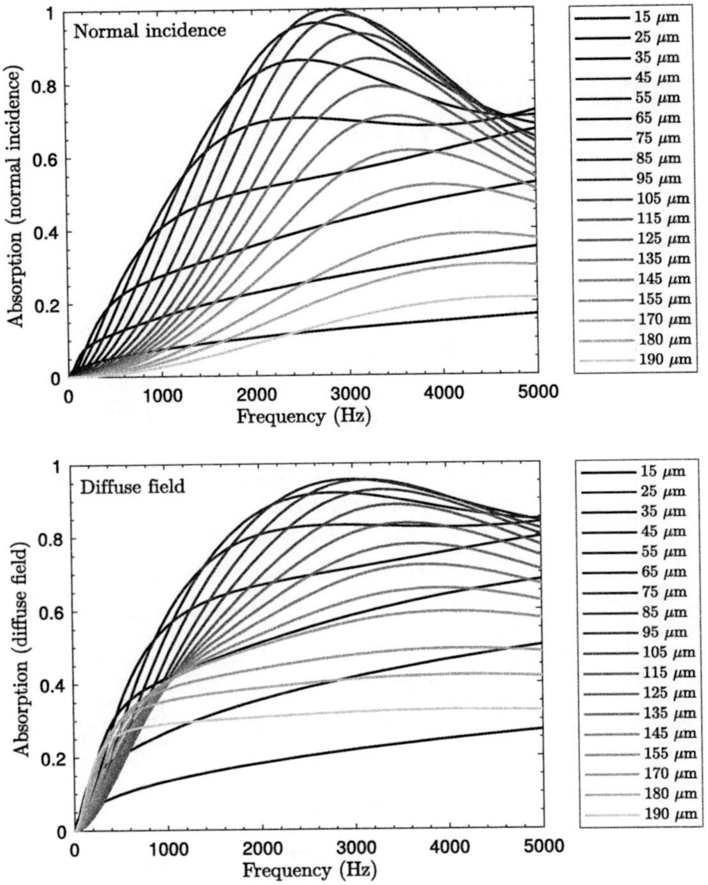

Fig. 9.8 Effect of the throat size on the sound absorption coefficient (top: normal incidence; bottom: diffuse field). Pore radius: $R_p = 230\ \mu\text{m}$, sample thickness $L = 25$ mm, rigid backing

with vertical bars for each sample thickness. The specific throat radii are increasing with the sample thickness.

Figure 9.10 displays the relative variations of all the studied macroscopic parameters with the throat radius, R_t, together with the selected rating in a DF. The variations are expressed as a percentage of the configuration maximizing the SAA rating in a DF. These results confirm the strong influence of the throat size on the airflow resistivity and its linear correlation with the viscous characteristic length. More detailed information can be found in Ref. [5].

A two-dimensional map of the $\text{SAA}_{125-4000}$ can be plotted as a function of the open porosity, ϕ, and the pore size, R_p, for each sample thickness and each excitation. 2D maps and the corresponding $\text{SAA}_{125-4000}$ ratings are presented in Fig. 9.11 for the NI and DF excitations with a thickness of 25 mm. The thick line shows the morphological

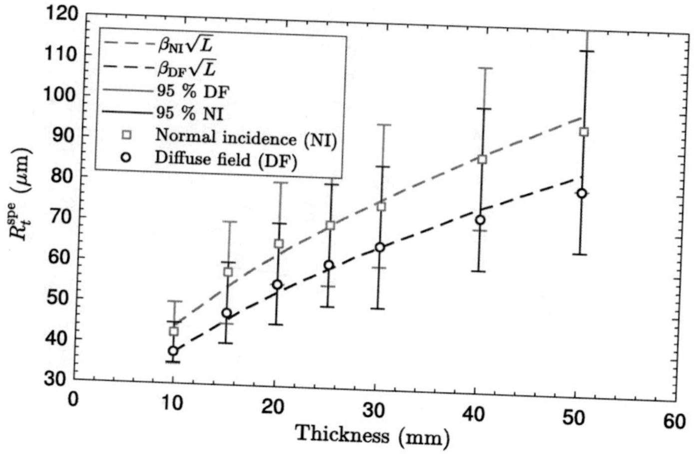

Fig. 9.9 Specific throat radii for NI (square) and DF excitations (circle) as a function of the sample thickness L. (Vertical bars: ranges of radii enabling to reach 95% of the maximum SAA rating for NI and DF; dashed lines: fits of type $\beta\sqrt{L}$ for NI and DF)

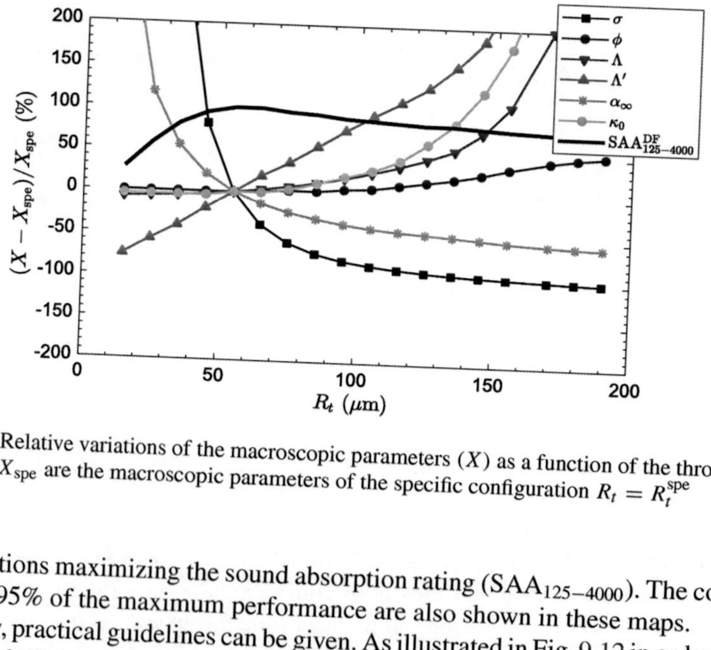

Fig. 9.10 Relative variations of the macroscopic parameters (X) as a function of the throat radius R_t, where X_{spe} are the macroscopic parameters of the specific configuration $R_t = R_t^{spe}$

configurations maximizing the sound absorption rating ($SAA_{125-4000}$). The contours enabling 95% of the maximum performance are also shown in these maps.

Finally, practical guidelines can be given. As illustrated in Fig. 9.12 in order to find the optimal thickness for a given microstructure or how to modify the microstructure for a targeted thickness [5].

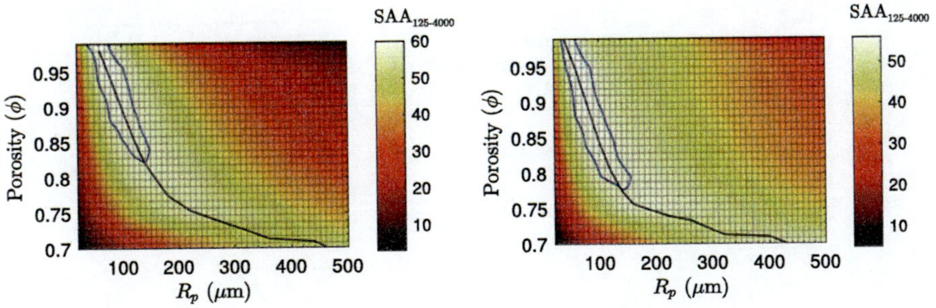

Fig. 9.11 Effect of the open porosity ϕ and pore size R_p on the sound absorption rating (left: NI; right: DF)—Sample (25 mm thick)

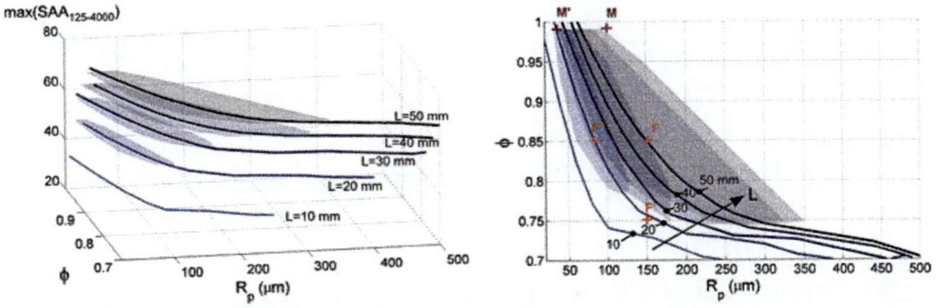

Fig. 9.12 Effect of the open porosity ϕ and pore size R_p on the sound absorption rating for various sample thicknesses with DF excitation (left: 3D; right: 2D)

9.2.4 *Airflow Resistive Screens: Tunable Absorption*

The control of the airflow resistance of microfiber non-wovens positioned on top of classical felts or foams weighting between 15 g/m^2 and 150 g/m^2 allows tuning the absorption coefficient towards the middle frequency range, without having a significant detrimental effect on the performance in the high frequency range [3].

The tunable absorption lay-up consists of two porous layers having different airflow resistances: the first one called "resistive screen" presents a much higher airflow resistance than the second one which is a classical open porous material with low tortuosity such as felt or melamine foam $1 < \alpha_\infty < 1.5$ (cf. Fig. 9.13). The high airflow resistance of the first layer increases the real part of the impedance in the

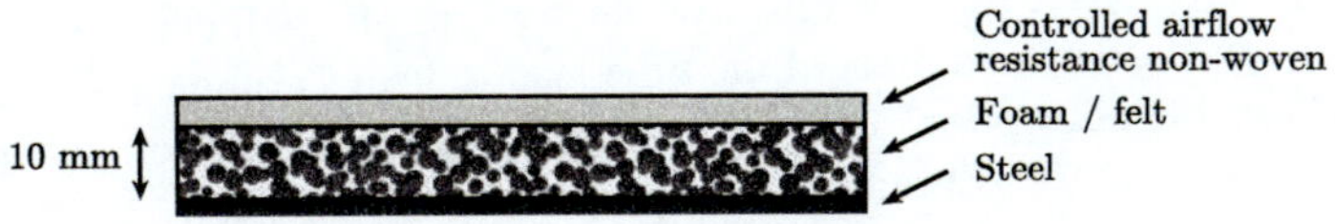

Fig. 9.13 Tunable absorption concept with airflow resistance non-wovens

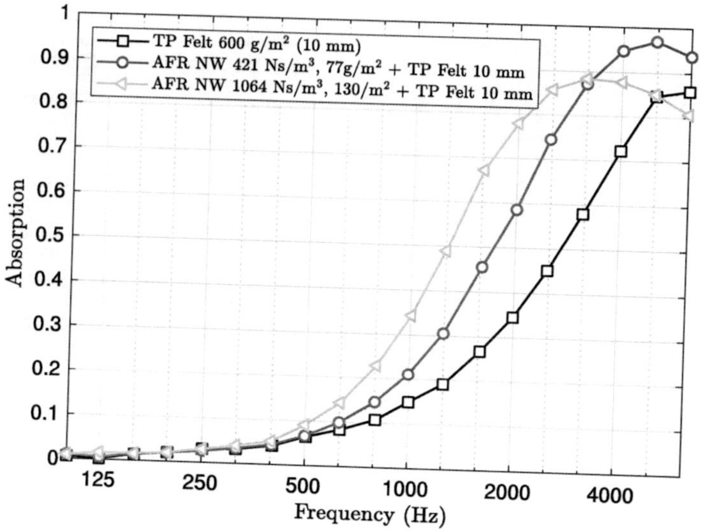

Fig. 9.14 Tunable absorption with airflow resistance non-wovens on thermoplastic felts: normal incidence absorption coefficient

middle frequency range and consequently improves the absorption performance at these frequencies: bi-permeable concept [3, 10]. Too highly tortuous semi-closed cell foams with almost closed membranes already present excellent absorption in the middle frequency range, the airflow resistance non-woven cannot improve such a saturated performance. On the contrary, it decreases the high frequency absorption coefficient, which was already affected by the high tortuosity value. In this case, the application of airflow resistance non-wovens should be avoided.

The control of the airflow resistance of the first layer allows tuning the absorption coefficient towards the middle frequencies, improving the absorption properties by up to 40% (cf. normal incidence absorption coefficient measurements as shown in Fig. 9.14). This "tunable absorption" can be realized with innovative light microfiber non-woven technologies weighting between 15 and 150 g/m^2 and covering the whole controlled airflow resistance range between 250 and 1500 Ns/m^3. These treatments remain highly porous, which makes them a real breakthrough solution for noise reduction [10].

In fact, it is quite easy to obtain high airflow resistance screens using compressed fiber felts, micro-perforated facings or resin bonded dense textiles. All these technologies lead to decreased porosity ($\phi \ll 0.8$) and much too small and/or almost closed pores which has a negative effect on the absorption coefficient especially in the high frequency range and most of the time lead to heavy complexes. The application of a "tunable absorption" concept can lead to better absorption performance with reduced space, to weight reduction up to 40%, or to both [10].

The use of foiled and perforated double porosity foams with a resonant foil membrane above the perforations, shows interesting absorption results in the middle fre-

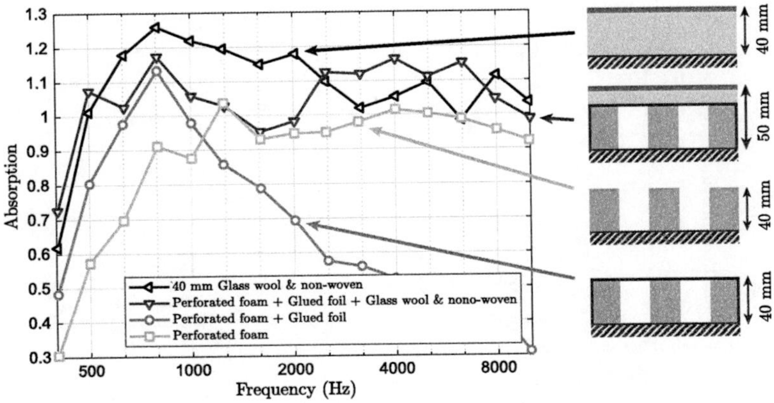

Fig. 9.15 Tunable absorption with airflow resistance non-wovens versus foiled double porosity material: diffuse field absorption coefficient [11]

quency range (cf. Fig. 9.15) [11]. In addition, covered by classical absorbing felt, it creates a broad band soundproofing system. The diffuse field absorption coefficient measurement values above 1 shown Fig. 9.15 obtained with 1.2 m^2 samples are due to the use of a "too small" reverberant room called Alpha Cabin having a 6.44 m^3 volume resulting in a 1.25 kHz Schroeder frequency [12].

However, it appears that a classical tunable absorber remains the best way to design a broadband absorbing device with thickness below 50 mm (cf. Fig. 9.15) [11]. This small thickness constraint is rather general for all transport industries because of lack of space generally linked to energy consumption.

It has been shown that perforated plates and screens (woven or non-woven) can be seen as thin layers of porous materials [13]. They are mainly controlled by the airflow resistivity, σ, (or the perforation radius) and the perforation rate, ϕ. The thickness of the screen is L and the one of the air cavity behind it is L_c. As shown in Fig. 9.16, a specific airflow resistance, maximizing the overall sound absorption, can be found for each perforation rate ϕ and thickness L independent of the cavity depth L_c [14].

This can be explained by the fact that if the airflow resistance is too low (e.g. large perforations), the acoustic waves can easily pass through the screen without enough dissipation. On the other hand, if the airflow resistivity is too high, the acoustic waves are reflected and thus little dissipation is added by the screen. This can be also explained using the sound absorption expression of a thin screen backed by an air cavity [14].

For thin porous layer backed by a cavity, the surface impedance Z_s can be approximated by:

$$Z_s \approx \sigma L + Z_b \tag{9.2}$$

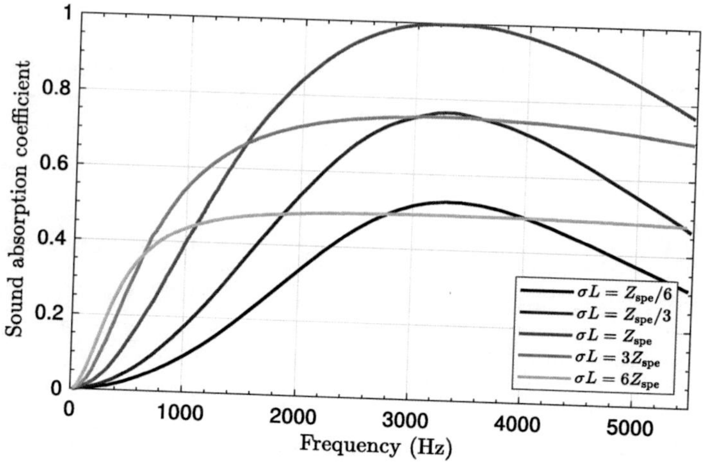

Fig. 9.16 Influence of the airflow resistance σL on the sound absorption coefficient (Normal incidence, $\phi = 0.2$, $L = 1$ mm backed by a 20 mm thick-air-cavity)

with $Z_b = -iZ_0 \cot(k_0 L_c)$ the backing impedance, i.e., a purely imaginary impedance for an air cavity.[1] The sound absorption coefficient α under normal incidence is expressed:

$$\alpha = 1 - \left| \frac{Z_s - Z_0}{Z_s + Z_0} \right|^2. \tag{9.3}$$

Thus, maximizing the sound absorption can be achieved by adapting the impedance:

$$Z_s = Z_0. \tag{9.4}$$

The airflow resistivity is linked to the hydraulic radius Λ:

$$\sigma = \frac{D\eta}{\phi \Lambda^2}, \tag{9.5}$$

with D the drag coefficient ($D = 8$ for circular hole, $D = 12$ for slit) and η the dynamic viscosity of air, then

$$\Lambda = R_{spe} = \sqrt{\frac{D\eta L}{\phi Z_0}} \approx 6 \cdot 10^{-4} \sqrt{\frac{L}{\phi}}. \tag{9.6}$$

The specific perforation radius is slightly lower under diffuse field than the one under normal incidence.

[1] A Fourier convention of $e^{i\omega t}$ was assumed here.

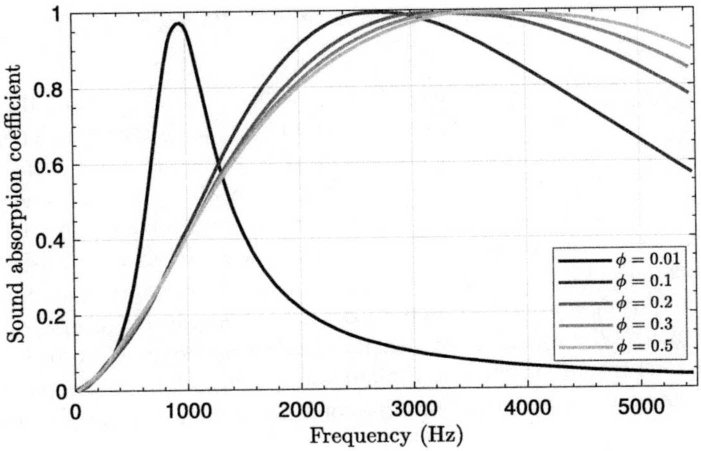

Fig. 9.17 Influence of the perforation rate (open porosity ϕ) on the sound on the sound absorption coefficient (Normal incidence, $L = 1$ mm backed by a 20 mm thick-air-cavity). The specific airflow resistance is adapted for each perforation rate

The effect of the opening rate is also of great importance. It allows controlling the frequency of the maximum sound absorption peak with the cavity depth (see Fig. 9.17). For porosity higher than 0.2, the frequency can be predicted as a quarter wave length in the double layer (screen + cavity) and from the Helmholtz resonator frequency for a porosity lower than 0.1 [14].

9.3 Insulation Optimization

For porous materials insulation optimization purposes, a complete poroelastic Biot-JCA model is necessary with the additional mechanical parameters: Young's modulus E, structural damping loss factor η_s and Poisson's ratio ν (density ρ as well of course). The Young's modulus is of particular interest if the poroelastic material is decoupling the mass barrier without any airgap. On the contrary, for double-wall constructions where airgaps are present, so where the barrier walls are self-supporting, the porous material will work effectively anyway whenever it is perfectly rigid or limp (extremely soft). The porous materials are not good insulators when used as a single layer, but they can be used to add viscothermal dissipation when coupled to single or double wall systems. Their elastic parameters can be of primary importance when they are bonded to the structure. More details will be given in the Chap. 10 about building applications.

9.3.1 Insulation Optimization: Fibrous Porous Material Case

For insulator poroelastic springs applications behind heavy layers, the optimized PET[2] or PET/PP[3] microfiber felts, described in Sect. 2.2.2, are too soft and decouple badly in the middle frequency range. In order to reach the rather ideal softness of "shoddy" materials while reducing weight ($E = 8$ kPa typically), one may introduce spring crimped or helicoidal hollow coarser fibers mixed with microfibers for improving the elastic properties, while keeping good viscothermal dissipation properties essential in the high frequency range [7]. On the contrary, glass fiber felts are too rigid for insulation automotive purposes (decoupling heavy layers), due to its too high Young's modulus at classical densities (which can be tuned a little bit somehow with glass composition, but in a limited way), restricting applications to absorption purposes only or integrated in self-supporting double wall air gaps without any gluing.

9.3.2 Insulation Optimization: Foam Porous Material Case

Let us concentrate here on the macro-stiffness control of foams realized through an adjustment of the Young's modulus of the constituting material depending on the targeted absorption or insulation application of the PUR[4] foam. Regarding a specific PUR foam index situation, which depends on the mixing of diols and triols and therefore on the resulting available –OH radicals, as well as on MDI diisocyanate molecule type with corresponding available N=C=O radicals, and while taking into account stoechiometric mixing ratios and water H_2O content, one can predict the number of "urethane" chemical bonds, which are directly linked to the stiffness of the foam (with "urea" secondary products bonds also) [9]. From this standpoint, it is then possible to scan above and below this stoechiometric value and get various low Young's modulus values here, with a certain chemically feasible variation (collapse, etc.).

9.3.3 The Standard Mass—Poroelastic Spring Insulator

The "old good" mass-poroelastic spring insulator is an excellent insulation technology allowing to reach very high transmission loss (TL) values in relatively small spaces. A 20 mm soft foam with a 3.5 kg/m^2 heavy layer simply positioned on a 0.8 mm steel plate reaches a transmission loss of 85 dB at 5 kHz, see Fig. 9.18. Doubling the mass of the heavy layer or the steel improves the transmission loss by 6

[2] PolyEthylene Terephthalate.
[3] PolyPropylene.
[4] Polyurethane.

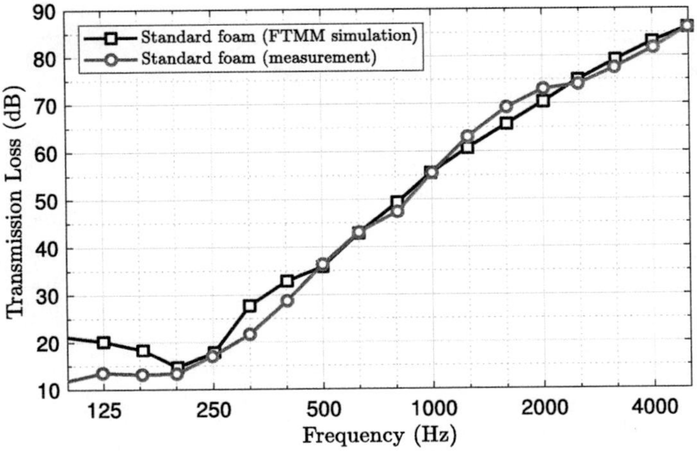

Fig. 9.18 Transmission loss of a standard mass-spring insulator: 3.5 kg/m^2–20 mm soft foam FTMM simulation vs. measurements

dB only. Thickness of the poroelastic spring and mass of the barriers are the macroscopic key lever and shows the limitation of this pure insulation concept for weight reduction.

The key transmission loss micro-structural optimization levers for a PUR injected foam used as poroelastic spring decoupler are:

- Airflow resistivity of the foam (which should be as high as possible while maintaining the other properties).
- Tortuosity of the foam (which should be as low as possible while maintaining the other properties [less important]).
- Young's modulus of the foam (which should be as low as possible while maintaining the other properties).
- Damping loss factor below the breathing frequency only (which should be as high as possible while maintaining viscothermal dissipation [less important]).

9.3.4 The Hybrid Stiff Insulator Concept

This hybrid stiff "green" light septum is simply a stiff compressed textile or felt glued on a light impervious layer, generally backed by an open porous foam or felt [15]. The light impervious layer acts as a mass in transmission loss of the stiff compressed textile or felt as long as the Young's modulus of the stiff upper felt is typically above $E > 1$ MPa, with a resonance dip in the curve Fig. 9.19 pushed above 4 kHz.

The advantage of this construction is that one maintains the absorption properties of the compressed textile or felt, as if they were positioned on a heavy layer and gets the typical response of a mass-spring system, i.e., transmission loss slopes of 18

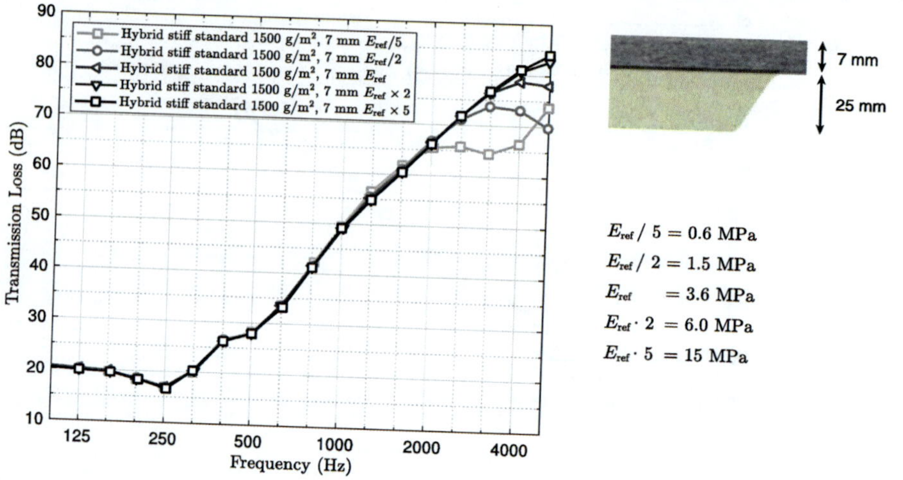

Fig. 9.19 Hybrid stiff concept Young's modulus dependency [15, 16]

dB/oct, up to a compressional resonance mode normal to the hard layer following the upper felt stiffness. If the impervious layer is not glued or missing, one loses the effect and recovers only a classical compressed textile or felt acoustic property (absorption type with 12dB/oct) [16]. This means that the Young's modulus and the mass per unit area are much more important here, than the airflow resistance of the compressed felt or textile (cf. Fig. 9.19).

9.4 Damping Optimization

9.4.1 Damping Optimization: Poroelastic Material Itself

For open cell porous materials described by the Biot-JCA theory [3], the damping should only be due to the structural losses within the solid phase. Of course, depending on the chemistry of the polymer foam, the viscoelastic properties of the skeleton will be observed to some extent, causing a strong or minor dependency of the damping with temperature.

We would like to emphasize the case where the closure rate of membranes at microstructure scale might induce additional strong viscous losses. In the dynamic case, the latter will cause an important variation of the damping with frequency as well. Indeed, the higher the frequency the higher the damping linked to these additional viscous effects [17]. This way one can increase the damping loss factor of standard PU viscoelastic foams formulations well above 0.5 and up to 0.7. The trouble is this almost closed cell foam does not present classical viscothermal dissipation and thus possesses almost no absorption as well as bad decoupling spring foam properties

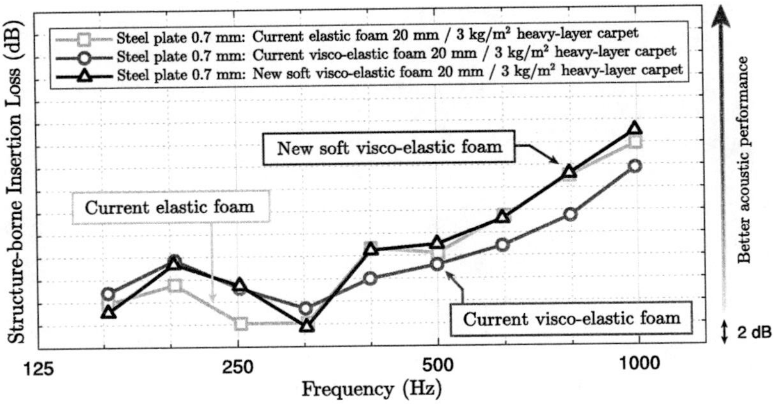

Fig. 9.20 Visco-elastic foams with various mechanical properties [18]

in the middle and high frequency ranges. The challenge is therefore to develop open cell viscoelastic foams with good structural damping above 0.33 and presenting still open cell microstructures. The result is an insertion loss increase below the breathing frequency of about 4 dB and classical poroelastic spring foam behaviour above (cf. Fig. 9.20) [18].

Note that the effect of the damping loss factor may have a significantly different effect depending on the type of the excitation. The damping loss factor has a limited influence when dealing with an air-borne excitation whereas it will influence the entire frequency range when dealing with structure-borne, turbulent boundary layer or rainfall excitation. This can be explained from the wavenumber point of view [19].

For fibrous materials, structural damping as well as friction between fibers with special slick treatments can eliminate the breathing frequency negative transmission loss (or insertion loss) values. Generally, fibrous materials achieve lower damping than foams, but the induced damping effect to structures in contact with them remains interesting.

9.4.2 Damping Optimization: Poroelastic Material Boundary Conditions

The boundary conditions at the interface between the poroelastic material and its supporting structure are essential for maximizing the induced damping effect to the structure. For thin air gap cases with no contact at all, the induced damping is not present. For sliding or even better fixed boundary conditions, the induced damping can be very significant [20], as illustrated in Fig. 9.21. Here, we show structure-borne transfer functions between acceleration and force on a bare or trimmed plate using poroelastic finite element simulation and measurements.

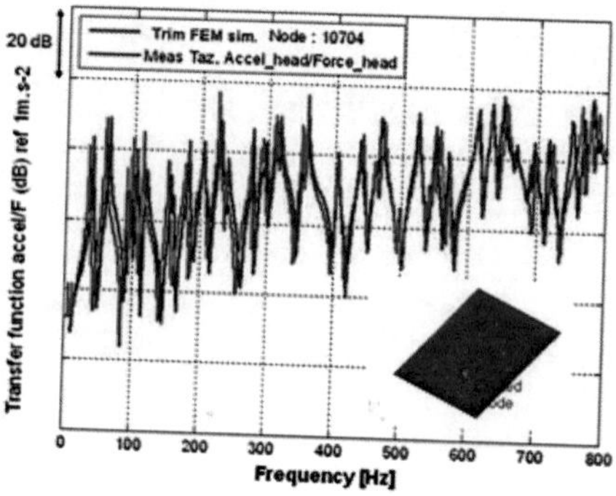

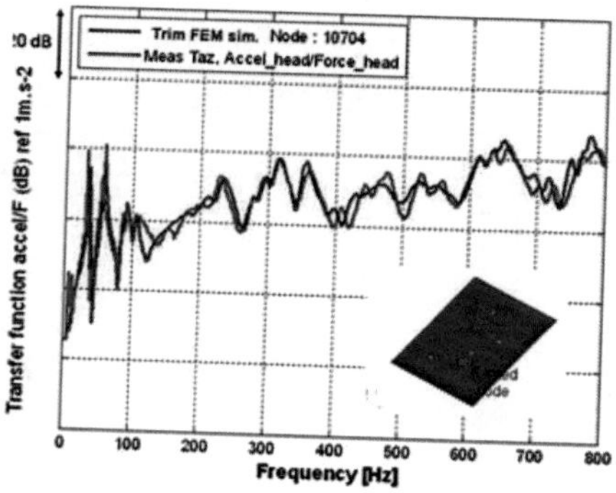

Fig. 9.21 (left) Bare structure and (right) trimmed structure [20]

9.5 Conclusions

A lot of "low-hanging fruits" technology transfers between automotive, building and aeronautic industries and beyond, of course, can be imagined. A first example is the tunable absorption concept, which is applied for over 10 years in the automotive industry with airflow resistive non-wovens and which is also applied for more than 5 years in the building industry using airflow resistive woven fabrics.

A second example is the hybrid stiff concept, i.e., a compressed stiff porous layer on an impervious layer, backed by a soft porous spring layer or an air gap. This configuration is more recent, applied presently in both automotive (dash and carpet insulators) and building industries (absorbing/insulating ceiling or walls). It could also be applied for aircraft interior linings with stiff compressed porous layers with nice woven fabrics facings and backfoiled. That would be lighter as well as more absorptive than grained plastic trims especially in the high frequency (5–7 mm porous layer absorption on a significant 2/3 of the interior passenger cavity).

A third example is the very low density glassfiber felt of 8 kg/m^3 or polymide foams of 6 kg/m^3 of the aeronautic industry that could be transferred to the automotive industry in particular for filling gaps between double wall constructions, etc.

In the following Chapters, we will present specific acoustic treatments and the particular challenges to control noise in the automotive, building and aeronautic industries.

References

1. M.A. Biot, Theory of propagation of elastic waves in a fluid-saturated porous solid. I. low-frequency range. J. Acoust. Soc. Am. **28**(2), 168–178 (1956)
2. M.A. Biot, Theory of propagation of elastic waves in a fluid-saturated porous solid. II. higher frequency range. J. Acoust. Soc. Am. **28**(2), 179–191 (1956)

3. J. Allard, N. Atalla, *Propagation of Sound in Porous Media: Modelling Sound Absorbing Materials 2e* (2009). Wiley
4. M. Tan Hoang, C. Perrot, Solid films and transports in cellular foams. J. Appl. Phys. **112**(5), 054911 (2012)
5. F. Chevillotte, C. Perrot, Effect of the three-dimensional microstructure on the sound absorption of foams: a parametric study. J. Acoust. Soc. Am. **142**(2), 1130–1140 (2017)
6. J. Manning, R. Panneton, Establishing relationships between acoustic and physical properties of shoddy-based fibre absorbers, in *INTER-NOISE and NOISE-CON Congress and Conference Proceedings*, vol. 2009 (Institute of Noise Control Engineering, 2009), pp. 1230–1235
7. A. Duval, F. Bonamy, R. Henry, *FR2836748 B1 Patent Family "mass-Spring Acoustic Insulation"* (Faurecia, France, 2002)
8. O. Doutres, N. Atalla, K. Dong, Effect of the microstructure closed pore content on the acoustic behavior of polyurethane foams. J. Appl. Phys. **110**(6), 064901 (2011)
9. A. Duval, M.T. Hoang, C. Perrot, V. Marcel, G. Bonnet, Chemistry-process morphology control of porous micro-structures: a bottom-up acoustic optimization approach, in *Light-Weighting and Acoustical Materials in Vehicles* (2013)
10. A. Duval, J.-F. Rondeau, G. Deshayes, F. Lhuillier, L. Bischoff, B. Teyssandier, *Generalized Light-Weight Concept: A Comprehensive Acoustic Package Weight Reduction Strategy* (Congrès SIA Confort Automobile et Ferroviaire, Le Mans, France, 2006)
11. J. Monet-Descombey, A. Duval, J.-F. Rondeau, G. Deshayes, Broadband thin absorbers, in *INTER-NOISE and NOISE-CON Congress and Conference Proceedings*, vol. 2011 (Institute of Noise Control Engineering, 2011), pp. 841–849
12. A. Duval, J.-F. Rondeau, L. Dejeager, F. Sgard, N. Atalla, Diffuse field absorption coefficient simulation of porous materials in small reverberant rooms: finite size and diffusivity issues, in *10ème Congrès Français d'Acoustique* (Lyon, France, 2010)
13. N. Atalla, F. Sgard, Modeling of perforated plates and screens using rigid frame porous models. J. Sound Vib. **303**(1–2), 195–208 (2007)
14. F. Chevillotte, Controlling sound absorption by an upstream resistive layer. Appl. Acoust. **73**(1), 56–60 (2012)
15. A. Duval, L. Bischoff, Stiff textiles or felts glued on light impervious layers: a new "green" light septum fiber technology, in *SAPEM 2011 Conference* (Ferrara, Italy, 2011)
16. A. Duval, J.-F. Rondeau, L. Dejaeger, F. Lhuillier, J. Monet-Descombey, Generalized light-weight concepts: a new insulator 3D optimization procedure, SAE Technical Paper (2013)
17. O. Danilov, F. Sgard, X. Olny, On the limits of an "in vacuum" model to determine the mechanical parameters of isotropic poroelastic materials. J. Sound Vib. **276**(3–5), 729–754 (2004)
18. L. Bischoff, C. Morgenstern, W. Berhard, A. Zopp, S. Schreck, Replacement of damping pads by using soft visco-elastic foam while maintaining high insulation properties, in *INTER-NOISE and NOISE-CON Congress and Conference Proceedings*, vol. 2012 (Institute of Noise Control Engineering, 2012) pp. 6177–6190
19. F. Chevillotte, F.-X. Bécot, L. Jaouen, Analysis of excitations from the wavenumber point of view, in *NOVEM Conference Proceedings, Dubrovnik, Croatia*, vol. 48988 (2015), pp. 1–2
20. A. Duval, L. Dejaeger, T. Reichensperger, L. Bischoff, N. Kobayashi, H. Yamaoka, Trim FEM simulation of a complete seat structure with foam cushions under structureborne and airborne excitations, in *INTER-NOISE and NOISE-CON Congress and Conference Proceedings, Ottawa, Canada*, vol. 2009 (Institute of Noise Control Engineering, 2009), pp. 3766–3775

Chapter 10
Industrial Applications II

Acoustic Package Optimization Methods in the Building Industry

Fabien Chevillotte

Abstract This chapter aims at presenting various building applications using porous materials. The role of porous materials and the relevant phenomena will be introduced for four typical applications, namely the acoustical correction, air-borne insulation, solid-borne insulation and ceiling applications. This chapter attempts to help the reader to identify the considered acoustical application, as well as in understanding the associated physical phenomena, in order to easily pinpoint the optimization key levers.

10.1 Introduction

Porous materials are widely used in building applications. Nevertheless, there are different applications and the relevant phenomena as well as their governing parameters differ. The purpose of this chapter is to give an overview of building applications and to attempt to explain how porous materials can help to improve the acoustical performance in these applications. Porous materials such as fibrous materials or foams are known as good "acoustical materials" but there is often a confusion between an absorbent material and an insulating material. Porous materials generally present good visco-thermal dissipation. This is enough to provide a good sound absorption performance but it is not enough to have good insulation properties when they are used alone.

Figure 10.1 compares the sound absorption coefficient in diffuse field (left) and the transmission coefficient (right) of a 46-mm thick glasswool (density $27\,\text{kg}\cdot\text{m}^{-3}$) and a 12.5-mm thick plasterboard. The single rating number R_w is also indicated for transmission properties [1]. One can note that the porous medium (glasswool) shows a good sound absorption improving with the frequency when compared to the plasterboard which is impervious and has no absorption property. On the contrary, the glasswool has a poor insulation performance compared to the plasterboard one.

F. Chevillotte (✉)
MATELYS - Research Lab., Bât. B, 7 Rue des Maraîchers,Vaulx-En-Velin, France
e-mail: fabien.chevillotte@matelys.com

N. Jiménez et al. (eds.), *Acoustic Waves in Periodic Structures, Metamaterials, and Porous Media*, Topics in Applied Physics 143,
https://doi.org/10.1007/978-3-030-84300-7_10

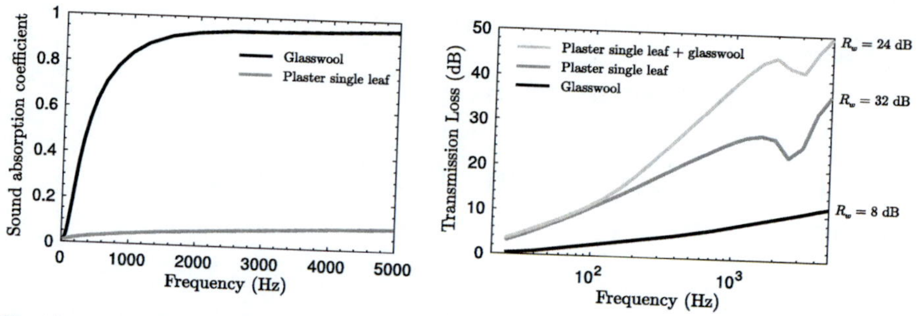

Fig. 10.1 Comparison of sound absorption coefficient (left) and transmission loss in diffuse field (right) of a 46-mm-thick glasswool and a 12.5 mm-thick-plasterboard

Nevertheless, the viscothermal dissipation of the porous medium can be useful to improve the insulation properties when assembled with other materials. The transmission loss of the glasswool added to the plasterboard is shown in the same figure. Note that gluing the porous material on the panel can affect the behaviour of the system and thus its transmission loss.

10.2 Acoustical Applications in the Building Industry

10.2.1 Acoustical Correction

The first building application, called "acoustical correction", consists of controlling the reverberation time according to the use of a room (e.g., living rooms, bedrooms, offices, classrooms, lobby, concert halls, etc.). In this application, the source and the receiver are placed in the same room, as shown in Fig. 10.2.

The reverberation time RT_{60} can be approximated by the Sabine's formula:

Fig. 10.2 Acoustical correction: direct propagation and multiple reflections of sound inside a room

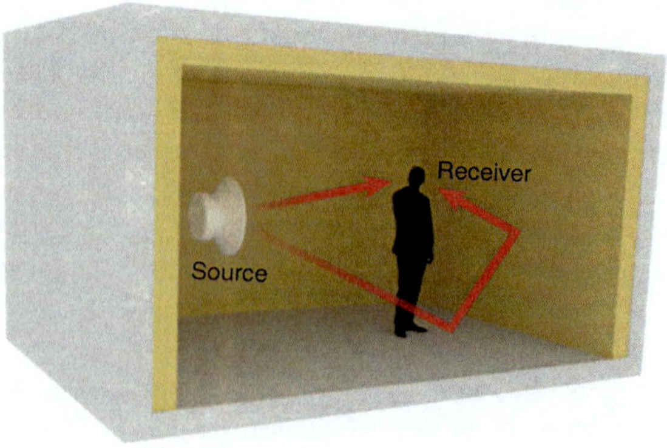

$$RT_{60} = \frac{24 \ln 10}{c_0} \frac{V}{A} \approx 0.161 \frac{V}{A}. \tag{10.1}$$

The reverberation time, RT_{60}, is defined as the required time to decrease the sound pressure level by 60 dB from the level of the excitation. Here, c_0 is the speed of sound, V the volume of the room, and A the equivalent absorption area defined as

$$A = \sum_i S_i \times \alpha_i, \tag{10.2}$$

where S_i and α_i are the surface and the sound absorption coefficient of surface element i, respectively.

For most applications, one generally tries to reduce the reverberation time. Looking at Eq. (10.1), this can be done by reducing the volume V, which is often not achievable, or by increasing the equivalent absorption area A. This latter is carried out by increasing the surface of absorbing materials or the absorption performance itself. The absorbing materials are always porous materials or an assembly of porous materials, perforated plates and/or screens.

10.2.2 Air-Borne Insulation

The second building application is the air-borne insulation. It can be insulation to interior noises (TV, radio, vacuum cleaner, etc.) as well as external noises (vehicles, roadworks, airport, etc.). The insulation can be achieved within the same housing or between different premises (housings, commercial surfaces, etc.).

10.2.2.1 Single Wall Partition

When dealing with air-borne insulation, we classically start with a single leaf partition. When considering a plane wave at oblique incidence θ, a drop in transmission loss happens at the so-called coincidence frequency. This decrease is due to the coincidence of the transverse acoustical wavenumber $k_t = k_0 \sin\theta$ of the incident wave and the natural bending wavenumber $k_b = \sqrt{\omega}\sqrt[4]{m'/D}$, with $k_0 = \omega/c_0$ the acoustical wavenumber (Fig. 10.3).

The coincidence frequency is given by

$$f_c' = \frac{1}{2\pi} \frac{c_0^2}{\sin^2\theta} \sqrt{\frac{m'}{D}}, \tag{10.3}$$

with the surface mass density

$$m' = \rho h, \tag{10.4}$$

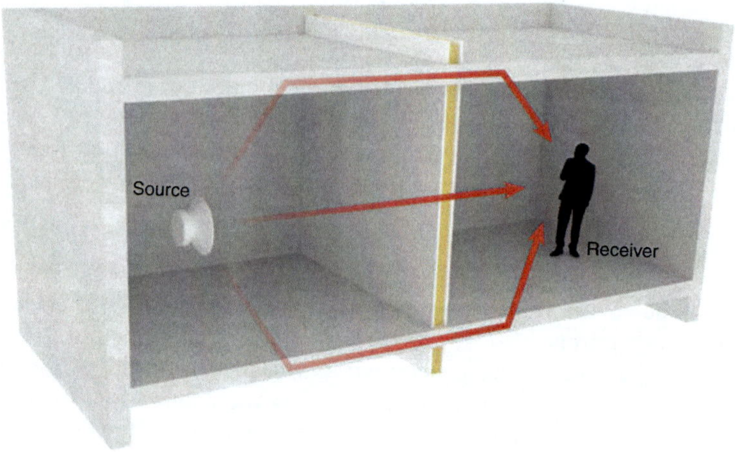

Fig. 10.3 Typical acoustical insulation problem and transmission pathways

and the bending stiffness

$$D = \frac{Eh^3}{12(1-\nu^2)}, \tag{10.5}$$

with ρ the mass density (kg/m^3), E the Young's modulus (Pa), ν the Poisson's ratio and h the thickness of the plate.

This coincidence frequency decreases while increasing the incidence angle (see Fig. 10.4). Under oblique plane wave, the transmission loss can be split in three zones controlled by different parameters:

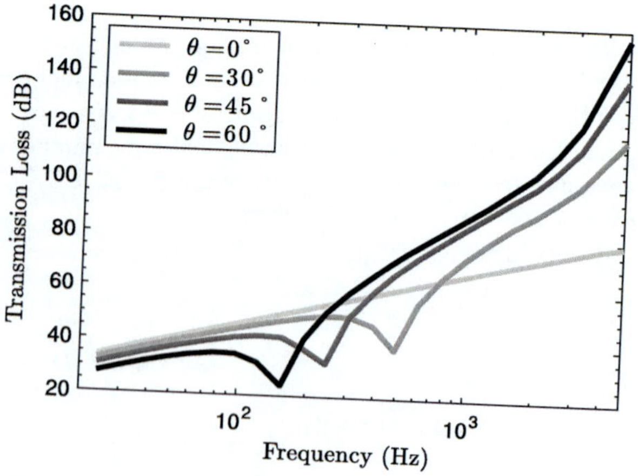

Fig. 10.4 Transmission loss of a 140 mm-thick single wall made of concrete for oblique plane wave excitation

(1) For $f < f_c'$: the mass law zone for frequencies lower than the coincidence one which is controlled by the surface mass density $m' = \rho h$ with a slope of 6 dB/octave.
(2) For $f \approx f_c'$: the coincidence zone which is mainly controlled by the damping loss factor η.
(3) For $f > f_c'$: a third zone for frequencies greater than the coincidence one which is controlled by the stiffness and a slope of 18 dB/octave.

When considering a diffuse field, the transmission loss can also be split in three zones but separated by the critical frequency f_c. The critical frequency is the lower coincidence frequency, corresponding to the greater incident angle ($\theta = 90°$), given by

$$f_c = \frac{c_0^2}{2\pi}\sqrt{\frac{m'}{D}}. \tag{10.6}$$

The typical transmission loss of a single panel (140 mm of concrete) submitted to a diffuse field is shown in Fig. 10.5. The transmission coefficient under diffuse field is computed thanks to an integration of oblique plane waves from 0 to 90°.

(1) For $f < f_c$: the first zone is still controlled by the mass which is controlled by the surfacic mass $m' = \rho h$ with a slope of 6 dB/octave (it is actually slightly lower around 5.5 dB/oct).
(2) For $f \approx f_c$: the critical zone is still controlled by the damping loss factor η.
(3) For $f > f_c$: a third zone which is controlled by the bending stiffness and the damping loss factor with slope of 9 dB/octave.

The slope goes from 18 dB/octave for a single incident angle to 9 dB/octave under diffuse field because the transmission loss in diffuse field integrates the stiffness effect and the coincidence frequencies of all incident angles.

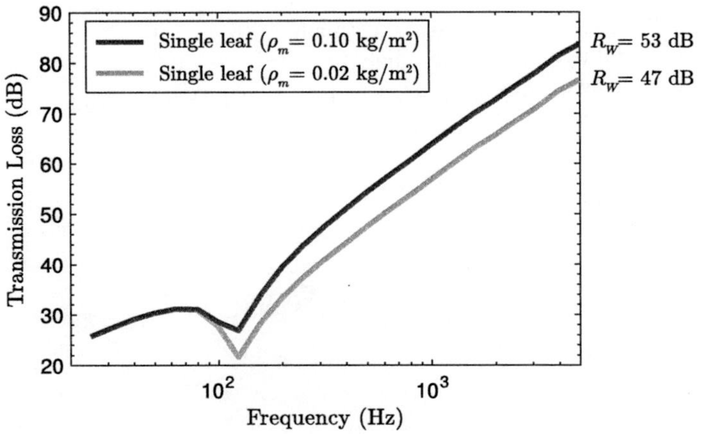

Fig. 10.5 Transmission loss of a 140 mm-thick single wall made of concrete for diffuse field excitation

The first way to enhance the transmission loss is to increase the surface mass density m'. This can be done by increasing the thickness which also implies an increase of the bending stiffness in power of 3 (see Eq. 10.5) or by increasing the mass density ρ. At the same time, the increase of mass density usually implies a stronger increase in Young modulus E. This means that increasing the mass generally implies a greater increase in bending stiffness and thus a decrease in the critical frequency. The decrease of the critical frequency strongly degrades the overall insulation performance. This degradation can be more important than the gain due to the increase of mass. Typical critical frequencies are 12 kHz for 1 mm-thick-steel plate, 2.5 kHz for 12.5 mm-thick-plasterboard 115 Hz for a 140 mm-thick-concrete slab.

10.2.2.2 Double Wall Partition

To get around this limitation, double wall partitions are usually employed (Fig. 10.6). Each panel has is own critical frequency. But thin plates enable to let the critical frequency relatively high. Nevertheless, another resonance appears at lower frequencies. It is a mass-spring-mass resonance and its characteristic frequency is called the breathing frequency, f_0, given by

$$f_0 = \sqrt{\frac{\mathrm{Re}\left(K_{eq}\right)}{4\pi^2 L_c}\left(\frac{1}{m_1'} + \frac{1}{m_2'}\right)}, \tag{10.7}$$

with K_{eq} the bulk modulus of the material filling the cavity, L_c the thickness of the cavity and m_i' the surface mass density of the i-th panel.

This resonance is a limitation, but the slope of the transmission loss is strongly increased right above f_0. The cavity can be filled with a porous material, usually a light glasswool, to increase the sound insulation at medium and high frequencies. The filling porous material mainly adds viscothermal dissipation which strongly improves the transmission loss above the breathing frequency.

The bulk modulus of air is the adiabatic value γP_0, with P_0 the atmospheric pressure and γ the heat capacity ratio ($\gamma = 1.4$ for air). When considering a typical

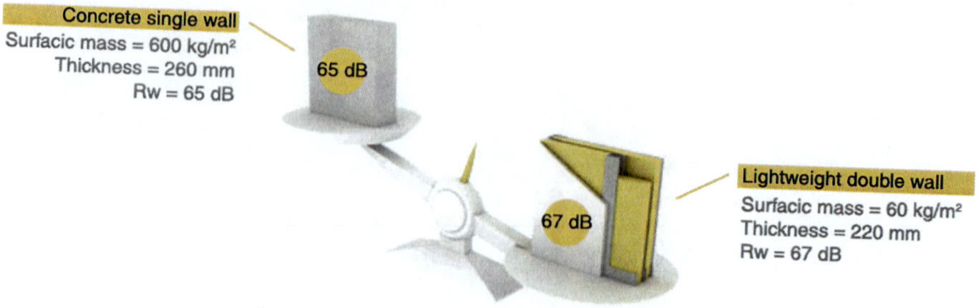

Fig. 10.6 Single leaf versus double leaf

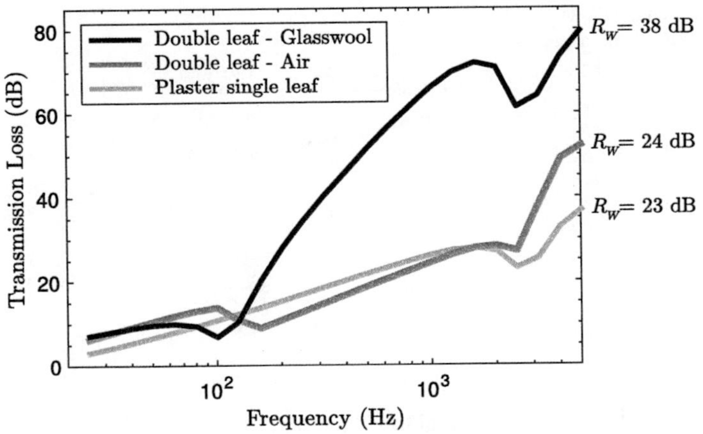

Fig. 10.7 Transmission loss of a double leaf partition (12.5 mm-thick-plasterboards and a 48 mm-thick-cavity filled with air or glasswool, stud-less) compared to a single leaf plasterboard of 12.5 mm

porous medium filling the cavity, the real part of the bulk modulus varies between the isothermal behaviour at low frequencies P_0/ϕ and the adiabatic one $\gamma P_0/\phi$, with ϕ the open porosity of the filling material [2]. Classical porous materials have a high porosity and an isothermal behaviour for frequencies around the breathing frequency. This means that filling the cavity with a porous such as a glasswool, decreases the stiffness of the cavity by a factor $\sqrt{1.4}$.

The improvement of viscothermal dissipation and the decrease of the breathing frequency are illustrated in Fig. 10.7 for a double wall partition made of two 12.5 mm-thick-plasterboards and a 48 mm-thick-cavity filled with air or glasswool. Note that additional phenomena such as double porosity [3] or adsorption/desorption can help to lower this bulk modulus [4].

Due to the self-supporting stiff walls, the Young's modulus of the porous material (glasswool here) is not important in this case, only viscothermal effects and density (so-called "limp" model) have to be taken into account. This situation becomes general whenever the porous material is decoupled from the walls with air gaps. Otherwise, full poroelastic models are compulsory and stiff decoupling porous materials have to be avoided.

10.2.2.3 Effect of Mechanical Links

To ensure the support of the partition, mechanical links have to be used between the two panels. This mechanical links, also called mounts or studs, introduce a solid-borne path which acts as an additional transmission path. The air-borne and the solid-borne paths are often computed separately and then combined [5] (see Fig. 10.8). Note that most of analytical models for solid-borne path computation are limited to two

Fig. 10.8 Schematic drawing of a double panel with mechanical links

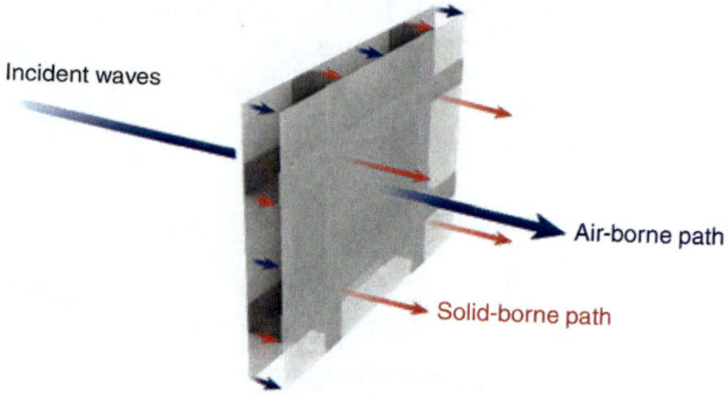

thin plates coupled by a mechanical link. A recent work has proposed a methodology to get around the thin plate assumption and to deal with more than one mechanical link [6].

The air-borne, solid-borne and total transmission loss coefficient are shown in Fig. 10.9 for the double wall partition used in the previous section (Fig. 10.7) with a mechanical link of stiffness $K_s = 10^6$ N/m and a loss factor $\eta = 0.08$ for the plasterboards. The solid-borne path is usually controlling the total transmission performance for frequencies higher 300 Hz for such a partition. The influence of the stud stiffness on the total transmission loss is illustrated in Fig. 10.10 for a given damping loss factor of the plasterboards ($\eta = 0.08$).

The influence of the loss factor on the total transmission loss is illustrated in Fig. 10.11 for a given stiffness of the mechanical links ($K_s = 10^6$ N/m). One can note that the influence of the damping loss factor of the plates is as important as the stiffness of the mechanical links. This means that the optimization of the stud stiffness has to be done for a given damping loss factor of the plates. Therefore, the loss factor of the plates is a major parameter. This can be explained using the wavenumber

Fig. 10.9 Influence of the mechanical link on the transmission loss of a double wall partition: air-borne, solid-borne and total contribution (12.5 mm-thick-plasterboards ($\eta = 0.08$) and a 48 mm-thick-cavity filled with a glasswool (27 kg.m^{-3}), stud stiffness $K_s = 1e6$ N/m)

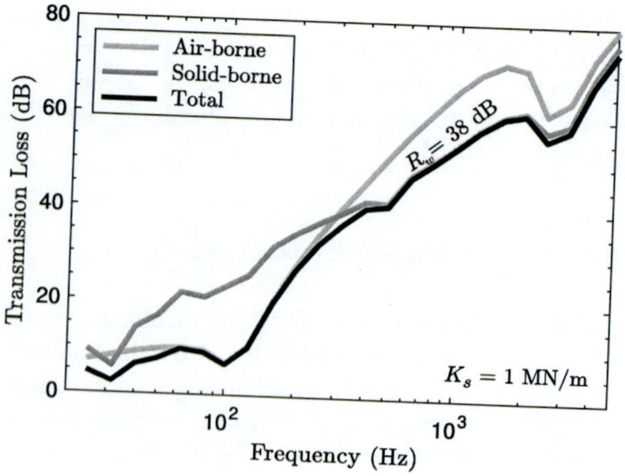

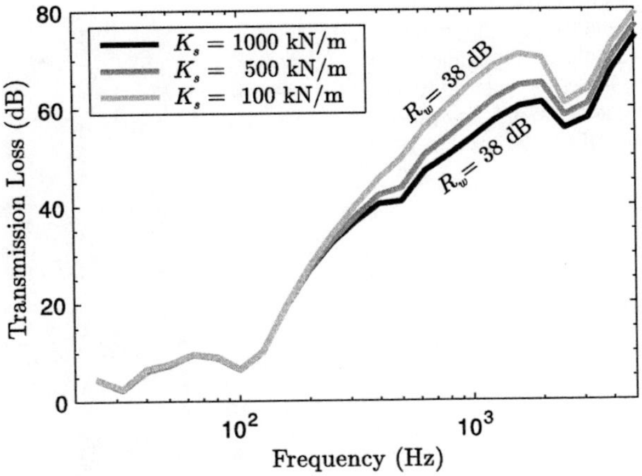

Fig. 10.10 Influence of stiffness of the mechanical link on the transmission loss of a double wall partition ($\eta = 0.08$)

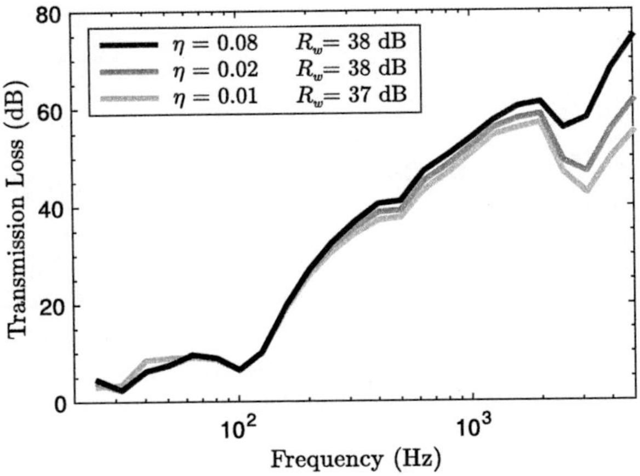

Fig. 10.11 Influence of damping loss factor of plates on the transmission loss of a double wall partition ($k_s = 10^6$ N/m)

analysis [7]. The solid-borne path presents a strong effect of the damping loss factor on the entire frequency range. This can be explained by the fact that the studs are excited by the first plate (emission side) and act as point forces exciting the second plate (reception side) and point forces excite a wide range of wavenumber. In this case there is always energy around the natural bending wavenumber of the second plate and a strong effect of the damping loss factor is observed.

This double wall partition problem is similar to the transmission problem presented in the part dealing with aeronautic applications (see Chap. 12) except that decouplers are employed to avoid the solid-borne path. Nevertheless, as a turbulent boundary layer excitation has to be considered (exciting a wide range of wavenumber [7]), the effect of the damping loss factor of the panels is also of primary importance.

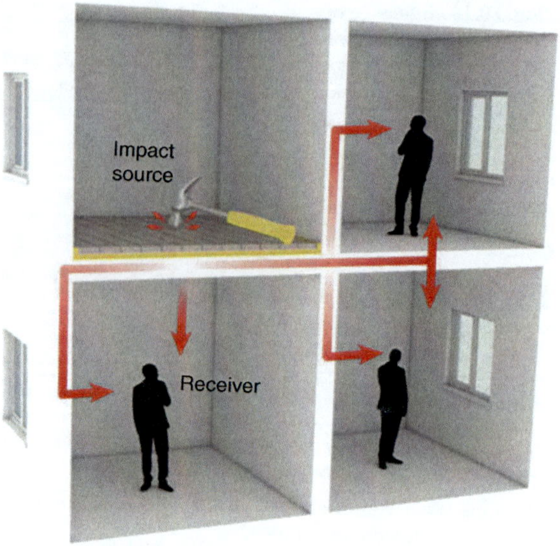

Fig. 10.12 Impact noise pathways in a typical building application. The underlay material (yellow) is added to increase the solid-borne insulation

10.2.3 Solid-Borne Insulation (Impact Noise)

Two floors are generally separated by a concrete slab, with a typical thickness around 140 mm. Its air-borne insulation is relatively good but floors can strongly radiate noise when submitted to impact source, e.g., falling objects, walking, etc. A spring-mass system can be added to increase the structure-borne insulation, as shown in Fig. 10.12. The mass is the surface floor (screed and tiles for instance) and the spring is an underlayer (porous or not). In this case, the porous layer must be sufficiently stiff to support the static load and at the same time sufficiently soft to shift the spring-mass resonance to lower frequencies in order to take advantage of the spring mass insulator. In this case, the performance is defined mainly by the elastic parameters of the porous layer. This is indeed the case presented in Sect. 9.3.3 previously.

Moreover, compared to an air-borne source, a mechanical source excites a wide range of wavenumbers, even at low frequencies, and the loss factor of the receiver panel (concrete slab in this case) is of primary importance [7]. The measurement of the impact noise is carried out thanks to a tapping machine [8, 9]. The tapping machine is placed on the upper floor (Fig. 10.12 top-left) and the sound pressure is measured in the lower room (e.g., at Fig. 10.12 bottom-left). First, one measures the normalized impact sound pressure level L_{0n} of the concrete slab and secondly the one of the concrete slab and the covering or the spring-mass system L_n. Then the reduction of impact sound pressure level is computed $\Delta L = L_{0n} - L_n$[10].

Normalized sound pressure levels are shown in Fig. 10.13 and the reduction of impact sound pressure level is shown in Fig. 10.14. The sound pressure level is strongly reduced at frequencies above the spring-mass resonance ($\approx 150 Hz$ in this configuration) but a slight increase happens at this resonance frequency. As this

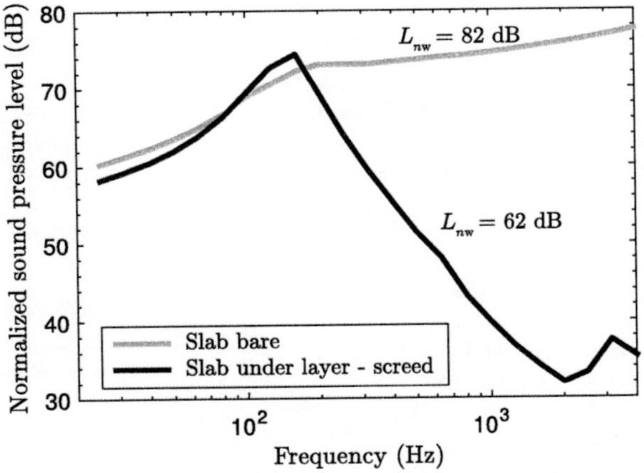

Fig. 10.13 Normalized sound pressure level with L_n and without under layer L_{n0}

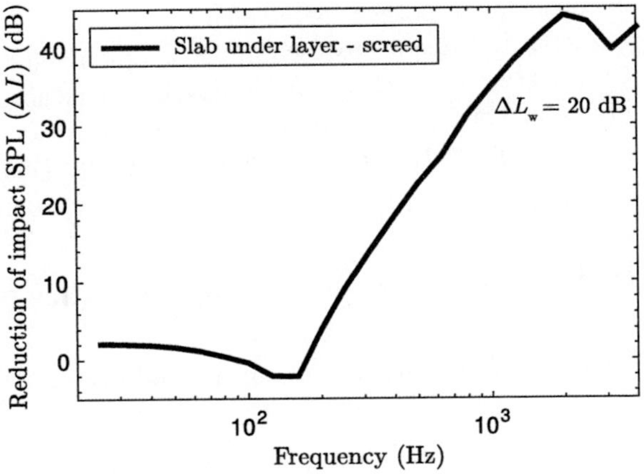

Fig. 10.14 Reduction of impact sound pressure level with under layer ΔL

resonance frequency is in the low frequency range, the single number rating ΔL_w is very sensitive to this phenomenon. When the underlayer is porous, one must be aware of the compressibility of the air which can become predominant when the mechanical stiffness is low.

10.2.4 Ceilings

Ceilings have a particular role since they are employed as sound absorbers but they can also introduce a lateral air-borne path between two rooms, as shown in Fig. 10.15. These parts have thus to be designed to control both the sound absorption in the emission room and to insure the air-borne sound insulation between two rooms. The

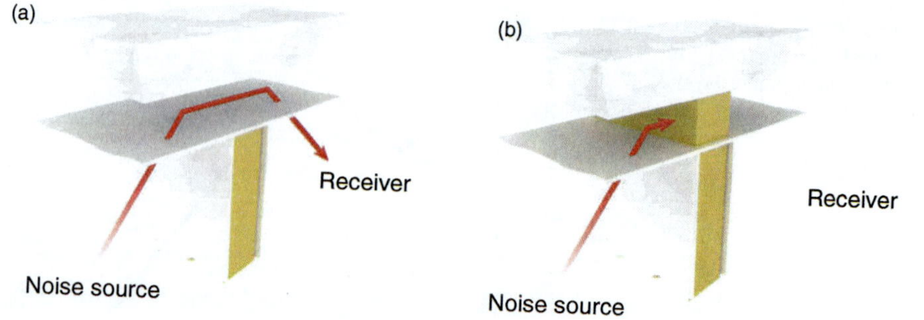

Fig. 10.15 Noise pathways in ceilings. **a** Noise propagating in the plenum introduces a lateral air-borne path. **b** Acoustic treatment of the plenum

treatment of the plenum is thus of primary importance to avoid lateral leaks and to attenuate the resonances of the plenum.

Acoustical treatments used in the automotive industry such as dash panels (see Chap. 11) are also designed to increase the sound absorption and the air-borne sound insulation. In both applications, a multilayer made of porous media, screens (or perforated plates) and/or impervious heavy layers can be employed.

10.3 New Trends in Building Acoustics

10.3.1 Thin and Aesthetic Absorbers

The new trends in building industry are varied. The first one is to use thin absorbers using textile or thin coatings. The idea is not to reach a sound absorption coefficient of 100% but rather to use thin treatments with lower absorption coefficient on large surfaces, such as the entire walls of a hotel lobby.

Another trend is to use additional suspended panels on ceiling or wall to increase the absorption area, e.g., in restaurants, offices, open-spaces, etc. Some panels can also include lighting. Absorption panels are used more and more often as aesthetic parts. Green materials and recycled ones are also often considered.

10.3.2 Low Frequency Performances and Non-conventional Phenomena

Similarly to other domains, the goal is to increase the low frequency performance of the solutions, for both absorption and insulation with a limited space, weight and cost. Acoustic metamaterials are of course studied. These materials attempt

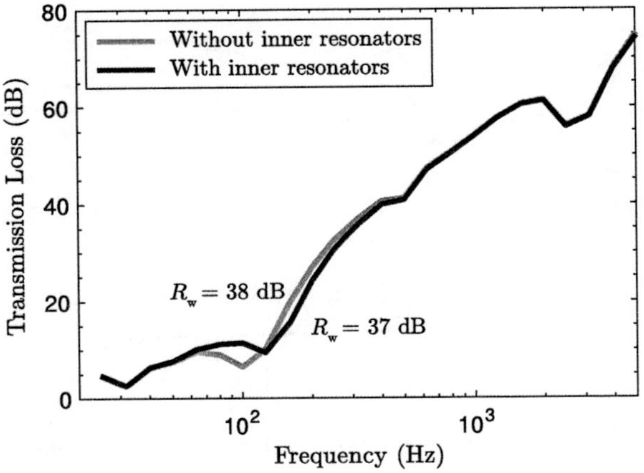

Fig. 10.16 Influence of inner resonators on the transmission loss of a double wall partition

to use additional physical phenomena to improve the sound absorbing or insulating performances. Resonating (acoustical, spring-mass or membrane-type resonators) or diffusion phenomena (pressure diffusion or multiple-scattering effects) are among them. In addition, the effect of the periodicity (Bragg's effect) can be used, but it is not required to take advantage of other non-conventional phenomena such as local resonances.

An example of porous medium with inner acoustical resonators (Helmholtz-type) in a double wall partition is studied here. The double wall partition is the same as the one studied in Sect. 10.2.2.3 with two plasterboards, mechanical links and a 48 mm-thick-cavity filled a glasswool and 20% of embedded resonators targeted around the breathing frequency.

The transmission loss values with and without resonators are compared in Fig. 10.16. One can note that the inner resonators enable to increase the transmission loss at the breathing frequency, which is shifted toward higher frequencies.

This can be explained looking at the bulk moduli shown in Fig. 10.17. Without inner resonators, the real part of the bulk modulus varies from the isothermal value P_0/ϕ (with $\phi \approx 1$ for a glasswool) at low frequencies to the adiabatic value $\gamma P_0/\phi$ at high frequencies. The imaginary part increases around the visco-inertial frequency (600 Hz). Looking at the bulk modulus of the glasswool with inner resonators, one can observe an increase of the imaginary part of the bulk modulus at the resonance frequency, which results in additional dissipation. On the other hand, one can notice a decrease of the real part of the bulk modulus below the resonance frequency and its increase right above it.

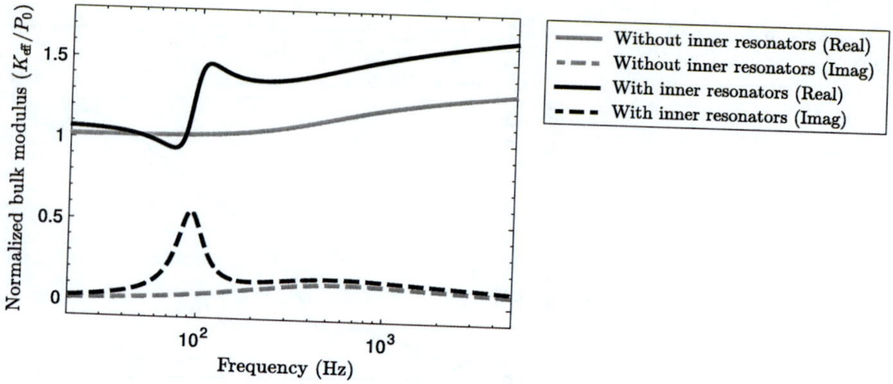

Fig. 10.17 Influence of inner resonators on the normalized Bulk modulus ($K/P0$)

As mentioned in Sect. 10.2.2.2, the breathing frequency increases with the real part of the bulk modulus (Eq. 10.7). Unfortunately, this phenomenon cancels the benefits of the additional dissipation of the resonators and the single number rating of the treatment with the resonators R_w is 1 dB lower that for the treatment without resonators. Obviously, different designs can be investigated but this type of drawback will always exist when the resonant phenomena are involved. This, however, does not happen for treatments based on diffusion phenomena.

10.3.3 *Rolling Noise*

Another trend is to divide buildings into commercial and habitable areas. Commercial shops typically occupy the ground floor, with private residencies on the upper floors. The delivery carts generate vibrations at low frequencies (100 Hz) that propagate easily throughout the building structure and on the upper floors, disturbing the habitants therein. Figure 10.18 shows that the sound pressure level due to rolling noise is far from the one generated with the tapping machine used for measuring impact noise according ISO standards [8, 9]. Moreover, one can note that typical decoupling spring-mass systems usually employed (see Sect. 10.2.3) are not suitable for rolling noise and can even amplify the rolling excitation when the spring-mass resonance coincides with the characteristic lobe of the rolling excitation. This problem is being addressed in current research using simulations and experiments [11].

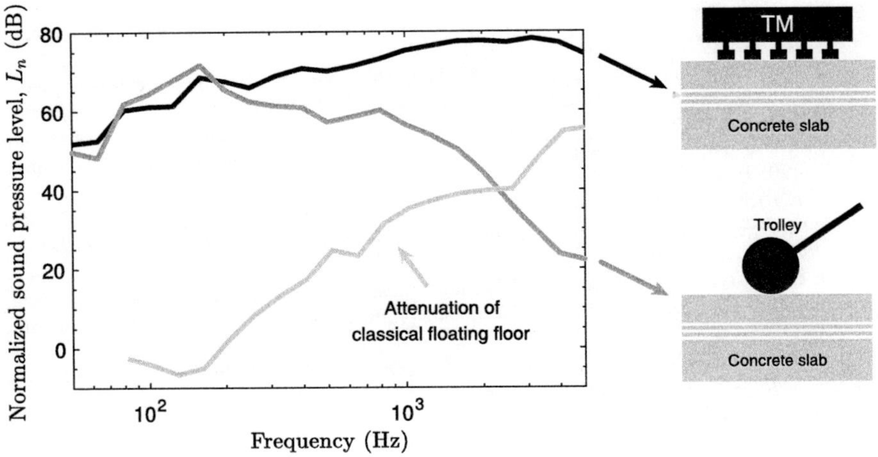

Fig. 10.18 Comparison of the sound pressure levels of tapping noise and rolling noise of a classical floating floor, as well as the attenuation of a typical floating floor

10.4 Conclusions

The porous media are employed to dissipate acoustical energy for absorption or insulation purposes. In double partition, porous media allow for an additional viscothermal dissipation at mid and high frequency and a decrease of the bulk modulus lowering the breathing frequency. When dealing with an acoustical problem in the building industry, we first have to identify which application is considered (acoustical correction, air-borne insulation, solid-borne insulation, ceiling) in order to understand which phenomena and therefore which parameters are important for optimization purposes.

Acknowledgements Pierre Leroy from Saint-Gobain-Isover is warmly thanked for the fruitful discussions and the review of this part.

References

1. NF EN ISO 717-1, Acoustics—rating of sound insulation in buildings and of building elements—part 1: air borne sound insulation, *International Standard Organisation* (2013)
2. Y. Champoux, J.-F. Allard, Dynamic tortuosity and bulk modulus in air-saturated porous media. J. Appl. Phys. **70**, 1975–1979 (1991)
3. X. Olny, C. Boutin, Acoustic wave propagation in double porosity media. J. Acoust. Soc. Am. **114**(1), 73–89 (2003)
4. R. Venegas, C. Boutin, Acoustics of sorptive porous materials. Wave Motion **828**, 135–174 (2017)
5. J. Davy, Sound transmission of cavity walls due to structure borne transmission via point and line connections. J. Acoust. Soc. Am. **132**(2), 814–821 (2012)

6. F. Chevillotte, F. Marchetti, On the modeling of multilayer systems with mechanical links, in *Proceedings of NOVEM 2018, Ibiza, Spain* (2018)
7. F. Chevillotte, F.-X. Bécot, L. Jaouen, Analysis of excitations from the wavenumber point of view, in *Proceedings of NOVEM 2015, Dubrovnik, Croatia* (2015)
8. NF EN ISO 10140-3, Acoustics—laboratory measurement of sound insulation of building elements—part 3: measurement of impact sound insulation, *International Standard Organisation* (2013)
9. NF EN ISO 10140-5, Acoustics—laboratory measurement of sound insulation of building elements—part 5: requirements for test facilities and equipment, *International Standard Organisation* (2013)
10. NF EN ISO 10140-1, Acoustics—laboratory measurement of sound insulation of building elements—part 1: application rules for specific products, *International Standard Organisation* (2013)
11. M. Edwards, F. Chevillotte, L. Jaouen, F.-X. Bécot, N. Totaro, Rolling noise in buildings, in *Proceedings of Internoise 2018, Chicago, USA* (2018)

Chapter 11
Industrial Applications III

Acoustic Package Optimization Methods in the Automotive Industry

Arnaud Duval

Abstract In this chapter we review the new up-coming challenges for noise treatments Research & Development in the automotive industry to improve the acoustic comfort in vehicles. After a description of the sources and their main transfer paths, noise treatments optimization key levers for both raw porous materials and multilayer lightweight treatments are presented and discussed ranging from basic 2D optimization at sample level to 3D at component level. A particular attention is given to newly emerging electric engines. Finally, the problematic of fast broadband vibroacoustic simulations with trims is analysed.

11.1 Introduction

The automotive industry is currently facing important challenges. One of them is dealing with strong environmental constraints linked to carbon dioxide (CO_2), nitrogen oxides ($NO_\times$) and other pollutants emission reduction. Another is the emergence of connected, autonomous, shared and electric vehicles (the famous "CASE" of Daimler's CEO). For the emission control, beyond aerodynamic, rolling and engine friction losses, one of the key levers is weight reduction with a simple rule of 10 kg corresponding to 1 g CO_2/km. The total weight of noise treatments including damping sheets exceeds 60 kg in a vehicle. Therefore, improvements offered by porous materials either through new treatments combining insulation, absorption and damping, in an innovative way, or through sole porous material or metamaterial use are of paramount importance for achievement of CO_2 emissions target.

For the electric vehicle, the distribution of principle contributing sources is changing drastically with a dominant broadband rolling noise centered in the mid-frequency combined with high frequency annoying tonal and broadband electric engine noise. Beyond typical noise, vibration and harshness (NVH) parts such as noise treatments or engine mounts, many other parts do also have an indirect but important acoustic

A. Duval (✉)
Trèves Products, Services & Innovation, 2-4 rue Emile Arquès, Reims, France
e-mail: arnaud.duval@treves-group.com

N. Jiménez et al. (eds.), *Acoustic Waves in Periodic Structures, Metamaterials, and Porous Media*, Topics in Applied Physics 143,
https://doi.org/10.1007/978-3-030-84300-7_11

function. These parts include pumps, heating, ventilating and air conditioning systems (HVAC), generators, cooling fans etc. The list can be very long indeed and very often combines acoustic and thermal management [1]. The electric vehicles (EV) and plug-in hybrid vehicles (PHEV) present unusual accessories annoyance as well as new power mode NVH transitions to manage. All this converges towards more and more accessories and engine encapsulation treatments where tuned porous multilayer materials excel in insulating and absorbing noises, but they are also prone to insulate heat when positioned in contact or near contact with the radiating surfaces.

11.2 Acoustic Package Optimization Methods in the Automotive Industry

11.2.1 Automotive Vehicle Main Noise Sources and Airborne Transfer Paths

The three main noise sources of an automotive vehicle are the engine noise (powertrain including the gearbox), rolling and aerodynamic noises. Figure 11.1 illustrates these three main noise sources as well as their major transfer paths. It demonstrates the relevance of considering exterior and interior noise treatments globally, each one having both an insulating (transmission loss: reflection of energy) and absorbing (absorption coefficient: dissipation of energy) role. A non-negligible induced damping effect from the noise treatments to the supporting structure may be observed even in the presence of damping sheets (pure bitumen dampers).

The traditional contribution of automotive sources versus speed is illustrated in Fig. 11.2 where the powertrain (engine with gearbox) is dominant up to 40 km/h,

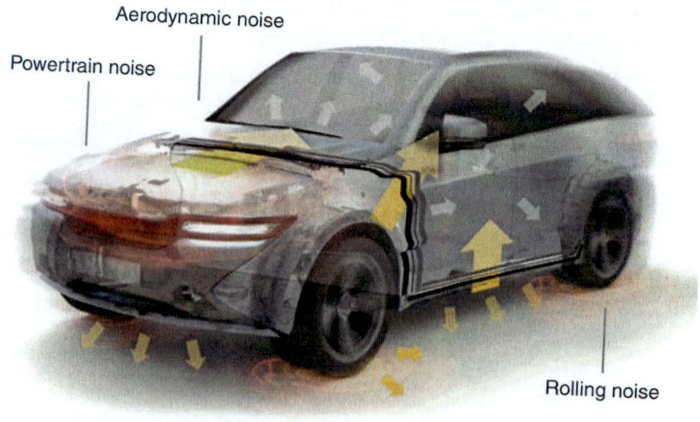

Fig. 11.1 Main sources and transfer paths in a vehicle

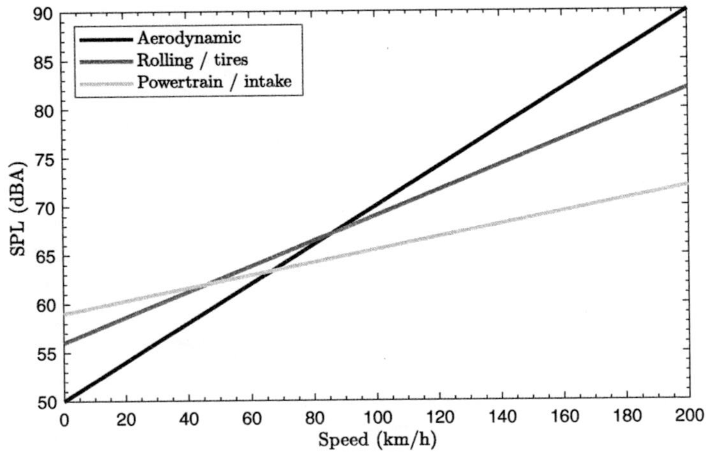

Fig. 11.2 Vehicle noise sources contribution depending on speed [2]

then the rolling noise takes over up to 90 km/h and, finally, above that speed the aerodynamic noise dominates. This is not the case anymore for electric vehicles, where the electric engine is so silent at low speed up to 30 km/h that artificial sounds with tonal content are necessary to alert pedestrians whether the electric vehicle is approaching or not, accelerating or not! New regulations define the type and the level of such sounds, also called acoustic vehicle alerting system (AVAS).

The main difference between an electric engine noise and a diesel engine noise is the very quiet low frequency range content of the electric engine noise: it is about 25 dB lower below 200 Hz. Indeed, the internal combustion engine (ICE) presents strong engine orders up to 250 Hz depending on the engine rotation speed, typically following the number of cylinders n divided by 2 for a 4 four strokes engine. Thus, in a 4 cylinders case, the dominating harmonics are the even orders (2, 4, 6, 8, ...). In the range of frequencies 800 Hz and 4000 Hz, which corresponds to the maximum sound pressure level (SPL), this difference is down by 10 dB compared to the diesel engine, as shown in Fig. 11.3.

This means that the low frequency masking of the internal combustion engine is not there anymore for the electric engine vehicle and that all automatic assistance: pumps, small electric motors, generators, air conditioning (HVAC) will have to be treated specifically either through additional interior noise insulation or dedicated encapsulation. This equipment is sometimes considered as the fourth noise source in an automotive vehicle. This dominating high frequency broadband content is reinforced with strong tonal "whistling" harmonics following the number of magnetic pair of poles, typical H24[1] and H48 for the traditional 8 pairs of poles construction. On top, the DC to AC current pulse width modulation (PWM) is also introducing

[1] Harmonics, engine orders.

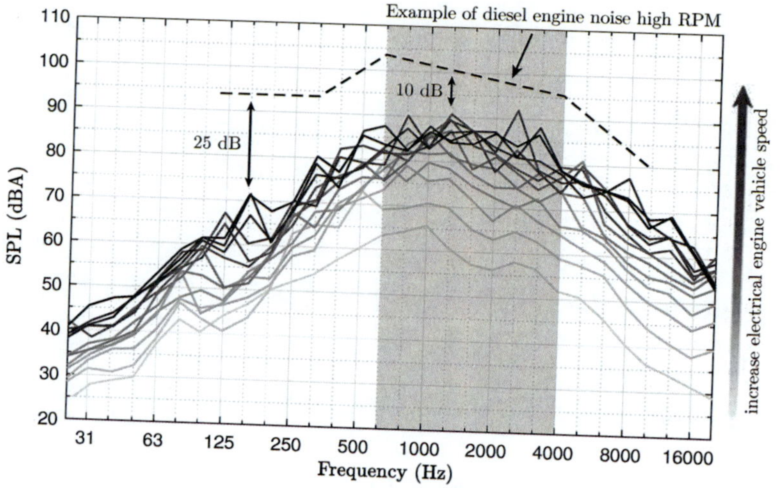

Fig. 11.3 Electric engine Sound Pressure Level in dB(A) in the engine compartment at various speed rolling conditions versus typical diesel engine SPL

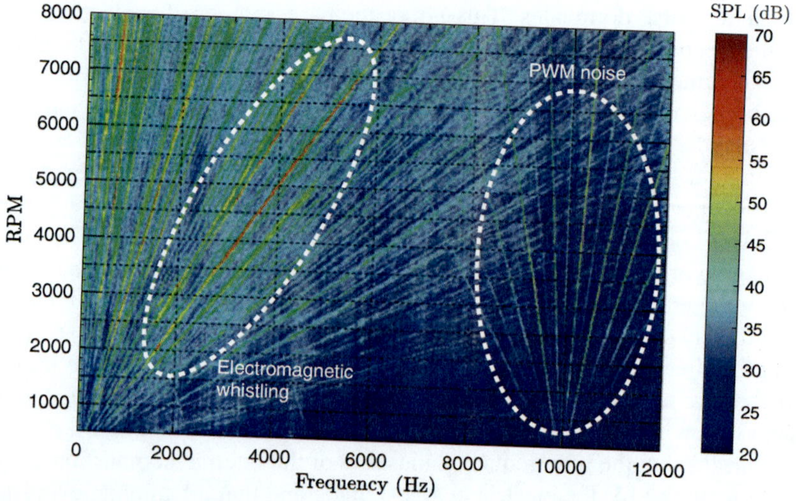

Fig. 11.4 Electric engine spectrogram in SPL (dB), RPM in function of frequency in the engine compartment (Courtesy of Renault)

harmonics in the high frequency range within a triangular shape centered around 10 kHz (cf. Fig. 11.4).

The aforementioned noise treatments put together allow recognizing an automotive vehicle (cf. Fig. 11.5). One can see engine encapsulation parts directly positioned on the powertrain having thermal and acoustic roles, body and underbody absorbers and shields having absorbing, insulating and aerodynamic roles, interior and trunk

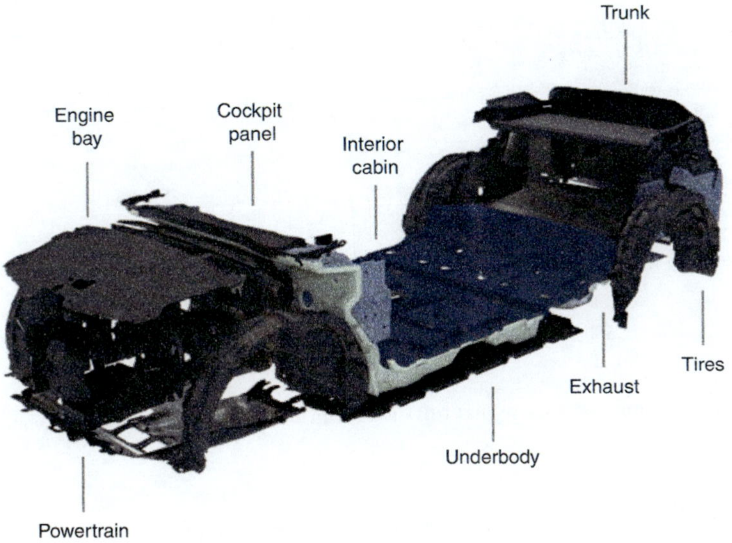

Fig. 11.5 Typical exterior and interior noise treatments (Courtesy of Trèves)

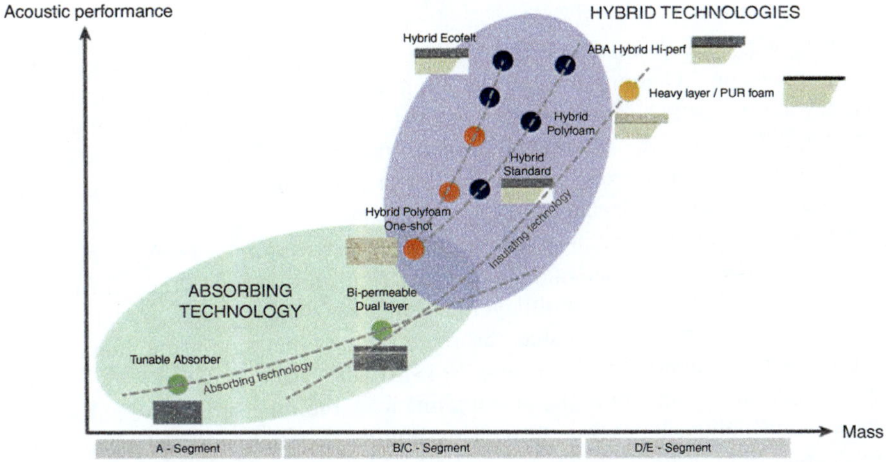

Fig. 11.6 Ranking of various typical absorbing, hybrid insulating/absorbing and insulating multi-layer noise treatment technologies

noise insulators having acoustical, mechanical and aspect roles. From a global acoustic package perspective, only damping sheets, sealing and absorbers in hollow body or passthroughs, door sealing and the headliner are missing in Fig. 11.5.

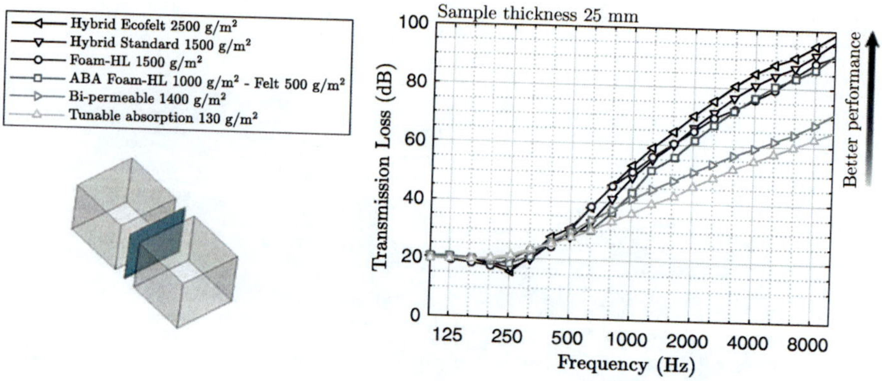

Fig. 11.7 Transmission loss results (dB) at 25 mm, using the Finite Transfer Matrix Method (FTMM)

11.2.2 Multilayer Noise Treatments Properties

11.2.2.1 Finite Transfer Matrix Method Comparison

The weight reduction challenge has taken a new shape in the past years due to high pressure on CO_2 emissions in the automotive industry. The new question is: what level of acoustic performance can one get with an insulator weighting globally less than 2500 g/m^2? The existing solutions at this weight being mainly dissipative (absorption) treatments give a satisfactory performance only if the passthroughs are poor and present critical leakages.

Respecting the less than 2500 g/m^2 weight target, a wide range of new or optimized concepts have been developed ranging from extremely absorbing to highly insulating noise treatments playing with multilayered insulators (typically three to four layers), in combination or not with tunable absorbers on the other side of the metal sheet (in the engine compartment for example) [3]. Each system answers to a specific passthrough quality situation, the best treatments combining broadband absorption with good insulation slopes (Fig. 11.6). This requires very light airflow resistive non-wovens and light airtight barriers in parallel with optimized poroelastic materials like foams or felts with adjusted stiffness. This approach shows with efficiency that optimizing a noise treatment without its environment may be misleading.

All aforementioned lightweight technologies have been developed at first using simplified Statistical Energy Analysis (SEA) simulation accounting for both insulation and absorption properties. When the absorption is localized in the reception cavity and when no spatial windowing is necessary, it is possible to use the famous coupled reverberant rooms equation,[2] instead of the reference SEA model (cf. Figs. 11.7, 11.8 and 11.9).

---

[2] $NR = TL + 10\log_{10}(A)$.

For the equivalent absorption area of the 1 m^3 reception cavity, the diffuse field absorption coefficient of the 1 m^2 sample is taken into account as well as the absorption coefficient due to the almost negligible viscothermal effects on the 5 m^2 resting "bare" boundaries. This absorption situation is representative of what is happening behind an instrument panel without additional absorption. Elsewhere, a full SEA or better full coupled ray-tracing model of the fully trimmed vehicle is necessary in order to simulate properly the absorption influence of a specific insulator on the sound pressure level at the ear location of the passengers [4].

This noise reduction analysis shows that good absorption performance cannot compensate for a strong lack of insulation performance. The Absorber-Barrier-Absorber (ABA) and hybrid stiff solutions are giving the best results for the 2500 g/m^2 total weight (1500 g/m^2 upper equivalent barrier). Note that the stiff upper layer of hybrid stiff treatments allows a full capture of the upper felt or polyfoam mass in the transmission loss performance unlike the soft upper felts of ABA lay-ups in the whole middle and high frequency 500 Hz upwards.

Once again, the heavier the better for the 2500 g/m^2 upper equivalent barrier weights like the Hybrid Ecofelt (or polyfoam) solutions. The main difficulty is to maintain the absorption performance at a high compression, see Fig. 11.8. Anyway, the super stiff hybrid Ecofelt does not present any high frequency dip in the transmission loss curve up to 10 kHz (Fig. 11.7). Nevertheless, above 4 kHz this potential dip would not have any influence on the global performance with leakages.

11.2.2.2 Finite Transfer Matrix Method Comparison with Leakages

Noise treatments in the automotive industry are generally very efficient as long as there are no leakages. This is the reason why the dash panel area is considered as the most severe one. Indeed, most of the critical passthroughs are concentrated there such as pedals, HVAC, steering column, cables etc. Nevertheless, the great majority of the

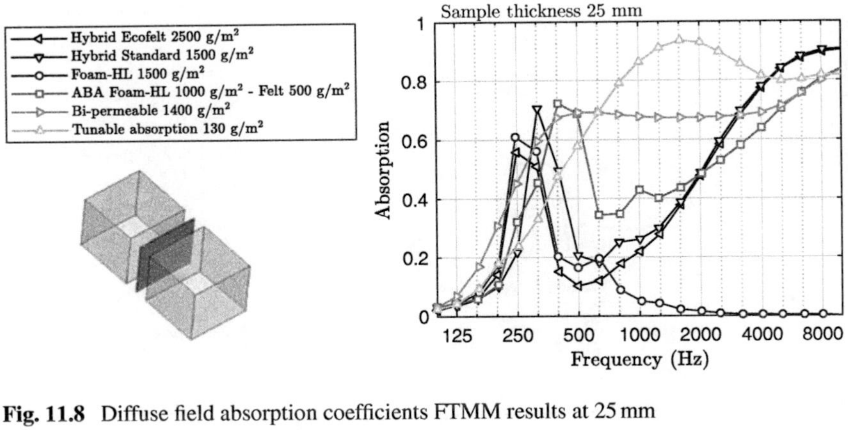

Fig. 11.8 Diffuse field absorption coefficients FTMM results at 25 mm

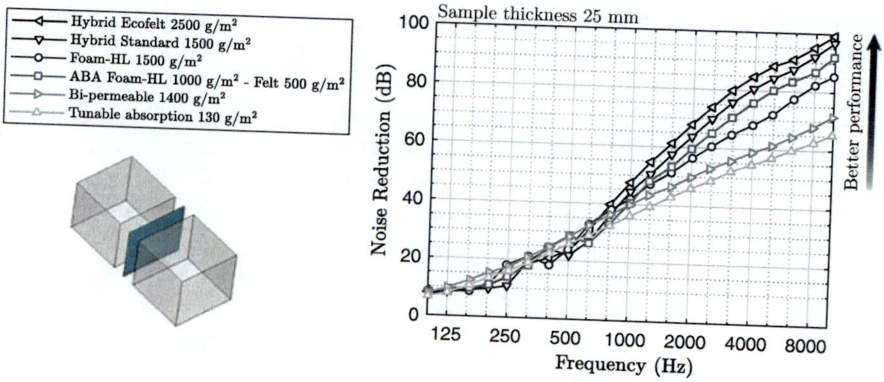

Fig. 11.9 Noise Reduction results: $NR = TL + 10\log_{10}(A)$ at 25 mm

passthroughs radiate behind the instrument panel, which builds an intermediate half-closed cavity having an insulation effect (masking) and where additional absorption works actually quite well (cf. Fig. 11.10) [3].

The concept of "airtight instrument panel" used by some premium German car-makers takes advantage of this instrument-panel insulation effect combined with conventional insulators. But it requires expensive sealing technologies especially in the under instrument panel driver area where the sealing must not block the movements of the pedals for obvious active safety reasons! Another classical solution is to use numerous absorption patches positioned under the instrument panel and on the instrument-panel undercover used in parallel with insulation concepts. The latter works quite well even if one is not profiting from the additional transmission loss of these absorption patches, the problem which can be solved by superimposing the absorption layer on the mass-spring insulation system (ABA or better hybrid-stiff concepts).

The best lightweight technology is in fact "pass-through dependent". The higher the leakages, the more effective are the absorption concepts shown on the left side

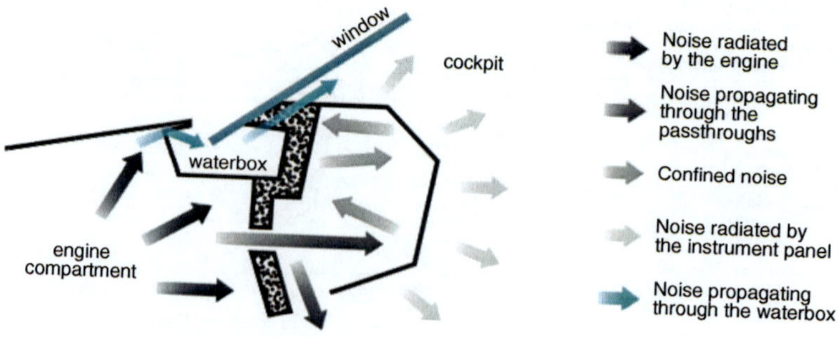

Fig. 11.10 Cockpit specific architecture with leakages and masking effects

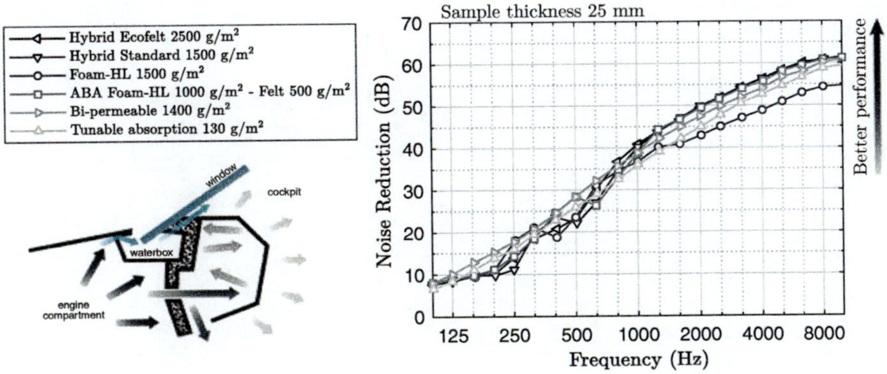

Fig. 11.11 Noise Reduction results: $NR = TL + 10\log(A)$ at 25 mm with 2% uncovered area

of Fig. 11.6. On the contrary, the lower leakages, the better is the performance of the insulation treatments, shown on the right side of Fig. 11.6. The "holy grail" of the noise treatment engineer is to maximize and broaden the working frequency range of both insulation and absorption at the same time while maintaining low weights. Using the TMM implementing the Biot-Allard theory, the noise reduction index, defined by $NR = TL + 10\log_{10}(A)$ with $A = \alpha_{\text{mat}} + 5\alpha_{\text{viscothermal}}$ is calculated without and with leakages represented as trim uncovered areas [3].

The resulting noise reduction curves show the minimal differences between the various solutions, as shown in Fig. 11.11, obtained when taking into account the leakages. Indeed, the ranking between insulating and absorbing technologies is still respected like in Fig. 11.9 but the differences in performance are reduced. Finally, the technology which is the most in difficulty with leakages is the traditional foam-heavy layer concept due to the absence of any absorption (only cavity boundaries viscothermal effects).

11.2.2.3 Body Mounted Engine Absorbers

In the last 20 years, the design of engine absorbers has been following a clear weight reduction path, starting from 60 kg/m³ for a cotton felt, down to 30 kg/m³ for a glass felt with microfibers of around 6 μm diameter, and, nowadays, down to 15 kg/m³ for polyurethane (PUR) slab foams with partially closed membranes (cf. Fig. 11.12). Globally, the absorption performance has been maintained and guaranteed by following the morphological paths described in Chap. 9: the homothetic reduction one for the fibrous materials and mainly the closure rate of membranes for the PUR slab foams. The cost has been driving evolution of the latter, as melamine foams are very close to glass felt from a morphological point of view: it is a 3D fibrous material with a super lightweight polymer skeleton presenting no membranes at all and having ligaments of 4–6 µm diameter. Indeed, closing the membranes of PUR with rather coarse cell size was cheaper than switching to a more expensive smaller cell size melamine foam technology, interesting anyway for high temperature resistance.

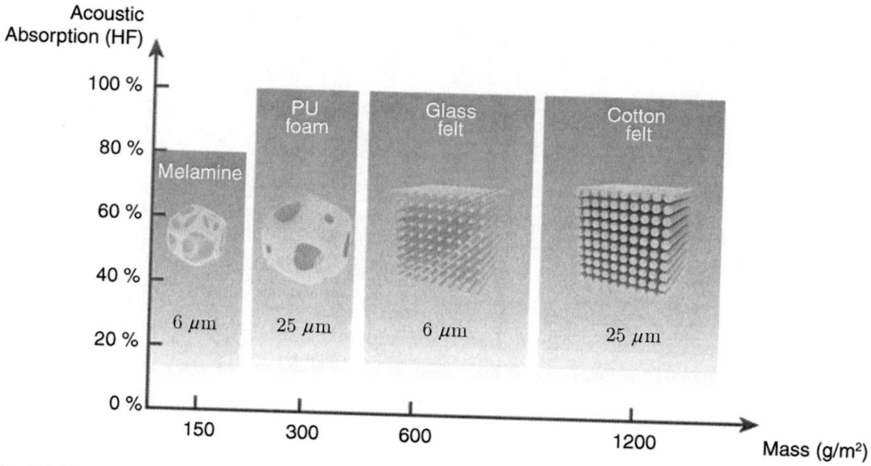

Fig. 11.12 Engine absorber ranking for a 20 mm blank with fiber or ligament diameter size versus surface weight. PU: polyurethane

As explained in the earlier chapter as well, the favorable increase of the airflow resistivity resulting from the closure rate of membranes was also combined with a strong increase of the tortuosity, which results in excellent absorption performance in the middle frequency range. This frequency range remains the most critical one for standard porous noise treatments with the double constraint of space and weight.

11.2.2.4 Automotive Insulator Optimization

In recent years, the link between the microstructure of porous media (foams, fibrous materials, and also some granular porous samples) and its acoustical macro-behaviour has been studied more and more starting from 2D models and progressing to 3D models [5, 6]. These studies allow computing the acoustic properties of porous materials from their microstructural characteristics. For this purpose, the method of Hoang and Perrot has been used to define the optimal microstructure morphology of a PUR foam, e.g. a periodic unit cell, corresponding to a targeted acoustic performance, based on three direct intrinsic measurements, namely: the porosity, the airflow resistivity and a ligament length (or cell size) [7].

Indeed, the optimization rule is to search in a morphology space where the microstructures are feasible for manufacturing afterwards [8, 9]. The final acoustic descriptor to optimize in the example demonstrated here was the Insertion Loss (IL) value in dB at 1000 Hz of a mass-spring insulator comprised of a soft porous foam covered by a heavy layer (EPDM-EVA barrier) [10].

Figure 11.13 shows two main microstructure optimization parameters of the surface response: the closure rate of the membranes and the dimensionless static permeability (normalized to the square of the cell dimension). This surface response

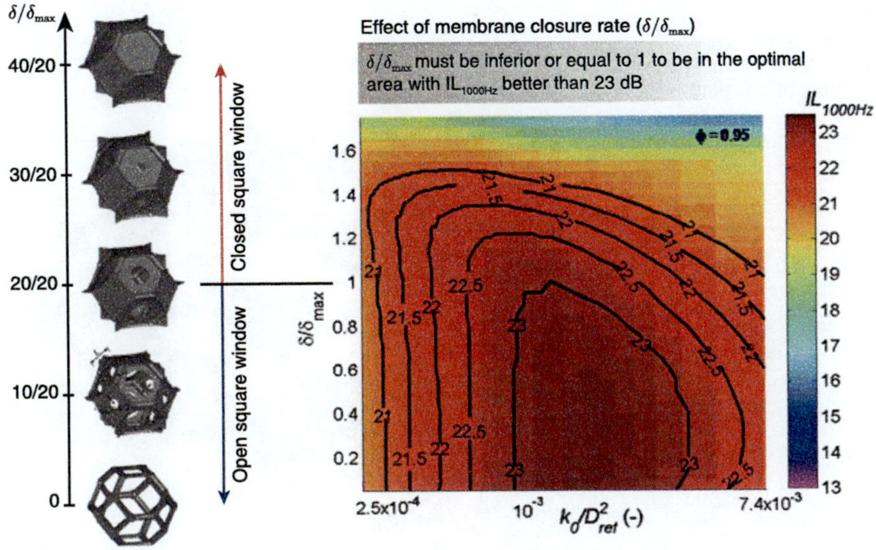

Fig. 11.13 Insertion Loss 1000 Hz surface response versus closure rate [10]

Insertion Loss analysis shows a potential for improvement by 1.5 dB compared to standard optimized foams. The corresponding optimal morphology may then be described and transferred to chemical and process engineers for the manufacturing phase. Beyond product-process know-how, both simulation and the design of experiments might be launched in order to obtain these optimal microstructures. Such optimization methodologies allow everybody in the industry to speak the same language and work together towards the development of the best acoustic material.

11.2.2.5 Engine Encapsulation

The new upcoming CO_2 emission regulations as well as exterior noise 68 dB(A) pass-by regulations for Europe to be adopted in 2024 in particular (70 dB(A) in 2020) lead to an increase of the engine treatments coverage need, from 15% now to 80% potentially in 2024 (for Internal Combustion Engine at least). Indeed, porous materials block the radiated heat when located close to the engine, maintaining the temperatures for an optimal energy consumption and, in addition, they guarantee efficient emission control technologies such as catalytic converters or particle filters. On top, engine encapsulation treatments are reducing the radiated noise through their insulation, absorption and induced damping properties.

Depending on the mounting constraints or decoupled attachment possibilities given by the various encapsulation technologies, the acoustic concepts will vary from pure insulation systems to a hybrid insulation – absorption systems and finally to pure absorption systems. Most of the times the hybrid solutions are the best.

Fig. 11.14 Engine encapsulation typical acoustic and thermal parts (courtesy of Trèves)

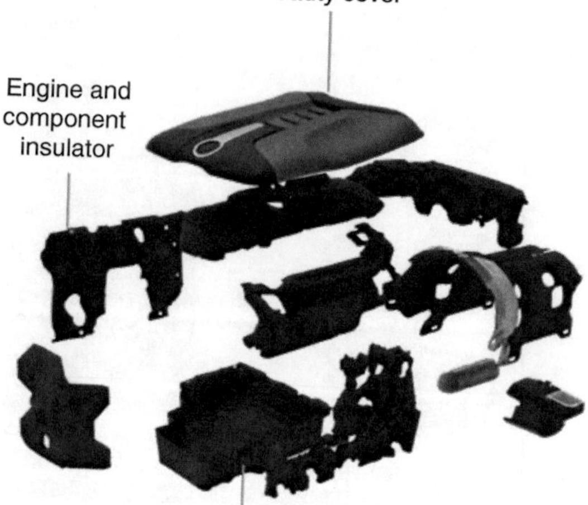

Figure 11.14 illustrates these acoustic and thermal encapsulation concepts from injected heavy layer back-foamed, to hybrid plastic GF/PP shells[3] back-foamed, to high density integral foams, micro-perforated aluminium foils on top of resin glass felts with or without slab foams behind called *dual layer*, etc. The trend is also to encapsulate the electric engines more and more, which has the consequence of leaving many small powered functions such as pumps, electric alternators, electric turbo-compressors radiate noise with the necessity to be encapsulated as well! Combined with the body mounted engine and underbody absorbers, these efficient engine encapsulation technologies allow an overall weight reduction of the interior noise treatments which, depending on the ICE or electric engine source, will be reconfigured for optimal performance focusing then more and more on rolling noise treatments.

11.3 Fast Broadband Highly Curved Insertion Loss Simulation of Automotive Trims

11.3.1 Trim Modelling Curvature Issue

For more than 25 years, the Transfer Matrix Method (TMM) implementation of the Biot-Allard theory for poroelastic materials has been used on a daily basis in the industry [11]. This multilayer simulation method is not only very fast (just a

[3] Glassfibre/polypropylene.

few seconds), but also very precise for flat samples (even for rather small sizes of around 1 m^2) thanks to the spatial windowing corrections often called Finite Transfer Matrix Method [11, 12]. The latter is valid not only for Transmission Loss coupled reverberant room suites simulations, but also for small reverberant room diffuse field absorption coefficient simulations [11–13]. Combined with SEA/virtual SEA or energy-based vibroacoustic approaches, the TMM is definitely a powerful tool for simulating noise treatments in the middle and high frequency range.

Some correlation issues appeared as early as 20 years ago especially when using 3D thickness maps for 3D insulator representation with SEA/TMM [14]. Early experimental investigations tended to prove that high curvatures were the main driving factor for the observed decrease of the Insertion Loss slopes and breathing frequency disappearance [15]. Poroelastic finite elements with the efficient (u, p) 4 d.o.f formulation and its commercial BEM-FEM software implementations seemed to be the modelling answer in order to capture this new 3D trim transmission loss physics in the low and middle frequency ranges [16–19]. The idea emerged to hybridize BEM-FEM with SEA using recomposed trim Insertion Losses gathered from a database of precomputed equivalent curvature cases. This method was carried out using poroelastic finite elements in the middle frequency range only and to extrapolate the Insertion Loss slopes to the high frequency [20, 21].

Indeed, one of the remaining difficulties was to perform accurate and fast broadband Insertion Loss simulation of any sound packages for curved surfaces. Analytical spectral approaches on cylindrical sandwiches [22–24] served as the base for the development of typical supporting structures: steel shell/poroelastic core/heavy layer shell model [25]. The aforementioned trimmed cylinder analytical model has been generalized through a unified cylindrical Transfer Matrix Method approach. This allows the construction of a vibroacoustic model for any multi-layered cylindrical systems of an arbitrary nature [26]. In particular, this system can be composed of orthotropic shells, 3D viscoelastic solids or 3D poroelastic media. The spectral nature of the approach makes it possible to rapidly cover the medium to high frequency ranges for industrial size systems. The objective of this work is to validate this recent spectral approach by application on a simplified dash insulator and comparing all aforementioned simulation methods with each other.

11.3.2 Cylindrical Transfer Matrix Method: A Spectral Approach

The theory of this recent fast broadband cylindrical Transfer Matrix Method approach (cf. Fig. 11.15) can be found in [26], including some experimental pure cylindrical validation cases as well as numerical validation cases from the literature.

Figure 11.16 presents a half-cylinder BEM-FEM poroelastic model (Rayon-VTM-TL software) which will be used as reference. It shows the lay-up from the BEM emission, the FEM steel structure, the poroelastic component, and the air gap

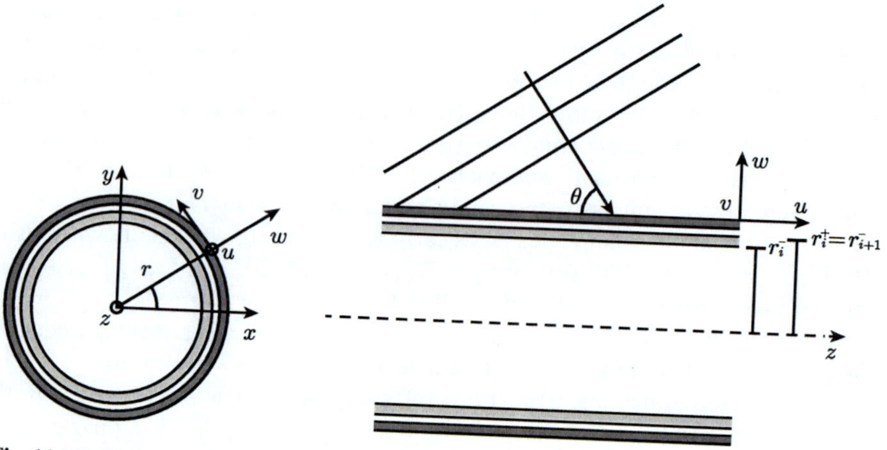

Fig. 11.15 Example of multilayered cylindrical system

Fig. 11.16 Curved Insertion Loss BEM-FEM poroelastic finite elements simulation model: hybrid stiff standard foam-felt lay-up 25 mm thick with a 32.5 cm or 25 cm half-cylinder radius

to the BEM reception. Whenever the mesh criteria are respected as stated in [27], the correlation quality of such BEM-FEM poroelastic model has been proven valid many times in the low and middle frequency ranges including highly curved dash insulators [11, 17–21].

The Insertion Loss simulation results comparison between the proposed spectral method, namely cylindrical Transfer Matrix Method (TMM), the classical flat TMM and flat and curved BEM-FEM poroelastic finite elements is presented in Fig. 11.17. They show excellent correlations above 250 Hz up to 2.5 kHz whenever the curvature is taken into account. Indeed, flat TMM and flat BEM-FEM poroelastic finite elements Insertion Loss simulations do correlate very well as Fig. 11.17 demonstrates, and it fits very well also with the measurements (as proven many times [11]). However, this flat method does not capture the Insertion Loss slope decrease for small curvature radii below 50 cm typically from globally 13 dB/oct down to 10 dB/oct

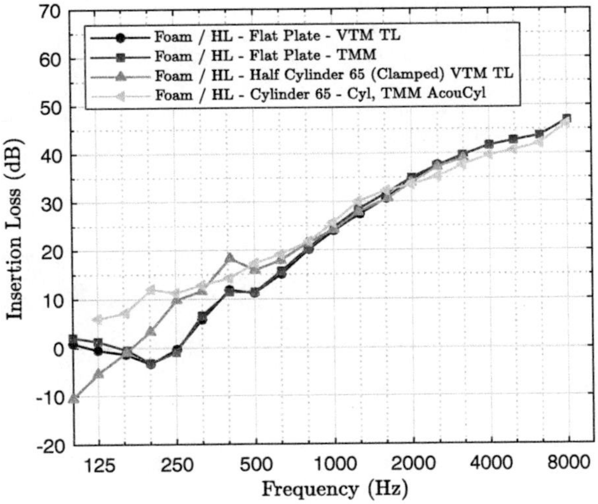

Fig. 11.17 Insertion Loss simulation comparison between flat TMM, the proposed cylindrical TMM approach and BEM-FEM poroelastic finite elements flat & curved (65 cm diameter) on a Foam 20 mm/Heavy Layer 3.5 kg/m^2 case [26]

[15, 19]. This Insertion Loss intersection between flat and curved Insertion Losses is seen to occur even earlier for hybrid stiff lay-ups than standard foam/heavy layer insulators. This phenomenon is amplified as the curvatures become smaller [28].

11.3.3 Curvature Radius Influence on Trim Insertion Loss Slopes

Figure 11.18 illustrates that the smaller the radius, the lower the Insertion Loss slope of a Foam 20 mm/Heavy layer 3.5 kg/m^2 trim. For a curved surface with radii between 40 cm downwards to 17.5 cm, the amplitude of the Insertion Loss increases by 8 dB between the ranges of 200 Hz and 630 Hz. The mismatch with the Insertion Loss of a flat surface is even higher and peaks at the breathing frequency of 200 Hz where a 15 dB difference can be observed. In the experience of the authors, these differences are less critical for radii above 50 cm in this specific case of soft poroelastic cores. This latter statement explains why the aeronautic industry never pointed out these curved trim modelling difficulties with main structure radii of about 3 m and above.

11.3.4 Trim Thickness Influence on Curved Insertion Loss Breathing Frequencies

The breathing frequency is the mass-spring-mass frequency resonance occurring when the upper mass and supporting steel structure vibrate in antiphase. This is a typical (1, −1) eigenmode (cf. Fig. 11.19 left), where the poroelastic foam is playing

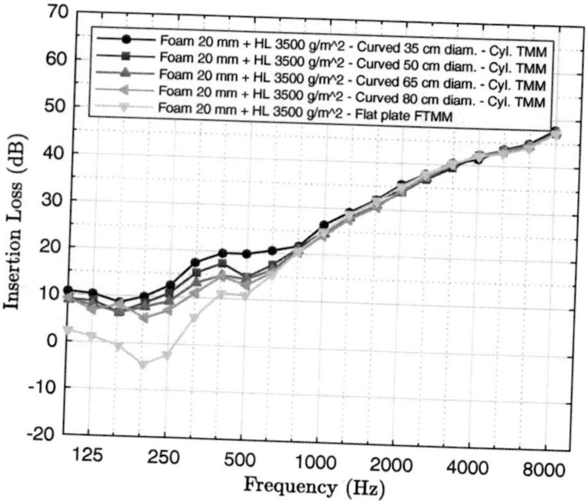

Fig. 11.18 Influence of the curvature radius on the Insertion Loss of a Foam 20 mm / Heavy Layer 3.5 kg/m^2 trim using Cylindrical TMM and Flat TMM

its decoupling role in the same way as a suspension would do for a pure vibratory problem. This is causing the dip in the Insertion Loss curve at 200 Hz for a 20 mm foam, 3.5 kg/m^2 heavy layer flat trim (cf. Figs. 11.17 and 11.18). This dip will shift towards higher frequencies as the foam thickness decreases.

The most remarkable aspect of curved-trim Insertion Losses is indeed the disappearance of the breathing frequency, where the mismatch between the Insertion Losses of curved and flat surfaces is at a maximum. In fact, up to 1.5 times the ring frequency of a cylinder, the membrane stiffness (in-plane waves) is dominating the vibroacoustic behaviour [22], thus shear coupling is dominating in the structure/porous/heavy layer or stiff felt lay-up: at this point there is no longer any breathing frequency (cf. Fig. 11.19 right). Taking into account the thickness distribution, typically every 5 mm, the aforementioned mismatch is migrating from 200 Hz for

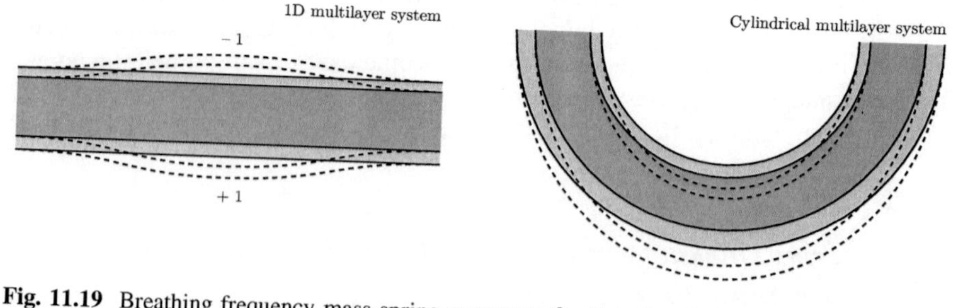

Fig. 11.19 Breathing frequency mass-spring resonance for flat trims (left)/In-plane wave dominating curved trimmed structure behaviour up to 1.5 times the ring frequency (right) (Foam/Heavy Layer trim)

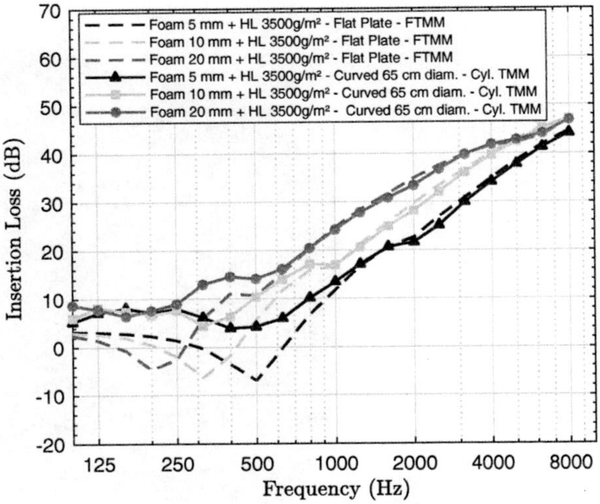

Fig. 11.20 Influence of the spring thickness on the Insertion Loss of a Foam/Heavy Layer 3.5 kg/m^2 trim using Cylindrical TMM and Flat TMM

20 mm spring thickness to 315 Hz for 10 mm, and to 500 Hz for 5 mm (cf. Fig. 11.20) [20], as a result of the generalized approach of modelling trims.

11.3.5 Multi-thickness Simplified 3D Dash Insulator Broadband Insertion Loss Simulation

Taking BEM-FEM poroelastic finite element simulation as a validated reference for complex 3D trim modelling, Fig. 11.21 presents a simplified curved dash insulator 0.8 mm thick steel validation case. As seen, it combines an upper truncated cylinder of 20 cm radius, a middle flat plate and a lower truncated cylinder of 32.5 cm radius. This kind of "S" shape simplified structure is very close to the actual shape of an automotive steel dash firewall with the upper cross-car beam, together with the generally flat middle area and the lower curved area overlapping with the front footwells carpet.

The advantage of this "S" shape is that the equivalent curvature approach (singly curved or flat here) [21], is applicable not only for the BEM-FEM poroelastic finite element Insertion Loss simulation, but also in a pure mode for both the cylindrical and flat transfer matrix methods, the models being exact for these shapes. Figure 11.22 shows the hybrid stiff localized lay-up on the "S" shape simplified dash panel and rather low multi-thickness repartition which is required to push towards the middle frequency range the breathing frequencies. This illustrates the modelling difficulties of curved trims in the middle frequency range [29].

Figure 11.23 shows the global Insertion Loss recomposition simulation results in the power sense using the various approaches described above and applied in the

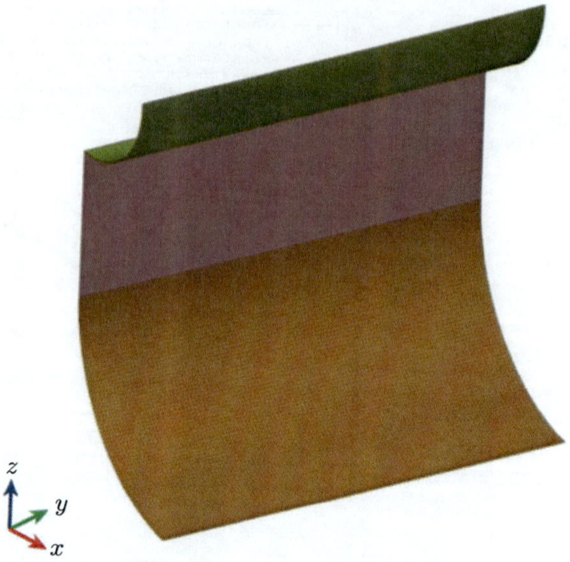

Fig. 11.21 Simplified dash insulator structure made of 0.8 mm thick steel: combination of a top truncated 20 cm diameter cylinder, middle flat plate and lower truncated 32.5 cm diameter cylinder

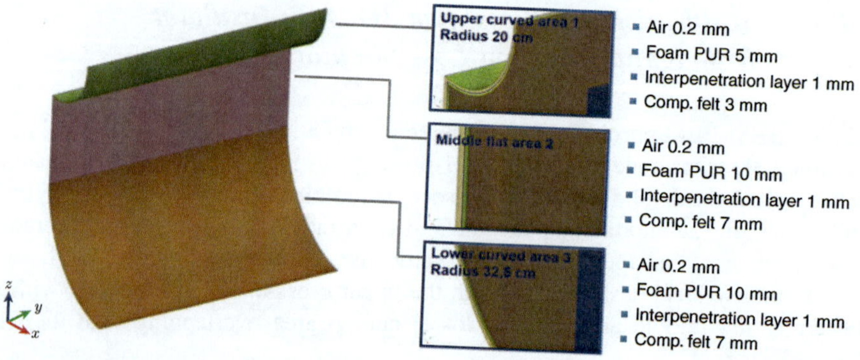

Fig. 11.22 Simplified S shape dash insulator trim: hybrid stiff localized lay-up and thickness repartition

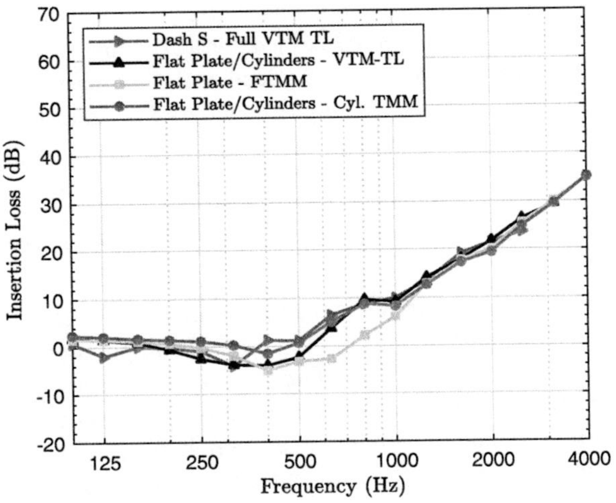

Fig. 11.23 Trimmed simplified dash insulator "S" shape hybrid stiff Foam/Felt application case following Fig. 11.22 hybrid stiff Foam/Felt lay-up and comparing all Insertion Loss simulation methods [29]

equivalent curvature approach to each subsystem like for Statistical Energy Analysis (SEA) sub-structuring:

- Reference full BEM-FEM poroelastic finite elements (Full VTM-TL with clamped surrounding boundary conditions like a real mounting in a transmission loss suite).
- Equivalent curvature BEM-FEM poroelastic finite elements (with free-free boundary conditions for the truncated cylinders).
- Flat TMM.
- Combined cylindrical and flat TMM (equivalent to free-free boundary conditions for the cylindrical TMM).

Once again, the Insertion Loss differences between simulation methods, whether or not accounting for the curvature, are significant with a maximum of 8 dB between 500 Hz and 1000 Hz. The correlation between the full BEM-FEM poroelastic finite elements and the combined cylindrical and flat TMM are excellent above 500 Hz up to 4000 Hz.

Figures 11.17 and 11.23 present a validation of the proposed cylindrical TMM, which has been integrated in the SEA+ software. This allows for better computational performance than reported in [26], with individual Insertion Losses computational times of less than 1 min., instead of a few hours with BEM-FEM poroelastic finite element simulations. Moreover, this Cylindrical TMM enables computing Insertion Losses with a broader frequency range from the middle frequencies above 250 Hz and up to 10 kHz in the high frequencies.

The limitation of the proposed approach lies in very complex shape supporting structures where full BEM-FEM poroelastic finite element models should work better in the low and middle frequency ranges. Other simplifications should be further investigated. This includes abrupt thickness changes in insulators, where small radii of curvature may lead to local radiation at the boundaries between constant-thickness regions [30]. Note poroelastic finite element methods capture these latter effects if the frequency range remains low. The equivalent curvature patch approach applied here cannot accurately predict the lateral coupling within the trim. To our knowledge and experience, the curvature effects are of primary importance and the abrupt changes of the trim thickness are of secondary importance.

11.3.6 *Summary of Curved Insertion Loss Simulations*

Integrated in a SEA/Virtual SEA framework [31, 32], the proposed cylindrical TMM spectral approach allows more rapid computing of the Insertion Loss of any singly-curved or flat multi-thickness noise treatment broadband. A 250 Hz, the Insertion Loss correlations are extremely promising using cylindrical TMM and, in addition, the computational times are extremely fast, e.g. less than one minute using the cylindrical TMM and 5–6 h using a poroelastic FEM. There is also a wider frequency range application achieved. These investigations prove additionally that the lower thicknesses of insulators, where late decoupling occurs (high breathing frequency mass-spring-mass resonance), do influence the Insertion Loss on the whole middle frequency range up 1000 Hz. This is mainly due to the disappearance of the breathing frequencies of curved trims versus flat trims. This cylindrical TMM simplifies enormously the SEA or energy based synthesis methods modelling effort, enabling easier and more accurate vibroacoustic simulations of a fully trimmed vehicle in the middle and high frequency range, especially in pre-project phases.

11.4 Conclusions

Optimizing porous noise treatments requires understanding the noise source characteristics and their transfer paths as well as identifying the dissipation mechanisms that should be fostered: absorption, insulation or vibration damping? The "ultimate target" of the NVH engineer remains to link chemistry, process, the resulting microstructure and finally the macro-acoustic parameters of porous noise treatments. The challenge consists in determining feasible and tunable microstructure morphological paths through a deep understanding of what each manufacturing process allows controlling. Simulating the processes such as foaming or thermoforming and linking the corresponding results with the performance of the future feasible acoustic package is now key. The computed Biot parameters, resulting from these micro-macro calculations, should be integrated in broadband vibroacoustic simulation models, for determining the best acoustic multilayer concepts and their optimal localization for weight reduction purpose or acoustic performance boost.

In automotive industry, apart from restricted available space issues, broadening the efficiency of noise treatments using lightweight metamaterials remains of strong interest. For example, porous metamaterials allowing an increase of the Transmission Loss in the low and middle frequency range without requiring any additional space nor weight are currently investigated carefully. This is also the reason why active noise cancellation using existing car loudspeakers is more often used as a low frequency complement (mainly treating engine noise) to the traditional middle and high frequency passive noise treatments using existing car loudspeakers. The real upcoming challenge is to design cost effective active noise cancellation systems to treat broadband rolling noise for frequencies up to 900 Hz or even 1000 Hz. This frequency range overlaps with traditional passive noise treatments between 250 Hz and 1000 Hz will lead to a new trade-off to define for complementary performances for active and passive noise reductions towards a better personalized acoustic comfort while reducing weight even further.

Acknowledgements Guillaume Crignon from the Trèves group is warmly thanked for the fruitful discussions as well as proofreading of this chapter.

References

1. L. Gagliardini, *Le livre blanc de l'acoustique en France en 2010* (Société Française d'Acoustique, 2010)
2. D. Ragot, F. Carta, " contribution to the definition of acoustic performances characteriza-tion cycles for electrical and hybrid motorizations, in *Congrès SIA Confort Automobile et Ferroviaire* (Le Mans, France, 2006)
3. A. Duval, J.-F. Rondeau, L. Bischoff, G. Deshayes, L. Dejaeger, " generalized light-weight concepts: improving the acoustic performance of less than 2500 g/m^2 insulators, in *SAE NVC Conference* (St Charles (MI), USA, 2009)
4. A. Duval, J.-F. Rondeau, L. Bischoff, C. Morgenstern, G. Deshayes, L. Gagliardini, "vehicle acoustic synthesis method 2nd generation: new developments with pu probes allowing to simulate unsteady operative conditions like run-ups, in *SAE NVC Conference* (St Charles (MI), USA, 2007)
5. C. Perrot, F. Chevillotte, R. Panneton, Bottom-up approach for microstructure optimization of sound absorbing materials. J. Acoust. Soc. Am. **124**(2), 940–948 (2008)
6. C. Perrot, F. Chevillotte, M. Tan Hoang, G. Bonnet, F.-X. Bécot, L. Gautron, A. Duval, Microstructure, transport, and acoustic properties of open-cell foam samples: experiments and three-dimensional numerical simulations. J. Appl. Phys. **111**(1), 014911 (2012)
7. M. Tan Hoang, C. Perrot, Solid films and transports in cellular foams. J. Appl. Phys. **112**(5), 054911 (2012)
8. O. Doutres, N. Atalla, K. Dong, Effect of the microstructure closed pore content on the acoustic behavior of polyurethane foams. J. Appl. Phys. **110**(6), 064901 (2011)
9. A. Duval, M.T. Hoang, C. Perrot, V. Marcel, G. Bonnet, Chemistry-process morphology control of porous micro-structures: a bottom-up acoustic optimization approach, in *SIA-SFA Conference: Light-Weighting and Acoustical Materials in Vehicles* (Compiègne, France, 2013)
10. M.T. Hoang, C. Perrot, V. Marcel, J.-F. Rondeau, Microstructure morphological optimization of foam-based lightweight insulators, in *9th Edition of the Automotive NVH Comfort Conference* (Le Mans, France, 2016)
11. J.-F. Allard, N. Atalla, *Propagation of Sound in Porous Media: Modelling Sound Absorbing Materials*, 2 edn. (Wiley, 2009)
12. M. Villot, C. Guigou-Carter, Using spatial windowing to take the finite size of plane structures into account in sound transmission, in *NOVEM Conference* (Biarritz, France, 2005)
13. A. Duval, J.-F. Rondeau, L. Dejeager, F. Sgard, N. Atalla, Diffuse field absorption coefficient simulation of porous materials in small reverberant rooms: finite size and diffusivity issues, in *10ème Congrès Français d'Acoustique* (Lyon, France, 2010)

14. G. Borello, Modélisation de la transmission acoustique de la partie avant d'un véhicule automobile par l'analyse statistique énergétique, in *Référence InterAC : RT9845, INTERAC* (1998)
15. A. Duval, J.-F. Rondeau, F. Lhuillier, G. Deshayes, L. Dejaeger, J. Pappaert, K. Kato, S. Peters, T. Keppens, Vehicle acoustic synthesis method: a hybrid approach to simulate interior noise of fully trimmed vehicles, in *Confort Automobile et Ferroviaire* (Le Mans, France, 2004)
16. N. Atalla, R. Panneton, P. Debergue, A mixed displacement-pressure formulation for poroelastic materials. J. Acoust. Soc. Am. **104**(3), 1444–1452 (1998)
17. A. Omrani, L. Mebarek, M.A. Hamdi, Transmission loss modeling of trimmed vehicle components, in *ISMA Conference* (Belgium, Leuven, 2006)
18. L. Mebarek, S. Lullier, Y. Kobayashi, D. Blanchet, Modelling of transmission loss for trimmed vehicle components, in *JSAE Annual Congress, Japan* (2007)
19. A. Duval, L. Dejaeger, J. Baratier, J.-F. Rondeau, Structureborne and airborne insertion loss simulation of trimmed curved and flat panels using rayon-vtm-tl: implications for the 3D design of insulator, in *Congrès SIA Confort Automobile et Ferroviaire* (Le Mans, France, 2008)
20. A. Duval, L. Dejaeger, C. Morgenstern, J.-F. Rondeau, Novel technique for the introduction of curved trims in sea/virtual sea models using poroelastic finite elements in the middle (and high) frequency range, in *Congrès SIA Confort Automobile et Ferroviaire* (Le Mans, France, 2010)
21. J.-F. Rondeau, A. Duval, J. Monet-Descombey, L. Dejaeger, Equivalent curvatures broadband insertion loss simulation technique coupling virtual SEA and BEM/FEM approaches, in *Proceedings of Internoise 2013, Innsbruck, Austria*, pp. 15–18 (2013)
22. C. Lesueur, *Rayonnement Acoustique Des Structures: Vibroacoustique, Interactions Fluide-Structure* (Eyrolles, 1988)
23. S. Ghinet, N. Atalla, H. Osman, Diffuse field transmission into infinite sandwich composite and laminate composite cylinders. J. Sound Vib. **289**(4–5), 745–778 (2006)
24. J. Magniez, J.-D. Chazot, M.A. Hamdi, B. Troclet, A mixed 3D-Shell analytical model for the prediction of sound transmission through sandwich cylinders. J. Sound Vib. **333**(19), 4750–4770 (2014)
25. J. Magniez, M.A. Hamdi, J.-D. Chazot, B. Troclet, A mixed "Biot-Shell" analytical model for the prediction of sound transmission through a sandwich cylinder with a poroelastic core. J. Sound Vib. **360**, 203–223 (2016)
26. C. Coguenanff, A. Duval, M. Goret, A spectral method for fast broadband insertion loss modeling of curved sound packages: correlation with poroelastic finite elements, in *INTER-NOISE and NOISE-CON Congress and Conference Proceedings, Chicago (IL), USA*, vol. 258, pp. 3023–3034 (Institute of Noise Control Engineering, 2018)
27. N. Dauchez, S. Sahraoui, N. Atalla, Convergence of poroelastic finite elements based on biot displacement formulation. J. Acoust. Soc. Am. **109**(1), 33–40 (2001)
28. A. Duval, G. Crignon, M. Goret, D. Lemaire, J.-B. Prunet, P. Chanudet, Ecofelt hybrid stiff nvh tunable insulator, in *Styrian Noise Vibration and Harshness Conference - SAE 2018–01-1494* (Graz, Austria, 2018)
29. A. Duval, M. Goret, G. Borello, Fast broadband curved insertion loss simulation of an inner dash insulator using a cylindrical transfer matrix method spectral approach, in *SAE NVC Conference - 2019–01-1583* (Grand Rapids (MI), USA, 2019)
30. S. Chaigne, A. Diallo, C. Capron, J.-P. Coyette, T. Leclercq, Vibro-Acoustic Analysis of Varying Thickness Trimmed Structure and Curved Trimmed Panels, in *Congrès SIA Confort Automobile et Ferroviaire* (Le Mans, France, 2004)
31. G. Borello, Predicting noise transmission in complex multilayers systems combining sea and transfer matrix method, in *International Congress on Sound and Vibration, ICSV23 Congress* (Greece, Athens, 2016)
32. G. Borello, L. Gagliardini, L. Houillon, L. Petrinelli, Virtual SEA: mid-frequency structure-borne noise modeling based on finite element analysis, in *SAE 2003–01-1555* (Traverse City (MI), USA, 2003)

Chapter 12
Industrial Applications IV

Acoustic Package Optimization Methods in the Aeronautic Industry

Israel Pereira, Sideto Futatsugi, and Maria L. V. Rodrigues

Abstract This chapter presents an overview of the challenges faced by the aeronautic industry on the search for a more efficient and comfortable cabin. In order to do that, the main noise sources are presented and characterized, together with a typical noise control treatment solution. Later, deeper focus on the porous materials use is described, presenting the difficulties for the characterization of such materials and consequently uncertainties on the project requirement's definition. Finally, some of the promising new studies on porous materials are discussed and their impact is evaluated in a case study.

12.1 Introduction

Air transportation became very popular during the last decades, being an affordable alternative for fast travel worldwide. With the popularization, new costumers' expectations arise and the decision on buying a flight ticket takes into account not only price but also the experience delivered. The importance of this experience increases on the business jet market where the expectations on the overall flight quality are higher.

One of the main topics related to the user experience is cabin comfort, in which noise and vibrations are two of the main factors that add to comfort perception [1, 2]. Quehl [2] showed that three of the top five attributes that compromise comfort during flight are related to noise and vibration. The proper estimation of cabin noise and

I. Pereira (✉) · S. Futatsugi
Embraer S.A, Av. Brigadeiro Faria Lima, 2170 São José dos Campos, Brazil
e-mail: israel.pereira@embraer.com.br

S. Futatsugi
e-mail: sideto.futatsugi@embraer.com.br

M. L. V. Rodrigues
Embraer R&D Portugal, Parque da Indústria Aeronáutica de Évora, Lote A-I, Herdade do Pinheiro e Casa Branca, 7005-797 Évora, Portugal
e-mail: mlvrodrigues@pt.embraer.com

N. Jiménez et al. (eds.), *Acoustic Waves in Periodic Structures, Metamaterials, and Porous Media*, Topics in Applied Physics 143,
https://doi.org/10.1007/978-3-030-84300-7_12

the design of the Noise Control Treatment (NCT) present big challenges due to the aircraft complexity, with its various noise sources that act on a broadband spectrum. The knowledge of the factors that act on cabin noise was significantly improved over the past decades [3] and the results of this progress can be seen on better acoustic comfort, mainly on business aircraft, and on the increase of the NCT efficiency, that allowed reducing the overall treatment weight keeping noise levels low.

Despite the technology evolution, the demand for higher efficiency on noise control continues and improvements on current solutions are reaching their limit on current form, therefore the development of innovative solutions is required. One field that presents strong innovation potential is the design of porous materials and a lot of efforts are seen on the topic in the past years. The main difficulties developing and applying these new solutions are due to tight weight and space restrictions, and also to the severe aeronautic certification requirements.

This document aims at presenting an overview of the challenges faced by the aeronautic industry on the search for a more efficient and comfortable cabin. In order to do that, the main noise sources are presented and characterized, together with a typical NCT solution. Later, deeper focus on the porous materials use is described, presenting the difficulties for the characterization of such materials and consequently uncertainties on the project requirement's definition. Finally, some of the promising new studies on porous materials are discussed and their impact is evaluated in a case study.

12.2 Description of Noise Sources

Although variations will happen due to constructive characteristics of each aircraft, typical cabin noise sources for a jet engine aircraft can be divided in noise coming from the Turbulent Boundary Layer (TBL), from the engine and from onboard systems (Fig. 12.1). TBL noise is the main responsible source for the mid-frequency noise, the engine for low and mid frequency and the systems will contribute mostly for higher frequencies.

12.2.1 Engine Sources

The engine is responsible for distinct noise sources, with different generation and transmission characteristics. Firstly, unbalance in the low and high pressure shafts (called N1 and N2 respectively), generates vibrations that are transmitted through the engine fixture structure into the aircraft cabin interior. The unbalance energy is converted into noise through the interior liners, floor and monument's (galleys, lavatories, closets, bar units, refreshment centers, etc.) radiation. This type of source has a low frequency spectral content and it is tonal, making this a point of attention due to the human sensitivity for this type of excitation.

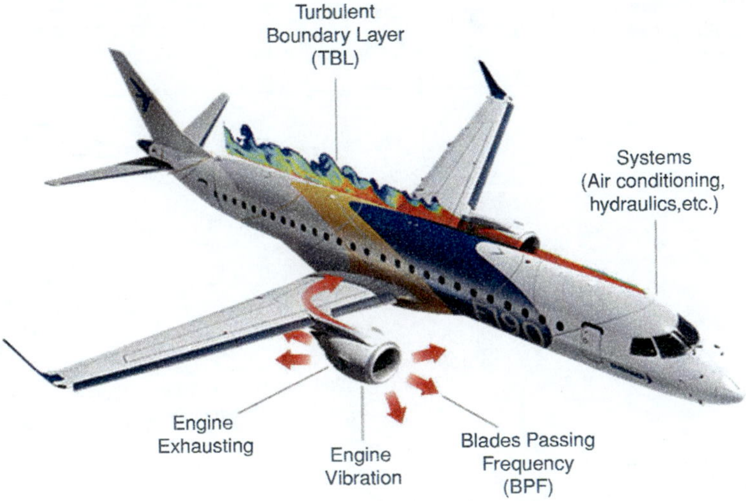

Fig. 12.1 Typical jet engine aircraft noise sources

Another noise source originated in the engine is the fan's Blades Passing Frequency (BPF). The excitation is also tonal, however the spectral content is much higher because it is the N1 rotational speed multiplied by the number of fan blades (usually some dozens of blades), and its higher harmonics.

Finally, another important engine noise source is caused by the exhaust gases turbulence. Unlike the aforementioned sources, this excitation has a broadband frequency spectrum, making the acoustic treatment strategy different from other engine sources. This kind of source may be relevant in wing mounted engines, impacting cabin zones located in the after portion of exhaust area.

12.2.2 Turbulent Boundary Layer (TBL)

The aerodynamic noise generated by the fuselage outboard pressure fluctuations due to the Turbulent Boundary Layer (TBL) is the main responsible for the cabin noise levels in the mid-frequencies, and for many cases, it is the main contributor for the overall noise level in the cabin.

In the TBL, there is a macroscopic movement of mass inside the layers of the boundary layer, thus generating an exchange of mass, momentum and energy. This pressure fluctuation travelling along the fuselage, dynamically excites the structure. There is part of energy directly radiated through the fuselage skin, and other part transmitted through the connection structures up to the interior parts, converting the vibratory energy into noise. Due to its nature, the TBL has got a broadband frequency content. The noise levels generated by the pressure fluctuation on the fuselage by

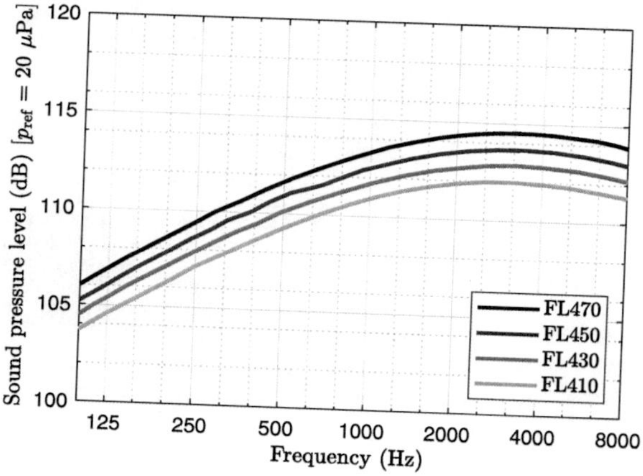

Fig. 12.2 Pressure level for different Flight Levels (FL). Mach = 0.8

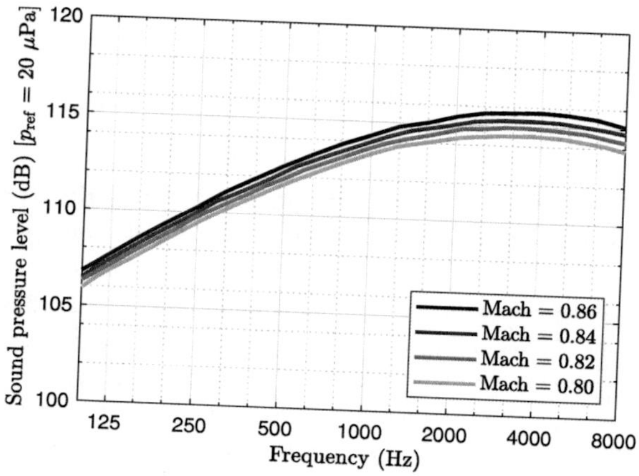

Fig. 12.3 Pressure level for different velocities. FL = 410

the TBL are directly associated with the aircraft flight altitude (Fig. 12.2), velocity (Fig. 12.3) and the distance from leading edge, i.e. TBL origin (Fig. 12.4).

The pressure estimation presented in results from Figs. 12.2, 12.3 to 12.4 were calculated [4] and will vary depending on other factors not listed here. But the sensitivity to velocity, altitude and distance will be similar.

12.2.3 Systems

Aircraft systems can generate several noise sources that may affect mainly the cabin noise perception (sound quality). The system's noise can be broadband, as those associated with the airflow inside the air-conditioning ducts, or tonal, as the noise generated by hydraulic pumps and cooling fans.

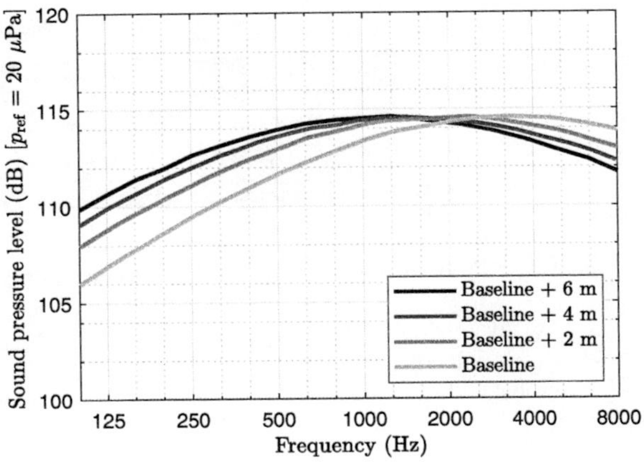

Fig. 12.4 Pressure level varying distance from TBL formation Mach = 0.8, FL = 410

Usually, the noise level generated by those systems does not impact the overall cabin noise level; however, the sound quality is affected if the proper countermeasure is not adopted. Furthermore, increasing the acoustic treatment performance for engine and TBL control, the systems noise tends to become more prominent, turning into the critical path for a more comfortable cabin.

12.3 The Use of Porous Materials

The choice of the vibroacoustic treatment strategy depends of the type of noise source, vibroacoustic energy path and the spectrum content. For instance, vibration caused by the engine N1 and N2 unbalance are typically attenuated using vibration isolators that act preventing the energy to be transmitted through the fuselage. For systems, the vibroacoustic control can be addressed in a variety of different strategies, depending on the system's specific characteristics. One of these strategies may be to act directly on the noise source, another may be isolating or attenuating the main energy path or, if those approaches are not feasible, the strategy can be the reallocation of the system far from the cabin.

The materials used to attenuate the energy that enters through the fuselage up to the interior liners is called Thermo-Acoustic Insulation System (TAIS). This set of materials actuates mainly in mid to high frequency and its typical components are presented in Fig. 12.5.

Viscoelastic material is applied in the fuselage skin in order to impose damping into the structure, reducing the skin vibration originated from the external sources. The Damping Loss Factor (DLF) is the key parameter for this material. Additionally, vibration isolators are employed to decouple the attachment between the primary structure (fuselage) and the interior liners, thus minimizing the structure-borne path. In order to attenuate airborne energy, porous materials are used, like fibers and foams.

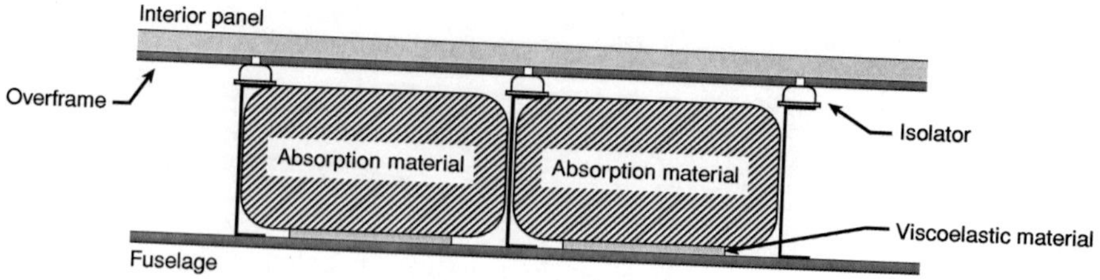

Fig. 12.5 Typical Thermo-Acoustic Insulation System (TAIS) configuration

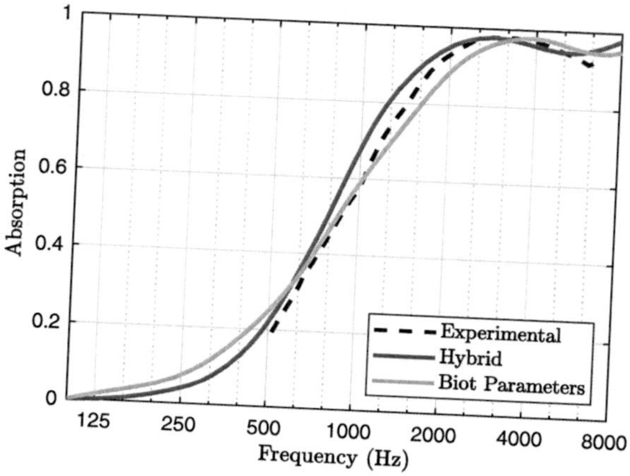

Fig. 12.6 Absorption estimation of a fiberglass sample using different methods

Those materials are lightweight with a high acoustic absorption capability in mid to high frequency.

The attenuation in low frequency range is limited due to the material thickness, and it is directly related to the distance available between the fuselage and the interior liners. Acoustic absorption coefficient is the key parameter. Finally an overframe (barrier) is used to avoid airborne noise propagation. Area density is the key parameter in this case.

In order to minimize the TAIS thickness and weight, a proper characterization of each material is mandatory. In this context, the use of porous materials presents a big challenge since those materials are well known for their inherent uncertainties and experimental characterization results dispersion [5].

Initially, the porous materials must be characterized looking for its acoustic absorptive properties. This is traditionally made using plane waves excitation on an impedance tube [6] but can also be made by characterizing the material's Biot parameters [7, 8]. Alternatively hybrid methods can be used when initial experimental data is used to estimate Biot parameters through algorithms. All these methodologies present many advantages and disadvantages, and the results can vary significantly among them, as it can be seen in Fig. 12.6, where the acoustic absorption of a fiberglass sample is estimated using all three methodologies: Impedance tube measurement; Biot parameters measurement; hybrid methodology.

Differences like seen in Fig. 12.6 can lead up to 0.5 dB variations on the cabin noise estimation, which seems not much when looking isolated, but as the complexity of the analysis increases, these uncertainties propagate and sum with other uncertainties sources resulting in low precision models. A simplified development workflow for the noise control treatment starts with the material characterization, followed by a local treatment design and finishing with a full cabin analysis. The degree of uncertainty of the results increases when progressing on this workflow since the integration become more complex.

Additionally to the uncertainties from the different available methodologies, experimental tests to obtain acoustic parameters from porous materials also present high dispersion, as it can be seen at the study presented in [5]. In this study acoustic properties of three different porous materials were measured by several laboratories and the results were significantly different among them.

After material characterization the noise control treatment composition must be evaluated, i.e. the integration of the noise control solutions with the fuselage and the interior panel. In this way, the Transmission Loss (TL) of the NCT package can be estimated and the acoustic requirements can be tuned. This composition analysis can be performed analytically, experimentally and numerically, the later using deterministic, statistical or hybrid models.

Given the low computational costs and the theory constrains, Statistical Energy Analysis (SEA) models are commonly used for mid to high frequency analysis of the treatment when applied in regions with large dimensions. Deterministic and hybrid models are applied in low frequencies or local analysis, when specific characteristics of certain treatment regions must be evaluated, like geometry modifications or leaks.

For the experimental tests, the main challenge is to obtain good repeatability for the TL measurements. When performed in a high quality laboratory, at similar atmosphere conditions and by the same operator, the results tend to have low variation but if any of these parameters vary, the comparison among tests become challenging and conclusions inaccurate.

12.4 Aeronautic Requirements

Additionally to technical challenges of the TAIS design process, aeronautic certification requirements and industry regulation significantly limit the materials that can be used in the treatment, making optimization even more challenging.

One of the most restrictive requirements is related to flammability. The materials are submitted to tight fire self-extinguish, flame propagation, radiated panel burn and, in some cases, flame penetration tests (FAR/EASA 25.853 and 25.856). Another requirement that porous materials shall comply is the humidity retention, that according to norm ASTM C1511 [9], a maximum 20 g of water can be absorbed for a 10" × 10" × 4" sample of the porous material after 15 min submerged.

A common strategy to help the porous materials to comply with the flammability and humidity requirements is using specific chemical treatments, such as flame retardants or inhibitors which usually deteriorate acoustic performance. However, additionally to the previously mentioned requirements, regulations like the European REACH (Registration, Evaluation, Authorization and restriction of CHemicals) restricts the use of various chemical components, even if it is only used during an intermediate step of the fabrication process. Therefore, both the materials that are part of the TAIS and chemical treatments that those may be submitted are restricted.

12.5 New Developments

With the challenges presented in previous sections, the search for new developments that could reduce technological restrictions of the aeronautic industry is fundamental. Porous materials have been studied and developments have been made aiming the acoustic performance improvement, mainly at low frequencies where those materials perform poorly.

An easy way to increase the acoustic absorption of porous materials at low frequencies is increasing the material thickness, like presented in Fig. 12.7, where the plane wave absorption of a fiber is numerically evaluated for 1 and 3 inches.

Unfortunately, increasing the material thickness is usually not possible due to space limitation between the fuselage and interior panel, therefore innovative solutions have been studied to increase the absorptions at low frequencies avoiding thickness increase. Some studies focus on optimizing the treatment layers so that the impedance mismatch result in high absorption at desired frequencies. One example can be seen on [10], where the author presented some variations at the layer distribution of different porous materials, aiming for higher absorption values at lower frequencies.

An evolution of that concept is the porous materials microstructure design through porosity optimization presented by [11] that showed significant gains at the acoustic

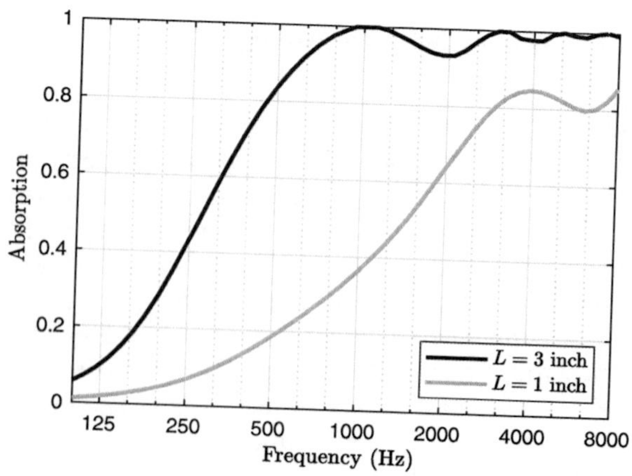

Fig. 12.7 Variation of the absorption with the porous material thickness

absorption characteristics by controlling the porous size of a foam sample. Another common research topic is the use of inclusions at porous materials in order to induce Bragg's scattering effect at the sample. These inclusions can significantly affect the energy attenuation as it can be seen in [12].

A straightforward expansion of the inclusions concept is the use of resonators inside the porous materials. This concept received attention for some time [13] and a recent example is presented in [14] where the author develop a study containing analytical, numerical and experimental analysis for resonators embedded in a porous material. The author concluded that significant TL gains may be expected at the resonator tuning frequency.

Recently the acoustic metamaterials concept have been explored [15] in which internal resonances are induced at the material in order to manipulate larger wavelengths than the expected for the base material, allowing elastic and acoustic energy attenuation at low frequencies. This concept has been expanded offering several different configurations such as membranes containing resonators [16] opening many possibilities to improve the aircraft acoustic treatment and performance.

12.6 Case Study

This section presents a typical scenario for an aircraft application. A simplified setup is presented in Fig. 12.8. The materials considered for this study are a 1.2 mm thick aluminium fuselage section, a 76.2 mm Fiberglass material with density (ρ) equal 8 kg/m^3, flow resistivity (σ) of 4000 N·s/m^4, porosity (ϕ) of 0.96, tortuosity (α_∞) equal 1, viscous (Λ) and thermal (Λ') lengths, respectively, 30 and 70 μm.

The overframe is a 0.5 mm solid material, with $\rho = 1000$ kg/m^3, tensile modulus (E) of 0.3 GPa and Poisson's ratio (ν) of 0.49. Finally, the interior panel is a composite material containing 2 external fiberglass plies (0.2 mm thick, $\rho = 1800$ kg/m^3, $E =$ 10 GPa, $\nu = 0.2$) and an interior core (8 mm thick, $\rho = 50$ kg/m^3, $E = 0.03$ GPa, $\nu =$ 0.2). Also, a thin air gap (0.01 mm) is placed between the elements so no structural connection is considered.

The TL for the configuration presented can be seen in Fig. 12.9, together with the results for intermediate configurations. The total thickness for this case study

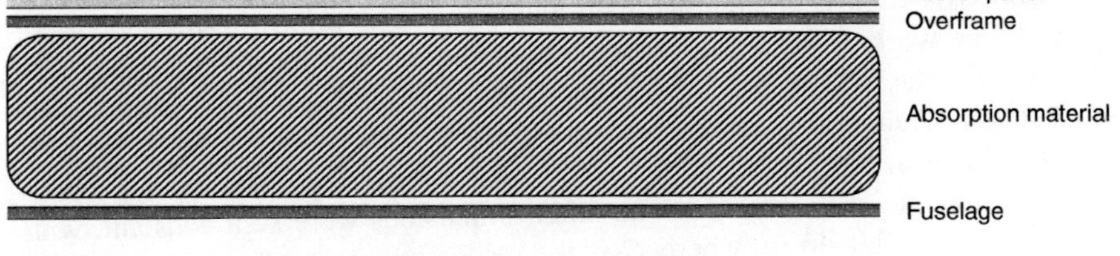

Fig. 12.8 NCT configuration considered for the case study

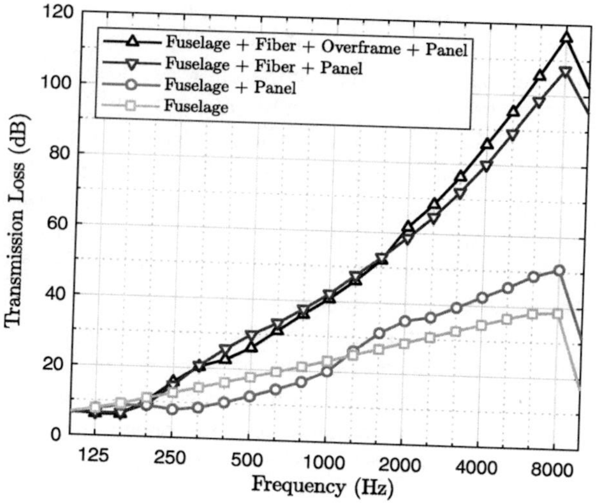

Fig. 12.9 Transmission Loss (TL) results for the case study configuration

solution is 76.7 mm. Note that solutions between 3 inches (76.2 mm) and 3.5 inches (89 mm) are typical in the aeronautic industry. The inclusion of the interior panel increases the TL at high frequencies, when comparing to the fuselage alone, but reduces it at lower frequencies due to the double wall effect.

The inclusion of the absorption fiber and the overframe significantly improves the TL at mid to high frequencies but little influence is seen at lower frequencies.

As seen in the Sect. 12.2.1 before, the aircraft engine can be responsible for low frequency excitation at the fuselage. This is also true for the TBL excitation, mainly at the aft part of the aircraft, i.e., the rear part, as it is shown in Fig. 12.4 where it is clear that the low frequency content importance increases progressively with the distance from the aircraft nose. Therefore, the results presented in Fig. 12.9 are not optimal for these sources and the design of a NCT that can act at lower frequencies is one of the main challenges to improve the aircraft acoustic comfort.

Even a simple addition of mass, which is far from desired in the aeronautic industry, may not be sufficient. For example, if the overframe density is twice the original, the TL in this case will only be affected above 1000 Hz, like shown in Fig. 12.10. That means that any mass to be added to the treatment must be optimized to give results at lower frequency.

The use of tuned resonators, as suggested by some of the papers discussed in previous sections, can provide attenuation at low frequencies even with limited space. Similar to the work presented in [14], the transmission loss for an embedded resonator array in the NCT was calculated and is presented in Fig. 12.11. For this analysis, 3 sizes of Helmholtz's resonators were used ($l_n = 0.007$ m, $l_n = 0.01$ m, $l_n = 0.012$ m), being l_n the length of the necks of the resonator, that way a broader frequency attenuation is achieved. The other dimensions were kept constant, with $D = 0.05$ m, $r = 0.0046$ m, where D is the resonators cavity diameter and r the radius of the resonators neck's transversal section.

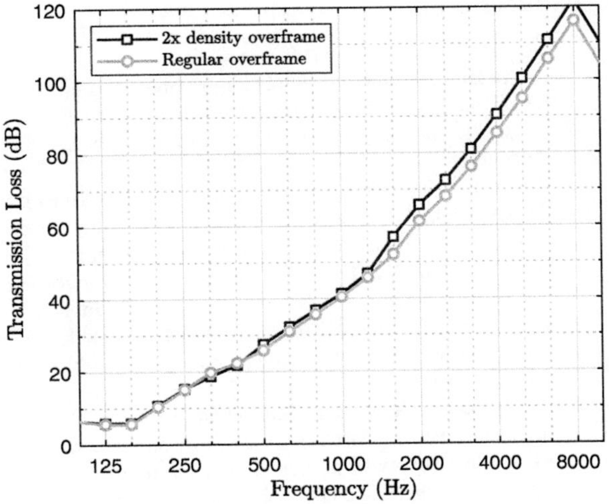

Fig. 12.10 TL results varing the density of the overframe

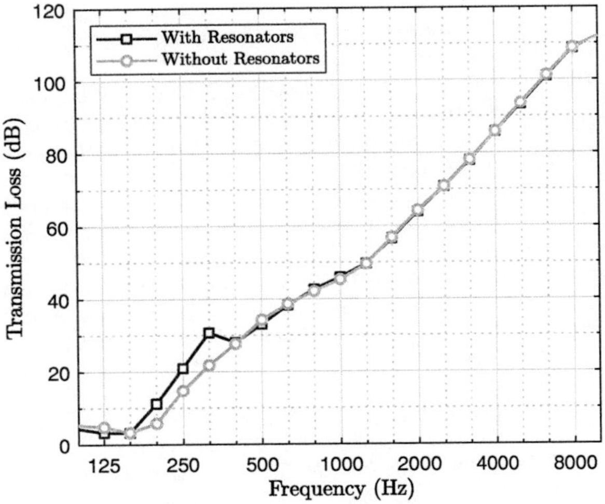

Fig. 12.11 TL results for the NCT with embedded resonators

It can be seen that up to 9 dB reduction can be obtained 300 Hz by using resonators but the influence of those are significant even in lower frequencies.

The use of resonators is an example of the potential that can be obtained by changing the current status of the noise control treatment applied to the aircraft. Variations of this concept like the use of microperforated layers, inclusions, porous design and acoustic metamaterials are solutions that present endless possibilities and can be a game changer for the industry, bringing a new perception of NVH comfort inside an aircraft.

References

1. M.M. Blok, P. Vink, I. Kamp, Comfortabel vliegen: comfort van het vliegtuiginterieur door de ogen van de gebruiker. Tijdschrift voor ergonomie **4**(32), 4–11 (2007)
2. J. Quehl, *Comfort Studies on Aircraft Interior Sound and Vibration*. Ph.D. thesis, Universität Oldenburg (2001)
3. J.F. Wilby, Aircraft interior noise. J. Sound Vib. **190**(3), 545–564 (1996)
4. D. Rennison, A.G. Piersol, J.F. Wilby, E.G. Wilby, A review of the acoustic and aerodynamic loads and the vibration response of the space shuttle orbiter vehicle–sts-1 dynamics verification assessment. *BBN Report 4438 for NASA, Jet Propulsion Laboratory* (1980)
5. K. V. Horoshenkov, A. Khan, F.-X. Bécot, L. Jaouen, F. Sgard, A. Renault, N. Amirouche, F. Pompoli, N. Prodi, P.E.A. Bonfiglio, Reproducibility experiments on measuring acoustical properties of rigid-frame porous media (round-robin tests). J. Acoust. Soc. Am. **122**(1), 345–353 (2007)
6. ISO 10534-2: Acoustics, Determination of sound absorption coefficient and impedance in impedance tubes—Part 2: transfer-function method (1998)
7. M.A. Biot, Theory of propagation of elastic waves in a fluid-saturated porous solid. I. Low-frequency range. J. Acoust. Soc. Am. **28**(2), 168–178 (1956)
8. M.A. Biot, Theory of propagation of elastic waves in a fluid-saturated porous solid. II. Higher frequency range. J. Acoust. Soc. Am. **28**(2), 179–191 (1956)
9. A. C1511, Standard test method for determining the water retention (repellency) characteristics of fibrous glass insulation (aircraft type) (2009)
10. E.A. Kulakauskas, Fábio Luis Val Quintans, Avaliação do uso de materiais porosos na perda de transmissão de painéis duplos, *Universidade Federal de Santa Catarina, MSc Thesis* (2016)
11. P. Mareze, *Análise da influência da microgeometria na absorção sonora de materiais porosos de estrutura rígida*. Ph.D. thesis, Universidade Federal de Santa Catarina (2013)
12. T. Weisser, J.-P. Groby, O. Dazel, F. Gaultier, E. Deckers, S. Futatsugi, L. Monteiro, Acoustic behavior of a rigidly backed poroelastic layer with periodic resonant inclusions by a multiple scattering approach. J. Acoust. Soc. Am. **139**(2), 617–629 (2016)
13. R.A. Prydz, L.S. Wirt, H.L. Kuntz, L.D. Pope, Transmission loss of a multilayer panel with internal tuned helmholtz resonators. J. Acoust. Soc. Am. **87**(4), 1597–1602 (1990)
14. O. Doutres, N. Atalla, H. Osman, Transfer matrix modeling and experimental validation of cellular porous material with resonant inclusions. J. Acoust. Soc. Am. **137**(6), 3502–3513 (2015)
15. Z. Liu, X. Zhang, Y. Mao, Y.Y. Zhu, Z. Yang, C.T. Chan, P. Sheng, Locally resonant sonic materials. Science **289**(5485), 1734–1736 (2000)
16. F. Langfeldt, J. Riecken, W. Gleine, O. Von Estorff, A membrane-type acoustic metamaterial with adjustable acoustic properties. J. Sound Vib. **373**, 1–18 (2016)

Correction to: Acoustic Wave Propagation in Viscothermal Fluids

Denis Lafarge

Correction to:
Chapter 6 in: N. Jiménez et al. (eds.), *Acoustic Waves in Periodic Structures, Metamaterials, and Porous Media*, Topics in Applied Physics 143,
https://doi.org/10.1007/978-3-030-84300-7_6

The book was inadvertently published with errors in chapter 6. These belated corrections have now been updated with the correction chapter.

The updated version of this chapter can be found at
https://doi.org/10.1007/978-3-030-84300-7_6

N. Jiménez et al. (eds.), *Acoustic Waves in Periodic Structures, Metamaterials, and Porous Media*, Topics in Applied Physics 143,
https://doi.org/10.1007/978-3-030-84300-7_13

Index

A
Absorption, 173
Absorption coefficient, 113
Absorption coefficient in diffuse field, 144
Acoustic barrier, 93
Adiabatic compressibility, 209
Adiabatic exponent, 209
Aerogels, 183
Anisotropic poroelasticity, 362
Antisymmetric modes, 186
Asymmetric absorbing panels, 193
Atomic chain, 4

B
Band gap, 7, 67, 154
Band structure, 75
Bending stiffness, 53
Biot's equations, 335
Biot's frequency, 173, 188
Bloch-Floquet theorem, 18, 153
Bloch parameters, 68
Bloch wave vector, 18
Bravais lattice, 8
Brillouin zone, 5, 74, 154
Bulk modulus, adiabatic, 209
Bulk modulus, isothermal, 209

C
Cauchy strain tensor, 336
Chirped absorber, 189
Circular membrane, 136
Clamped circular plate, 133
Coincidence frequency, 139
Complex frequency plane, 171
Compressibility modulus, 336
Constitutive laws, 221
Convective derivative, 219
Critical coupling, 171
Critical frequency, 139
Cylindrical duct, 119

D
Dead-end porosity, 180
Decorated membranes, 181
Degenerate resonances, 196
Delany-Bazley model, 122
Diatomic chain, 6
Diffusion coefficient, 117
Dipolar resonances, 186
Direct lattice, 5
Dispersion relation, 5, 153
Double wall sound proofing, 145
Dynamic tortuosity, 337

E
Effective bulk modulus, 109
Effective characteristic impedance, 108
Effective density, 86, 109
Effective parameters, 108
Effective wavenumber, 108
Elastic moduli tensor, 15
Elastic plates, 133, 181
Elastic wall, 138
End correction, 160
Energy balance, 219
Equi-frequency contour, 37
Evanescent waves, 3, 71

N. Jiménez et al. (eds.), *Acoustic Waves in Periodic Structures, Metamaterials, and Porous Media*, Topics in Applied Physics 143,
https://doi.org/10.1007/978-3-030-84300-7

Extended Plane Wave Expansion (EPWE) method, 73

F

Fabry–Pérot resonance, 184
Filling fraction, 25
Finite element method, 334
Flexural waves, 53

G

Gradded absorber, 189
Gradient refractive index lenses, 61, 86
Grating metaporous, 175

H

Half-wavelength resonator, 129
Helmholtz-membrane resonator, 196
Helmholtz resonator, 130, 179
Homogenization limit, 86
Hooke's law, 15

I

Ideal gas law, 208
Impedance matching, 173

J

Johnson-Champoux-Allard (JCA) model, 124
Johnson-Champoux-Allard-Lafarge (JCAL) model, 125

K

Kinematic viscosity, 228

L

Localized modes, 89
Locally resonant elements, 126
Locally resonant sonic crystals, 87
Luneburg lens, 62

M

Mass conservation, 218
Mass-spring-mass resonance, 147
Mass-spring resonators, 57
Membranes, 136
Metamaterials, 87, 156, 168
Metaporous absorbers, 175, 359
Metasurfaces, 178
Micro-perforated plates, 84, 133
Miki model, 124
Modified PWE method, 34
Momentum balance, 219
Monoatomic chain, 4
Monopolar resonances, 186
Multilayer structure, 141
Multiple scattering, 43, 44
Multiple Scattering Theory (MST), 50, 77

N

Narrow slits, 118
Negative bulk modulus, 156
Negative mass-density, 158
Non-reciprocal systems, 112

O

Oblique incidence, 142, 148

P

Paris' formula, 144
Perfect absorption, 171
Periodic structure, 3
Phononic crystal plates, 37
Phononic crystals, 22, 151
Plane Wave Expansion (PWE) method, 15, 71
Poroelastic material, 336
Poroelastic plates, 183
Porous materials, 121, 141, 173
Prandtl number, 118
Primitive vectors, 8

Q

Quarter-wavelength resonance, 168
Quarter-wavelength resonator, 129, 179

R

Rainbow-trapping absorbers, 194
Random-incidence absorption coefficient, 144
Reciprocal lattice, 5, 9
Reciprocal system, 112
Rectangular duct, 120
Reflection problem, 170
Refractive lenses, 61

S

Scattering matrix, 109, 169
Slits, 118
Slow sound, 178
Sonic-crystal metaporous, 175
Sonic crystals, 33, 65
Specific acoustic impedance, 114
Specific heat coefficient, 209
Speed of sound, 209
Split-ring metaporous, 176
Squared clamped elastic plate, 134
Squared membrane, 137
Strain tensor, 15
Stress tensor, 336
Structure factor, 25, 74
Supercell, 37, 76
Symmetric modes, 186
Symmetric system, 113, 170

T

Thermal expansion coefficient, 208
Thermoviscous fluid, 117
Thermoviscous losses, 118
Time-reversal symmetry, 171
Transfer Matrix Method (TMM), 104
Transmission matrix, 107
Transmission problems, 110

V

Viscous and thermal boundary layers, 83, 118
Viscous drag tensor, 338

W

Wall, 138
Waveguides, 89
Wigner-Seitz cell, 8
Wood's anomaly, 176